Comets as Tracers of Solar System Formation and Evolution

Kathleen Mandt • Olivier Mousis •
Dominique Bockelée-Morvan • Christopher Russell
Editors

Comets as Tracers of Solar System Formation and Evolution

Previously published in *Space Science Reviews* Volume 197, Issues 1–4, 2015

Springer

Editors
Kathleen Mandt
Space Science and Engineering
Southwest Research Institute
San Antonio, TX, USA

Dominique Bockelée-Morvan
Observatoire de Paris
LESIA/CNRS, UPMC
Meudon, France

Olivier Mousis
Aix Marseille Université
CNRS, LAM
Marseille, France

Christopher Russell
Institute of Geophysics & Planetary Physics
University of California Los Angeles
Los Angeles, CA, USA

ISBN 978-94-024-1102-7

Library of Congress Control Number: 2017942719

Cover Image: Illustration of the stages of solar system formation leading up to the present era where comets, like the Rosetta target 67-P/Churyumov-Gerasimenko shown here, are used to evaluate the conditions that prevailed during the formation of the solar system. The process begins when a dark cloud core experiences gravitational collapse. A protostar forms in the center and a disk of gas and dust forms around the protostar. Eventually the material of the disk forms the planets, moons and small bodies leading to the solar system as it appears today.
Credit line: Cover art created by Elsa M. Mandt

Printed on acid-free paper

This Springer imprint is published by Springer Nature
The registered company is Springer Science+Business Media B.V.
The registered company address is: Van Godewijckstraat 30, 3311 GX Dordrecht, The Netherlands

Contents

On the Evolution of Comets
A. Guilbert-Lepoutre · S. Besse · O. Mousis · M. Ali-Dib · S. Höfner · D. Koschny · P. Hager: Space Science Reviews 2015, 2015:271–296 (15 April 2015)
DOI:10.1007/s11214-015-0148-9

Constraints from Comets on the Formation and Volatile Acquisition of the Planets and Satellites
K.E. Mandt · O. Mousis · B. Marty · T. Cavalié · W. Harris · P. Hartogh · K. Willacy: Space Science Reviews 2015, 2015:297–342 (21 May 2015)
DOI:10.1007/s11214-015-0161-z

Space Sci Rev (2015) 197:1–1
DOI 10.1007/s11214-015-0216-1

Dedication to Claudia Alexander

Published online: 23 November 2015

As described in the Introduction, the aim of the International Cometary Workshop was to bring together the cream of the crop of cometary experts to discuss the role of comets in the understanding the origin and evolution of the solar system. The meeting was a great success, thanks not only to the work of the science and local organising committee led by Kathleen Mandt and Olivier Mousis, but also to the enthusiastic support of Claudia Alexander, the Project Scientist for NASA's role in the ESA Rosetta mission. Claudia Alexander passed away in July 2015, leaving behind a sense of great sadness for all involved with Rosetta, but also a strong legacy for the planetary science and space physics community.

Claudia worked at NASA's Jet Propulsion Laboratory. Following her work as project manager for NASA's Galileo mission, she was appointed Project Scientist for NASA's participation in the ESA Rosetta mission. I met Claudia very soon after I was assigned as the ESA project Scientist of Rosetta in 2013, and it was clear immediately that I had met a powerhouse in the community, a freight train of enthusiasm and somebody I knew I could immediately consider a friend and great colleague. In the months following Claudia and I worked at the project science tasks of optimising and stimulating science output of the mission, with Claudia helping me get up to speed with the new community I had joined. One highlight she flagged early on in our interactions was the International Cometary Workshop and I was very excited to attend. During the ICW meeting in the middle of a presentation Claudia strode up to me, with her signature beaming smile and declared "This is soooo great!! There is a real buzz! Can you imagine what we will be discussing next year!" It was indeed a great meeting and set the momentum for the cometary community for what was to come with Rosetta and we are still immersed in that scientific excitement.

A number of aspects of the success of Rosetta are only there because of Claudia, her work, her drive and her personality, her way of telling the world how important our work in planetary science and exploration is. In honour of her great contributions to the cometary research community, we dedicate this special issue to Claudia.

M. Taylor, Rosetta Project Scientist, ESTEC, ESA

Space Sci Rev (2015) 197:3–4
DOI 10.1007/s11214-015-0217-0

Dedication to Claudia Alexander

Published online: 23 November 2015

As the planetary sciences editor for Space Science Reviews and a co-editor of this volume, I wish to join with Matt Taylor in the dedication of the volume to Claudia Alexander, one of the most outgoing personalities in the field of planetary sciences and one whose presence will be greatly missed. Claudia received her B.S. degree from UC Berkeley and then joined UCLA to pursue graduate studies in planetary and space physics and I became her first graduate advisor. Claudia received her M.S. in space physics but her true love was cometary science. After a stint at JPL she enrolled in the University of Michigan where she received her doctorate in cometary science, after which she returned to JPL. She and I teamed up shortly thereafter to propose (unsuccessfully) an ion-propelled mission to comet Tempel 2. This effort may have been a stepping stone on her path to Rosetta. In 2005 Claudia became the first[1] of my students to deliberately crash a spacecraft, when as project manager she sent Galileo into the Jovian atmosphere. Claudia always took time to point out that crashing Galileo into Jupiter was not an accident.

Claudia was a strong and very positive personality. She loved life and was a cheerleader for the teams in which she worked. She enjoyed travelling and chronicled her journeys with her camera. She was an avid educator and one of her passions was the creation of children's books, two (Alexander 2015a, 2015b) of which she published shortly before her death. I was honored to write a foreword to one of these books and in the process read the book carefully. She was clearly a gifted educator. The book was engaging; it explained important planetary concepts to young minds; and it was well executed as was everything in her life. She will be greatly missed by not just the cometary community but by the planetary community at large. I join with Matt Taylor in dedicating this special issue to Claudia.

C.T. Russell
Planetary Sciences Editor, Space Science Reviews

[1]Rick Elphic followed her in 2015 by crashing LADEE into the lunar surface. Linda Spilker is expecting to dispose of Cassini in the Saturnian atmosphere in 2017.

References

C. Alexander, Which of the Mountains is Greatest of All? (Red Phoenix Books, 2015a), 130 pp.
C. Alexander, Venus the Morning Star (Red Phoenix Books, 2015a), 88 pp.

Space Sci Rev (2015) 197:5–7
DOI 10.1007/s11214-015-0215-2

PREFACE

Comets as Tracers of Solar System Formation and Evolution

Kathleen E. Mandt[1] **· Olivier Mousis**[2] **· Dominique Bockelée-Morvan**[3] **· Christopher T. Russell**[4]

Published online: 2 November 2015

The dawn of the space age has brought on a new era for exploring comets. Starting in 1985, a total of ten spacecraft have flown by seven different comets at high relative velocities reaching distances as close as 180 km from the surface. These missions made a variety of observations, including the first ever in situ measurements of cometary plasma allowing detailed studies of the interaction of the cometary coma and the solar wind. The International Cometary Explorer (ICE) achieved the first cometary flyby passing within 7800 km of comet 21P/Giacobini-Zinner on 11 September 1985. This same spacecraft passed through the tail of comet 1P/Halley in May 1986, millions of kilometers from the nucleus. Five other spacecraft, Vega 1 & 2, Sakigake, Suisei and Giotto also visited comet 1P/Halley in 1986, with Giotto passing through the coma within 600 kilometers of the nucleus on 14 March 1986. Giotto was the first European deep space mission, provided the closest images of comet 1P/Halley and carried mass spectrometers that measured the composition of ions and neutrals in Halley's coma. Giotto went on to conduct a second flyby of a comet, passing within 200 kilometers from the nucleus of comet 26P/Grigg-Skjellerup on 10 July 1992. The Deep Space 1 mission, primarily designed to test new technology, flew within 2000 kilometers of comet 19P/Borrelly on 22 September 2001. The Stardust mission is the only mission so far to return samples of cometary dust collected during a flyby within 240 kilometers of the surface of comet 81P/Wild 2 on 2 January 2004 and return the samples to Earth for laboratory analysis. Stardust later flew within 180 kilometers of Comet 9P/Tempel 1 on 14 February 2011. The Deep Impact mission sent an impactor into Comet 9P/Tempel 1 on 4 July 2005,

✉ C.T. Russell
ctrussel@igpp.ucla.edu

1 Space Science and Engineering Division, Southwest Research Institute, 6220 Culebra Rd, San Antonio, TX 78238, USA

2 Aix Marseille Université, CNRS, LAM (Laboratoire d'Astrophysique de Marseille) UMR 7326, 13388, Marseille, France

3 LESIA, Observatoire de Paris, LESIA/CNRS, UPMC, Université Paris-Diderot, 92195 Meudon, France

4 Institute of Geophysics and Planetary Physics/UCLA, Los Angeles, CA 90095-1567, USA

excavating dust and allowing studies of dust and subsurface composition as well as physical properties of the nucleus. This spacecraft was later renamed the EPOXI mission and conducted a flyby of comet 103P/Hartley 2 on 4 November 2010.

Although much was learned from the in situ measurements made during these cometary flybys, they were limited to single high-speed passes through the coma at one point during the comet's orbit around the Sun. In order to provide a more detailed study of a comet, the Rosetta mission was designed to escort a comet during a large portion of its orbit around the Sun and deposit a lander. The Rosetta orbiter has twelve science instruments and the Philae lander has nine additional instruments. Rosetta was named after the famous "Rosetta Stone", an ancient tablet that allowed archeologists to translate Egyptian hieroglyphics providing a breakthrough in understanding of ancient Egyptian civilization. Rosetta was launched in 2004 and arrived at its target, comet 67P/Churyumov-Gerasimenko on 6 August 2014. The spacecraft was the first to ever orbit a comet, obtaining orbits down to 6 km from the surface. This is also the first mission to successfully land a probe on a comet, which was achieved on 12 November 2014.

In order to prepare the community for the task of placing Rosetta observations within the greater context of using comets as tracers of solar system formation and evolution, a workshop was held in Toulouse, France 1-3 April 2014. More than 100 scientists from 18 different countries participated in this three-day science program, which was comprised of themed sessions, and featured a mixture of invited reviews, invited and contributed talks and posters. Invited speakers included many of the community leaders in cometary science, measurements and technology development. The workshop covered topics ranging from the dynamical and chemical evolution of the solar nebula during formation, to the techniques for measuring the composition of comets. Of particular focus was the role that measurements made by Rosetta could be expected to play in understanding the origin of solar system bodies. The purpose of this special issue is to build upon the results of this workshop, providing a formal record of the state of knowledge leading into the Rosetta mission.

This special volume contains eight papers that cover a range of subjects relating the current understanding of comets to solar system formation and evolution. The first four papers summarize what is known about cometary composition and physical properties of cometary nuclei. Anita Cochran and co-authors in "The composition of comets" review the current state of knowledge of the composition of the gas and dust in the comae of comets based on spectral observations made in numerous wavelength ranges and in situ measurements with mass spectrometers. Observations of how coma gas composition changes with distance from the Sun are also reviewed. Various taxonomic studies are discussed, along with the limitations of taxonomic descriptions. The paper concludes with a description of the Rosetta instrument capabilities for measuring the composition of the gas and dust of the comet 67P/Churyumov-Gerasimenko, as well as the implications for measurements made early in the mission. In "Cometary isotopes measurements", Dominique Bockelée-Morvan and co-authors focus on isotopic measurements made in cometary comae, micrometeorites, interplanetary dust particles and Stardust grains. These measurements cover D/H, $^{14}N/^{15}N$, $^{16}O/^{18}O$, $^{12}C/^{13}C$, and $^{32}S/^{34}S$ ratios in cometary grains and gases and are critical for evaluating the origin of cometary material, and the physical and chemical conditions in the early Solar Nebula. This paper also presents the capabilities of Rosetta instruments in measuring isotopic ratios. In "The subsurface structure and density of cometary nuclei", Philippe Lamy and co-authors review the knowledge gained of cometary interior structure from ground-based radar studies of the first few meters of cometary nuclei that pass close to Earth. What little information is available suggests significant variation in subsurface properties of cometary nuclei from one comet to the next. Murthy Gudipati and co-authors outline how laboratory studies have contributed to the current understanding of comets in "Laboratory studies towards understanding comets". Their paper presents recent advancements in understanding of comets thanks to laboratory studies, outlines the need for new laboratory simulations, and makes predictions for future explorations. Laboratory studies are a critical aspect of cometary science because they help to improve understanding of the cometary interior, surface, coma and tail.

These contributions are followed by two papers that summarize understanding of the chemistry and dynamics involved in the formation of the solar system. In "The composition of the protosolar disk and the formation conditions of comets", Karen Willacy and co-authors summarize our understanding of the composition, structure and evolution of protoplanetary disks by reviewing the chemical processes that began in the parent molecular cloud core and continued through the collapse of that core to form the protosun and the solar nebula, as well as during the evolution of the solar nebula itself. Studies of protoplanetary disks around young stars similar to the proto-Sun are help to understand the evolution of disks similar to the solar nebula while they are in the process of evolving to form their own planetary systems. Through their work we learn about the physical and chemical conditions under which comets formed, and about the types of dynamical processing that shaped the solar system we see today. In "Dynamics of the formation of the solar system", Luke Dones and co-authors discuss the evolution of our understanding of the origins of the three known cometary reservoirs: the Oort Cloud for long-period, the Kuiper Belt for ecliptic comets, and the asteroid belt in the case of main-belt comets.

The last two contributions discuss the implications of cometary studies for understanding the formation and evolution of the solar system. In "On the evolution of comets", Aurélie Guilbert-Lepoutre and co-authors evaluate how the observed physical, chemical and orbital characteristics of comets could have evolved over the more than four billion years due to the time-varying radiative and collisional environment of the solar system. To start, the composition and internal structure of comets inherited from their formation stage may be very diverse. Further evolution of comet nuclei involving some possible processing from radiogenic heating, space weathering and large- and small-scale collisions might have modified their primordial structures and compositions with various degrees. In "Constraints from comets on the formation and volatile acquisition of the planets and satellites", Kathleen Mandt and co-authors evaluate the dual role comets play in understanding the formation and evolution of the solar system. First, the composition of comets provides information about the origin of the giant planets and their moons because they are believed serve as a record of conditions during the early stages of solar system formation. After the solar system formed, the orbits were perturbed allowing them to travel into the inner solar system, impact the planets, and thus contribute to the volatile inventory of planetary atmospheres.

The success of this volume is due to many people; but first of all, the editors wish to thank the authors who had the difficult job of distilling the myriad papers on our current understanding of comets into highly readable articles. The editors also benefited from an excellent group of referees who acted as a test readership, refining the manuscripts provided by the authors. These referees included Jérôme Aléon, the late Claudia J. Alexander to whom this volume is dedicated, Claude Arpigny, Michael Belton, Guillermo Manuel Munoz Caro, Dale P. Cruikshank, Jacques Crovisier, Walter F. Huebner, Wing-Huen Ip, Michiel Hogerheijde, Paul Kamoun, Klaus Pontoppidan, Dina Prialnik, Francois Robert Philippe Rousselot, Diane Wooden.

Equally important has been the strong support this project has received at Springer and the extra effort expended by Jennifer Satten, Jurgita Vanagaite and Nirmala Kumar. At UCLA we were skillfully assisted by Marjorie Sowmendran who acted as the interface between the authors, editors, referees and the publisher.

Space Sci Rev (2015) 197:9–46
DOI 10.1007/s11214-015-0183-6

The Composition of Comets

Anita L. Cochran[1] · Anny-Chantal Levasseur-Regourd[2] · Martin Cordiner[3] · Edith Hadamcik[4] · Jérémie Lasue[5] · Adeline Gicquel[3,6] · David G. Schleicher[7] · Steven B. Charnley[3] · Michael J. Mumma[3] · Lucas Paganini[3] · Dominique Bockelée-Morvan[8] · Nicolas Biver[8] · Yi-Jehng Kuan[9]

Received: 26 November 2014 / Accepted: 6 July 2015 / Published online: 25 July 2015

Abstract This paper is the result of the International Cometary Workshop, held in Toulouse, France in April 2014, where the participants came together to assess our knowledge of comets prior to the ESA Rosetta Mission. In this paper, we look at the composition of the gas and dust from the comae of comets. With the gas, we cover the various taxonomic studies that have broken comets into groups and compare what is seen at all wavelengths. We also discuss what has been learned from mass spectrometers during flybys. A few caveats for our interpretation are discussed. With dust, much of our information comes from flybys. They include in situ analyses as well as samples returned to Earth for laboratory measurements. Remote sensing IR observations and polarimetry are also discussed. For both gas and dust, we discuss what instruments the Rosetta spacecraft and Philae lander will bring to bear to improve our understanding of comet 67P/Churyumov-Gerasimenko as "ground-truth" for our previous comprehensive studies. Finally, we summarize some of the initial Rosetta Mission findings.

Keywords Comets · Origins · Taxonomy · Rosetta

✉ A.L. Cochran
anita@astro.as.utexas.edu

1 McDonald Observatory, The University of Texas, Austin, TX, USA

2 LATMOS, UPMC, Paris, France

3 NASA Goddard Space Flight Center, Greenbelt, MD, USA

4 LATMOS, UPMC, Guyancourt, France

5 IRAP, UPS, Toulouse, France

6 MPS, Göttingen, Germany

7 Lowell Observatory, Flagstaff, AZ, USA

8 Observatoire de Paris, LESIA, Meudon, France

9 National Taiwan Normal University, Taipei, Taiwan, ROC

1 Introduction

Comets are leftovers from when our Solar System formed. These icy bodies were formed out of the solar nebula in the outer regions of our Solar System. Many of these bodies were swept up and incorporated into the giant planets. The remnants of planetary formation were either ejected from the nascent Solar System, were gravitationally perturbed to the Oort cloud or remained in reservoirs past the orbit of Neptune. The comets we see today have undergone little change from their primordial states. Comets, therefore, represent important objects to study in order to determine constraints suitable for models of the early solar nebula.

Whipple (1950) first described comets as "dirty snowballs" that are composed of a mixture of ices and dust. The mass is approximately equally divided into these components. As the comets approach the Sun, they are heated and the ices sublime. Since comets are small (generally a few to 10s of km), the resultant gas is not bound to the nucleus and expands outwards into the vacuum of space. When it leaves the nucleus, the gas carries with it some of the solid particles (refractory grains, refractory organics, icy grains, aggregates of ice and dust). The resultant material forms the cometary comae (and sometimes tails) that are so prominent when a comet is in the inner Solar System. We do not observe the nucleus composition directly, except when observing from a spacecraft flying past or during a rendezvous mission. What we generally observe is the gas in the coma and the light reflected off the refractories and we *infer* composition from those.

The gases that come directly from the nucleus first flow through a region near the nucleus where the gas densities are sufficiently high that collisions, and thus chemical reactions, can take place. This so-called "collisional" zone is <1000 km for all but the most productive comets. With typical outflow velocities near 1 AU of 850 m/sec, this means that the outflowing gas clears this region in the first 20 minutes of outflow. Once out of the collisional zone, photochemical processes change the composition of the gases.

The comets were formed over a wide range of heliocentric distances and conditions. Petit et al. (2015, this issue) discuss the dynamics of the formation. In this review, we will discuss what we have learned about the chemical composition of the gas and dust using studies at all wavelengths. Our conclusions will be tempered by some caveats based on some current observations. The subject of isotope ratios, however, will be covered in the review by Bockelée-Morvan et al. (2015, this issue). Finally, we will discuss what observations the Rosetta Mission can make to improve our understanding, We include some early results of this mission.

2 Gas Composition

The spectra of comets are mostly composed of emissions from gas-phase molecules superposed onto a continuum that results from sunlight reflecting off dust. Many of the emissions arise from resonance fluorescence of the gas, but prompt emissions also play an important role. Optical spectra of comets have been obtained since the 1860s, while UV, radio and IR spectra have been obtained since the 1970s.

Optical spectra of comets contain few parent species. The spectra are mostly bands of fragment molecules (often called daughters though they can be grand-daughters or chemical products). Thus, the parent species must be inferred using lifetimes and chemical reaction networks. The IR and radio spectra (including mm and sub-mm) contain a mixture of parent and daughter species. Many of the molecules only possess transitions in the IR because they do not have permanent dipole moments. In the UV, we see parents and daughters, as well as a number of important atomic features, such as H, C and O.

Early on, it was noted that the spectra of different comets seemed similar (rarely was an unusual feature noted in the spectra). The strengths of the emission lines relative to the continuum did seem to vary and the relative strengths of bands changed with heliocentric distance. This led to the question of whether all comets shared the same composition or whether there were distinctly different classes of comets. This question is important for our understanding of the homogeneity of the solar nebula at the epoch(s) and the region(s) where comets formed. Thus were born some large comparative studies of comets at all wavelengths in order to understand the chemical homogeneity or diversity of comets. Of course, even if comets formed identically, their spectra might have been altered by activity or irradiation over their lifetime. Thus, we need to look for clues of evolution in the spectra.

2.1 Optical Observations

Since optical observations of comets have existed for far longer than spectra at other wavelengths, there are many more comets observed in the optical than at any other wavelengths. The optical observations consist of high-resolution and low-resolution spectra, as well as photometry obtained with narrow-band filters that isolate the molecular bands. Optical observations are also the most sensitive to faint comets, meaning that the optical observations can be gathered for more comets and at larger heliocentric distances than the other wavelength observations. Additionally, the typical photometric aperture or long slit used is larger than those used in the IR, allowing observations of a larger portion of the coma. The larger the aperture, the less sensitive are the observations to outflow velocity or short term temporal variability (which instead can be measured at other wavelengths). Feldman et al. (2004) discuss how to turn optical observations of fragment species into knowledge of the composition of comets.

The spectral observations have the advantage that they can easily isolate the molecular features from the continuum, but suffer from small apertures. Some observers use long-slit spectrographs that allow for measurements of the coma as a function of cometocentric distance. This is an improvement over small apertures and allows direct measurement of scale lengths for decay, but the long slit only allows for sampling in specific directions within the coma. Spectroscopic observations often have significant spectral grasp, so that multiple species can be observed simultaneously, minimizing worries of temporal variability. Large surveys with low-resolution optical spectra have been carried out since the 1970s (Newburn and Spinrad 1984, 1989; Cochran 1987; Fink and Hicks 1996; Fink 2009; Langland-Shula and Smith 2011; Cochran et al. 2012) with well over 150 comets observed spectroscopically.

High-resolution studies have tended to focus more on individual cometary properties and are excellent for detailed work. Examples of the types of studies performed are $^{12}C/^{13}C$ (Danks et al. 1974; Kleine et al. 1995), $^{14}N/^{15}N$ (Arpigny et al. 2003; Manfroid et al. 2005, 2009) and O ($^1S/^1D$) (Cochran and Cochran 2001; Cochran 2008; Festou and Feldman 1981; Capria et al. 2005, 2008; McKay et al. 2012, 2013; Decock et al. 2013, 2015).

Photometry with filters allows for much larger apertures and is therefore much more sensitive than spectroscopy (Schleicher and Farnham 2004). Continuum removal is done with observations through filters that isolate continuum regions. Large photometric studies of comets have been carried out (A'Hearn and Millis 1980; A'Hearn et al. 1995; Schleicher and Bair 2014), with more than 160 comets observed. Photometry suffers from the problem of not being able to measure weak features (e.g. CH or NH_2) because there is a practical limit to how accurately the underlying continuum can be determined and removed.

With photometry, the different filters are observed sequentially, thus temporal variations could change our understanding of the relative abundance of species. However, the cycle time to observe all filters is typically much shorter than the expected timescale for temporal variability. With long-slit spectra, the spectra may be obtained at multiple position angles to sample any inhomogeneities of the coma. At that point, temporal variations must be considered. Typically, the time between position angles is much longer than between filters with photometry.

Without prior knowledge of any asymmetry, single-aperture photometry and small-aperture and long-slit spectroscopy all must assume that the distribution of the gas in the coma is symmetric. This is rarely the case. Examples of asymmetry can be seen in Cochran et al. (2012, see Fig. 5). While not able to probe the morphology of the coma, these observations are, however, measuring the bulk composition of the nucleus. It is this composition that is the necessary measurement to aid our understanding of the conditions in the solar nebula, while morphology tells us more about the physical condition of the nucleus today. However, morphology also may tell us about inhomogeneities of the nucleus, lending clues to the size and variation in composition of the presumed cometesimals from which the nucleus was assembled.

To some extent, this assumption of symmetry can be mitigated with an integral field unit (IFU) spectrograph, where spectra can be obtained from many positions in the coma at the same time and with spectral resolution sufficient to isolate bands. IFU spectroscopy is not as readily available as other instrument types and most IFU spectrographs have very small fields-of-view. Alternatively, the coma can be imaged with narrow-band filters to measure the morphology of the coma.

Early studies of cometary diversity were carried out by A'Hearn and Millis (1980), Newburn and Spinrad (1984) and Cochran (1987). These papers included measurements of daughters such as CN, C_2, C_3, and O I. All the authors noted that the ratios of species appeared very similar for most comets but there were occasional comets that seemed depleted in C_2 or C_3. However, the sample sizes in these first papers were small and it was not apparent what cometary properties were common to the unusual comets.

Since 1995, several large studies have been published with the total number of comets studied by all groups around 250 (A'Hearn et al. 1995; Fink 2009; Langland-Shula and Smith 2011; Cochran et al. 2012; Schleicher and Bair 2014). Many of these comets were observed by more than one group. Additional species were added in these surveys, including OH, NH, NH_2, and CH.

From these studies, it has become apparent that the majority of comets really do have very similar relative compositions. However, 25–30 % of the comets look different than the rest. As noted in the earlier studies, many of the comets that are different are depleted in C_2 and C_3.

Unfortunately, intercomparing surveys is difficult, since the various authors have used different g-factors and scale lengths in their reductions, as well as different cut-offs for the definition of depletion. However, the surveys are mostly consistent in comet-by-comet comparisons when they do overlap on the same comet, agreeing which comets show differences. The inconsistent reductions, however, are why ever larger surveys by a single group can make the most progress.

Figure 1 demonstrates that the depletions are not a subtle effect by comparing observations of comets 21P/Giacobini-Zinner (GZ) and 8P/Tuttle when both were near 1 AU from the Sun and 0.5 AU from Earth. The Tuttle spectrum demonstrates the full, normal, range of observed molecules. GZ is the prototypical depleted comet and it is clear that there is much less C_3/CN and C_2/CN in GZ than in Tuttle.

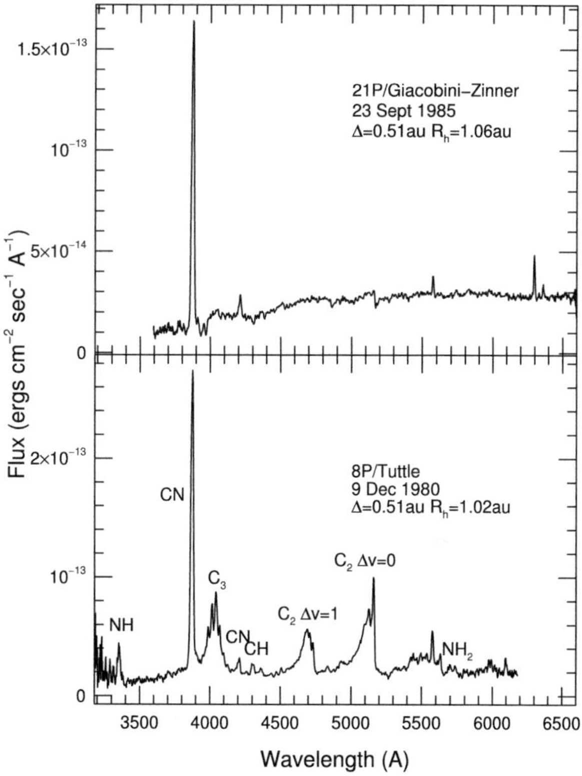

Fig. 1 Spectra of 21P/Giacobini-Zinner and 8P/Tuttle obtained at McDonald Observatory are shown, scaled to the CN band at 3880 Å. The comets were observed at comparable heliocentric and geocentric distances. Inspection of the figure shows that the well-defined C_2 and C_3 bands seen in Tuttle are nearly absent in Giacobini-Zinner's spectrum. (Spectra courtesy of A. Cochran)

After the earliest studies, it seemed that the depletion occurred primarily in Jupiter Family comets. However, the studies of A'Hearn et al. (1995), Fink (2009), Langland-Shula and Smith (2011) and Cochran et al. (2012) had sufficiently large samples that it became apparent that the relationship between dynamical state and depletion was not so simple. These studies showed that depleted comets can be from any dynamical type: Jupiter Family, Halley Type, Long-period (dynamically new and not). However, it is more common for depleted comets to be Jupiter Family comets; Cochran et al. (2012) found that two thirds of the depleted comets were Jupiter Family while one third were long period. In addition, a higher percentage of the Jupiter Family comets (37 %) were depleted than of the long period comets (18.5 %). Thus, the depletion is either pointing to different evolutionary states or it is pointing to mixing of the formational reservoirs.

Cochran et al. (2012) pointed out that the picture is a little more subtle than just "carbon-chain depleted" or "typical". The definitions for carbon-chain depletion typically only account for C_2 relative to OH or CN and it is these C_2-depleted comets that represent 25–30 % of all comets. However, Cochran et al. found that only $\sim$ 10 % of comets are depleted in both C_2 *and* C_3. They also found some number of comets with different patterns for some other molecules, but the statistics did not warrant breaking the comets into additional groups.

Schleicher and Bair (2014) have analyzed photometry of 167 comets observed and reduced in a consistent manner. Of that sample, 101 were of sufficient quality that they were able to be included in the analysis of the statistics of the sample. They find the typical carbon-chain depletions found by other authors. However, their statistics are better than previous studies and they now claim that there are 7 classes of comets: 1) typical (still $\sim$ 70 %

of comets); 2) comets depleted in both C_2 and C_3 with the depletion as strong as GZ or stronger; 3) comets depleted in both C_2 and C_3 but weaker than the GZ depletion; 4) comets depleted in C_2 but not C_3 (in agreement with Cochran et al. 2012); 5) comets depleted in C_3 but not C_2; 6) comets depleted in NH but not depleted in any carbon species; and 7) comets depleted in CN relative to C_2 and C_3 but still depleted in C_2 and C_3 compared to OH (a class of 1–96P/Machholz 1; Schleicher 2008).

With the most recent findings, especially that certain depletions are not confined to a single dynamical type, the question of formation vs. evolution is important. One of the best constraints on this question is comet 73P/Schwassmann-Wachmann 3 (SW3). In 1995, comet SW3 underwent a splitting event into three pieces; subsequently, those pieces split again. SW3 is a strongly C_2 and C_3 depleted comet and thus has a distinctive "fingerprint". If the depletion were just an evolutionary effect, from multiple perihelion passages, we would expect it to be confined mostly to the surface and the interior would appear typical. However, observations of the distinct pieces during the 2006 apparition showed that all the pieces had identical depletions both in the parents observed in the IR and the daughters observed in the optical and IR (Kobayashi et al. 2007; Schleicher and Bair 2011; Jehin et al. 2008). In addition, there was no change from measurements of SW3 obtained before the splitting in 1995 (Schleicher and Bair 2011). This has been interpreted as strong evidence that C_2 and C_3 depletions are primarily from the formation of the comets, not from their subsequent evolution.

2.2 UV Observations

The UV part of the spectrum of comets is not as well studied as the optical because the Earth's atmosphere precludes observations from ground-based telescopes (the OH $(0,0)$ band at 3080 Å is technically in the UV but can be observed from the ground; however the atmospheric extinction at this wavelength is typically around ~ 1.5 mags/airmass). Thus, UV observations require a spacecraft-borne telescope or a rocket experiment. The electronic bands of some diatomic molecules are visible in the FUV; atomic daughter products that result from solar UV dissociation are also seen.

In the UV, transitions of CO, H, H_2, O, C, OH, CS and S_2 have been regularly detected. Most of these species are products of a parent molecule, though the processes to produce them are dependent on the transition. Whether S_2 is a parent or a daughter is still unknown; regardless S_2 is very short-lived. CO is seen to have two different band series, the fourth positive bands and the Cameron bands. The fourth positive bands are the result of solar pumped fluorescence of CO. The Cameron bands result from either electron impact on CO or dissociative excitation of CO_2, the latter being more important. Thus, these two bands can yield information on two possible parents.

The UV era of cometary observations started in 1970 with the Orbiting Astronomical Observatory (OAO-2) observations of comet Bennett (Code et al. 1972). In these observations OH and Ly α, both daughters of H_2O, were detected.

Significant progress in our understanding of the UV spectra of comets came with the International Ultraviolet Explorer (IUE) satellite. Ultimately, IUE was used to observe more than 50 comets between 1978 and 1996. Festou (1998) summarized the findings from these observations. He concluded that "with the exception of the S_2 molecule, new and long-period comets have a chemical composition that does not differ by more than a factor of two from that of periodic comets" (but see discussion below).

Since 1990, the UV has been covered with the Hubble Space Telescope (HST) and a variety of instruments. Also operational during this period, the Far Ultraviolet Spectroscopic

Explorer (FUSE) mission operated from 1999–2007 and the Galaxy Evolution Explorer (GALEX) operated from 2003. Of these three missions, HST has been the workhorse, having been used to observe many comets.

FUSE was used to observe the FUV from 900–1200 Å in four comets. In that bandpass, emissions due to H I, O I, CO (including three new Hopfield-Birge band systems), H_2 and N I were detected. A He I resonance transition at 584 Å was seen in second order and was taken as an indicator of solar wind charge transfer in the coma. Ar and N_2 were searched for but not detected (Feldman 2005; Feldman et al. 2009). In addition, FUSE detected many lines that were initially unidentified, but were later determined to be H_2 (Liu et al. 2007).

The GALEX satellite uses time-tagged images and grism mode to obtain both FUV and NUV observations. The FUV observations have concentrated on C I. The NUV is used to observe OH, CS and CO^+ (Morgenthaler et al. 2009). Comets fill the field-of-view of the GALEX detector so it is difficult to estimate the background. Morgenthaler et al. (2011) describe the process of extracting the C I lifetime from observations of comet C/2004 Q2 (Machholz). They found that either the CO lifetime is shorter than previously thought or another shorter-lived species must be contributing to the production of C I in the inner coma.

Using the FIMS/SPEAR instrument on the Korean STSAT-1, Lim et al. (2014) observed comet C/2001 Q4 (NEAT) in the FUV. They observed the CO fourth positive bands as well as C I and S I atomic features. They concluded that the production rate of C is $\sim 60\ \%$ of CO, agreeing with results obtained for other comets, e.g. Morgenthaler et al. (2011) for comet C/2004 Q2 (Machholz). For comet NEAT, Lim et al. found that S I production was about 1 % of the water production rate. It was assumed the parent was H_2S.

Meier and A'Hearn (1997) had already published a study of S I in 19 comets observed with IUE and HST. They also observed CS in these spectra, presumably a daughter of CS_2. Typical values for CS_2 derived are not sufficient to supply all of the S. Meier and A'Hearn pointed out that all the transitions they see are optically thick. This means that simple fluorescence models are not sufficient and they developed a radiative transfer/collisional model to interpret their spectra. For the comets they studied, Meier and A'Hearn derived sulfur abundances of 0.1 % to 1 % relative to water.

HST has observed 18 comets with a variety of instruments. The comets are a mixture of Jupiter Family and long-period comets. Comet 103P/Hartley 2 (the target of the EPOXI mission) was observed with HST during three different apparitions (1991, 1998, 2010), the only comet observed over more than one apparition. With most of these observations, the CO fourth positive bands and the OH bands are observed simultaneously, thus enabling a measure of the $CO{:}H_2O$ for many comets. Lupu et al. (2007) presented models of the CO fourth positive bands in four comets observed with HST. These observations showed a large range of Q_{CO}/Q_{H_2O}, from $< 1\ \%$ for C/2000 WM1 (LINEAR) to $> 20\ \%$ for comet C/1996 B2 (Hyakutake).

Since then, Weaver et al. (2011) have reported that comet 103P/Hartley 2 had a CO/H_2O production rate ratio of 0.15–0.45 %, amongst the most carbon monoxide poor comets measured. The temporal variation in CO abundance was seen to correlate with the rotation of the comet. It should be remembered that A'Hearn et al. (2011) detected strong CO_2 (CO_2/H_2O $\sim 20\ \%$) with the EPOXI spacecraft at a similar time. Thus, Hartley 2 is CO_2 rich while CO poor.

Weaver et al. noted that CO/H_2O varies by a factor of 50 across all comets observed to date using UV data. This is in agreement with IR observations (e.g. Paganini et al. 2014b). Though the most CO-rich comets are long period comets, not all long period comets are high in CO (e.g. C/2000 WM1 (LINEAR) is very low). Even though CO can be produced as a daughter product of species such as CO_2 or H_2CO, Weaver et al. concluded that the HST high spatial resolution suggests that the CO observed by HST originates from the nucleus.

The Solar Wind ANistropies (SWAN) all-sky Ly α camera on the Solar and Heliospheric Observatory (SOHO) mission has been used regularly to monitor comets since its launch in 1995. Hydrogen is a dissociation product of many species, the most notable of these H_2O. To date, SWAN has been used to observe over 60 comets. These comets are either observed as routine interlopers in the all-sky images or can be specifically targeted for monitoring. With its orbit at the L1 Lagrangian point, SOHO with SWAN is able to observe comets at very small heliocentric distances as well as north and south of the ecliptic. When a comet is very near to the Sun, the lifetime of H_2O against photodissociation is extremely short and water dissociates completely within the collisional zone. Examples of comets observed close to the Sun are C/2002 V1 (NEAT), C/2002 X5 (Kudo–Fujikawa), 2006 P1 (McNaught) and 96P/Machholz 1 (Combi et al. 2011). The closest passage to the Sun by a comet observed by SOHO came in fall 2013 when comet C/2012 S1 (ISON) broke up within a few solar radii, hours before perihelion (Combi et al. 2014).

While the SWAN observations tell us only about the water production rate of comets and not of other species, they are extremely valuable for being able to monitor the production rates of comets as they approach and recede from the Sun. Combi et al. (2011) pointed out that there are two types of behaviors: comets whose water production rate increases with a steep slope as they approach the Sun and those with a more shallow slope. However, they pointed out that all of the steep-sloped comets are Jupiter Family comets of the type that have typical perihelia near 1 AU. Indeed, the only periodic comets in the moderate-slope group are 153P/Ikeya-Zhang and 96P/Machholz 1. Neither is a Jupiter Family comet. Ikeya-Zhang has a ~ 350 year period. Machholz 1 is in a highly inclined (58°) orbit.

2.3 IR Observations

The lack of a permanent dipole moment in many of the cometary parent molecules means that these molecules cannot be observed in the optical or radio; they have strong vibrational transitions in the IR. Studies of these spectral regions were not possible until the 1970s when IR detectors were introduced to astronomy. The first detectors were not very sensitive and were also very small.

The IR offers some difficult challenges not seen in the optical part of the spectrum. The Earth's atmosphere has significant opacity in the IR due to molecules such as CO_2 and H_2O. Thus, there are broad regions of the atmosphere that are opaque to the cometary photons. This means that we can only study molecules that occur in "clean" regions of the spectrum. It also argues for higher spectral resolution to avoid bad regions of the spectrum. Even the clean regions have significant, and variable, opacity so the atmosphere must be carefully measured, with equal amounts of observation time on sky and comet, and the atmosphere must be modeled and removed. Figure 2 demonstrates the importance of high spectral resolution for IR work.

Most of the IR observations are in the 1–5 μm region, with most of the interest in the 2.5–5 μm bandpass. IR observations are typically limited by the regions in the Earth's atmosphere that are most transparent. Doppler shifts of cometary lines can move them from the good region into the more opaque regions of the bandpass. Limitations on time available to observe and the spectral grasp can lead to different molecules being detected for different comets.

A variety of species in the IR spectra of comets have been detected, including H_2O, CH_4, C_2H_2, C_2H_6, H_2CO, NH_3, CH_3OH, HCN, etc. All of these are seen in molecular clouds. However, there are still some species seen in molecular clouds and not comets. In addition to the parent species seen in the IR, fragments such as OH, NH_2 and CN are observed.

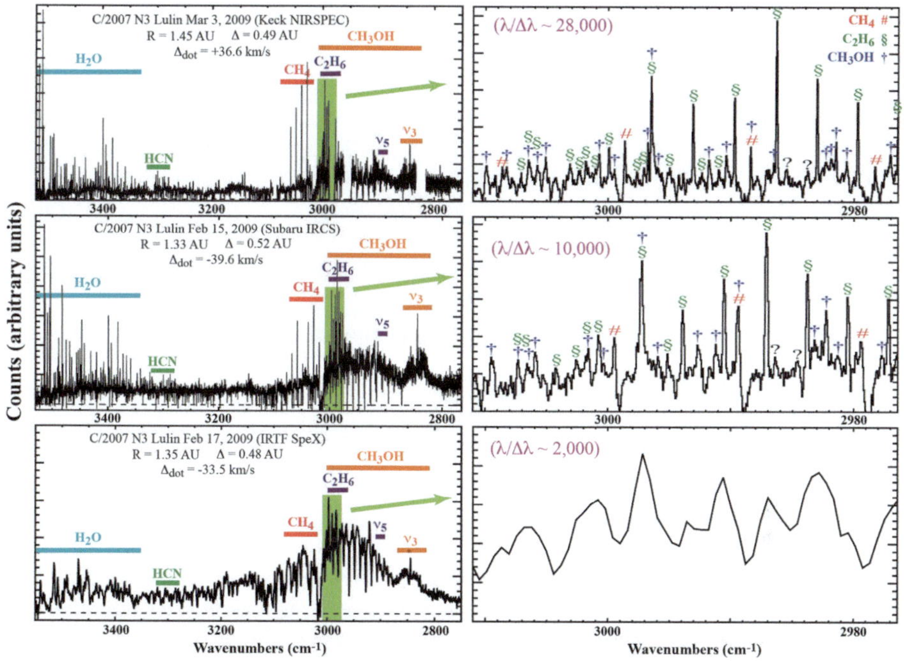

Fig. 2 Direct comparison of high and moderate-resolution spectra for C/2007 N3 Lulin. *Left panels* show the complete spectra (before telluric correction) from ∼ 2.83–3.64 μm (∼ 3550–2750 cm^{-1}). *Right panels* show a zoomed-in portion (∼ 3.32–3.36 μm) of these full spectra, as indicated by *green boxes* in the *left figures*. This illustrates two important points: (1) High-resolution spectra are necessary for the complete chemical interpretation of moderate-resolution spectra. (2) IR spectra are rich in cometary emissions underlining the importance of spectral surveys. Note that these spectra were obtained on different dates so the geocentric Doppler-shift is different (particularly for the NIRSPEC spectrum) so some lines are not present on all dates due to telluric extinction. (Spectrum courtesy N. Dello Russo)

Dello Russo et al. (2013) include a high-resolution atlas of comet 103P/Hartley 2. The first such infrared spectra of a comet obtained with a cross-dispersed echelle spectrometer appeared in Mumma et al. (2001) for comet C/1999 H1 (Lee). A complete spectral atlas for comet Lee appeared in Dello Russo et al. (2006).

Determining production rates from IR spectra requires taking into account the various excitation processes. These are discussed in detail in Bockelée-Morvan et al. (2004). As with the optical models, differences in model parameters can make intercomparing results a bit problematic. In recent years, intense activity has led to greatly improved fluorescence models for primary volatiles in comets and improved transmittance models for the terrestrial atmosphere (e.g. DiSanti et al. 2013; Gibb et al. 2013; Kawakita and Mumma 2011; Lippi et al. 2013; Radeva et al. 2011; Villanueva et al. 2012).

The sample of comets observed in the IR is much smaller than the sample in the optical because the IR observations require bigger telescopes and the IR cannot be used to observe the fainter comets. In addition, IR observations are generally obtained over a smaller range of heliocentric distances than, for example, optical data, because the comets are fainter far from the Sun. This is rapidly changing. Paganini et al. (2013) detected CO in the unusual comet, 29P/Schwassmann-Wachmann 1 at 6.26 AU. At this point, about three dozen comets have been observed in some part of the IR.

IR observations are scattered over many papers that concentrate on a single or a few comets, with brief mention of groupings of comets. Since 2013, papers have been published on 2P/Encke (Radeva et al. 2013), 21P/Giacobini-Zinner (DiSanti et al. 2013), 81P/Wild 2 (Dello Russo et al. 2014b), 103P/Hartley 2 (Kawakita et al. 2013; Dello Russo et al. 2013; Bonev et al. 2013), C/2003 K4 (LINEAR) (Paganini et al. 2015) C/2009 P1 (Garradd) (DiSanti et al. 2014b), C/2010‘G2 (Hill) (Kawakita et al. 2014), C/2012 F6 (Lemmon) (Paganini et al. 2014a), C/2012 S1 (ISON) (Bonev et al. 2014), and C/2013 R1 (Lovejoy) (Paganini et al. 2014b). Inspection of this list shows that it is a mix of long-period and Jupiter Family comets. However, it has been much harder to observe Jupiter Family comets in the IR since they generally are at the faint end of detectability.

From the IR observations, there is an emerging picture that there are three different kinds of comets: typical, organic enriched and organic severely depleted (cf. Mumma and Charnley 2011). However, as Mumma and Charnley pointed out, taxonomies based on optical fragments and IR primary species are not always in agreement. As an example, 8P/Tuttle, a Halley Type Comet, is found to be typical in C_2/CN in the optical but peculiar in the IR because C_2H_2 and HCN are severely depleted and C_2H_6 is low, in contrast to CH_3OH that is enriched (Bonev et al. 2008).

The limited windows through the atmosphere that ground-based IR observations must use mean that important parents cannot be observed from the ground. The most important is CO_2. CO_2 is important because it represents a significant fraction of the ice (around 20 % is not uncommon). CO_2 is also much more volatile than H_2O. Thus, while the activity of comets is controlled by the sublimation of water inside $\sim$3 AU, outside of this region, it is CO_2 or CO that control the activity (CO can either be a daughter or a parent). Observations obtained with the AKARI spacecraft demonstrate this point, showing that the CO_2 abundance relative to H_2O is much higher outside of 2.5 AU (Ootsubo et al. 2012). Some comets were observed both outside and within $\sim$2.5 AU and the driver of activity is shown to change. No evidence was seen for a difference between Jupiter Family and long-period comets.

Reach et al. (2013) used the Spitzer Space Telescope to determine the amount of CO_2 relative to H_2O. While many of the Spitzer-observed comets look to have abundances similar to the AKARI comets, Reach et al. found a number of comets that were CO_2 poor. Those CO_2 poor comets are a mix of optically classified typical and depleted comets. Reach et al. concluded that more of the CO_2 poor comets were depleted than typical comets but they did not have taxonomic classifications on several of the CO_2 poor comets. With recent photometric observations, it turns out that the CO_2 poor comet group contains 3 depleted comets, 5 typical comets and 1 unclassifiable comet (Schleicher 2015, personal communication). It should be noted that while the AKARI observations are spectral, the Spitzer observations were imaging observations. The filter that contains the CO_2 also contains a considerable quantity of dust that was measured with a nearby continuum filter. Depending on the dust properties, it is possible that some of the noted depletions are instead dust contamination.

While the AKARI and Spitzer spacecraft have been used to look at a large number of comets, these spacecraft are available for observations for only a limited amount of time. McKay et al. (2012, 2013) and Decock et al. (2013, 2015) have used discussions of branching ratios for the processes to produce [O I] (e.g. Festou and Feldman 1981) and have developed procedures and empirical release rates to use observations of the forbidden oxygen lines in the optical to *infer* the abundance of CO or CO_2. This is done by looking at the ratio of the $O(^1S)$ (or green) line to the sum of the $O(^1D)$ (or red) pair of lines. Unfortunately, this method relies on lifetimes and branching ratios that are not well determined in the lab and calibration against the IR is sorely needed.

A key use of IR spectra has been to map the ortho/para ratio of the gas for a variety of molecules. If a molecule has H atoms that are located symmetrically, then these molecules may have states that have different spin orientations of the hydrogen. As an example, for H_2O, the two H atoms can have their spin orientation the same or opposite one another. The first of these orientations is known as the "ortho" state, while the second is the "para" state. A unique feature of the spin states is that it is not possible to flip the state by radiative or collisional processes. Thus, the ortho/para ratio (OPR) of the coma gas measures the ortho/para ratio of the gas reservoir when the gas was frozen into the nuclear ices. Measurement of the rotational distribution in each spin species can lead to a measurement of the ortho/para ratio and thus to the nuclear spin temperature. Since most of these molecules reach equilibrium at some temperature, the ortho/para ratio measures the spin temperature of the ice (but see below). Bockelée-Morvan et al. (2004) discuss how these conversions are achieved.

Mumma et al. (1987) were the first to measure an OPR in H_2O for a comet. Since then, about a dozen OPRs in water have been measured. Mumma and Charnley (2011) summarized the values through 2011. Recently, Bonev et al. (2013) and Kawakita et al. (2013) used this technique to study the spin temperature in H_2O for comet 103P/Hartley 2. As discussed in these papers, they found a range of ratios that did not always agree within the error bars. However, in general, it appeared that T_{spin} was below 40 K.

Shinnaka et al. (2011) presented OPRs in NH_2 (optical) and NH_3 (IR) for 15 comets. They found the same value for most comets, with SW3 standing out as being different. Occasionally, OPRs can be measured in other species (e.g. CH_4, cf. Mumma and Charnley 2011). These observations, however, are rarer because they rely on being able to observe the comets at critical Doppler shifts. In eight comets, spin temperatures for NH_3 and H_2O were relaxed and in agreement; spin temperatures for methane also agreed when co-measured with H_2O or NH_3 (e.g. Mumma and Charnley 2011).

The use of OPRs was believed to be a sensitive measure of the formational temperature of the ice. However, recent laboratory experiments (Fillion et al. 2012) pointed out that the speed of desorption has a major affect on this ratio. Thus, it remains to be seen how useful OPRs are as constraints on temperature in the solar nebula vs. as taxonomic tools.

2.4 Radio Observations

Pure rotational transitions for molecules with dipole moments occur in the radio portion of the spectrum. These simpler transitions make it easier to detect complex volatiles. In addition, radio spectroscopy has very high spectral resolution, allowing for accurate measurements of spectral line shapes and Doppler shifts in the cometary coma. Thus, with radio observations it is possible to measure the line-of-sight outflow velocity of the coma. Radio telescope beams tend to be very large. Thus, the radio observations measure the total abundance of a species but have to assume the outflow is symmetric. Mapping may be obtained by shifting the beam center, though the additional positions are generally limited in number. The mapping can be improved upon with interferometry (see Sect. 2.4.1).

The earliest radio observations were made of the daughter species, OH, at 18 cm. Since 1973, there have been more than 50 comets observed with the Nançay telescope (Crovisier et al. 2002; Colom et al. 2011).

There are more than a dozen species seen in the radio in comets. Species such as CO, H_2CO, CH_3OH, CH_3CN, HC_3N, HCN, NH_3, HCOOH, HNCO, and H_2S are thought to be primary (parent) volatiles. OH, CS, and HNC are likely to be product species. As with the IR, all of these species cannot be observed simultaneously so we have detections of these species in some comets but not others. Thus, though more than 40 comets have been observed in the radio, very few species have been observed in more than 10 comets (Fig. 1 of Crovisier et al. 2009 gives an indication of the number of molecules seen per comet for different types of comets).

The comet with the record for the most species observed is C/1995 O1 (Hale-Bopp). This comet's brightness allowed for the detection of molecules that had never before (or since) been seen and for measurement of the behavior of the species with heliocentric distance (Biver et al. 1997, 2002; Bockelée-Morvan et al. 2000). In addition to the molecules mentioned above, SO_2, H_2CS, NH_2CHO, CH_3CHO, and $HCOOCH_3$ were detected. The most complex molecule detected in the spectrum of Hale-Bopp was HO–CH_2CH_2–OH (ethylene glycol) (Crovisier et al. 2004).

Hale-Bopp was an exceptionally bright comet, with a very high water production rate. Indeed, it was the brightest comet since the radio and IR techniques have become mature and routine. Therefore, it was possible not only to observe more molecules in Hale-Bopp than in other comets, but also to observe them over a larger range of heliocentric distances. Most species were observed out to distances of 3–4 AU. However, HCN and CH_3OH were both observed to 6 AU and CO was observed at 14 AU, some four years after perihelion (Biver et al. 2002). Figure 5 of Biver et al. shows the trends of the production rates of nine molecular species as a function of heliocentric distance, both pre- and post-perihelion. It is obvious from this figure that at large heliocentric distances the CO is controlling the cometary activity, while within about 3 AU the H_2O sublimation is driving the activity.

Crovisier et al. (2009) explored the radio-derived abundances of several species relative to water for about 30 comets. They found no evidence for the three categories of typical, depleted and enhanced comets seen in the IR, nor did they detect a difference between Jupiter Family and other dynamical types of comets. They noted that there is a wide spread of abundances relative to water for most species. The exception to this is for their observations of HCN, which, according to Crovisier et al., show a narrow range of values. Figure 3 shows the range of abundance values, relative to water, found in the IR and radio for a number of comets.

2.4.1 *The Rise of Radio Interferometry*

Combining the benefits of high spectral resolution with simultaneous spatial mapping, radio interferometry is a powerful technique for probing the distribution and kinematics of cometary gas and dust (e.g. Blake et al. 1999; Boissier et al. 2014). Currently in Cycle 2 Early Science mode, the Atacama Large Millimeter/submillimeter Array (ALMA) is a state-of-the-art radio/sub-mm interferometer under construction at 5 km altitude in Chile's Atacama Desert. Once completed in 2015–2016, using 66 antennae separated on baselines up to 16 km, extremely detailed mapping of molecular line and continuum emission from cometary comae will be possible across the frequency range 84–950 GHz, with an angular resolution up to $\sim 0.005''$. Cordiner et al. (2014) first demonstrated the power of ALMA for quantitative measurements of the distributions of molecules and dust in the inner comae of typical bright comets, with spectrally and spatially-resolved maps of HCN, HNC, H_2CO and 0.9 mm dust continuum in C/2012 F6 (Lemmon) and C/2012 S1 (ISON).

Observations of these two Oort cloud comets were made using the ALMA Band 7 receivers, covering frequencies between 338.6 and 364.6 GHz (0.82–0.89 mm) using 28–30 12-m antennae (with baselines 15–2700 m, which provided an angular resolution of approximately $0.5''$). Comet Lemmon was observed post-perihelion on 2013 June 1–2 at heliocentric distance $r_H = 1.75$ AU and ISON was observed pre-perihelion on 2013 November 15–17 at $r_H = 0.54$ AU. The spectral resolution was about $0.42\ \mathrm{km\,s^{-1}}$. For further details see Cordiner et al. (2014).

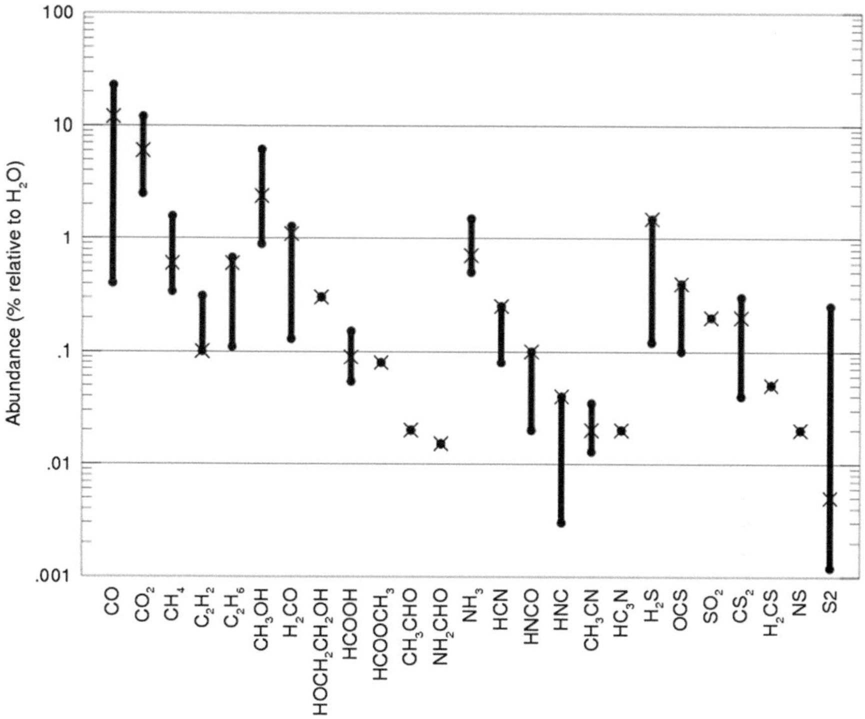

Fig. 3 The abundances for many of the IR- and radio-observed species are shown as a percentage relative to water. The *bar* shows the range of values observed for each species. The "×" marks the value determined for comet C/1995 O1 (Hale-Bopp). CO, CH_3OH, H_2CO, HCN, H_2S and CS_2 were all observed in at least 10 comets; all other species where observed in fewer. The species with a single value (an X-dot) were only observed in Hale-Bopp (based on Fig. 1 of Bockelée-Morvan 2011)

Figure 4 shows spectrally-integrated flux contour maps for the observed molecules in each comet. Dramatic differences are evident between different molecular species observed in the same comet, and between the same species observed in the two comets. By eye, the HCN distributions in both comets appear quite rotationally-symmetric about the central peak. For Lemmon, no offset between the HCN and continuum peaks is distinguishable, whereas ISON's HCN peak (indicated with a white '+') is offset 80 km eastward from the continuum peak (white '×'). Shown in Fig. 4b, the HNC map for ISON exhibits a wealth of remarkably extended spatial structure, with at least three streams (identified at $> 6\sigma$ confidence), emanating away from the main peak (indicated with white dashed arrows). The majority of HNC emission from both comets is asymmetric, originating predominantly in the anti-sunward hemispheres of their comae.

Formaldehyde also shows strikingly different distributions for comets Lemmon and ISON (Figs. 4c and 4f), highlighting the complex origin of this species. Lemmon has a remarkably flat and extended H_2CO map, as demonstrated by the size of the region traced by the 40 % contour compared with the other maps. By contrast, the H_2CO distribution for comet ISON is dominated by a strong, much more compact central peak, and has a relatively symmetrical contour pattern, similar to that of HCN.

Using the method of Boissier et al. (2007, 2014), Cordiner et al. (2014) modeled the interferometric visibility amplitudes for the observed species in order to determine their

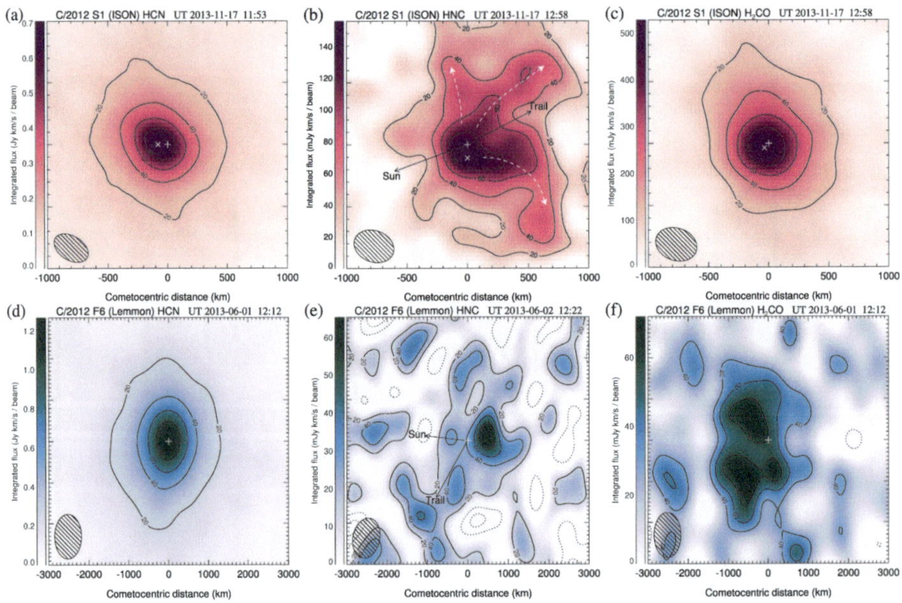

Fig. 4 Contour maps of spectrally-integrated molecular line flux observed in comets S1/ISON (*top row*) and F6/Lemmon (*bottom row*). Contour intervals in each map are 20 % of the peak flux. The 20 % contour has been omitted from panel (**f**) for clarity. Negative contours are *dashed*. On panel (**b**), *white dashed arrows* indicate putative HNC streams/jets. The RMS noise (σ, in units of $\mathrm{mJy\,beam^{-1}\,km\,s^{-1}}$) and contour spacings ($\delta$) in each panel are as follows: (**a**) $\sigma = 10.9$, $\delta = 25.2\sigma$, (**b**) $\sigma = 13.6$, $\delta = 2.3\sigma$, (**c**) $\sigma = 11.0$, $\delta = 9.6\sigma$, (**d**) $\sigma = 13.1$, $\delta = 19.4\sigma$, (**e**) $\sigma = 14.8$, $\delta = 0.9\sigma$, (**f**) $\sigma = 13.7$, $\delta = 1.1\sigma$. The peak position of the (simultaneously observed) 0.9 mm continuum is indicated with a *white* '+'. For ISON, a *white* '×' indicates the integrated molecular emission peak position, offset from the continuum (dust) peak in each case. Sizes (FWHM) and orientations of the point-spread functions are indicated in *lower-left* (*hatched ellipses*); observation dates and times are also given. See Cordiner et al. (2014) for further details

origins within the coma. In both comets, HCN was found to originate from (or within a hundred km of) the nucleus, with a spatial distribution largely consistent with spherically-symmetric, uniform outflow. By contrast, the HNC and H_2CO distributions were consistent with the release of these species as coma products. The H_2CO parent scale length was found to be a few thousand km in Lemmon (at $r_H = 1.75$ AU) and only a few hundred km in ISON (at $r_H = 0.54$ AU), consistent with destruction of the H_2CO precursor by photolysis or thermal degradation at a rate which scales in proportion to the solar radiation flux. The scale length for the putative parent of HNC in comet ISON was found to be ~ 1000 km.

The release of HNC and H_2CO as product species implies the existence of organic precursor materials in the coma, which undergo sublimation, photochemical and/or thermal degradation to produce the observed molecules in the gas phase. Heating or photolysis of materials such as grains, polymers or other macro-molecules, and their subsequent breakdown at distances $\sim$ 100–10,000 km from the nucleus presents the most compelling hypothesis for the origin of the observed H_2CO and HNC. The presence (and composition) of the hypothesized macro-molecular precursors will be measured by the COSIMA instrument on the Rosetta spacecraft during its encounter with comet 67P/Churyumov-Gerasimenko in 2014–2015 (Kissel et al. 2007; Le Roy et al. 2012).

As ALMA's construction and commissioning are completed over the next two years, the array will grow larger and even more sensitive, enabling the distributions and kinematics of

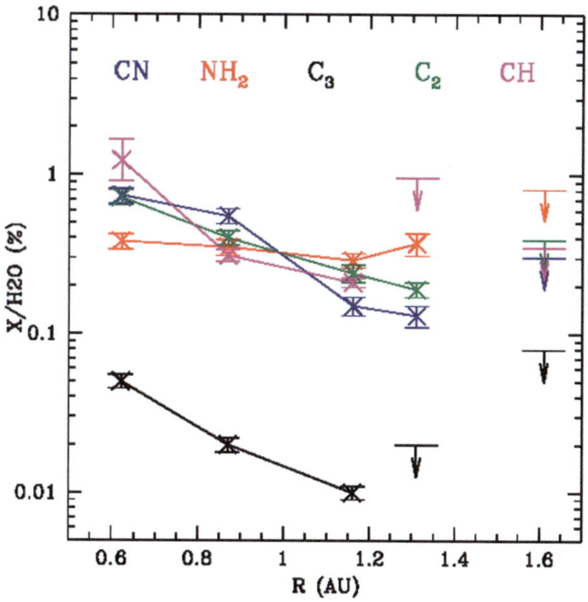

Fig. 5 The mixing ratios of various fragment species are shown as a function of heliocentric distance for comet C/2012 S1 (ISON). The data were obtained with various telescopes but all are high spectral resolution observations. NH_2 shows no change relative to H_2O but the other species increased their fraction with respect to H_2O as the comet went from 1 AU to 0.6 AU

individual coma gases to be resolved on smaller scales than have ever been achieved before; detections of new species should also be possible. Such observations are expected to yield further fundamental new insights into our understanding of cometary compositions.

2.5 Caveat: What About Comets Near the Sun?

The taxonomic studies discussed above generally refer to comets that are at 1 AU or larger heliocentric distances. While we have evidence for a change of the volatile species that controls activity at distances greater than $\sim$3 AU (generally CO or CO_2) compared with within $\sim$3 AU (H_2O), we rarely get to study comets within the orbit of the Earth. Those that we do get to study are generally outside the orbit of Venus. And it is rarer still to observe a comet outside 1 AU and follow it inside the orbit of Mercury.

In fall 2013, we got just such an opportunity with comet C/2012 S1 (ISON). ISON was the very rare Sun-grazing comet that was discovered well in advance of its perihelion passage (and subsequent destruction in this case). At larger heliocentric distances, optical observations showed this comet to be typical in some species, depleted in others and even enhanced in some species. IR observations of C_2H_2 and HCN seemed typical; CH_3OH, C_2H_6 and CH_4 were depleted; NH_3 was enhanced (Dello Russo et al. 2014a; DiSanti et al. 2014a). Stranger still, DiSanti et al. noted that the H_2CO started out depleted, trended to typical and ended up enriched as the comet went from >1 AU to $\ll 1$ AU. Figure 5 shows what happened in optical spectra as a function of time (McKay et al. 2014). With the exception of NH_2, the normal fragment molecules appear to have increased in production relative to water as the comet moved well inside of 1 AU. Opitom et al. (2014) saw similar behavior for C_2 and CN production relative to OH using the TRAPPIST telescope monitoring the comet from 1.3 $\rightarrow$ 0.3 AU inbound. In addition, examination of the top row of the ALMA data in Fig. 4 shows that there are large differences in the distribution of the different species.

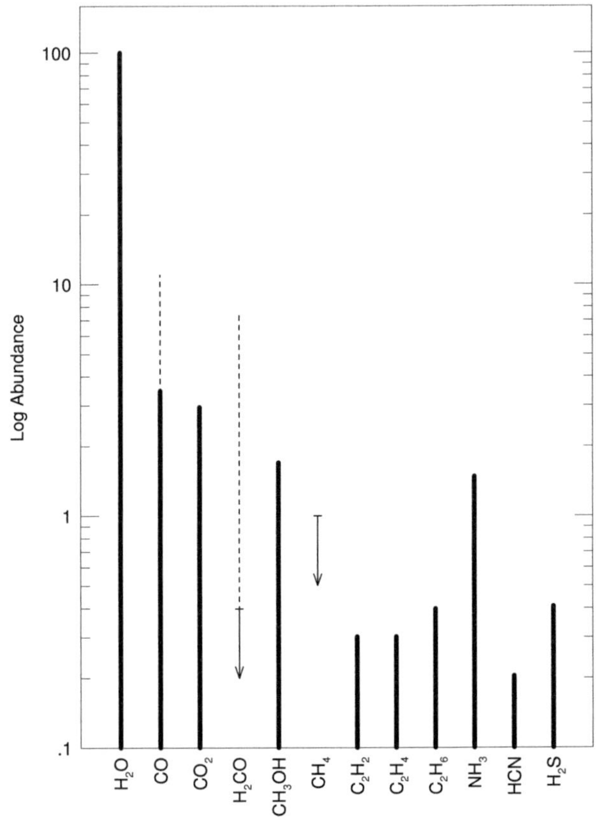

Fig. 6 The abundances of various species that were detected with the mass spectrometers onboard the Giotto spacecraft are shown. *Heavy solid lines* show the abundances for species detected in comet 1P/Halley. Upper limits for the coma are shown with *downward arrows*. The *lighter dashed lines* show the additional contribution of the extended sources in the coma. However, note that Giotto's closest approach to the nucleus was 1100 km so some species could be produced by chemistry in the first 1100 km of outflow, while others may be released entirely or in part from the nucleus. Based on Eberhardt (1999)

2.6 Mass Spectrometer Measurements

All of the above discussion was based on remote-sensing observations of comets and includes observations of several hundred comets. Ideally, one would like to take samples of the ices back to the laboratory for measurement, especially if they could be kept in the ice state so that the structure could be preserved and chemistry prevented. However, sampling and transporting a sample back to the lab is a process that is still on the horizon. Alternatively, one can take the laboratory to the comet. The Rosetta mission is doing just that, with samples of the coma from the main spacecraft and samples of the ice from the Philae lander.

Prior to Rosetta, in situ direct measurements of the cometary material with a mass spectrometer have only been accomplished once, for comet 1P/Halley. The Giotto spacecraft was successful in sampling the coma during its flyby and analyzing the composition. The results of this have been discussed by Eberhardt (1999). It was the Giotto mass spectrometer that first measured many primary volatiles leading to a detailed comparison of relative abundances. Figure 6 graphically shows the values, taken from Table 1 of Eberhardt, for the abundances of the ten molecules detected as direct contributions from the nucleus for comet Halley. There are also two upper limits. In addition, this experiment showed that some species, CO and H_2CO, that were directly produced at the nucleus, also resulted from an extended source (however, note that the spacecraft was never closer to the nucleus than 1100 km so had no spatial information closer than that). CO obviously could be a parent as well as a daughter of CO_2, but the H_2CO extended source is rarely seen in comets (but see

Sect. 2.4.1 above). However, as Eberhardt pointed out, the lack of detection of two sources from the ground-based observations is more a function of the difficulty of differentiating the sources at the spatial scale of the ground-based observations. The first measurements of the D/H ratio in H_2O were obtained (with IMS and NMS on-board Giotto), providing D/H values of about 3×10^{-4} (Balsiger et al. 1995; Eberhardt et al. 1995).

While these mass spectrometer observations of Halley are direct measurements, it should be remembered that they were still measuring the coma gas after it had flowed off the nucleus. Thus, some of the species had already undergone some chemistry and photo processes. In addition, the nature of a mass spectrometer means that there will be break-up of some of the species going through the instrument. Thus, to interpret the measurements of the complex mass spectrum, one needs a detailed model of the *expected* abundances to match. This is especially true of species that are available at the percent level or lower. In addition, depending on the mass resolution of the instrument, there is a limit on differentiating species of similar mass. An important example is CO vs. N_2, both at 28 amu. The Rosina spectrograph onboard Rosetta has improved resolution from the Giotto instruments but still relies on a model to interpret some of its results.

2.7 Gas Composition: Rosetta Capabilities and Early Results

There are instruments on both the Rosetta orbiter and Philae lander designed to determine the composition of the coma. These range from remote sensing cameras with filters or spectrometers to coma-sniffing mass spectrometers.

On the orbiter, ROSINA (Rosetta Orbiter Spectrograph for Ion and Neutral Analysis; Balsiger et al. 2007) is a mass spectrometer with three parts: DFMS (Double Focusing Mass Spectrometer, a high mass resolution spectrometer); RTOF (Reflectron Time-of-Flight Mass Spectrometer, a high mass range spectrometer) and COPS (pressure sensor to measure the total and ram pressure). DFMS can be used in restricted ranges to separate critical species such as ^{13}C vs. CH or CO vs. N_2. RTOF can identify organic molecules, such as PAHs. MIRO (Microwave Instrument for the Rosetta Orbiter; Gulkis et al. 2007) is used to monitor H_2O, CO and CO_2 abundances. Alice (Stern et al. 2007), a UV imaging spectrometer, monitors coma features such as the CO fourth-positive system, Ly α, O, as well as scans the surface. VIRTIS (Visible and Infrared Thermal Imaging Spectrometer, Coradini et al. 2007) is used to determine the composition of the ices on the surface of the nucleus. These instruments will be used over a wide range of heliocentric distances as the spacecraft flies alongside the comet from rendezvous through perihelion.

On the lander, samples are analyzed with COSAC (COmetary SAmpling and Composition Experiment; Goesmann et al. 2007) and Ptolemy (Wright et al. 2007). COSAC is an evolved gas analyzer that will concentrate on elemental and molecular composition. Ptolemy is an evolved gas analyzer that will emphasize measuring the isotopes. The length of time that the lander can be used to sample the gas depends upon power replenishment for the batteries. On the original descent to the surface, the lander bounced several times and was only able to send back a limited amount of data. However, the lander has communicated with the ground again after the comet got closer to the Sun, starting in June 2015.

Early results from these instruments have shown a wealth of detail. MIRO detected the ground-state rotation line of H_2O in June 2014, when the comet was 3.9 AU from the Sun (Gulkis et al. 2015). This was a blue-shifted asymmetric line, indicating that it came from the day side of the comet. MIRO observations showed that the surface was very insulating.

Early ROSINA observations showed that the density of gas varied with the rotation period and the latitude. CO_2 and H_2O varied with different rotational phases; the CO_2 and H_2O

arise from different places on the nucleus (Hässig et al. 2015). The ratio between H_2O, CO and CO_2 varies quite substantially along the spacecraft trajectory. ROSINA was also used to detect various isotopes of H and O (Altwegg et al. 2015). The D/H was higher than other comets. ROSINA was able to detect both CO and N_2, with modeling needed to derive abundances (Rubin et al. 2015). They found that N_2/CO is severely depleted relative to the protosolar values.

Together, the orbiter and lander observations provide "ground-truth" about comet 67P/Churyumov-Gerasimenko (CG). We already know from ground-based observations that CG is one of the carbon-chain depleted comets. It will take all of the observations of comets detailed in previous sections to place CG into context.

3 Solid Particle Composition and Properties

This section summarizes our understanding of the composition of solid particles in cometary comae. It is derived from flybys performed prior to the Rosetta rendezvous, from Earth (or Earth's orbit) remote spectroscopic and polarimetric observations of various comets, e.g., comet 1P/Halley, the first to experience a close spacecraft flyby, comet C/1995 O1 (Hale-Bopp), a very active Oort cloud comet, and some rather bright Jupiter family comets (JFCs). Solid particles are ejected with gases whenever the surface of a nucleus is close enough to the Sun to allow ice sublimation. Solid components in cometary comae appear to be mostly composed of refractory materials such as silicates, organics and amorphous carbon. Refractory dust particles present mixtures of different mineralogies. Icy grains, which rapidly evaporate, and semi-refractory carbonaceous components that suffer photolysis, may also be found in inner cometary comae.

3.1 Evidence from Cometary Flybys

3.1.1 1P/Halley Flybys: First Discoveries on Cometary Dust Composition

Major discoveries came from the dust mass spectrometers on board Vega 1 and 2 (PUMA) and Giotto (PIA) spacecraft in March 1986. The relative velocity of the probes and comet Halley on its retrograde orbit was so high (about 70 km s^{-1}) that only information on the atomic composition could be obtained (Kissel et al. 1986a,b). From spectrometric analyses (mostly PUMA 1), dust particles were found to consist not only of a mixture of major rock-forming elements (e.g., silicates, metals, sulphides, with Mg, Si, Ca, Fe), but also, quite unexpectedly, of carbon-hydrogen-oxygen-nitrogen compounds, the so-called "CHONs". The presence of large polymers of organic molecules was also suspected from the positive ion cluster composition analyzer (with RPA on-board Giotto), the most refractory polymers being likely part of dust particles (e.g., Krueger et al. 1991), and from the three-channel spectrometer (TKS) on board Vega 2 (e.g., Moreels et al. 1994). It was suggested that some gases detected in the coma were degradation products of polymers of formaldehyde on cometary solid particles (Cottin et al. 2004).

The very low geometric albedo of the nucleus surface, of about 4 % (e.g., Keller et al. 1986), was also a clue to a highly porous surface, with its slightly reddish color suggesting the presence of dark organic compounds. Moreover, local data on dust impacts (from the DID instrument) and on light scattered by dust (from the OPE instrument) could be compared along the Giotto trajectory (Levasseur-Regourd et al. 1999). These two in situ data sets, with their similarities and correlations, offered a diagnostic test of the validity of cometary coma models. Significant increases in local intensity and decreases in polarization for distances to the nucleus below 2000 km suggested the presence of small icy particles in the innermost coma (Levasseur-Regourd et al. 1999). Fits of the data (in the 2000–10000 km nucleus distance range) with a dust dynamical model indicate that the grain size distribution index was of about −2.6, that the density of dust particles was very low, of about 100 $kg\,m^{-3}$, and that their geometric albedo was about 4 %, suggesting that the dust particles were both very fluffy and dark (Fulle et al. 2000).

Rocky and carbonaceous materials were mixed together on a very fine scale and isotopic abundances appeared to be solar (Jessberger 1999).

3.1.2 81P/Wild 2 Flyby: Dust Sample Collection

Stardust collected about 4 mg of cometary material from comet 81P/Wild 2 in January 2004, capturing dust particles ejected into the coma using aerogel, with a comet-spacecraft relative speed of 6.1 $km\,s^{-1}$. After delivery of the samples to Earth in January 2006, continued analyses provided valuable insights into the properties and composition of solid cometary material. Dust impact data gathered (by DFMI) during the flyby indicated that dust particles were fragile aggregates fragmenting as they were evolving in the inner coma of the comet (Tuzzolino et al. 2004). Chemical composition (from the CIDA dust mass spectrometer) indicated the predominance of organic matter with light elements present as gas phases, whereas the dust was rich in nitrogen-containing species, with the presence of some sulfur ions (Kissel et al. 2004).

Analysis of the Stardust aerogels tracks showed the typical size range of dust particles to be between 5 and 25 μm. These particles are a mixture of compact and cohesive grains (65 %) and friable less cohesive aggregated structures (35 %) with constituent grains of a size less than 1 μm and a size distribution consistent with the ones derived in the comae of comets (Hörz et al. 2006). The collected particles are chemically heterogeneous at the largest scale (of the order of 1 μm). The mean elemental composition suggests a CI-like composition consistent with a bulk solar system composition for primitive material (e.g., Brownlee et al. 2012). The majority of the Stardust particles appear primarily composed of ferromagnesian silicates with a larger range of Mg–Fe content than other comets (Zolensky et al. 2008), Fe–Ni sulphides and Fe–Ni metal. The range of olivine and low-Ca pyroxene compositions indicates a wide range of formation conditions, reflecting a large scale mixing between the inner and outer protoplanetary disk (Zolensky et al. 2006). No hydrous phases of silicates were detected, which suggests a lack of aqueous processing of Wild 2 dust (Keller et al. 2006; Zolensky et al. 2011). Mg-carbonates were detected in Stardust samples, and they may be produced by large KBO collisions or by nebular condensates (Flynn et al. 2008). The material accreted includes Al-rich and Si-rich chondrule fragments together with some CAI-like fragments. These materials, combined with fine-grained components in the tracks, are analogous to components in unequilibrated chondrite meteorites and clustered interplanetary dust particles collected in Earth's stratosphere, so-called "IDPs" (Joswiak et al. 2012).

Organics found in comet 81P/Wild 2 dust samples show a heterogeneous and unequilibrated distribution in abundance and composition. Their characterization suggests that amorphous carbon and organic carbon are dominant (Matrajt et al. 2008). The carbon isotopic composition of glycine strongly favors an non-terrestrial origin for this compound (i.e. not contamination), making it the first amino acid detected in cometary material (Elsila et al. 2009).

H, C, N and O isotopic compositions are heterogeneous in the samples, but extreme anomalies are rare, indicating that Wild 2 is not a pristine aggregate of presolar materials. Few grains with ^{17}O and ^{16}O isotopic anomalies were found, showing them to be a minor phase of the comet, and indicating the presence of both presolar material and material formed at high temperature in the inner solar system and transported to the Kuiper belt before comet accretion (McKeegan et al. 2006). The presence of deuterium and ^{15}N excesses also suggests that some organics have an interstellar/protostellar heritage (Sandford et al. 2006). The oxygen isotopic compositions are consistent with chondritic material trends and the presence of one additional ^{16}O enriched reservoir (Nakashima et al. 2012). The ^{26}Al–^{26}Mg isotope presents no evidence of radiogenic ^{26}Mg, thus indicating that part of the cometary grains have formed more than 1.7 million years after the oldest solids in the solar system, the calcium- and aluminum-rich inclusions (CAIs). This material must have been incorporated into the comet several million years after the CAIs formed (Matzel et al. 2010). One particular Wild 2 chondrule-like fragment, called Iris, is similar to CR chondrite materials, and was age-dated to more than 1.8 million years after CAI formation (Ogliore et al. 2012). Ogliore et al. concluded that comet formation probably occurred after Jupiter formation and about 2 million years after CAIs.

It nevertheless needs to be noted that, due to the fragility of the dust particles and the relatively high speed of collection by Stardust, the analyses of the composition of the particles and their interpretation is perturbed by the mode of collection, which induced partial melting of some of the material (e.g., preferential loss of light elements combined with possible amorphization and graphitization, Spencer and Zare 2007; Fries et al. 2009).

3.1.3 9P/Tempel 1 and 103P/Hartley 2: Icy Grains and Aggregates

The nucleus of comet 9P/Tempel 1 was hit in July 2005 by a projectile released by the Deep Impact spacecraft (A'Hearn et al. 2005). It allowed comparison of the properties of materials found in the coma, possibly originating from a processed surface, to sub-surface materials released by the impact. After the impact, there was an enhanced UV scattering that persisted for about 20 minutes. Strong absorptions in IR spectra near 3 μm were noticed, providing evidence for icy grains in the plume (Schulz et al. 2006; Sunshine et al. 2007). Before impact, the C_2H_6/H_2O ratio in the coma was depleted by a factor of three relative to organics-normal comets but after impact it increased to the organics-normal value ($\sim$0.6 %) suggesting that the hypervolatile ethane was more depleted in the surface layer than in the excavated material (Mumma et al. 2005).

The Deep Impact spacecraft was retargeted to comet 103P/Hartley 2, which it flew by in November 2010 as part of the EPOXI mission. The nucleus was discovered to be bi-lobed in shape, with knobby terrains on the lobes and relatively smooth regions on the waist, to have an average geometric albedo of about 4 %, and to present a clustering of jets on parts of the lobes (A'Hearn et al. 2011). At closest approach, many images suggested the presence of large chunks of icy particles, with sizes possibly reaching tens of centimeters. Further analyses confirmed the existence of such aggregates, sublimating on their sunny side and presenting densities below 100 $kg\,m^{-3}$ (Kelley et al. 2013). The velocity of the individual icy grains is smaller than the expansion velocities expected for 1 μm pure water ice particles, confirming that they are probably components of aggregates (Protopapa et al. 2014).

Similarities between the material excavated from Tempel 1 and the outgassing from Hartley 2 suggest icy aggregates built of relatively pure icy grains with sizes of about one micron. Quite comparable values are also derived for near-infrared remote observations of the coma

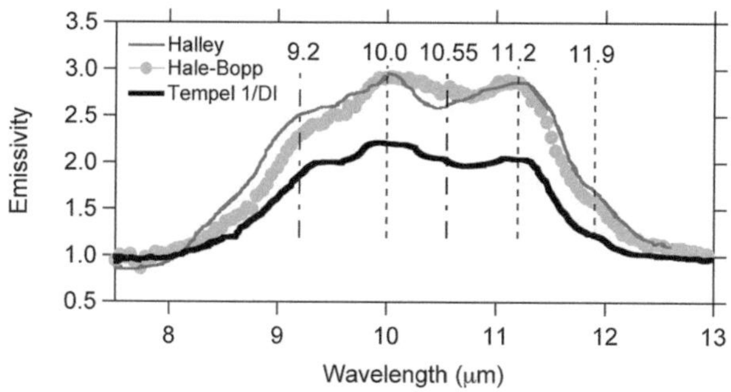

Fig. 7 Comparison of silicate features, as observed for 1P/Halley and C/1995 O1 Hale-Bopp (adapted from Hanner 1999), and for 9P/ Tempel 1 after the Deep Impact event and release of subsurface material (adapted from Kelley and Wooden 2009). In each case the total flux is divided by the black body contribution. The main features are at 10 μm, 11.2 μm and 11.9 μm for crystalline olivine, and at 9.2 μm and 10.55 μm for crystalline ortho-pyroxene

of comet 17P/Holmes after its October 2007 outburst, with solid particles in the coma suggested to consist of both refractory dust and cold ice grains, not in thermal contact (Yang et al. 2009; Beer et al. 2009).

3.2 Results from Remote Observations

Solid particles in cometary comae have been studied for several tens of comets bright enough to be observed from Earth or near-Earth-based observatories. Results stem from infrared spectroscopic observations (in the 8–13 μm atmospheric window or over a larger range from space observatories), as well as from linear polarimetric observations (including polarimetric imaging techniques). After the results from Halley flybys and before space missions to JFCs, the very active Oort cloud comet C/1995 O1 (Hale-Bopp) increased our understanding of the composition of the solid component of comae. Its activity was higher than for all other comets previously observed, and significant similarities in the IR dust spectrum were found with comet Halley (Fig. 7; Hanner 1999).

3.2.1 Infrared Spectroscopy: Silicates

Silicate components of the dust contribute to a weak continuum emission and strong resonances in the 10 μm domain for Hale-Bopp. Different silicates were identified in spectra: amorphous pyroxene, amorphous olivine and crystalline olivine. Closer to the Sun, Mg-rich crystalline pyroxene was also identified. The very significant strength observed for the features in C/1995 O1 (Hale-Bopp) could be due to a large number of submicron-sized grains (0.2 μm or smaller). The smaller Mg-rich crystals could come from the jets, although no differences in silicate mineralogy were observed between the different regions, e.g., inner coma, coma and jets (Hayward et al. 2000). Other Oort cloud comets present similar spectra with Mg-rich grains (Hanner 1999; Wooden et al. 2004), but with lower strength of the features.

Comet 9P/Tempel 1 was extensively observed before and after the Deep Impact event. Before impact, spectra contained a broad feature in the 8–12 μm range, attributed to amorphous silicates. Spectra of JFCs with low dust production rates are generally found to have weak silicate features, discernible in high signal-to-noise Spitzer IRS spectra of comets (Kelley and Wooden 2009). These weak silicate features, attributable to the coma dust population, are better revealed after subtraction of the relatively strong thermal contribution from the nucleus (Hanner 1999; Sitko et al. 2004; Harker et al. 2005, 2007; Kelley and Wooden 2009; Woodward et al. 2011). Weak silicate features arise from a deficiency of small grains rather than a lack of silicate material (Hanner 1999; Woodward et al. 2011). After impact on Tempel 1, spectra showed a very significant feature due to crystalline Mg-silicate (see Fig. 7), similar to those observed for Hale-Bopp (Harker et al. 2005, 2007; Lisse et al. 2007). These features progressively disappeared as the plume dissipated away from the nucleus. Grains first ejected from Tempel 1 were mainly submicron-sized amorphous-carbon grains and submicron-sized silicates (Sugita et al. 2005; Harker et al. 2007). The second part of the ejecta mainly contained Mg-rich crystal silicates with submicron sizes, originating in the sub-surface layer. It may be added that hydrated minerals, which have implications for the location of their formation region (Lisse et al. 2007; Kelley and Wooden 2009) have also been detected. Amorphous carbon (submicron-sized) grains were suggested to come from a carbonaceous layer formed by cosmic irradiation during the extremely long time spent by the nucleus in the transneptunian region (Sugita et al. 2007; Furusho et al. 2007). This hypothesis is compatible with the observations of the impact crater by the Stardust-NExT mission (Schultz et al. 2013).

Both Tempel 1 and Hale-Bopp have high-crystalline silicate fractions, but there are important differences in their precise mineralogy (Lisse et al. 2007; Kelley and Wooden 2009). In comet Hale-Bopp and many Oort cloud comets, only Mg-rich grains are present. Mg–Fe crystals are also found in Tempel 1 and JFCs. The presence of Mg-silicates requires significant processing in the protosolar nebula (Wooden 2002). Annealing by heating may be an important process to convert amorphous grains to crystals. New observations and modeling are necessary to understand the transport and formation of the different classes of comets.

3.2.2 Infrared Spectroscopy: Organics and Icy Particles

Refractory organics, which mainly contribute to the featureless emission and possibly to a 3.4 μm emission (Green et al. 1992), are more difficult to detect. They can be mixed with rocky components, and constitute a matrix gluing the different minerals, as in IDPs (Flynn et al. 2003, 2013; Flynn 2011). They can also coat the dust grains, a process possibly having taken place in the interstellar medium (Hanner and Bradley 2004; Greenberg and Hage 1990; Ehrenfreund et al. 2003). When particles are ejected in the coma, they get heated, leading to changes in their optical properties and even to evaporation and destruction. Distributed sources may appear (Bockelée-Morvan et al. 2004; Cottin et al. 2004) and fragmentation of dust particles is likely to occur.

Remote detection of water ice is extremely difficult, with an icy grain halo likely to be limited to a few hundred kilometers at heliocentric distances below 2.5 AU (Hanner 1981; Beer et al. 2006, 2008), meaning that high spatial resolution is required to attempt any detection. Water ice has three absorption bands in the near-infrared, centered at 1.5, 2.0, and 3.0 μm. First detections of water ice were obtained on comet Hale-Bopp from the UKIRT telescope with the comet at about 7 AU from the Sun (Davies et al. 1997) and from the ISO space observatory at 2.9 AU (Lellouch et al. 1998). Infrared spectra of comets Hale-Bopp and C/2002 T7 (LINEAR) were reproduced with an intimate mixture of water ice and silicate spherical grains, assuming water ice to be in an amorphous state (Davies et al. 1997; Kawakita et al. 2004). In the infrared spectra of comets 17P/Holmes in its outburst and C/2011 L4 (PanSTARRS), water ice bands were detected at 2.0 μm and reproduced with submicron icy grains (Yang et al. 2009, 2014). Remote observations of Tempel 1 from the XMM-Newton observatory provided evidence for ice particles in the inner coma after the impact (Schulz et al. 2006). Observations of color changes after the Deep Impact encounter in the visible and near IR domains also suggested an increase in small grains and in ice relative to refractory dust in the coma (Knight et al. 2007; Schleicher et al. 2006). Indeed, remote observations from the Spitzer observatory suggested a very high ice-to-dust ratio of about 10 in the excavated material (which greatly exceeds the gas-to-dust production rate ratio of about 0.5 measured for the background coma), although a ratio in the 1 to 3 range cannot be excluded if a large amount of material fell back to the surface and sublimated (Gicquel et al. 2012).

3.2.3 Color and Polarimetric (Linear and Circular) Observations

Observations of solar light scattered by solid particles in cometary comae depend on the observational conditions (e.g., phase angle, wavelength) and on the particles' physical properties (e.g., complex refractive indices and composition, size, shape, geometric albedo). Differences in physical properties between the particles in different regions of the coma are pointed out by differences in the linear polarization of the scattered light and by spectral variations in brightness and polarization. At visible wavelengths, grains or particles with sizes of between 0.5 and 10 μm seem to dominate the scattered light, affecting the degree of linear polarization and color variation in the brightness and in the polarization.

The color of the dust is associated with the variation of the scattered intensity with wavelength. In the visible domain, the scattered light is generally redder than the solar continuum (Jewitt and Meech 1986; Kolokolova et al. 2004). The reddening slope decreases towards the near-infrared. The color depends on the size distribution of the grains and aggregates and on their refractive indices, mainly for grains larger than the wavelength. Some local variations of colors have been observed in comae. In comet Hale-Bopp, the color in the curved jets was less red than the average background, possibly because of an increase in the number of submicron-sized grains (Furusho et al. 1999). At the beginning of the disruption of comet C/1999 S4 (LINEAR), the color was bluer while fast moving small grains were detected, although the color of the largest fragments was red and became less red when the fragmenting particles went farther away from the nucleus (e.g., Hadamcik and Levasseur-Regourd 2003a). To interpret the slightly bluer color in the visible and near infrared in the dust ejecta after the Deep Impact event, small submicron-sized grains were suggested, as well as the presence of pure or mixed water ice crystals or the sublimation of organics (Hodapp et al. 2007; Fernández et al. 2007; Beer et al. 2009). Generally, bluer color requires relatively transparent grain materials such as water ice, Mg-rich silicates, or perhaps non-radiation damaged nor heated organics (Kiselev et al. 2004; Hadamcik and Levasseur-Regourd 2009; Zubko et al. 2011, 2012; Hadamcik et al. 2014).

The solar light scattered by cometary comae is partially linearly polarized. Polarimetric properties vary with size and size distribution, morphology and structure, as well as with complex refractive indices, of the scattering particles. Observations can thus point out changes in dust properties and provide information on composition through numerical and experimental simulations (see e.g. for reviews, Kolokolova et al. 2015). Evidence for changes in local polarization was established by OPE/Giotto during the Halley flyby, with an increase in jets and a decrease near the nucleus at constant large phase angle and wavelength (Levasseur-Regourd et al. 1999). Such behavior could also be monitored by remote CCD imaging polarimetry on comet Levy (Renard et al. 1992) and later on other comets at different phase angles (e.g., Jones and Gehrz 2000; Hadamcik and Levasseur-Regourd 2003b, Hadamcik et al. 2013; Deb Roy et al. 2015; Hines et al. 2014).

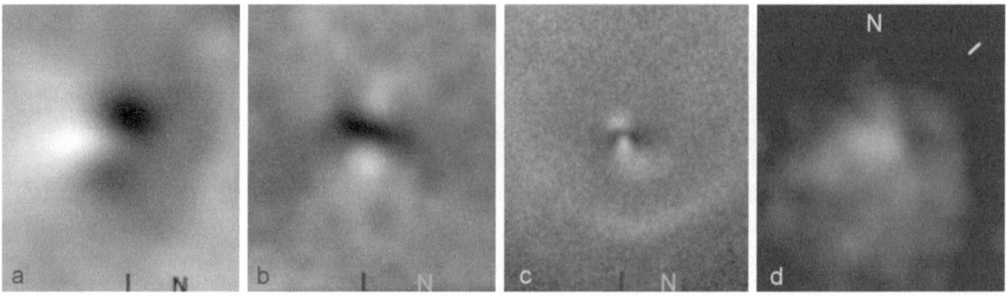

Fig. 8 Evidence for different properties of solid particles in different regions of cometary comae, from linear polarization (P) maps at a given phase angle (α) (**a**) Comet C/1990 K1 Levy at $\alpha = 18°$, in a field of view of 4000 km by 4000 km (*black*, P = −2.6 %; *white*, P = 0.5 %). (**b**) Comet C/1995 O1 Hale-Bopp at $\alpha = 7°$, in a field of 38000 km by 38000 km (*black*, P = −6 %; *white*, P = 2 %). (**c**) Comet Hale-Bopp, now at $\alpha = 44°$, in a field of 82000 km by 82000 km (*black*, P = 9 %; *white*, P = 18 %). (**d**) Comet 9P/Tempel 1, before Deep Impact event, without any conspicuous feature, at $\alpha = 41°$, in a field of 70000 km × 70000 km (*black*, P = 1 %; *white*, P = 10 %). The solar direction is indicated by a *black* or *white tick*

Three main regions of polarization have been detected, as illustrated in Fig. 8: the background coma, jet-like features (in active comets) with a higher polarization than in the coma, and (for some comets only) a circumnucleus polarimetric halo, clearly detected on polarimetric images of comets Levy and Hale-Bopp at different phase angles. The polarization in the halo, if any, is more negative than the surroundings for phase angles smaller than about 20° (see Fig. 8a and b) and is lower than the surrounding coma for larger phase angles (see Fig. 8c and d).

The averaged polarization, measured through different increasing aperture sizes (e.g., Kiselev et al. 2001; Hadamcik et al. 2013), provides the whole coma polarization when using large apertures (Hadamcik and Levasseur-Regourd 2003b), for a given phase angle and wavelength. Polarimetric phase curves show a small negative branch (down to about −1 to −2 %), and inversion angle near 20°, and a broad positive branch with a maximum (up to about 15 to 30 %) by 90° to 100° (e.g. Hadamcik and Levasseur-Regourd 2003b). From synthetic phase curves built for all observed comets in the same wavelength range, two classes may be characterized (see Fig. 9): the active comets with generally well focused jets that present a high maximum in polarization, and the comets that present a lower maximum in polarization (Levasseur-Regourd et al. 1996). The latter are less active or may show an important decrease of polarization away from the photometric center, such as comet Tempel 1 before the Deep Impact event (Hadamcik et al. 2007a) or comet C/1996 Q1 (Tabur) (Kiselev et al. 2001).

For a given phase angle above 20°, the whole coma polarization slightly increases with the wavelength, at least up to about 2 μm. It seems to decrease for wavelengths greater than 2 μm, possibly due to some thermal emission (Kolokolova et al. 2004), although observations and thermal models of cometary comae show that the thermal emission has its onset between 3 μm and 4 μm for heliocentric distances in the 1–3 AU range (Sitko et al. 2004; Harker et al. 2002, 2004). A decrease of polarization with increasing wavelength in the visible has been observed for some comets such as 21P/Giacobini-Zinner (Kiselev et al. 2000) or during some event such as the complete disruption of comet C/1999 S4 (LINEAR) (Hadamcik and Levasseur-Regourd 2003a). In the circumnucleus polarimetric halo region, such a spectral gradient inversion was noticed through in situ observations of Halley and remote observations of Hale-Bopp, possibly providing clues to different properties for freshly ejected dust particles.

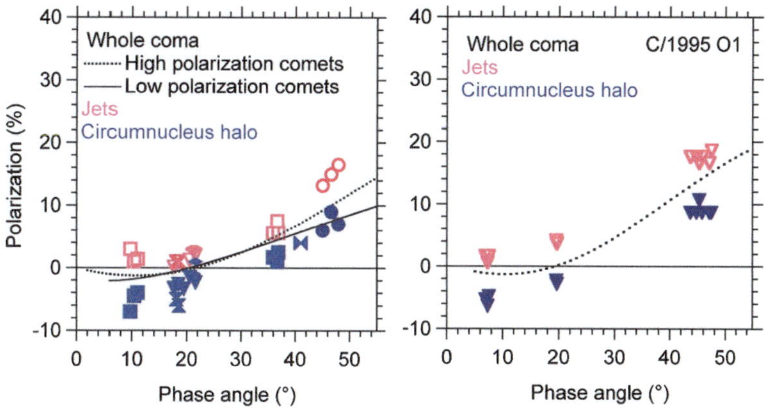

Fig. 9 Evidence for changes in linear polarization, and thus in solid particle properties, within specific coma features (jets, circumnucleus polarimetric halo) observed at different phase angles. *Left*, comets with either high or low maxima in polarization; *right*, comet C/1995 O1 Hale-Bopp. While synthetic phase curves can be derived for whole cometary comae, the polarization is higher in jets (*unfilled symbols*) and lower in the circumnucleus halo (*filled symbols*), pointing out changes in composition and sizes (updated from Hadamcik and Levasseur-Regourd 2003b)

In addition to its linear polarization, the light scattered by cometary dust appears to present a weak circular polarization signal. The origin of circular polarization is still debated but could arise from the alignment of dust particles, light scattering by asymmetric particles, or the presence of prebiotic homochiral organic molecules in cometary dust. More observations are necessary to better constrain its origin. A discussion can be found in Kiselev et al. (2014).

3.2.4 *Properties of Solid Particles Inferred from Light Scattering Measurements, with Emphasis on Composition*

Numerous authors have tried to constrain the scattering properties of cometary dust. From experimental simulations, Hadamcik et al. (2007b) established that fluffy Mg–Fe–SiO and C mixtures could represent satisfactory cometary dust analogs. From numerical simulations of polarization data and silicate emission features, Kolokolova et al. (2007) concluded that the dust in comets with a high maximum in polarization consists of highly porous aggregates that may be associated with fresh dust as in new comets. The dust in comets with a lower maximum polarization consists of less porous particles which may be associated with more highly processed dust such as expected for the surfaces of JFC comets. Compact porous particles also dominated the coma of long period comet C/2007 N3 (Lulin), as evidenced by a negative branch that extended to near-IR wavelengths (Woodward et al. 2011). Finally, whenever polarimetric phase curves are obtained on a large enough range of phase angles in different wavelengths (as possible mostly for Halley and Hale-Bopp), numerical and experimental simulations may be used to infer some average properties, such as size and size distribution, ratio between transparent and absorbing materials, and ratio between fluffy aggregates and compact particles (Levasseur-Regourd et al. 2008; Lasue et al. 2009; Zubko et al. 2012; Hines et al. 2014).

The circumnucleus polarimetric halo provides an example of the complexity of interpretations. First of all, it is not always observed; it can be hidden by jets or not continuously present or may not exist at all. Secondly, its interpretation is controversial. The deep negative branches observed for C/1995 O1 (Hale-Bopp) and 81P/Wild 2 seem to be possible only if transparent particles exist in this region. Suggestions are large agglomerates of water ice grains, an excess of Mg-rich silicate grains as compared to dark carbonaceous compounds, and relatively transparent organics, possibly covering silicate grains (Flynn et al. 2003, 2013). These organics, heated in the coma after ejection, may become darker by carbonization and can release gases as an extended source (Hadamcik et al. 2014). Various compositions and evolutions may actually co-exist in the innermost comae. Additionally, a smaller polarization may originate in some depolarization induced by multiple scattering, as was certainly the case in the plume ejected from Tempel 1 after impact. Nevertheless, the polarimetric spectral gradient, possibly negative or neutral near the nucleus, and positive after an evolution of the composition, could provide further clues.

There can be a strong synergy between the dust particle properties derived from studies of color and linear polarization with their properties assessed by studies of coma dynamics and thermal emission, where comet Hale-Bopp is an excellent example. As stated in Sect. 3.2.3, Hale-Bopp's curved jets or arcs were less red than the average background, possibly because of an increase in the number of submicron-sized grains. Compared to the background coma, Hale-Bopp arcs also showed higher linear polarization (Jones and Gehrz 2000), higher dust color temperatures (Hayward et al. 2000), strong silicate features (Hayward et al. 2000) indicative of a high concentration of submicron silicate crystals as well as micron-sized porous particles (Harker et al. 2002, 2004). The coma dynamical models showed that arc/jet particles needed to be small (submicron) and relatively transparent in order that solar radiation pressure not smear out the arcs nor compress the arc spacing (Hayward et al. 2000). Ground truth, as expected from a rendezvous mission such as Rosetta, monitoring the dust properties of a given comet at different distances from the nucleus and from the Sun, coupled with remote observations, is certainly of major importance for future interpretations of spectroscopic and polarimetric observations of various comets.

3.3 Sun-Grazing Comets

When a comet gets very close to the Sun, it is heated sufficiently that the refractory particles can start to sublimate, releasing metals such as Na and K in the coma. Observations of these metals can give key insights into the atomic abundances in cometary dust that can only otherwise be gleaned with dust analyzers in situ. However, these lines only become prominent when a comet is very close to the Sun, challenging our ability to observe the comet from the ground or with space telescopes owing to the small solar elongation. Indeed, these comets are so close to the Sun that the observations must be obtained during the daytime. As noted above, comet C/2012 S1 (ISON) was one such Sun-grazing comet that was extensively observed, though it faded substantially when closest to the Sun, limiting our ability to detect the metal lines.

Comet C/1965 S1 (Ikeya-Seki) was one of the brightest Sun-grazing comets to be studied spectroscopically. Observations were obtained from Kitt Peak National Observatory (Slaughter 1969; Arpigny 1979), Lick Observatory (Preston 1967), Sacramento Peak (Curtis 1966), Haute Provence (Dufay et al. 1965), Radcliffe (Thackeray et al. 1966), and possibly others, when the comet was as close to the Sun as ~ 14 solar radii.

In addition to the normal cometary gas emissions, emission lines due to Na I, K I, Ca I, Ca II, Cr I, Mn I, Fe I, Ni I, Cu I, Co I, and V I were observed. From analysis of the Fe lines, a Boltzmann temperature of ~ 4500 K was determined (Slaughter 1969; Preston

1967). Arpigny (1979) noted that Mg, Al, Si ad Ti were not detected. Compared with solar abundances, the abundances of some of these metals were underabundant, while others were overabundant (Arpigny 1979; Preston 1967). However, arguments were presented (Arpigny 1979) that the relative *elemental* abundances were probably close to solar or meteoritic values.

3.4 Dust Composition: Rosetta Capabilities and Early Results

The instruments on the Rosetta mission that could provide direct information on the composition of solid particles for 67P/Churyumov-Gerasimenko (CG) are mostly COSIMA and VIRTIS on the main Rosetta spacecraft, to be used during the whole rendezvous mission, and COSAC at Philae landing. The COmetary Secondary Ion Mass Analyser, COSIMA, is the first instrument applying SIMS (secondary ion mass spectrometry) technique to in situ analysis. It collects dust particles at the Rosetta location on targets handled by a manipulation unit, in order to obtain time-of-flight spectra over a mass range from 1 to 3500 amu, leading to the composition of the dust particles, whether they are organic or inorganic (Kissel et al. 2007). The Visual IR Thermal Imaging Spectrometer, VIRTIS, provides imaging spectroscopy in the 0.25–5.0 μm spectral range, in order to map composition and evolution of dust jets in the coma, as well as to derive the composition of the dust grains in the inner coma and on the surface (Coradini et al. 2007). By August 2014, strong hints of carbon-bearing compounds, with spectral features compatible with complex macromolecular carbonaceous materials, were already noted on the nucleus surface. (http://blogs.esa.int/rosetta/2014/09/08/virtis-maps-comet-hot-spots/). The COmetary SAmpling and Composition experiment, COSAC, relies on a multi-column enantio-selective gas chromatograph, coupled to a linear time-of-flight mass spectrometer to derive the composition of volatile species collected at the surface and at about 20 cm below the surface with the Philae lander (Goesmann et al. 2007).

Interpretation of data, e.g., from dust experiments GIADA (Colangeli et al. 2007), COSIMA (Kissel et al. 2007) and MIDAS (Riedler et al. 2007) on Rosetta, as well as from CONSERT (Kofman et al. 2007), APSX (Klingelhöfer et al. 2007) or Ptolemy (Wright et al. 2007) on Philae, should also significantly contribute to a better understanding of the dust composition, and on its evolution i) from the subsurface to the surface of the nucleus and to the coma and ii) from large and decreasing solar distances before perihelion passage to increasing solar distances after perihelion passage.

Preliminary results from the Rosetta mission have already established that the surface of CG is rich in non-volatile organics, likely complex mixtures of various carbon-hydrogen (aromatic and aliphatic), oxygen-hydrogen (carboxylic or alcoholic) and nitrogen-hydrogen groups (Capaccioni et al. 2015). The largest particles (with sizes greater than 50 μm) are demonstrated to be fluffy and fragile aggregates, and found to present sodium-rich surfaces (Schulz et al. 2015). While the presence of such aggregates had been suspected from simulations (e.g., Hadamcik et al. 2007a; Levasseur-Regourd et al. 2008), the Rosetta ground-truth now allows us to speculate about the early Solar System evolution. A significant amount of non-volatile organics could have enriched the surface of terrestrial planets, with fluffy particles more resistant to atmospheric ablation than compact ones. More results and discoveries are soon expected, from Rosetta pre-perihelion (at solar distances less than 3 AU) and post-perihelion measurements, and from Philae measurements.

4 Summary

We have learned a great deal about comets from the studies outlined in this paper. However, the picture is not yet fully coherent since we find different stories when we change wavelength. The Rosetta mission will help us to gain detail and to compare with the data that have come before. However, it is obvious that all that can be learned from the enigmatic comets has not yet been found.

Acknowledgements The authors would like to thank Dr. Paul Feldman, Dr. Jacques Crovisier, Dr. Neil Dello Russo and Dr. Adam McKay for helpful comments and making data available for our use. Partial support from CNES (the French Space Agency) is acknowledged.

References

M.F. A'Hearn, R.L. Millis, Abundance correlations among comets. Astron. J. **85**, 1528–1537 (1980)

M.F. A'Hearn, R.L. Millis, D.G. Schleicher, D.J. Osip, P.V. Birch, The ensemble properties of comets: results from narrowband photometry of 85 comets, 1976–1992. Icarus **118**, 223–270 (1995)

M.F. A'Hearn, M.J.S. Belton, W.A. Delamere, J. Kissel, K.P. Klaasen, L.A. McFadden, K.J. Meech, H.J. Melosh, P.H. Schultz, J.M. Sunshine, P.C. Thomas, J. Veverka, D.K. Yeomans, M.W. Baca, I. Busko, C.J. Crockett, S.M. Collins, M. Desnoyer, C.A. Eberhardy, C.M. Ernst, T.L. Farnham, L. Feaga, O. Groussin, D. Hampton, S.I. Ipatov, J.-Y. Li, D. Lindler, C.M. Lisse, N. Mastrodemos, W.M. Owen, J.E. Richardson, D.D. Wellnitz, R.L. White, Deep impact: excavating comet Tempel 1. Science **310**, 258–264 (2005)

M.F. A'Hearn, M.J.S. Belton, W.A. Delamere, L.M. Feaga, D. Hampton, J. Kissel, K.P. Klaasen, L.A. McFadden, K.J. Meech, H.J. Melosh, P.H. Schultz, J.M. Sunshine, P.C. Thomas, J. Veverka, D.D. Wellnitz, D.K. Yeomans, S. Besse, D. Bodewits, T.J. Bowling, B.T. Carcich, S.M. Collins, T.L. Farnham, O. Groussin, B. Hermalyn, M.S. Kelley, M.S. Kelley, J.-Y. Li, D.J. Lindler, C.M. Lisse, S.A. McLaughlin, F. Merlin, S. Protopapa, J.E. Richardson, J.L. Williams, EPOXI at comet Hartley 2. Science **332**, 1396–1400 (2011)

K. Altwegg, H. Balsiger, A. Bar-Nun, J.J. Berthelier, A. Bieler, P. Bochsler, C. Briois, U. Calmonte, M. Combi, J. De Keyser, P. Eberhardt, B. Fiethe, S. Fuselier, S. Gasc, T.I. Gombosi, K.C. Hansen, M. Hässig, A. Jäckel, E. Kopp, A. Korth, L. LeRoy, U. Mall, B. Marty, O. Mousis, E. Neefs, T. Owen, H. Rème, M. Rubin, T. Sémon, C.-Y. Tzou, H. Waite, P. Wurz, 67P/Churyumov-Gerasimenko, a Jupiter family comet with a high D/H ratio. Science **347**, A387 (2015)

C. Arpigny, Relative abundances of the heavy elements in comet Ikeya-Seki (1965 VIII), in *Liege International Astrophysical Colloquia*, ed. by A. Boury, N. Grevesse, L. Remy-Battiau (1979), pp. 189–197

C. Arpigny, E. Jehin, J. Manfroid, D. Hutsemékers, R. Schulz, J.A. Stüwe, J.-M. Zucconi, I. Ilyin, "Anomalous" nitrogen isotope ratio in comets. Science **301**, 1522–1524 (2003)

H. Balsiger, K. Altwegg, J. Geiss, D/H and $^{18}O/^{16}O$ ratio in the hydronium ion and in neutral water from in situ ion measurements in comet 1P/Halley. J. Geophys. Res. **100**, 5827–5834 (1995)

H. Balsiger, K. Altwegg, P. Bochsler, P. Eberhardt, J. Fischer, S. Graf, A. Jäckel, E. Kopp, U. Langer, M. Mildner, J. Müller, T. Riesen, M. Rubin, S. Scherer, P. Wurz, S. Wüthrich, E. Arijs, S. Delanoye, J. de Keyser, E. Neefs, D. Nevejans, H. Rème, C. Aoustin, C. Mazelle, J.-L. Médale, J.A. Sauvaud, J.-J. Berthelier, J.-L. Bertaux, L. Duvet, J.-M. Illiano, S.A. Fuselier, A.G. Ghielmetti, T. Magoncelli, E.G. Shelley, A. Korth, K. Heerlein, H. Lauche, S. Livi, A. Loose, U. Mall, B. Wilken, F. Gliem, B. Fiethe, T.I. Gombosi, B. Block, G.R. Carignan, L.A. Fisk, J.H. Waite, D.T. Young, H. Wollnik, Rosina-Rosetta orbiter spectrometer for ion and neutral analysis. Space Sci. Rev. **128**, 745–801 (2007)

E. Beer, M. Podolak, D. Prialnik, The contribution of icy grains to the activity of comets. I. Grains lifetime and distribution. Icarus **180**, 473–486 (2006)

E. Beer, D. Prialnik, M. Podolak, The contribution of grains to the activity of comets. II. The brightness of the coma. Icarus **195**, 340–347 (2008)

E. Beer, D.H. Wooden, R. Schulz, The grain evolution model for icy grains ejected from 9P/Tempel 1 by Deep Impact, in *Deep Impact as a World Observatory Event: Synergies in Space, Time and Wavelengths*, ed. by H.U. Kaüfl, C. Sterken (Springer, Berlin, 2009), pp. 59–67

N. Biver, D. Bockel'ee-Morvan, P. Colom, J. Crovisier, J.K. Davies, W.R.F. Dent, D. Despois, E. Gerard, E. Lellouch, H. Rauer, R. Moreno, G. Paubert, Evolution of the outgassing of comet Hale-Bopp (C/1995 O1) from radio observations. Science **275**, 1915–1918 (1997)

N. Biver, D. Bockelée-Morvan, P. Colom, J. Crovisier, F. Henry, E. Lellouch, A. Winnberg, L.E.B. Johansson, M. Gunnarsson, H. Rickman, F. Rantakyrö, J.K. Davies, W.R.F. Dent, G. Paubert, R. Moreno, J. Wink, D. Despois, D.J. Benford, M. Gardner, D.C. Lis, D. Mehringer, T.G. Phillips, H. Rauer, The 1995–2002 long-term monitoring of comet C/1995 O1 (Hale-Bopp) at radio wavelength. Earth Moon Planets **90**, 5–14 (2002)

G.A. Blake, C. Qi, M.R. Hogerheijde, M.A. Gurwell, D.O. Muhleman, Sublimation from icy jets as a probe of the interstellar volatile content of comets. Nature **398**, 213–216 (1999)

D. Bockelée-Morvan, An overview of comet composition, in *IAU Symposium 280*, ed. by J. Cernicharo, R. Bachiller (2011), pp. 261–274

D. Bockelée-Morvan, D.C. Lis, J.E. Wink, D. Despois, J. Crovisier, R. Bachiller, D.J. Benford, N. Biver, P. Colom, J.K. Davies, E. Gérard, B. Germain, M. Houde, D. Mehringer, R. Moreno, G. Paubert, T.G. Phillips, H. Rauer, New molecules found in comet C/1995 O1 (Hale-Bopp). Investigating the link between cometary and interstellar material. Astron. Astrophys. **353**, 1101–1114 (2000)

D. Bockelée-Morvan, J. Crovisier, M.J. Mumma, H.A. Weaver, The composition of cometary volatiles, in *Comets II*, ed. by M.C. Festou, H.U. Keller, H.A. Weaver (University of Arizona Press, Tucson, 2004), p. 391

J. Boissier, D. Bockelée-Morvan, N. Biver, J. Crovisier, D. Despois, B.G. Marsden, R. Moreno, Interferometric imaging of the sulfur-bearing molecules H_2S, SO, and CS in comet C/1995 O1 (Hale-Bopp). Astron. Astrophys. **475**, 1131–1144 (2007)

J. Boissier, D. Bockelée-Morvan, N. Biver, P. Colom, J. Crovisier, R. Moreno, V. Zakharov, O. Groussin, L. Jorda, D.C. Lis, Gas and dust productions of comet 103P/Hartley 2 from millimetre observations: interpreting rotation-induced time variations. Icarus **228**, 197–216 (2014)

B.P. Bonev, M.J. Mumma, Y.L. Radeva, M.A. DiSanti, E.L. Gibb, G.L. Villanueva, The peculiar volatile composition of comet 8P/Tuttle: a contact binary of chemically distinct cometesimals? Astrophys. J. Lett. **680**, L61–L64 (2008)

B.P. Bonev, G.L. Villanueva, L. Paganini, M.A. DiSanti, E.L. Gibb, J.V. Keane, K.J. Meech, M.J. Mumma, Evidence for two modes of water release in comet 103P/Hartley 2: distributions of column density, rotational temperature, and ortho-para ratio. Icarus **222**, 740–751 (2013)

B.P. Bonev, M.A. DiSanti, G.L. Villanueva, E.L. Gibb, L. Paganini, M.J. Mumma, The inner coma of comet C/2012 S1 (ISON) at 0.53 AU and 0.35 AU from the Sun. Astrophys. J. Lett. **796**, L6 (2014)

D. Brownlee, D. Joswiaka, G. Matrajt, Overview of the rocky component of Wild 2 comet samples: insight into the early solar system, relationship with meteoritic materials and the differences between comets and asteroids. Meteorit. Planet. Sci. **47**, 453–470 (2012)

F. Capaccioni, A. Coradini, G. Filacchione, S. Erard, G. Arnold, P. Drossart, M.C. De Sanctis, D. Bockelee-Morvan, M.T. Capria, F. Tosi, C. Leyrat, B. Schmitt, E. Quirico, P. Cerroni, V. Mennella, A. Raponi, M. Ciarniello, T. McCord, L. Moroz, E. Palomba, E. Ammannito, M.A. Barucci, G. Bellucci, J. Benkhoff, J.P. Bibring, A. Blanco, M. Blecka, R. Carlson, U. Carsenty, L. Colangeli, M. Combes, M. Combi, J. Crovisier, T. Encrenaz, C. Federico, U. Fink, S. Fonti, W.H. Ip, P. Irwin, R. Jaumann, E. Kuehrt, Y. Langevin, G. Magni, S. Mottola, V. Orofino, P. Palumbo, G. Piccioni, U. Schade, F. Taylor, D. Tiphene, G.P. Tozzi, P. Beck, N. Biver, L. Bonal, J.-P. Combe, D. Despan, E. Flamini, S. Fornasier, A. Frigeri, D. Grassi, M. Gudipati, A. Longobardo, K. Markus, F. Merlin, R. Orosei, G. Rinaldi, K. Stephan, M. Cartacci, A. Cicchetti, S. Giuppi, Y. Hello, F. Henry, S. Jacquinod, R. Noschese, G. Peter, R. Politi, J.M. Reess, A. Semery, The organic-rich surface of comet 67P/Churyumov-Gerasimenko as seen by VIRTIS/Rosetta. Science **347**, 628 (2015)

M.T. Capria, G. Cremonese, A. Bhardwaj, M.C. De Sanctis, O (^{1}S) and O (^{1}D) emission lines in the spectrum of 153P/2002 C1 (Ikeya-Zhang). Astron. Astrophys. **442**, 1121–1126 (2005)

M.T. Capria, G. Cremonese, A. Bhardwaj, M.C. De Sanctis, E. Mazzotta Epifani, Oxygen emission lines in the high resolution spectra of 9P/Tempel 1 following the Deep Impact event. Astron. Astrophys. **479**, 257–263 (2008)

A.L. Cochran, Another look at abundance correlations among comets. Astron. J. **93**, 231–238 (1987)

A.L. Cochran, Atomic oxygen in the comae of comets. Icarus **198**, 181–188 (2008)

A.L. Cochran, W.D. Cochran, Observations of O (^{1}S) and O (^{1}D) in spectra of C/1999 S4 (LINEAR). Icarus **154**, 381–390 (2001)

A.L. Cochran, E.S. Barker, C.L. Gray, Thirty years of cometary spectroscopy from McDonald Observatory. Icarus **218**, 144–168 (2012)

A.D. Code, T.E. Houck, C.F. Lillie, Ultraviolet observations of comets, in *Scientific Results from the Orbiting Astronomical Observatory (OAO-2)* (1972), p. 109

L. Colangeli, J.J. Lopez-Moreno, P. Palumbo, J. Rodriguez, M. Cosi, V. Della Corte, F. Esposito, M. Fulle, M. Herranz, J.M. Jeronimo, A. Lopez-Jimenez, E.M. Epifani, R. Morales, F. Moreno, E. Palomba, A. Rotundi, The Grain Impact Analyser and Dust Accumulator (GIADA) experiment for the Rosetta Mission: design, performances and first results. Space Sci. Rev. **128**, 803–821 (2007)

P. Colom, J. Crovisier, N. Biver, D. Bockelée-Morvan, Observations of the 18-cm lines of OH in comets with the Nançay radio telescope, in *EPSC-DPS Joint Meeting 2011* (2011), p. 837

M.R. Combi, Z. Boyd, Y. Lee, T.S. Patel, J.-L. Bertaux, E. Quémerais, J.T.T. Mäkinen, SOHO/SWAN observations of comets with small perihelia: C/2002 V1 (NEAT), C/2002 X5 (Kudo-Fujikawa), 2006 P1 (McNaught) and 96P/Machholz 1. Icarus **216**, 449–461 (2011)

M.R. Combi, N. Fougere, J.T.T. Mäkinen, J.-L. Bertaux, E. Quémerais, S. Ferron, Unusual water production activity of comet C/2012 S1 (ISON): outbursts and continuous fragmentation. Astrophys. J. Lett. **788**, L7 (2014)

A. Coradini, F. Capaccioni, P. Drossart, G. Arnold, E. Ammannito, F. Angrilli, A. Barucci, G. Bellucci, J. Benkhoff, G. Bianchini, J.P. Bibring, M. Blecka, D. Bockelee-Morvan, M.T. Capria, R. Carlson, U. Carsenty, P. Cerroni, L. Colangeli, M. Combes, M. Combi, J. Crovisier, M.C. De Sanctis, E.T. Encrenaz, S. Erard, C. Federico, G. Filacchione, U. Fink, S. Fonti, V. Formisano, W.H. Ip, R. Jaumann, E. Kuehrt, Y. Langevin, G. Magni, T. McCord, V. Mennella, S. Mottola, G. Neukum, P. Palumbo, G. Piccioni, H. Rauer, B. Saggin, B. Schmitt, D. Tiphene, G. Tozzi, VIRTIS: an imaging spectrometer for the Rosetta Mission. Space Sci. Rev. **128**, 529–559 (2007)

M.A. Cordiner, A.J. Remijan, J. Boissier, S.N. Milam, M.J. Mumma, S.B. Charnley, L. Paganini, G. Villanueva, D. Bockelée-Morvan, Y.-J. Kuan, Y.-L. Chuang, D.C. Lis, N. Biver, J. Crovisier, D. Minniti, I.M. Coulson, Mapping the release of volatiles in the inner comae of comets C/2012 F6 (Lemmon) and C/2012 S1 (ISON) using the Atacama Large Millimeter/Submillimeter Array. Astrophys. J. Lett. **792**, L2 (2014)

H. Cottin, M.-C. Gazeau, Y. Benilan, F. Raulin, Origin of cometary extended sources from degradation of refractory organics on grains: polyoxymethylene as formaldehyde parent molecule. Icarus **167**, 397–416 (2004)

J. Crovisier, P. Colom, E. Gérard, D. Bockelée-Morvan, G. Bourgois, Observations at Nançay of the OH 18-cm lines in comets. The data base. Observations made from 1982 to 1999. Astron. Astrophys. **393**, 1053–1064 (2002)

J. Crovisier, D. Bockelée-Morvan, N. Biver, P. Colom, D. Despois, D.C. Lis, Ethylene glycol in comet C/1995 O1 (Hale-Bopp). Astron. Astrophys. **418**, L35–L38 (2004)

J. Crovisier, N. Biver, D. Bockelée-Morvan, J. Boissier, P. Colom, D.C. Lis, The chemical diversity of comets: synergies between space exploration and ground-based radio observations. Earth Moon Planets **105**, 267–272 (2009)

G.W. Curtis, Daylight observations of the 1965 F comet at the Sacramento Peak Observatory. Astron. J. **71**, 194–196 (1966)

A.C. Danks, D.L. Lambert, C. Arpigny, The $^{12}C/^{13}C$ ratio in comet Kohoutek (1973f). Astrophys. J. Lett. **194**, 745–751 (1974)

J.K. Davies, T.L. Roush, D.P. Cruikshank, M.J. Bartholomew, T.R. Geballe, T. Owen, C. de Bergh, The detection of water ice in comet Hale-Bopp. Icarus **127**, 238–245 (1997)

P. Deb Roy, H.S. Das, B.J. Medhi, Imaging polarimetry of comet C/2012 L2. Icarus **245**, 241–246 (2015)

A. Decock, E. Jehin, D. Hutsemékers, J. Manfroid, Forbidden oxygen lines in comets at various heliocentric distances. Astron. Astrophys. **555**, A34 (2013)

A. Decock, E. Jehin, P. Rousselot, D. Hutsemékers, J. Manfroid, S. Raghuram, A. Bhardwaj, B. Hubert, Forbidden oxygen lines at various nucleocentric distances in comets. Astron. Astrophys. **573**, A1 (2015)

N. Dello Russo, M.J. Mumma, M.A. DiSanti, K. Magee-Sauer, E.L. Gibb, B.P. Bonev, I.S. McLean, L.-H. Xu, A high-resolution infrared spectral survey of Comet C/1999 H1 Lee. Icarus **184**, 255–276 (2006). doi:10.1016/j.icarus.2006.04.020. http://adsabs.harvard.edu/abs/2006Icar..184..255D

N. Dello Russo, R.J. Vervack, H.A. Weaver, C.M. Lisse, H. Kawakita, H. Kobayashi, A.L. Cochran, W.M. Harris, D. Bockelée-Morvan, N. Biver, J. Crovisier, A.J. McKay, A high-resolution infrared spectral survey of 103P/Hartley 2 on the night of the EPOXI closest approach. Icarus **222**, 707–722 (2013)

N. Dello Russo, R. Vervack, H. Weaver, C. Lisse, H. Kawakita, H. Kobayashi, A. McKay, A. Cochran, W. Harris, N. Biver, D. Bockelée-Morvan, J. Crovisier, E. Jehin, M. DiSanti, The evolving chemical composition of C/2012 S1 (ISON) as it approached the Sun, in *Asteroids, Comets, Meteors 2014*, Helsinki, Finland (2014a)

N. Dello Russo, R.J. Vervack, H. Kawakita, H. Kobayashi, H.A. Weaver, W.M. Harris, A. Cochran, N. Biver, D. Bockelée-Morvan, J. Crovisier, The volatile composition of 81P/Wild 2 from ground-based high-resolution infrared spectroscopy. Icarus **238**, 125–136 (2014b)

M.A. DiSanti, B.P. Bonev, G.L. Villanueva, M.J. Mumma, Highly depleted ethane and mildly depleted methanol in comet 21P/Giacobini-Zinner: application of a new empirical ν_2-band model for CH_3OH near 50 K. Astrophys. J. Lett. **763**, 1 (2013)

M. DiSanti, B. Bonev, E. Gibb, G. Villanueva, L. Paganini, M. Mumma, J. Keane, K. Meech, G. Blake, A. McKay, The chemical composition of comet C/2012 S1 (ISON) between 1.2 AU and 0.35 AU from the Sun, in *Asteroids, Comets, Meteors 2014*, Helsinki, Finland (2014a)

M.A. DiSanti, G.L. Villanueva, L. Paganini, B.P. Bonev, J.V. Keane, K.J. Meech, M.J. Mumma, Pre- and post-perihelion observations of C/2009 P1 (Garradd): evidence for an oxygen-rich heritage? Icarus **228**, 167–180 (2014b)

J. Dufay, P. Swings, C. Fehrenbach, Spectrographic observations of comet Ikeya-Seki (1965f). Astrophys. J. Lett. **142**, 1698 (1965)

P. Eberhardt, Comet Halley's gas composition and extended sources: results from the neutral mass spectrometer on Giotto. Space Sci. Rev. **90**, 45–52 (1999)

P. Eberhardt, M. Reber, D. Krankowsky, R.R. Hodges, The D/H and $^{18}O/^{16}O$ ratios in water from comet P/Halley. Astron. Astrophys. **302**, 301–316 (1995)

P. Ehrenfreund, H.J. Fraser, J. Blum, J.H.E. Cartwright, J.M. García-Ruiz, E. Hadamcik, A.C. Levasseur-Regourd, S. Price, F. Prodi, A. Sarkissian, Physics and chemistry of icy particles in the universe: answers from microgravity. Planet. Space Sci. **51**, 473–494 (2003)

J. Elsila, D. Glavin, J. Dworkin, Cometary glycine detected in samples returned by Stardust. Meteorit. Planet. Sci. **44**, 1323–1330 (2009)

P.D. Feldman, Spectroscopy of comets with the Far Ultraviolet Spectroscopic Explorer satellite. Phys. Scr. T **119**, 7–12 (2005)

P.D. Feldman, A.L. Cochran, M.R. Combi, Spectroscopic investigations of fragment species in the coma, in *Comets II*, ed. by M.C. Festou, H.U. Keller, H.A. Weaver (University of Arizona Press, Tucson, 2004), pp. 425–447

P.D. Feldman, R.E. Lupu, S.R. McCandliss, H.A. Weaver, The far-ultraviolet spectral signatures of formaldehyde and carbon dioxide in comets. Astrophys. J. Lett. **699**, 1104–1112 (2009)

Y.R. Fernández, C.M. Lisse, M.S. Kelley, N. Dello Russo, A.T. Tokunaga, C.E. Woodward, D.H. Wooden, Near-infrared light curve of comet 9P/Tempel 1 during Deep Impact. Icarus **187**, 220–227 (2007)

M.C. Festou, 18 years of systematic UV studies of comets with IUE, in *Ultraviolet Astrophysics Beyond the IUE Final Archive*, ed. by W. Wamsteker, R. Gonzalez Riestra, B. Harris (1998), pp. 45–51

M.C. Festou, P.D. Feldman, The forbidden oxygen lines in comets. Astron. Astrophys. **103**, 154–159 (1981)

J.-H. Fillion, M. Bertin, A. Lekic, A. Moudens, L. Philippe, X. Michaut, Understanding the relationship between gas and ice: experimental investigations on ortho-para ratios, in *European Conference on Laboratory Astrophysics*, ed. by C. Stehlè, C. Joblin, L. d'Hendecourt (2012), pp. 307–314

U. Fink, A taxonomic survey of comet composition 1985–2004 using CCD spectroscopy. Icarus **201**, 311–334 (2009)

U. Fink, M.D. Hicks, A survey of 39 comets using CCD spectroscopy. Astrophys. J. Lett. **459**, 720–743 (1996)

G. Flynn, Organic grain coatings in primitive interplanetary dust: a timescale for formation of Solar System organic matter, in *Workshop on Formation of the First Solids in the Solar System*, Kauai, Hawaii (2011), p. 9143. LPI Contrib. 1639

G.J. Flynn, L.P. Keller, M. Feser, S. Wirick, C. Jacobsen, The origin of organic matter in the Solar System: evidence from the interplanetary dust particles. Geochim. Cosmochim. Acta **67**, 4791–4806 (2003)

G.J. Flynn, H. Leroux, K. Tomenka, I. Ohnishi, T. Mikouchi, S. Wirick, L.P. Keller, C. Jacobsen, S.A. Sandford, Carbonate in comets: a comparison of comets 1P/Halley, 9P/Tempel 1, and 81P/Wild 2, in *Lunar Planet. Sci. Conf. 39* (2008). Abstract 1979

G. Flynn, S. Wirick, L.P. Keller, Organic grain coatings in primitive interplanetary dust particles: implications for grain sticking in the solar nebula. Earth Planets Space **65**, 1159–1166 (2013)

M. Fries, M.J. Burchell, A. Kearsley, A. Steele, Capture effects in carbonaceous material: a Stardust analogue study. Meteorit. Planet. Sci. **44**, 1465–1474 (2009)

M. Fulle, A. Levasseur-Regourd, N. McBride, E. Hadamcik, In-situ dust measurements from within the coma of 1P/Halley: first order approximation with a dust dynamical model. Astron. J. **119**, 1968–1977 (2000)

R. Furusho, B. Suzuki, N. Yamamoto, H. Kawakita, T. Sasaki, Y. Shimizu, T. Kurakami, Imaging polarimetry and color of the inner coma of comet Hale-Bopp (C/1995 O1). Publ. Astron. Soc. Jpn. **51**, 367–473 (1999)

R. Furusho, Y. Ikeda, D. Kinoshita, W.-H. Ip, H. Kawakita, T. Kasuga, Y. Sato, H.C. Lin, M.S. Chang, Z.H. Lin, J.-I. Watanabe, Imaging polarimetry of comet 9P/Tempel 1 before and after the Deep Impact. Icarus **190**, 454–458 (2007)

E.L. Gibb, B.P. Bonev, G. Villanueva, M.A. DiSanti, M.J. Mumma, Solar fluorescence model of CH_3D as applied to comet emission. J. Mol. Spectrosc. **291**, 118–124 (2013). doi:10.1016/j.jms.2013.05.007. http://adsabs.harvard.edu/abs/2013JMoSp.291..118G

A. Gicquel, D. Bockelée-Morvan, V.V. Zakharov, M.S. Kelley, C.E. Woodward, D.H. Wooden, Investigation of dust and water ice in comet 9P/Tempel 1 from Spitzer observations of the Deep Impact event. Astron. Astrophys. **542**, A119 (2012)

F. Goesmann, H. Rosenbauer, R. Roll, C. Szopa, F. Raulin, R. Sternberg, G. Israel, U. Meierhenrich, W. Thiemann, G. Munoz-Caro, COSAC, the cometary sampling and composition experiment on Philae. Space Sci. Rev. **128**, 257–280 (2007)

S.F. Green, J.K. Davies, T.R. Geballe, T. Brooke, A.T. Tokunaga, A strong 3.4 micron emission feature in comet Austin 1989c1, in *Asteroids, Comets, Meteors 1991*, ed. by A.W. Harris, E. Bowell (1992), pp. 211–214

J.M. Greenberg, J.I. Hage, From interstellar dust to comets—a unification of observational constraints. Astrophys. J. Lett. **361**, 260–274 (1990)

S. Gulkis, M. Frerking, J. Crovisier, G. Beaudin, P. Hartogh, P. Encrenaz, T. Koch, C. Kahn, Y. Salinas, R. Nowicki, R. Irigoyen, M. Janssen, P. Stek, M. Hofstadter, M. Allen, C. Backus, L. Kamp, C. Jarchow, E. Steinmetz, A. Deschamps, J. Krieg, M. Gheudin, D. Bockelée-Morvan, N. Biver, T. Encrenaz, D. Despois, W. Ip, E. Lellouch, I. Mann, D. Muhleman, H. Rauer, P. Schloerb, T. Spilker, MIRO: Microwave Instrument for Rosetta Orbiter. Space Sci. Rev. **128**, 561–597 (2007)

S. Gulkis, M. Allen, P. von Allmen, G. Beaudin, N. Biver, D. Bockelée-Morvan, M. Choukroun, J. Crovisier, B.J.R. Davidsson, P. Encrenaz, T. Encrenaz, M. Frerking, P. Hartogh, M. Hofstadter, W.-H. Ip, M. Janssen, C. Jarchow, S. Keihm, S. Lee, E. Lellouch, C. Leyrat, L. Rezac, F.P. Schloerb, T. Spilker, Subsurface properties and early activity of comet 67P/Churyumov-Gerasimenko. Science **347**, 709 (2015)

E. Hadamcik, A.C. Levasseur-Regourd, Dust coma of comet C/1999 S4 (LINEAR): imaging polarimetry during nucleus disruption. Icarus **166**, 188–194 (2003a)

E. Hadamcik, A. Levasseur-Regourd, Imaging polarimetry of cometary dust: different comets and phase angles. J. Quant. Spectrosc. Radiat. Transf. **79–80**, 661–678 (2003b)

E. Hadamcik, A. Levasseur-Regourd, Optical properties of dust from Jupiter family comets. Planet. Space Sci. **57**, 1118–1132 (2009)

E. Hadamcik, A.C. Levasseur-Regourd, V. Leroi, D. Bardin, Imaging polarimetry of the dust coma of Comet Tempel 1 before and after Deep Impact at Haute-Provence Observatory. Icarus **190**, 459–468 (2007a). doi:10.1016/j.icarus.2007.05.010. http://adsabs.harvard.edu/abs/2007Icar..190..459H

E. Hadamcik, J.-B. Renard, F.J.M. Rietmeijer, A.C. Levasseur-Regourd, H.G.M. Hill, J.M. Karner, J.A. Nuth, Light scattering by fluffy Mg–Fe–SiO and C mixtures as cometary analogs (PROGRA^2 experiment). Icarus **190**, 660–671 (2007b)

E. Hadamcik, A.K. Sen, A.C. Levasseur-Regourd, R. Gupta, J. Lasue, R. Botet, Dust in comet 103P/Hartley 2 coma during EPOXI mission. Icarus **222**, 774–785 (2013)

E. Hadamcik, J.-B. Renard, A. Buch, N. Carrasco, N. Johnson, J. Nuth, Linear polarization of light scattered by cometary analogs: new samples, in *Asteroids, Comets and Meteors 2014*, Helsinki, Finland (2014)

M.S. Hanner, On the detectability of icy grains in the comae of comets. Icarus **47**, 342–350 (1981)

M.S. Hanner, The silicate material in comets. Space Sci. Rev. **90**, 99–108 (1999)

M.S. Hanner, J.P. Bradley, Composition and mineralogy of cometary dust, in *Comets II*, ed. by M.C. Festou, H.U. Keller, H.A. Weaver (University of Arizona Press, Tucson, 2004), pp. 555–564

D.E. Harker, D.H. Wooden, C.E. Woodward, C. Lisse, Grain properties of comet C/1995 O1 (Hale-Bopp). Astrophys. J. Lett. **580**, 579–597 (2002)

D.E. Harker, D.H. Wooden, C.E. Woodward, C. Lisse, Erratum: grain properties of comet C/1995 O1 (Hale-Bopp) (Ap.J., 580, 579 [2002]). Astrophys. J. Lett. **615**, 1081 (2004)

D.E. Harker, C.E. Woodward, D. Wooden, The dust grains from 9P/Tempel 1 before and after the encounter with Deep Impact. Science **310**, 278–280 (2005)

D.E. Harker, C.E. Woodward, D.H. Wooden, R.S. Fisher, C.A. Trujillo, Gemini-N mid-IR observations of the dust properties of the ejecta excavated from comet 9P/Tempel 1 during Deep Impact. Icarus **190**, 432–453 (2007)

M. Hässig, K. Altwegg, H. Balsiger, A. Bar-Nun, J.J. Berthelier, A. Bieler, P. Bochsler, C. Briois, U. Calmonte, M. Combi, J. De Keyser, P. Eberhardt, B. Fiethe, S.A. Fuselier, M. Galand, S. Gasc, T.I. Gombosi, K.C. Hansen, A. Jäckel, H.U. Keller, E. Kopp, A. Korth, E. Kührt, L. Le Roy, U. Mall, B. Marty, O. Mousis, E. Neefs, T. Owen, H. Rème, M. Rubin, T. Sémon, C. Tornow, C.-Y. Tzou, J.H. Waite, P. Wurz, Time variability and heterogeneity in the coma of 67P/Churyumov-Gerasimenko. Science **347**, 276 (2015)

T.L. Hayward, M.S. Hanner, Z. Sekanina, Thermal infrared imaging and spectroscopy of comet Hale-Bopp (C/1995 O1). Astrophys. J. Lett. **538**, 428–455 (2000)

D.C. Hines, G. Videen, E. Zubko, K. Muinonen, Y. Shkuratov, V.G. Kaydash, M. Knight, M.L. Sitko, C.M. Lisse, M. Mutchler, D. Hammer, P.A. Yanamandra-Fisher, Hubble Space Telescope pre-perihelion ACS/WFC imaging polarimetry of comet ISON (C/2012 S1) at 3.81 AU. Astrophys. J. Lett. **780**, L32–L38 (2014)

K.W. Hodapp, G. Aldering, K.J. Meech, A.L. Cochran, P. Antilogus, E. Pécontal, W. Chickering, N. Blanc, Y. Copin, D.K. Lynch, R. Rudy, S. Mazuk, C.C. Venturini, R.C. Puetter, R.B. Perry, Visible and near-

infrared spectrophotometry of the Deep Impact ejecta of comet 9P/Tempel 1. Icarus **187**, 185–198 (2007)
F. Hörz, R. Bastien, J. Borg, J.P. Bradley, J.C. Bridges, D.E. Brownlee, M.J. Burchell, M. Chi, M.J. Cintala, Z.R. Dai, Z. Djouadi, G. Dominguez, T.E. Economou, S.A.J. Fairey, C. Floss, I.A. Franchi, G.A. Graham, S.F. Green, P. Heck, P. Hoppe, J. Huth, H. Ishii, A.T. Kearsley, J. Kissel, J. Leitner, H. Leroux, K. Marhas, K. Messenger, C.S. Schwandt, T.H. See, C. Snead, F.J. Stadermann, T. Stephan, R. Stroud, N. Teslich, J.M. Trigo-Rodríguez, A.J. Tuzzolino, D. Troadec, P. Tsou, J. Warren, A. Westphal, P. Wozniakiewicz, I. Wright, E. Zinner, Impact features on Stardust: implications for comet 81P/Wild 2 dust. Science **314**, 1716 (2006)
E. Jehin, J. Manfroid, H. Kawakita, D. Hutsemékers, M. Weiler, C. Arpigny, A. Cochran, O. Hainaut, H. Rauer, R. Schulz, J.-M. Zucconi, Optical spectroscopy of the B and C fragments of comet 73P/Schwassmann-Wachmann 3 at the ESO VLT. LPI Contributions 1405, (2008), p. 8319. http://adsabs.harvard.edu/abs/2008LPICo1405.8319J
E. Jessberger, Rocky cometary particulates: their elemental, isotopic, and mineralogical ingredients. Space Sci. Rev. **90**, 91–97 (1999)
D. Jewitt, K.J. Meech, Cometary grain scattering versus wavelength, or, "What color is comet dust?". Astrophys. J. Lett. **310**, 937–952 (1986)
T.J. Jones, R.D. Gehrz, Infrared imaging polarimetry of comet C/1995 O1 (Hale-Bopp). Icarus **143**, 338–346 (2000)
D.J. Joswiak, D.E. Brownlee, G. Matrajt, A.J. Westphal, C.J. Snead, Z. Gainsforth, Comprehensive examination of large mineral and rock fragments in Stardust tracks: mineralogy, analogous extraterrestrial materials, and source regions. Meteorit. Planet. Sci. **47**, 471–524 (2012)
H. Kawakita, M.J. Mumma, Fluorescence excitation models of ammonia and amidogen radical (NH_2) in comets: application to comet C/2004 Q2 (Machholz). Astrophys. J. Lett. **727**, 91 (2011)
H. Kawakita, J.-i. Watanabe, T. Ootsubo, R. Nakamura, T. Fuse, N. Takato, S. Sasaki, T. Sasaki, Evidence of icy grains in comet C/2002 T7 (LINEAR) at 3.52 AU. Astrophys. J. Lett. **601**, L191–L194 (2004)
H. Kawakita, H. Kobayashi, N. Dello Russo, R.J. Vervack, M. Hashimoto, H.A. Weaver, C.M. Lisse, A.L. Cochran, W.M. Harris, D. Bockelée-Morvan, N. Biver, J. Crovisier, A.J. McKay, Parent volatiles in comet 103P/Hartley 2 observed by Keck II with NIRSPEC during the 2010 apparition. Icarus **222**, 723–733 (2013)
H. Kawakita, N. Dello Russo, R. Vervack Jr., H. Kobayashi, M.A. DiSanti, C. Opitom, E. Jehin, H.A. Weaver, A.L. Cochran, W.M. Harris, D. Bockelée-Morvan, N. Biver, J. Crovisier, A.J. McKay, J. Manfroid, M. Gillon, Extremely organic-rich coma of comet C/2010 G2 (Hill) during its outburst in 2012. Astrophys. J. Lett. **788**, 110 (2014)
H.U. Keller, C. Arpigny, C. Barbieri, R.M. Bonnet, S. Cazes, M. Coradini, C.B. Cosmovici, W.A. Delamere, W.F. Huebner, D.W. Hughes, C. Jamar, D. Malaise, H.J. Reitsema, H.U. Schmidt, W.K.H. Schmidt, P. Seige, F.L. Whipple, K. Wilhelm, First Halley multicolour camera imaging results from Giotto. Nature **321**, 320–326 (1986)
L.P. Keller, S. Bajt, G.A. Baratta, J. Borg, J.P. Bradley, D.E. Brownlee, H. Busemann, J.R. Brucato, M. Burchell, L. Colangeli, L. D'Hendecourt, Z. Djouadi, G. Ferrini, G. Flynn, I.A. Franchi, M. Fries, M.M. Grady, G.A. Graham, F. Grossemy, A. Kearsley, G. Matrajt, K. Nakamura-Messenger, V. Mennella, L. Nittler, M.E. Palumbo, F.J. Stadermann, P. Tsou, A. Rotundi, S.A. Sandford, C. Snead, A. Steele, D. Wooden, M. Zolensky, Infrared spectroscopy of comet 81P/Wild 2 samples returned by Stardust. Science **314**, 1728 (2006)
M.S. Kelley, D.H. Wooden, The composition of dust in Jupiter-family comets inferred from infrared spectroscopy. Planet. Space Sci. **57**, 1133–1145 (2009)
M.S. Kelley, D.J. Lindler, D. Bodewits, M.F. A'Hearn, C.M. Lisse, L. Kolokolova, J. Kissel, B. Hermalyn, A distribution of large particles in the coma of comet 103P/Hartley 2. Icarus **222**, 634–652 (2013)
N.N. Kiselev, K. Jockers, V.K. Rosenbush, F.P. Velichko, T. Bonev, N. Karpov, Anomalous wavelength dependence of polarization of comet 21P/Giacobini-Zinner. Planet. Space Sci. **48**, 1005–1009 (2000)
N.N. Kiselev, K. Jockers, V.K. Rosenbush, P.P. Korsun, Analysis of polarimetric, photometric, and spectroscopic observations of comet C/1996 Q1 (Tabur). Sol. Syst. Res. **35**, 480–495 (2001)
N.N. Kiselev, N. Nikolai, K. Jockers, T. Bonev, CCD imaging polarimetry of comet 2P/Encke. Icarus **168**, 385–391 (2004)
N.N. Kiselev, V. Rosenbush, A.C. Levasseur-Regourd, L. Kolokolova, Polarimetry of comets, in *Polarization of Stars and Planetary Systems*, ed. by L. Kolokolova, J. Hough, A.C. Levasseur-Regourd (Cambridge University Press, Cambridge, 2014), pp. 379–404
J. Kissel, D.E. Brownlee, K. Buchler, B.C. Clark, H. Fechtig, E. Grün, K. Hormung, E.B. Igenbergs, E.K. Jessberger, F.R. Krueger, H. Kuczera, J.A.M. McDonnell, G.M. Morfill, J. Rahe, G.H. Schwehm, Z. Sekanina, N.G. Utterback, H.J. Volk, H.A. Zook, Composition of comet Halley dust particles from Giotto observations. Nature **321**, 336–337 (1986a)

J. Kissel, R.Z. Sagdeev, J.L. Bertaux, V.N. Angarov, J. Audouze, J.E. Blamont, K. Buckler, E.N. Evlanov, H. Fechtig, M.N. Fomenkova, H. von Hoerner, N.A. Inogamov, V.N. Khromov, W. Knabe, F.R. Krueger, Y. Langevin, V.B. Leonas, A.C. Levasseur-Regourd, G.G. Managadz, S.N. Podkolzin, V.D. Shapiro, S.R. Tabaldyev, B.V. Zubkov, Composition of comet Halley dust particles from Vega observations. Nature **321**, 280–282 (1986b)

J. Kissel, F.R. Krueger, J. Silén, B.C. Clark, The cometary and interstellar dust analyzer at comet 81P/Wild 2. Science **304**, 1774–1776 (2004)

J. Kissel, K. Altwegg, B.C. Clark, L. Colangeli, H. Cottin, S. Czempiel, J. Eibl, C. Engrand, H.M. Fehringer, B. Feuerbacher, M. Fomenkova, A. Glasmachers, J.M. Greenberg, E. Grün, G. Haerendel, H. Henkel, M. Hilchenbach, H. von Hoerner, H. Höfner, K. Hornung, E.K. Jessberger, A. Koch, H. Krüger, Y. Langevin, P. Parigger, F. Raulin, F. Rüdenauer, J. Rynö, E.R. Schmid, R. Schulz, J. Silén, W. Steiger, T. Stephan, L. Thirkell, R. Thomas, K. Torkar, N.G. Utterback, K. Varmuza, K.P. Wanczek, W. Werther, H. Zscheeg, COSIMA high resolution time-of-flight secondary ion mass spectrometer for the analysis of cometary dust particles onboard Rosetta. Space Sci. Rev. **128**, 823–867 (2007)

M. Kleine, S. Wyckoff, P.A. Wehinger, B.A. Peterson, The carbon isotope abundance ratio in comet Halley. Astrophys. J. Lett. **439**, 1021–1033 (1995)

G. Klingelhöfer, J. Brückner, C. D'Uston, R. Gellert, R. Rieder, The Rosetta Alpha Particle X-ray Spectrometer (APXS). Space Sci. Rev. **128**, 383–396 (2007)

M.M. Knight, K.J. Walsh, M.F. A'Hearn, R.A. Swaters, B.A. Zauderer, N.H. Samarasinha, R. Vázquez, H. Reitsema, Ground-based visible and near-IR observations of comet 9P/Tempel 1 during the Deep Impact encounter. Icarus **187**, 199–207 (2007)

H. Kobayashi, H. Kawakita, M.J. Mumma, B.P. Bonev, J.-i. Watanabe, T. Fuse, Organic volatiles in comet 73P-B/Schwassmann-Wachmann 3 observed during its outburst: a clue to the formation region of the Jupiter-family comets. Astrophys. J. Lett. **668**, L75–L78 (2007)

W. Kofman, A. Herique, J.-P. Goutail, T. Hagfors, I.P. Williams, E. Nielsen, J.-P. Barriot, Y. Barbin, C. Elachi, P. Edenhofer, A.-C. Levasseur-Regourd, D. Plettemeier, G. Picardi, R. Seu, V. Svedhem, The COmet Nucleus Sounding Experiment by Radiowave Transmission (CONSERT): a short description of the instrument and of the commissioning stages. Space Sci. Rev. **128**, 413–432 (2007)

L. Kolokolova, M.S. Hanner, A.C. Levasseur-Regourd, B.A.S. Gustavson, Physical properties of cometary dust from light scattering and thermal emission, in *Comets II*, ed. by M.C. Festou, H.U. Keller, H.A. Weaver (University of Arizona Press, Tucson, 2004), pp. 577–604

L. Kolokolova, H. Kimura, N. Kiselev, V. Rosenbush, Two different evolutionary types of comets proved by polarimetry and infrared properties of their dust. Astron. Astrophys. **463**, 1189–1196 (2007)

L. Kolokolova, J. Hough, A.C. Levasseur-Regourd, *Polarization of Stars and Planetary Systems* (Cambridge University Press, Cambridge, 2015)

F.R. Krueger, A. Korth, J. Kissel, The organic matter of comet Halley as inferred by joint gas phase and solid phase analyses. Space Sci. Rev. **56**, 167–175 (1991)

L.E. Langland-Shula, G.H. Smith, Comet classification with new methods for gas and dust spectroscopy. Icarus **213**, 280–322 (2011)

J. Lasue, A.C. Levasseur-Regourd, E. Hadamcik, G. Alcouffe, Cometary dust properties retrieved from polarization observations: application to C/1995 O1 Hale-Bopp and 1P/Halley. Icarus **199**, 129–144 (2009)

L. Le Roy, G. Briani, C. Briois, H. Cottin, N. Fray, L. Thirkell, G. Poulet, M. Hilchenbach, On the prospective detection of polyoxymethylene in comet 67P/Churyumov-Gerasimenko with the COSIMA instrument onboard Rosetta. Planet. Space Sci. **65**, 83–92 (2012)

E. Lellouch, J. Crovisier, T. Lim, D. Bockelée-Morvan, K. Leech, M.S. Hanner, B. Altieri, B. Schmitt, F. Trotta, H.U. Keller, Evidence for water ice and estimate of dust production rate in comet Hale-Bopp at 2.9 AU from the Sun. Astron. Astrophys. **339**, L9–L12 (1998)

A.C. Levasseur-Regourd, E. Hadamcik, J.B. Renard, Evidence of two classes of comets from their polarimetric properties at large phase angles. Astron. Astrophys. **313**, 327–333 (1996)

A.C. Levasseur-Regourd, N. McBride, E. Hadamcik, M. Fulle, Similarities between in situ measurements of local dust scattering and dust flux impact data within the coma of 1P/Halley. Astron. Astrophys. **348**, 636–641 (1999)

A.C. Levasseur-Regourd, M. Zolensky, J. Lasue, Dust in cometary comae: present understanding of the structure and composition of dust particles. Planet. Space Sci. **56**, 1719–1724 (2008)

Y.-M. Lim, K.-W. Min, P.D. Feldman, W. Han, J. Edelstein, Far-ultraviolet observations of comet C/2001 Q4 (NEAT) with FIMS/SPEAR. Astrophys. J. Lett. **781**, 80 (2014)

M. Lippi, G.L. Villanueva, M.A. DiSanti, H. Böhnhardt, M.J. Mumma, B.P. Bonev, D. Prialnik, A new model for the ν_1 vibrational band of HCN in cometary comae, with application to three comets. Astron. Astrophys. **551**, A51 (2013)

C.M. Lisse, K.E. Kraemer, J.A. Nuth, A. Li, D. Joswiak, Comparison of the composition of the Tempel 1 ejecta to the dust in comet C/Hale Bopp 1995 O1 and YSO HD 100546. Icarus **187**, 69–86 (2007)

X. Liu, D.E. Shemansky, J.T. Hallett, H.A. Weaver, Extreme non-LTE H_2 in comets C/2000 WM1 (LINEAR) and C/2001 A2 (LINEAR). Astrophys. J. Suppl. Ser. **169**, 458–471 (2007)

R.E. Lupu, P.D. Feldman, H.A. Weaver, G.-P. Tozzi, The fourth positive system of carbon monoxide in the Hubble Space Telescope spectra of comets. Astrophys. J. Lett. **670**, 1473–1484 (2007)

J. Manfroid, E. Jehin, D. Hutsemékers, A. Cochran, J.-M. Zucconi, C. Arpigny, R. Schulz, J.A. Stüwe, Isotopic abundance of nitrogen and carbon in distant comets. Astron. Astrophys. **432**, L5–L8 (2005)

J. Manfroid, E. Jehin, D. Hutsemékers, A. Cochran, J.-M. Zucconi, C. Arpigny, R. Schulz, J. Stüwe, I. Liyin, The CN isotopic ratios in comets. Astron. Astrophys. **503**, 613–624 (2009)

G. Matrajt, M. Ito, S. Wirick, S. Messenger, D.E. Brownlee, D. Joswiak, G. Flynn, S. Sandford, C. Snead, A. Westphal, Carbon investigation of two Stardust particles: a TEM, NanoSIMS, and XANES study. Meteorit. Planet. Sci. **43**, 315–334 (2008)

J.E.P. Matzel, H.A. Ishii, D. Joswiak, I.D. Hutcheon, J.P. Bradley, D. Brownlee, P.K. Weber, N. Teslich, G. Matrajt, K.D. McKeegan, G.J. MacPherson, Constraints on the formation age of cometary material from the NASA Stardust Mission. Science **328**, 483 (2010)

A.J. McKay, N.J. Chanover, J.P. Morgenthaler, A.L. Cochran, W.M. Harris, N.D. Russo, Forbidden oxygen lines in comets C/2006 W3 Christensen and C/2007 Q3 Siding Spring at large heliocentric distance: implications for the sublimation of volatile ices. Icarus **220**, 277–285 (2012)

A.J. McKay, N.J. Chanoveer, J.P. Morgenthaler, A.L. Cochran, W.M. Harris, N. Dello Russo, Observations of the forbidden oxygen lines in DIXI target comet 103P/Hartley. Icarus **222**, 684–690 (2013)

A. McKay, A. Cochran, N. Dello Russo, H. Weaver, R. Vervack, W. Harris, H. Kawakita, M. DiSanti, N. Chanover, Z. Tsvetanov, Evolution of fragment-species production in comet C/2012 S1 (ISON) from 1.6 AU to 0.4 AU, in *Asteroids, Comets, Meteors 2014*, Helsinki, Finland (2014)

K.D. McKeegan, J. Aléon, J. Bradley, D. Brownlee, H. Busemann, A. Butterworth, M. Chaussidon, S. Fallon, C. Floss, J. Gilmour, M. Gounelle, G. Graham, Y. Guan, P.R. Heck, P. Hoppe, I.D. Hutcheon, J. Huth, H. Ishii, M. Ito, S. Jacobsen, A. Kearsley, L.A. Leshin, M.-C. Liu, A. Lyon, K. Marhas, B. Marty, G. Matrajt, A. Meibom, S. Messenger, S. Mostefaoui, S. Mukhopadhyay, K. Nakamura-Messenger, L. Nittler, R. Palma, R.O. Pepin, D.A. Papanastassiou, F. Robert, D. Schlutter, C.J. Snead, F. Stadermann, R. Stroud, P. Tsou, A. Westphal, E.D. Young, K. Ziegler, L. Zimmermann, E. Zinner, Isotopic compositions of cometary matter returned by Stardust. Science **314**, 1724 (2006)

R. Meier, M.F. A'Hearn, Atomic sulfur in cometary comae based on UV spectra of the S I triplet near 1814 Å. Icarus **125**, 164–194 (1997)

G. Moreels, J. Clairemidi, P. Hermine, P. Brechignac, P. Rousselot, Detection of a polycyclic aromatic molecule in comet P/Halley. Astron. Astrophys. **282**, 643–656 (1994)

J.P. Morgenthaler, W.M. Harris, M.R. Combi, P.D. Feldman, H.A. Weaver, The GALEX comets, in *AAS/Division for Planetary Sciences Meeting Abstracts* (2009)

J.P. Morgenthaler, W.M. Harris, M.R. Combi, P.D. Feldman, H.A. Weaver, GALEX FUV observations of comet C/2004 Q2 (Machholz): the ionization lifetime of carbon. Astrophys. J. Lett. **726**, 8 (2011)

M.J. Mumma, S.B. Charnley, The chemical composition of comets—emerging taxonomies and natal heritage. Annu. Rev. Astron. Astrophys. **49**, 471–524 (2011)

M.J. Mumma, H.A. Weaver, H.P. Larson, The ortho-para ratio of water vapor in comet P/Halley. Astron. Astrophys. **187**, 419–424 (1987)

M.J. Mumma, I.S. McLean, M.A. DiSanti, J.E. Larkin, N. Dello Russo, K. Magee-Sauer, E.E. Becklin, T. Bida, F. Chafee, A.R. Conrad, D.F. Figer, A.M. Gilbert, J.R. Graham, N.A. Levenson, R.E. Novak, D.C. Reuter, H.I. Teplitz, M.K. Wilcox, L.-H. Xu, A survey of organic volatile species in comet C/1999 H1 (Lee) using NIRSPEC at the Keck Observatory. Astrophys. J. Lett. **546**, 1183–1193 (2001)

M.J. Mumma, M.A. DiSanti, K. Magee-Sauer, B.P. Bonev, G.L. Villanueva, H. Kawakita, N. Dello Russo, E.L. Gibb, G.A. Blake, J.E. Lyke, R.D. Campbell, J. Aycock, A. Conrad, G.M. Hill, Parent volatiles in comet 9P/Tempel 1: before and after impact. Science **310**, 270–274 (2005). doi:10.1126/science.1119337. http://adsabs.harvard.edu/abs/2005Sci...310..270M

D. Nakashima, T. Ushikubo, D.J. Joswiak, D.E. Brownlee, G. Matrajt, M.K. Weisberg, M.E. Zolensky, N.T. Kita, Oxygen isotopes in crystalline silicates of comet Wild 2: a comparison of oxygen isotope systematics between Wild 2 particles and chondritic materials. Earth Planet. Sci. Lett. **357**, 355–365 (2012)

R.L. Newburn, H. Spinrad, Spectrophotometry of 17 comets. I. The emission features. Astron. J. **89**, 289–309 (1984)

R.L. Newburn, H. Spinrad, Spectrophotometry of 25 comets: post-Halley updates for 17 comets plus new observations for eight additional comets. Astron. J. **97**, 552–569 (1989)

R.C. Ogliore, G.R. Huss, K. Nagashima, A.L. Butterworth, Z. Gainsforth, J. Stodolna, A.J. Westphal, D. Joswiak, T. Tyliszczak, Incorporation of a late-forming chondrule into comet Wild 2. Astrophys. J. Lett. **745**, L19 (2012)

T. Ootsubo, H. Kawakita, S. Hamada, H. Kobayashi, M. Yamaguchi, F. Usui, T. Nakagaw, M. Ueno, M. Ishiguro, T. Sekiguchi, J. Watanabe, I. Sakon, T. Shimonishi, T. Onaka, AKARI near-infrared spectroscopic survey for CO_2 in 18 comets. Astrophys. J. Lett. **752**, 15 (2012)

C. Opitom, E. Jehin, J. Manfroid, D. Hutsemakers, M. Gillon, TRAPPIST monitoring of C/2012 S1 (ISON) and C/2013 R1 (Lovejoy), in *Asteroids, Comets, Meteors 2014*, Helsinki, Finland (2014)

L. Paganini, M.J. Mumma, H. Boehnhardt, M.A. DiSanti, G.L. Villanueva, B.P. Bonev, M. Lippi, H.U. Käufl, G.A. Blake, Ground-based infrared detections of CO in the centaur-comet 29P/Schwassmann-Wachmann 1 at 6.26 AU from the Sun. Astrophys. J. **766**, 100 (2013). doi:10.1088/0004-637X/766/2/100. http://adsabs.harvard.edu/abs/2013ApJ...766..100P

L. Paganini, M.A. DiSanti, M.J. Mumma, G.L. Villanueva, B.P. Bonev, J.V. Keane, E.L. Gibb, H. Boehnhardt, K.J. Meech, The unexpectedly bright comet C/2012 F6 (Lemmon) unveiled at near-infrared wavelengths. Astron. J. **147**, 15 (2014a)

L. Paganini, M.J. Mumma, G.L. Villanueva, J.V. Keane, G.A. Blake, B.P. Bonev, M.A. DiSanti, E.L. Gibb, K.J. Meech, C/2013 R1 (Lovejoy) at IR wavelengths and the variability of CO abundances among Oort Cloud comets. Astrophys. J. Lett. **791**, 122 (2014b)

L. Paganini, M.J. Mumma, G.L. Villanueva, M.A. DiSanti, B.P. Bonev, The volatile composition of comet C/2003 K4 (LINEAR) at near-IR wavelengths—Comparisons with results from the Nancay radio observatory and from the Odin, Spitzer, and SOHO space telescopes. Ap. J. (2015, in press)

G.W. Preston, The spectrum of comet Ikeya-Seki (1965f). Astrophys. J. Lett. **147**, 718–742 (1967)

S. Protopapa, J.M. Sunshine, L.M. Feaga, M.S.P. Kelley, M.F. A'Hearn, T.L. Farnham, O. Groussin, S. Besse, F. Merlin, J.-Y. Li, Water ice and dust in the innermost coma of comet 103P/Hartley 2. Icarus **238**, 191–204 (2014)

Y.L. Radeva, M.J. Mumma, G.L. Villanueva, M.F. A'Hearn, A newly developed fluorescence model for C_2H_6 ν_5 and application to cometary spectra acquired with NIRSPEC at Keck II. Astrophys. J. Lett. **729**, 135 (2011)

Y.L. Radeva, M.J. Mumma, G.L. Villanueva, B.P. Bonev, M.A. DiSanti, M.F. A'Hearn, N. Dello Russo, High-resolution infrared spectroscopic measurements of comet 2P/Encke: unusual organic composition and low rotational temperatures. Icarus **223**, 298–307 (2013)

W.T. Reach, M.S. Kelley, J. Vaubaillon, Survey of cometary CO_2, CO, and particulate emissions using the Spitzer Space Telescope. Icarus **226**, 777–797 (2013)

J.B. Renard, A.C. Levasseur-Regourd, A. Dolfus, Polarimetric CCD imaging of comet Levy. Ann. Geophys. **10**, 288–292 (1992)

W. Riedler, K. Torkar, H. Jeszenszky, J. Romstedt, H.S.C. Alleyne, H. Arends, W. Barth, J.V.D. Biezen, B. Butler, P. Ehrenfreund, M. Fehringer, G. Fremuth, J. Gavira, O. Havnes, E.K. Jessberger, R. Kassing, W. Klöck, C. Koeberl, A.C. Levasseur-Regourd, M. Maurette, F. Rüdenauer, R. Schmidt, G. Stangl, M. Steller, I. Weber, MIDAS—the Micro-Imaging Dust Analysis System for the Rosetta Mission. Space Sci. Rev. **128**, 869–904 (2007)

M. Rubin, K. Altwegg, H. Balsiger, A. Bar-Nun, J.-J. Berthelier, A. Bieler, P. Bochsler, C. Briois, U. Calmonte, M. Combi, J. De Keyser, F. Dhooghe, P. Eberhardt, B. Fiethe, S.A. Fuselier, S. Gasc, T.I. Gombosi, K.C. Hansen, M. Hässig, A. Jäckel, E. Kopp, A. Korth, L. Le Roy, U. Mall, B. Marty, O. Mousis, T. Owen, H. Rème, T. Sémon, C.-Y. Tzou, J.H. Waite, P. Wurz, Molecular nitrogen in comet 67P/Churyumov-Gerasimenko indicates a low formation temperature. Science **348**, 232–235 (2015)

S.A. Sandford, J. Aléon, C.M.O. Alexander, T. Araki, S. Bajt, G.A. Baratta, J. Borg, J.P. Bradley, D.E. Brownlee, J.R. Brucato, M.J. Burchell, H. Busemann, A. Butterworth, S.J. Clemett, G. Cody, L. Colangeli, G. Cooper, L. D'Hendecourt, Z. Djouadi, J.P. Dworkin, G. Ferrini, H. Fleckenstein, G.J. Flynn, I.A. Franchi, M. Fries, M.K. Gilles, D.P. Glavin, M. Gounelle, F. Grossemy, C. Jacobsen, L.P. Keller, A.L.D. Kilcoyne, J. Leitner, G. Matrajt, A. Meibom, V. Mennella, S. Mostefaoui, L.R. Nittler, M.E. Palumbo, D.A. Papanastassiou, F. Robert, A. Rotundi, C.J. Snead, M.K. Spencer, F.J. Stadermann, A. Steele, T. Stephan, P. Tsou, T. Tyliszczak, A.J. Westphal, S. Wirick, B. Wopenka, H. Yabuta, R.N. Zare, M.E. Zolensky, Organics captured from comet 81P/Wild 2 by the Stardust spacecraft. Science **314**, 1720 (2006)

D.G. Schleicher, The extremely anomalous molecular abundances of comet 96P/Machholz 1 from narrowband photometry. Astron. J. **136**, 2204–2213 (2008)

D.G. Schleicher, A.N. Bair, The composition of the interior of comet 73P/Schwassmann-Wachmann 3: results from narrowband photometry of multiple components. Astron. J. **141**, 177 (2011)

D.G. Schleicher, A.N. Bair, Compositional taxonomy for comets: background and motivation, in *Asteroids, Comets, Meteors 2014*, Helsinki, Finland (2014)

D.G. Schleicher, T.L. Farnham, Photometry and imaging of the coma with narrowband filters, in *Comets II*, ed. by M.C. Festou, H.U. Keller, H.A. Weaver (University of Arizona Press, Tucson, 2004), pp. 449–469

D.G. Schleicher, K.L. Barnes, N.F. Baugh, Photometry and imaging results for comet 9P/Tempel 1 and Deep Impact: gas production rates, postimpact light curves, and ejecta plume morphology. Astron. J. **131**, 1130–1137 (2006)

P.H. Schultz, B. Hermalyn, J. Veverka, The Deep Impact crater on 9P/Tempel 1 from Stardust-NExT. Icarus **222**, 502–515 (2013)

R. Schulz, A. Owens, P.M. Rodriguez-Pascual, D. Lumb, C. Erd, J.A. Stüwe, Detection of water ice grains after the Deep Impact onto comet 9P/Tempel 1. Astron. Astrophys. **448**, L53–L56 (2006)

R. Schulz, M. Hilchenbach, Y. Langevin, J. Kissel, J. Silen, C. Briois, C. Engrand, K. Hornung, D. Baklouti, A. Bardyn, H. Cottin, H. Fischer, N. Fray, M. Godard, H. Lehto, L. Le Roy, S. Merouane, F.-R. Orthous-Daunay, J. Paquette, J. Rynö, S. Siljeström, O. Stenzel, L. Thirkell, K. Varmuza, B. Zaprudin, Comet 67P/Churyumov-Gerasimenko sheds dust coat accumulated over the past four years. Nature **518**, 216–218 (2015)

Y. Shinnaka, H. Kawakita, H. Kobayashi, E. Jehin, J. Manfroid, D. Hutsemékers, C. Arpigny, Ortho-to-para abundance ratio (OPR) of ammonia in 15 comets: OPRs of ammonia versus $^{14}N/^{15}N$ ratios in CN. Astrophys. J. Lett. **729**, 81 (2011)

M.L. Sitko, D.K. Lynch, R.W. Russell, M.S. Hanner, 3–14 micron spectroscopy of comets C/2002 Q4 (Hoenig), C/2002 V1 (NEAT), C/2002 X5 (Kudo-Fujikawa), C/2002 Y1 (Juels-Holvorcem), and 69P/Taylor and the relationships among grain temperature, silicate band strength, and structure among comet families. Astrophys. J. Lett. **612**, 576–587 (2004)

C.D. Slaughter, The emission spectrum of comet Ikeya-Seki 1965-f at perihelion passage. Astron. J. **74**, 929–943 (1969)

M.K. Spencer, R.N. Zare, Comment on "Organics captured from comet 81P/Wild 2 by the Stardust spacecraft". Science **317**, 1680 (2007)

S.A. Stern, D.C. Slater, J. Scherrer, J. Stone, M. Versteeg, M.F. A'Hearn, J.L. Bertaux, P.D. Feldman, M.C. Festou, J.W. Parker, O.H.W. Siegmund, Alice: the Rosetta ultraviolet imaging spectrograph. Space Sci. Rev. **128**, 507–527 (2007)

S. Sugita, T. Ootsubo, T. Kadono, M. Honda, S. Sako, T. Miyata, I. Sakon, T. Yamashita, H. Kawakita, H. Fujiwara, T. Fujiyoshi, N. Takato, T. Fuse, J. Watanabe, R. Furusho, S. Hasegawa, T. Kasuga, T. Sekiguchi, D. Kinoshita, K.J. Meech, D.H. Wooden, W.H. Ip, M.F. A'Hearn, Subaru telescope observations of deep impact. Science **310**, 274–278 (2005). doi:10.1126/science.1119091. http://adsabs.harvard.edu/abs/2005Sci...310..274S

S. Sugita, T. Kadono, S. Sako, T. Ootsubo, M. Honda, H. Kawakita, R. Furusho, J. Watanabe, Mid-IR observations of Deep Impact reveal the primordial origin of a surface of comet 9P/Tempel 1, in *Lunar and Planetary Science Conference* (2007), p. 1911

J.M. Sunshine, O. Groussin, P.H. Schultz, M.F. A'Hearn, L.M. Feaga, T.L. Farnham, K.P. Klaasen, The distribution of water ice in the interior of comet Tempel 1. Icarus **190**, 284–294 (2007)

A.D. Thackeray, M.W. Feast, B. Warner, Daytime spectra of comet Ikeya-Seki near perihelion. Astrophys. J. Lett. **143**, 276–279 (1966)

A.J. Tuzzolino, T.E. Economou, B.C. Clark, P. Tsou, D.E. Brownlee, S.F. Green, J.A.M. McDonnell, N. McBride, M.T.S.H. Colwell, Dust measurements in the coma of Comet 81P/Wild 2 by the dust flux monitor instrument. Science **304**, 1776–1780 (2004)

G.L. Villanueva, M.A. DiSanti, M.J. Mumma, L.-H. Xu, A quantum band model of the ν_3 fundamental of methanol (CH_3OH) and its application to fluorescence spectra of comets. Astrophys. J. Lett. **747**, 37 (2012)

H.A. Weaver, P.D. Feldman, M.F. A'Hearn, N. Dello Russo, S.A. Stern, The carbon monoxide abundance in comet 103P/Hartley 2 during the EPOXI flyby. Astrophys. J. Lett. **734**, L5 (2011)

F.L. Whipple, A comet model: the acceleration of comet Encke. Astrophys. J. Lett. **111**, 375–394 (1950)

D.H. Wooden, Comet grains: their IR emission and their relation to ISM grains. Earth Moon Planets **89**, 247–287 (2002)

D.H. Wooden, C.E. Woodward, D.E. Harker, Discovery of crystalline silicates in comet C/2001 Q4 (NEAT). Astrophys. J. Lett. **612**, L77–L80 (2004)

C.E. Woodward, T.J. Jones, B. Brown, E.L. Ryan, M. Krejny, L. Kolokolova, M.S. Kelley, D.E. Harker, M.L. Sitko, Dust in comet C/2007 N3 (Lulin). Astron. J. **141**, 181–190 (2011)

I.P. Wright, S.J. Barber, G.H. Morgan, A.D. Morse, S. Sheridan, D.J. Andrews, J. Maynard, D. Yau, S.T. Evans, M.R. Leese, J.C. Zarnecki, B.J. Kent, N.R. Waltham, M.S. Whalley, S. Heys, D.L. Drummond, R.L. Edeson, E.C. Sawyer, R.F. Turner, C.T. Pillinger, Ptolemy—an instrument to measure stable isotopic ratios of key volatiles on a cometary nucleus. Space Sci. Rev. **128**, 363–381 (2007)

B. Yang, D. Jewitt, S. Bus, Comet 17P/Holmes in outburst: the near-infrared spectrum. Astron. J. **137**, 4538–4546 (2009)

B. Yang, J. Keane, K. Keech, T. Owen, R. Wainscoat, Multi-wavelength observations of comet C/20122 L4 (Pan-STARRS). Astrophys. J. Lett. **784**, L23–L27 (2014)

M.E. Zolensky, T.J. Zega, H. Yano, S. Wirick, A.J. Westphal, M.K. Weisberg, I. Weber, J.L. Warren, M.A. Velbel, A. Tsuchiyama, P. Tsou, A. Toppani, N. Tomioka, K. Tomeoka, N. Teslich, M. Taheri, J. Susini, R. Stroud, T. Stephan, F.J. Stadermann, C.J. Snead, S.B. Simon, A. Simionovici, T.H. See, F. Robert, F.J.M. Rietmeijer, W. Rao, M.C. Perronnet, D.A. Papanastassiou, K. Okudaira, K. Ohsumi, I. Ohnishi, K. Nakamura-Messenger, T. Nakamura, S. Mostefaoui, T. Mikouchi, A. Meibom, G. Matrajt, M.A. Marcus, H. Leroux, L. Lemelle, L. Le, A. Lanzirotti, F. Langenhorst, A.N. Krot, L.P. Keller, A.T. Kearsley, D. Joswiak, D. Jacob, H. Ishii, R. Harvey, K. Hagiya, L. Grossman, J.N. Grossman, G.A. Graham, M. Gounelle, P. Gillet, M.J. Genge, G. Flynn, T. Ferroir, S. Fallon, D.S. Ebel, Z.R. Dai, P. Cordier, B. Clark, M. Chi, A.L. Butterworth, D.E. Brownlee, J.C. Bridges, S. Brennan, A. Brearley, J.P. Bradley, P. Bleuet, P.A. Bland, R. Bastien, Mineralogy and petrology of comet 81P/Wild 2 nucleus samples. Science **314**, 1735 (2006)

M. Zolensky, K. Nakamura-Messenger, F. Rietmeijer, H. Leroux, T. Mikouchi, K. Ohsumi, S. Simon, L. Grossman, T. Stephan, M. Weisberg, M. Velbel, T. Zega, R. Stroud, K. Tomeoka, I. Ohnishi, N. Tomioka, T. Nakamura, G. Matrajt, D. Joswiak, D. Brownlee, F. Langenhorst, A. Krot, A. Kearsley, H. Ishii, G. Graham, Z.R. Dai, M. Chi, J. Bradley, K. Hagiya, M. Gounelle, J. Bridges, Comparing Wild 2 particles to chondrites and IDPs. Meteorit. Planet. Sci. **43**, 261–272 (2008)

M. Zolensky, D. Frank, L. Le, Olivine and pyroxene compositions in fine-grained chondritic materials, in *Lunar Planet. Sci. Conf. 42* (2011), abstract 1898

E. Zubko, R. Furusho, K. Kawabata, T. Yamamoto, K. Muinonen, G. Videen, Interpretation of photopolarimetric observations of comet 17P/Holmes. J. Quant. Spectrosc. Radiat. Transf. **112**, 1848–1863 (2011)

E. Zubko, K. Muinonen, Y. Shkuratov, E. Hadamcik, A.C. Levasseur-Regourd, Evaluating the carbon depletion found by the Stardust mission in comet 81P/Wild 2. Astron. Astrophys. **544**, L8 (2012)

Space Sci Rev (2015) 197:47–83
DOI 10.1007/s11214-015-0156-9

Cometary Isotopic Measurements

Dominique Bockelée-Morvan[1] · Ursina Calmonte[2] · Steven Charnley[3] · Jean Duprat[4] · Cécile Engrand[4] · Adeline Gicquel[5] · Myrtha Hässig[6] · Emmanuël Jehin[7] · Hideyo Kawakita[8] · Bernard Marty[9] · Stefanie Milam[3] · Andrew Morse[10] · Philippe Rousselot[11] · Simon Sheridan[10] · Eva Wirström[12]

Received: 11 December 2014 / Accepted: 5 May 2015 / Published online: 16 May 2015

Abstract Isotopic ratios in comets provide keys for the understanding of the origin of cometary material, and the physical and chemical conditions in the early Solar Nebula. We review here measurements acquired on the D/H, $^{14}N/^{15}N$, $^{16}O/^{18}O$, $^{12}C/^{13}C$, and $^{32}S/^{34}S$ ratios in cometary grains and gases, and discuss their cosmogonic implications. The review includes analyses of potential cometary material available in collections on Earth, recent measurements achieved with the Herschel Space Observatory, large optical telescopes,

✉ D. Bockelée-Morvan
dominique.bockelee@obspm.fr

1 Observatoire de Paris, CNRS, UPMC, Université Paris-Diderot, 5 place Jules Janssen, 92195 Meudon, France

2 Physics Institute, Space Research and Planetary Science, University of Bern, Sidlerstrasse 5, 3012 Bern, Switzerland

3 Astrochemistry Laboratory, Code 691 NASA Goddard Space Flight Center Greenbelt, 8800 Greenbelt Rd, Greenbelt, MD 20771, USA

4 Centre de Sciences Nucléaires et de Sciences de la matière (CSNSM), Université Paris-Sud, UMR 8609-CNRS/IN2P3, 91405 Orsay, France

5 Max Planck Institute for Solar System Research, Justus-von-Liebig-Weg 3, 37077 Göttingen, Germany

6 Southwest Research Institute, San Antonio, TX, USA

7 Département d'Astrophysique, de Géophysique et d'Océanographie de Université de Liège, Place du 20 Août 7, 4000 Liège, Belgium

8 Koyama Astronomical Observatory, Kyoto Sangyo Univ., Motoyama, Kamigamo, Kita, Kyoto 603-8555, Japan

9 CRPG-CNRS, Université de Lorraine, 15 Rue Notre Dame des Pauvres, 54500 Vandoeuvre-lès-Nancy, France

10 Space Sciences, The Open University, Walton Hall, Milton Keynes MK7 6AA, UK

11 Observatoire des Sciences de l'Univers THETA, Institut UTINAM—UMR CNRS 6213, University of Franche-Comté, BP 1615, 25010 Besançon Cedex, France

12 Onsala Space Observatory, Chalmers University of Technology, 439 92 Onsala, Sweden

and Rosetta, as well as recent results obtained from models of chemical-dynamical deuterium fractionation in the early solar nebula. Prospects for future measurements are presented.

Keywords Comets · Isotopes

1 Introduction

The solar system formed about 4.6 billion years ago from an infall of matter inside a molecular cloud. A wealth of processes happened, involving chemistry, and dynamics at all scales before the solar system reached its present state with its cortege of planets and small bodies (e.g. Ceccarelli et al. 2014; Mandt et al. 2015, this issue). The view that comets are among the most primitive materials in the solar system stems from their primitive element abundances, similar to that of the bulk solar system for most elements (as deduced from analyses of 1P/Halley and 81P/Wild 2 for rock forming elements), and their extraordinary composition richness in elements in the gas phase, which resembles that of interstellar clouds (Jessberger and Kissel 1991; Bockelée-Morvan et al. 2000; Cochran et al. 2015, this issue; Brownlee 2014). Comets formed in the outer skirts of the solar nebula and trapped volatile species formed either in the presolar cloud or in situ. However, the discovery that comets incorporated high-temperature components formed in the neighborhood of the forming Sun, points out to a large diversity of formation conditions for the comet material, which in turn provides constraints on mixing processes in the solar nebula (Brownlee et al. 2006; Bockelée-Morvan et al. 2002; Wooden 2008).

The isotopic abundances are key tracers for reconstructing the origin and history of cometary material. Indeed, isotopic fractionation is sensitive to environmental conditions, such as temperature, density, radiation, and composition. The strongest variations among volatile element isotope ratios are observed for hydrogen. Therefore deuterium fractionation is an Ariadne's thread for understanding the very first steps of solar system formation, starting with the pre-collapse phase to the formation of planetesimals (Ceccarelli et al. 2014). Key information can also be inferred from the large nitrogen isotopic variations observed among solar system objects, although they are still difficult to interpret.

In this chapter, we review isotopic measurements acquired so far on comets, including on dust particles collected by the Stardust mission (Brownlee 2014). Different lines of evidence indicate that samples from comets have been recovered on Earth. Such samples are of uttermost interest since they allow extremely precise elemental, mineralogical and isotopic analyses at different scales down to the nanometer. Hence this review also present isotopic properties of interplanetary dust particles (IDPs) and micrometeorites of probable cometary origin. To put the measurements into context, we summarize recent studies on interstellar deuterium and nitrogen fractionation, initiated by recent measurements in interstellar sources. Finally, we present prospectives for new isotopic measurements in comets, especially with instruments onboard the Rosetta/Philae space mission.

2 Isotopic Measurements in Cometary Volatiles

Because of the faint signatures of deuterated species, D/H ratios were only obtained in a handful of bright comets. Table 1 lists measurements of the D/H ratio in cometary water acquired so far. It is worth remembering that the D/H ratio is equal to half the HDO/H_2O ratio, because water contains two atoms of hydrogen.

Table 1 D/H ratios in cometary molecules

Species	Parent molecule	D/H ratio[a]	Comet	Reference
H_2DO^+	H_2O	$(3.08 \pm 0.5) \times 10^{-4}$	1P/Halley	Balsiger et al. (1995)
		$(3.06 \pm 0.34) \times 10^{-4}$	1P/Halley	Eberhardt et al. (1995)
		$(2.1 \pm 0.3) \times 10^{-4}$	1P/Halley	Brown et al. (2012)
HDO	H_2O	$(2.9 \pm 1.0) \times 10^{-4}$	C/1996 B2 (Hyakutake)	Bockelée-Morvan et al. (1998)
		$(3.3 \pm 0.8) \times 10^{-4}$	C/1995 O1 (Hale-Bopp)	Meier et al. (1998a)
		$(4.09 \pm 1.45) \times 10^{-4}$	8P/Tuttle	Villanueva et al. (2009)
		$(1.61 \pm 0.24) \times 10^{-4}$	103P/Hartley 2	Hartogh et al. (2011)
		$(2.06 \pm 0.22) \times 10^{-4}$	C/2009 P1 (Garradd)	Bockelée-Morvan et al. (2012)
		$(5.3 \pm 0.7) \times 10^{-4}$	67P/Churyumov-Gerasimenko	Altwegg et al. (2014)
		$< 2.5 \times 10^{-4}$[b]	153P/Ikeya-Zhang	Biver et al. (2006)
		$< 5.6 \times 10^{-4}$	C/2007 N3 (Lulin)	Gibb et al. (2012)
		$< 2 \times 10^{-4}$	45P/Honda-Mrkos-Pajdušáková	Lis et al. (2013)
OD	H_2O	$(2.5 \pm 0.7) \times 10^{-4}$	C/2002 T7 (LINEAR)	Hutsemékers et al. (2008)
D	H_2O	$(4.6 \pm 1.4) \times 10^{-4}$	C/2001 Q4 (NEAT)	Weaver et al. (2008)
DCN	HCN	$(2.3 \pm 0.4) \times 10^{-3}$	C/1995 O1 (Hale-Bopp)	Meier et al. (1998b)
		$(2.3 \pm 0.6) \times 10^{-3}$	C/1995 O1 (Hale-Bopp)	Crovisier et al. (2004)
		$< 1 \times 10^{-2}$	C/1996 B2 (Hyakutake)	Bockelée-Morvan et al. (1998)
		$< 1 \times 10^{-2}$	103P/Hartley 2	Gicquel et al. (2014)
HDCO	H_2CO	$< 2 \times 10^{-2}$	1P/Halley	Balsiger et al. (1995)
		$< 5 \times 10^{-2}$	C/1995 O1 (Hale-Bopp)	Crovisier et al. (2004)
CH_3D	CH_4	$< 1 \times 10^{-1}$	153P/2002 C1 (Ikeya-Zhang)	Kawakita et al. (2003)
		$< 1 \times 10^{-2}$	C/2001 Q4 (NEAT)	Kawakita et al. (2005)
		$< 5 \times 10^{-3}$	C/2004 Q2 (Machholz)	Bonev et al. (2009)
		$< 6.4 \times 10^{-3}$	C/2004 Q2 (Machholz)	Kawakita and Kobayashi (2009)
		$< 7.5 \times 10^{-3}$	C/2007 N3 (Lulin)	Gibb et al. (2012)
NH_2D	NH_3	$< 4 \times 10^{-2}$	C/1995 O1 (Hale-Bopp)	Crovisier et al. (2004)
ND	NH_3	$< 6 \times 10^{-3}$	C/1996 B2 (Hyakutake)	Meier et al. (1998c)
CH_3OD	CH_3OH	$< 1 \times 10^{-2}$	1P/Halley	Eberhardt et al. (1994)
		$< 3 \times 10^{-2}$	C/1995 O1 (Hale-Bopp)	Crovisier et al. (2004)
CH_2DOH	CH_3OH	$< 8 \times 10^{-3}$	C/1995 O1 (Hale-Bopp)	Crovisier et al. (2004)
HDS	H_2S	$< 2 \times 10^{-1}$	C/1995 O1 (Hale-Bopp)	Crovisier et al. (2004)
		$< 8 \times 10^{-3}$	17P/Holmes	Biver et al. (2008)
CD		$< 3 \times 10^{-2}$	C/1996 B2 (Hyakutake)	Meier et al. (1998c)

[a]Value for the parent molecule, except for ND and CD, where it refers to ND/NH, and CD/CH values

[b]Revised value using updated H_2O production rates from Biver et al. (2007)

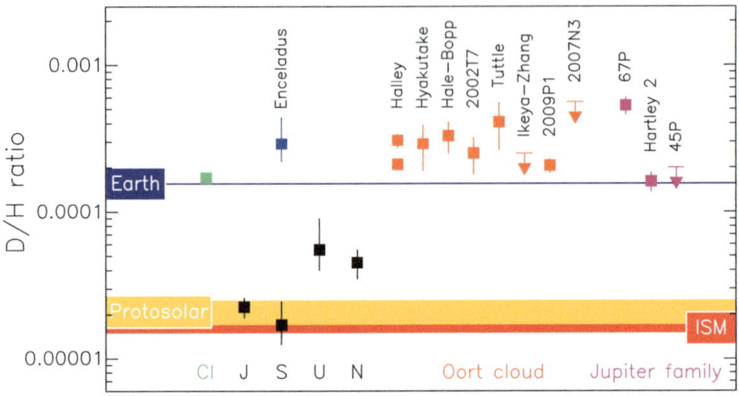

Fig. 1 D/H ratio in the water of comets compared to values in carbonaceous meteorites (CI), Earth's oceans (VSMOW), and Enceladus. Displayed data for planets, the interstellar medium, and the protosolar nebula refer to the value in H_2. Adapted from Lis et al. (2013)

2.1 Deuterium

2.1.1 Water

The first measurements of the D/H ratio in cometary H_2O were obtained in comet 1P/Halley from mass-resolved ion-spectra of H_3O^+ acquired with the Ion Mass Spectrometer (IMS) (Balsiger et al. 1995) and the Neutral Mass Spectrometer (NMS) (Eberhardt et al. 1995) instruments onboard the European Giotto spacecraft. These independent data provided a consistent D/H value of $\sim 3 \times 10^{-4}$, which corresponds to twice the Vienna Standard Mean Ocean Water VSMOW of 155.76 ± 0.1 ppm (Table 1, Balsiger et al. 1995; Eberhardt et al. 1995). However, Brown et al. (2012) reexamined the NMS measurements, reevaluating the D/H ratio in comet 1P/Halley to $(2.1 \pm 0.3) \times 10^{-4}$.

From observations undertaken with the Caltech Submillimeter Observatory and the James Clerk Maxwell telescope (JCMT), HDO was detected in the bright long-period comets C/1996 B2 (Hyakutake) and C/1995 O1 (Hale-Bopp) from its 1_{01}–0_{00} line at 464.925 GHz (Bockelée-Morvan et al. 1998; Meier et al. 1998a). The derived D/H values in these two comets are in agreement with the determinations in comet Halley (Table 1, Fig. 1). However, observations of the HDO 1_{10}–1_{01} transition at 509.292 GHz in the Halley-type comet 153P/Ikeya-Zhang yielded D/H $< 2.5 \times 10^{-4}$, suggesting for the first time an hydrogen isotopic diversity in the Oort-cloud population (Biver et al. 2006).

Using the new high resolution Cryogenic Infrared Echelle Spectrograph (CRIRES) of the Very Large Telescope (VLT), Villanueva et al. (2009) observed in early 2008 the HDO ro-vibrational transitions near 3.7 μm in the Halley-family comet 8P/Tuttle originating from the Oort cloud. Twenty three lines were co-added to get a marginal detection of HDO, from which a formal value of D/H of $(4.09 \pm 1.45) \times 10^{-4}$ was derived.

In cometary atmospheres, water photodissociates into mainly OH and H. The OD/OH ratio was measured to be $(2.5 \pm 0.7) \times 10^{-4}$ in the Oort-Cloud comet C/2002 T7 (LINEAR) through ground-based observations of the OH $A^2\Sigma^+$–$X^2\Pi_i$ ultraviolet bands at 310 nm obtained with the VLT feeding the Ultraviolet-Visual Echelle Spectrograph (UVES) (Hutsemékers et al. 2008). No individual OD line was detected, but a marginal 3σ detection of OD was obtained by co-adding the brightest lines. Atomic deuterium (D) emission was also discovered during ultraviolet observations of comet C/2001 Q4 (NEAT) in April 2004 using the Space Telescope Imaging Spectrograph (STIS) of the Hubble Space Telescope (Weaver et al. 2008). The Lyman-α emission from both D and atomic hydrogen were detected, from which a preliminary value $D/H = (4.6 \pm 1.4) \times 10^{-4}$ was derived, assuming that H_2O is the dominant source of the observed D and H, as it is likely the case. The strength of the optical method is that both normal and rare isotopologues have lines in the same spectral interval and are observed simultaneously, avoiding problems related to comet variable activity.

The most recent spectroscopic D/H measurements in cometary water were acquired using the ESA Herschel space observatory. The HDO 1_{10}–1_{01} transition at 509.292 GHz was observed using the Heterodyne Instrument for the Far-Infrared (HIFI) in the Jupiter-family comets 103P/Hartley 2 and 45P/Honda-Mrkos-Pajdušáková (Hartogh et al. 2011; Lis et al. 2013), and in the Oort-cloud comet C/2009 P1 (Garradd) (Bockelée-Morvan et al. 2012). Observations of HDO were interleaved with observations of the H_2O and $H_2^{18}O$ 1_{10}–1_{01} lines. Since the H_2O ground state rotational lines in comets are optically thick, optically thin lines of $H_2^{18}O$ provide, in principle, a more reliable reference for the D/H determination. The $HDO/H_2^{18}O$ was measured to be 0.161 ± 0.017 for comet Hartley 2, i.e., consistent with the VSMOW value for this ratio of 0.1554 ± 0.0001. The $HDO/H_2^{18}O$ value of 0.215 ± 0.023 for comet Garradd suggests a significant difference (3-σ) in deuterium content between the comet Garradd and Hartley 2. Hartogh et al. (2011) derived a D/H ratio of $(1.61 \pm 0.24) \times 10^{-4}$ for comet Hartley, assuming an $H_2^{16}O/H_2^{18}O$ ratio of 500 ± 50 which encompasses the VSMOW value and values measured in cometary water (Jehin et al. 2009, see Sect. 2.3). For comet Garradd, the derived D/H ratio is $(2.06 \pm 0.22) \times 10^{-4}$ based on the $HDO/H_2^{16}O$ production rate ratio, and $(2.15 \pm 0.32) \times 10^{-4}$, using the same method as Hartogh et al. (2011). Herschel observations in the Jupiter family comet 45P/Honda-Mrkos-Pajdušáková were unsuccessful in detecting the HDO 509 GHz line, but resulted in a 3σ upper limit for the D/H ratio of 2.0×10^{-4} which is consistent with the value measured in comet 103P/Hartley 2 and excludes the canonical pre-Herschel value measured in Oort-cloud comets of $\sim 3 \times 10^{-4}$ at the 4.5σ level (Lis et al. 2013).

Whereas the Herschel measurements were suggesting a Earth-like value for Jupiter family comets, the ROSINA mass-spectrometer onboard Rosetta (Balsiger et al. 2007) measured in comet 67P/Churymov-Gerasimenko a D/H value of $(5.3 \pm 0.7) \times 10^{-4}$, i.e., more than three times the VSMOW value (Altwegg et al. 2014).

2.1.2 HCN

The first detection of DCN (J:5-4; 362.046 GHz) was obtained by Meier et al. (1998b) in the bright Oort comet C/1995 O1 (Hale-Bopp) using the JCMT, where a D/H ratio of $(2.3 \pm 0.4) \times 10^{-3}$ was derived. Crovisier et al. (2004) obtained a marginal 3σ detection of DCN (J:2-1; 217.2385 GHz) with the IRAM 30m telescope toward the same comet and reported a D/H ratio consistent with Meier et al. (1998b). The deuterium enrichment in HCN is 7 times larger than that measured in H_2O in comet Hale-Bopp and most Oort cloud comets. (Sect. 2.1.1).

Bockelée-Morvan et al. (1998) derived a 3σ upper limit of DCN (J:5-4; 362.046 GHz) in the bright long-period comet C/1996 (Hyakutake) by using the 10.4m Leighton telescope of the Caltech Submillimeter Observatory (CSO). They obtained a $D/H < 1.0 \times 10^{-2}$. Gicquel et al. (2014) measured for the first time an upper limit for DCN (J:4-3; 362.0465 GHz) in a Jupiter-Family comet (103P/Hartley 2) using the JCMT. They concluded to a D/H ratio $< 1.0 \times 10^{-2}$. Blake et al. (1999) deduced a larger D/H ratio in jets from interferometric radio data, which might suggest variations in the coma. The $(D/H)_{HCN}$ ratios deduced from single-dish observations are summarized in Table 1.

By comparison, a range of (0.4–7.0) × 10^{-2} for the DCN/HCN ratio was determined in the interstellar medium (ISM) (Roberts et al. 2002; Jørgensen et al. 2004). The low DCN/HCN value in comets compared to the ISM can suggest some reprocessing in the Solar Nebula that occurred before the incorporation of these molecules into comets (Mousis et al. 2000), as it will be discussed in depth in Sect. 4.2. Alternatively, the D/H ratios in both water and HCN in comet Hale-Bopp may reflect an ion-chemistry at about 30–35 K which took place either in the presolar cloud or in the outer regions of the solar nebula (Meier et al. 1998b).

2.1.3 CH_4

Because methane has four identical H-atoms to be substituted with deuterium, the abundance ratio of singly deuterated methane (CH_3D) relative to CH_4 is enhanced, i.e., $D/H = 0.25 \times (CH_3D/CH_4)$.

Constraints on the D/H ratio in methane were obtained from observations of vibrational bands in the near-IR. The strongest vibrational band of CH_3D in cometary atmospheres is the ν_4 band around 3000 cm^{-1}, which is close to the ν_3 vibrational band of CH_4. They are close to each other and this situation allows us to measure emission lines of CH_4 and CH_3D simultaneously.

The first trial to detect CH_3D in comet 153P/Ikeya-Zhang provided a 2σ upper limit for the D/H ratio of 0.075 (Kawakita and Watanabe 2003). A more stringent upper limit of 0.01 was later reported in comet C/2001 Q4 (NEAT) (Kawakita et al. 2005). In comet C/2004 Q2 (Machholz), 3σ upper limits of 0.003 and 0.005 were reported by Kawakita and Kobayashi (2009) and Bonev et al. (2009), respectively. More recently, Gibb et al. (2012) derived $D/H < 0.007$ (3σ) in comet C/2007 N3 (Lulin).

Chemical network models for low temperature environments (e.g., Aikawa and Herbst 1999) predict large CH_3D/CH_4 abundance ratios, up to 0.1–0.2 at 10 K (i.e, D/H up to 0.025–0.05). Upper limits measured in comets cannot be explained by methane deuteration at such low temperatures.

2.1.4 Other Molecules

Some constraints on the level of deuteration in other cometary molecules have been obtained. Using Hale-Bopp data in the millimeter range, Crovisier et al. (2004) obtained D/H upper limits for NH_3, H_2CO, and CH_3OH (both CH_3OD, and CH_3DOH), which range typically from 0.01 to 0.05. An upper limit of $D/H < 0.008$ was obtained for H_2S in comet 17P/Holmes (Biver et al. 2008). Measurements for some of these molecules will be possible from Rosetta (see Sect. 5.1 and Table 6), or will need to await for a very bright comet observed with ALMA (see Sect. 5.2).

2.2 Nitrogen

Figure 2 summarizes nitrogen isotopic ratios measured in comets, and how they compare to values measured in other primitive Solar System materials.

2.2.1 HCN and CN

The nitrogen isotopic ratio $^{14}N/^{15}N$ was measured for the first time in comet C/1995 O1 (Hale-Bopp) in 1997 in HCN as well as in the CN violet band. The $^{14}N/^{15}N$ values derived from the HCN emission at sub-millimetre range (323 ± 46, Jewitt et al. 1997; 330 ± 98

Table 2 Isotopic ratios in comets: Carbon, oxygen, nitrogen and sulfur

Isotopic ratio	Species	Value	Comet	References
$^{12}C/^{13}C$	[C_2]	93 ± 10	4 comets	Wyckoff et al. (2000)
		85 ± 20	C/2002 T7	Rousselot et al. (2012)
		80 ± 20	C/2001 Q4	Rousselot et al. (2012)
		94 ± 33	C/2012 S1	Shinnaka et al. (2014a)
	[CN]	91 ± 4	21 comets	Manfroid et al. (2009)
		95 ± 15	103P/Hartley 2	Jehin et al. (2011)
		95 ± 25	C/2012 F6	Decock et al. (2014)
	[HCN]	111 ± 12	C/1995 O1	Jewitt et al. (1997)
		114 ± 26	17P/Holmes	Bockelée-Morvan et al. (2008)
$^{14}N/^{15}N$	[HCN]	205 ± 70	C/1995 O1	Bockelée-Morvan et al. (2008)
		139 ± 26	17P/Holmes	Bockelée-Morvan et al. (2008)
	[CN]	141 ± 29	21 comets	Manfroid et al. (2009)
		220 ± 40	73P (fragment C)	Manfroid et al. (2009)
		165 ± 40	17P/Holmes	Bockelée-Morvan et al. (2008)
		155 ± 25	103P/Hartley 2	Jehin et al. (2011)
	[NH_2]	90–190	12 comets	Rousselot et al. (2014)
		139 ± 38	C/2012 S1	Shinnaka et al. (2014b)
		140 ± 30	C/2012 F6	Decock et al. (2014)
$^{16}O/^{18}O$	[H_2O]	518 ± 45	1P/Halley	Balsiger et al. (1995)
		470 ± 40	1P/Halley	Eberhardt et al. (1995)
		530 ± 60	4 comets	Biver et al. (2007)
		425 ± 55	C/2002 T7	Hutsemékers et al. (2008)
		300 ± 150	C/2012 F6	Decock et al. (2014)
		523 ± 32	C/2009 P1	Bockelée-Morvan et al. (2012)
$^{32}S/^{34}S$	[CS]	27 ± 3	C/1995 O1	Jewitt et al. (1997)
	[CS]	16 ± 3	17P/Holmes	Biver et al. (2008)
	[S^+]	23 ± 6	1P/Halley	Altwegg (1996)
	[H_2S]	16 ± 3	C/1995 O1	Crovisier et al. (2004)

Ziurys et al. 1999), were consistent with the telluric value (272, Anders and Grevesse 1989) while the ratio derived from optical high resolution spectroscopy of CN (140 ± 35, Arpigny et al. 2003) was widely discordant. Subsequent observations of a dozen comets of various origins lead to $^{14}N/^{15}N = 147.8 \pm 5.7$ from CN (Jehin et al. 2009). The discrepancy between CN and HCN, a presumed parent, was eventually solved with the quasi simultaneous sub-millimeter and optical observations of HCN and CN in comet 17P/Holmes performed a couple of days after its huge and bright outburst in October 2007 (Bockelée-Morvan et al. 2008). The nitrogen isotopic ratios derived from the two techniques are found in agreement with each other and with the values found in other comets. The reanalysis of the Hale-Bopp sub-millimeter data (Bockelée-Morvan et al. 2008) also gave much lower values for the nitrogen ratio and established that HCN has most probably the same non-terrestrial nitrogen isotopic composition as CN. No other measurements in HCN have been made due to the difficulty of detection of the faint $HC^{15}N$ lines in the radio range but the CN bright violet band $(0, 0)$ at 3880 Å was observed in about 20 comets since Hale-Bopp with high resolution optical spectrometers like the Ultraviolet and Visual Echelle Spectrograph (UVES) mounted at the ESO Very Large Telescope (VLT), the High Resolution Echelle Spectrometer

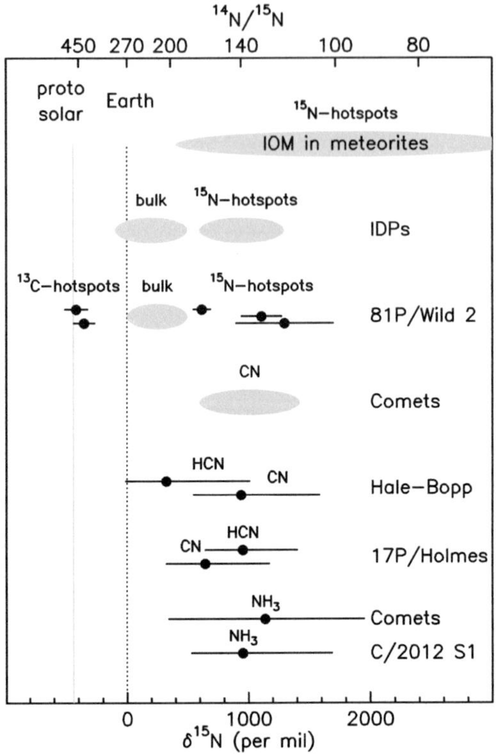

Fig. 2 N isotope measurements in primitive matter of the Solar System, given either as $^{14}N/^{15}N$ ratios or $\delta^{15}N$ values ($\delta^{15}N = (^{14}N/^{15}N)/(^{14}N/^{15}N)_{Earth} - 1$). Shown are the protosolar value measured by the Genesis mission (Marty et al. 2011), values in HCN and CN for comet 17P/Holmes (Bockelée-Morvan et al. 2008), comet Hale-Bopp (Arpigny et al. 2003; Bockelée-Morvan et al. 2008) and other comets (Jehin et al. 2009), in NH_3 from NH_2 observations (Rousselot et al. 2014; Shinnaka et al. 2014b), in dust particles of comet 81P/Wild 2 collected by the Stardust mission (McKeegan et al. 2006), in interplanetary dust particles IDPs (Floss et al. 2006), and in the insoluble organic matter (IOM) in carbonaceous meteorites (Busemann et al. 2006). So-called hotspots are regions that present strong isotopic enrichments relative to the surrounding material. ^{13}C hotspots refer to 81P/Wild 2 submicrometre grains enriched in ^{13}C with respect to the terrestrial $^{12}C/^{13}C$ ratio and depleted in ^{15}N. *Points with error bars* correspond to single measurements. When many measurements are available, *ellipses* encompassing the range of measured values are drawn. Adapted from Bockelée-Morvan (2011)

(HIRES) at Keck, the High Dispersion Spectrograph (HDS) at Subaru and the 2D Coudé at the Harlan telescope at the McDonald Observatory (Table 2). They provided measurements of $^{14}N/^{15}N$ for comets of various origins (new Oort-cloud comets, Halley type comets, and Jupiter family comets), for comets at various heliocentric distances (from 0.6 AU to 3.7 AU), at different distances from the nucleus, during comet outbursts or splitting events (Jehin et al. 2004; Manfroid et al. 2005; Hutsemékers et al. 2005; Jehin et al. 2006, 2008, 2009; Manfroid et al. 2009; Jehin et al. 2011; Decock et al. 2014). All the measurements are providing the same value for the nitrogen ratio within the errors but one comet, the carbon poor and split comet 73P/Schwassmann-Wachmann 3 (Jehin et al. 2008; Shinnaka et al. 2011). Whilst the value for the $^{12}C/^{13}C$ ratio (see Sect. 2.3.1), derived from C_2, CN and HCN, is in very good agreement with the solar and terrestrial value of 89, the nitrogen isotopic ratio, derived from CN and HCN, is very different from the telluric value of 272. The large ^{15}N excess in cometary volatiles relative to the Earth atmospheric value indicates that N-bearing volatiles in the solar nebula underwent important nitrogen isotopic fractionation at some stage of the Solar System formation or earlier in the proto-solar cloud (Rodgers and Charnley 2008).

2.2.2 NH_3

Ammonia molecules represent typically a little bit less than about 1 % (relative to water) of the molecules observed in cometary coma (e.g. Biver et al. 2012), only about ten molecules have a similar or higher abundance. Photolysis of ammonia by solar UV radiation in the coma produces the NH_2 radical with 95 % efficiency (Huebner et al. 1992). This radical has a long lifetime against photodissociation and has many emission lines in the optical wavelength range.

Direct measurement of the $^{14}N/^{15}N$ isotopic ratio in ammonia, through inversion (in radio), rotational (submillimetre), or vibrational (near-infrared) transitions, are usually difficult due to a lack of sensitivity for the $^{15}NH_3$ detection at the available facilities. Indirect measurement of the $^{14}N/^{15}N$ isotopic ratio in ammonia, however, can be done by using NH_2 emission lines in the optical range because these lines have a strong intensity and are easy to observe in this spectral range. The main problem for doing such a measurement has been, for a long time, the lack of accurate wavelengths measurements. Up to recently only one paper had been published mentioning isotope shifts between $^{14}NH_2$ and $^{15}NH_2$ (Dressler and Ramsay 1959) but this shift was presented in a general plot that was by far not accurate enough for searching for the faint $^{15}NH_2$ emission lines.

To efficiently search for these lines it was necessary to perform new laboratory experiments. Such experiments have been conducted in the Advanced Infrared Line Exploited for Spectroscopy (AILES) of synchrotron SOLEIL (Source Optimisé de Lumière d'Energie Intermédiaire du LURE) and permitted us to establish a list of wavelengths for $^{15}NH_2$ emission lines accurate to about 0.01 Å for the ((0, 10, 0)–(0, 0, 0)) and ((0, 9, 0)–(0, 0, 0)) bands of the $\tilde{A}^2A_1 - \tilde{X}^2B_1$ system. These wavelengths led to the first clear detection of $^{15}NH_2$ in comets by Rousselot et al. (2014) made by combining spectra of 12 comets taken by the UVES spectrograph at the 8-m ESO Very Large Telescope ($\lambda/\Delta\lambda \simeq 80000$) located in Chile. The inferred $^{14}N/^{15}N$ ratio is 127 in ammonia (in the range 90–190 taking into account uncertainties).

This work assumed two reasonable assumptions: (i) a similar photodissociation efficiency for $^{14}NH_3$ and $^{15}NH_3$ to produce, respectively, $^{14}NH_2$ and $^{15}NH_2$ radicals, (ii) similar transition probabilities for both $^{14}NH_2$ and $^{15}NH_2$. For the second point the average of many spectra obtained at different heliocentric velocities, i.e. with different Doppler shifts with respect to the solar absorption lines, avoided the issue (in the absence of a complete fluorescence modelling) of being affected by the Swing effects and helped to provide a more robust measurement.

Immediately after the first report another measurement was published, based on a high-resolution ($\lambda/\Delta\lambda = 72000$) spectrum obtained on comet C/2012 S1 (ISON) with the 8-m SUBARU telescope and the High Dispersion Spectrograph (Shinnaka et al. 2014b). The measured value of $^{14}N/^{15}N = 139 \pm 38$ is consistent with the previous measurement, and confirms that ammonia in comets is highly fractionated in ^{15}N, comparably to CN and HCN.

2.3 Carbon, Oxygen, and Sulfur

2.3.1 Carbon

There are two stable carbon isotopes: ^{12}C and ^{13}C. The ratio $^{12}C/^{13}C$ has been measured in different solar system objects. The terrestrial value is 89 and is similar to the one detected in other solar system objects, such as the Sun, Venus, the Moon, Mars, Jupiter, Saturn, Titan, Neptune, meteorites or the IDPs (Woods and Willacy 2009). This ratio is larger than the value of $\sim$68 measured in the local interstellar medium, that may result from a ^{13}C enrichment since the formation of the Solar System (Milam et al. 2005).

For comets, historically the first $^{12}C/^{13}C$ ratio determinations were obtained from the $^{12}C^{13}C$ emission spectrum in the optical range for four bright comets: Ikeya 1963 I (Stawikowski and Greenstein 1964), Tago-Sato-Kosaka 1969 IX (Owen 1973), Kohoutek 1973 XII (Danks et al. 1974), and Kobayashi-Berger-Milon 1975 IX (unpublished data from Lambert and Danks mentioned by Lambert and Danks 1983; Vanysek 1977; Wyckoff et al. 2000). The emission lines used for theses measurements, that belong to the $(1,0)$ Swan bands, are strongly blended with NH_2 emission lines. For this reason the CN B-X $(0,0)$ band was later favored for measuring the $^{12}C/^{13}C$ ratio. It has been used for the first time for comet 1P/Halley (Wyckoff and Wehinger 1988; Wyckoff et al. 1989; Kleine et al. 1995). Since then the CN B-X (0,0) band has been used with success for measuring the $^{12}C/^{13}C$ ratio in many comets of different origins, with an average ratio of 91.0 ± 3.6 (Manfroid et al. 2009). Recently emission lines of the $(2,1)$ and $(1,0)$ bandheads of $^{12}C^{13}C$ have been used to measure the $^{12}C/^{13}C$ ratio in comets C/2002 T7 (LINEAR) and C/2001 Q4 (NEAT) (Rousselot et al. 2012).

Some other determinations of the $^{12}C/^{13}C$ ratio have been performed with radio spectroscopy of $H^{13}CN$. These measurements have been performed for comets C/1995 O1 (Hale-Bopp) (Jewitt et al. 1997; Ziurys et al. 1999; Bockelée-Morvan et al. 2008) and 17P/Holmes (Bockelée-Morvan et al. 2008). All determinations are consistent with the terrestrial value. Measurements made in the dust samples collected by the Stardust spacecraft are presented in Sect. 3.

2.3.2 Oxygen

The first value of the oxygen isotopic ratio in a comet was obtained by the in-situ measurements aboard the Giotto spacecraft and gave $^{16}O/^{18}O = 495 \pm 37$ for H_3O^+ in comet 1P/Halley (Eberhardt et al. 1995; Balsiger et al. 1995). Seventeen years after the Giotto flyby, a deep integration spectrum of the bright comet 153P/2002 C1 (Ikeya-Zhang) with the sub-millimeter satellite Odin led to the detection of the $H_2^{18}O$ line at 548 GHz (Lecacheux et al. 2003). Subsequent observations resulted in the determination of $^{16}O/^{18}O = 530 \pm 60$, 530 ± 60, 550 ± 75 and 508 ± 33 in the Oort cloud comets Ikeya-Zhang, C/2001 Q4, C/2002 T7 and C/2004 Q2 respectively (Biver et al. 2007). Within the error bars, all these measurements are consistent with Halley and the terrestrial value ($^{16}O/^{18}O_{VSMOW} = 499$), although marginally higher.

A few ^{18}OH lines belonging to the ultraviolet OH system at 310 nm were detected in the spectra of the bright comets C/2002 T7 (LINEAR) and C/2012 F6 (Lemmon) with UVES at the ESO VLT and allowed for the first time the determination of the $^{16}O/^{18}O$ ratio from ground-based observations. From the $^{16}OH/^{18}OH$ ratio, $^{16}O/^{18}O = 425 \pm 55$ and 350 ± 150 were derived respectively (Hutsemékers et al. 2008; Decock et al. 2014). These values are equal, within the uncertainties, to the terrestrial value and to the ratio measured in other comets, although marginally smaller. An enrichment of ^{18}O in comets has been predicted by some models of the pre-solar nebula by CO self-shielding to explain the so-called"oxygen anomaly" i.e. the fact that oxygen isotope variations in meteorites cannot be explained in any easy way (Yurimoto and Kuramoto 2004). Recently H_2O, and $H_2^{18}O$ 1_{10}–1_{01} lines were observed with high signal-to-noise ratio in the Oort Cloud comet C/2009 P1 (Garradd) using the HIFI instrument onboard the IR Herschel space telescope (Bockelée-Morvan et al. 2012). The measured oxygen ratio of 523 ± 32 is consistent with the terrestrial value.

2.3.3 Sulfur

A few measurements have been acquired for the $^{32}S/^{34}S$ ratio in cometary gases. The first investigation was made in comet 1P/Halley through in situ mass spectrometry (S^+ ion, Altwegg 1996). Using millimetre spectroscopy, determinations were made for CS (Jewitt et al. 1997; Biver et al. 2008), a photo-dissociation product of CS_2, and for H_2S (Crovisier et al. 2004). Inferred values are given in Table 2. They are consistent within 2σ with the terrestrial value of 23.

3 Measurements in Cometary Dust and Analogs

Samples from comets have been identified on Earth. Dust particles directly sampled from comet 81P/Wild 2 have been collected by the Stardust mission and returned to laboratories in 2006. For others, a cometary origin is inferred from different lines of evidence. Such samples provide unique information since extremely precise elemental, mineralogical and isotopic analyses at different scales down to the nanometer permit direct comparison with asteroidal material delivered by meteorites. However, they do not allow access to the composition of ices, because these phases, if any, were lost during processing on parent bodies, during atmospheric entry, or during sampling in the case of Stardust grains.

At least three types of samples with certain to probable cometary origin are available in laboratories:

- The NASA Discovery Stardust mission collected micro-grains of cometary material in the tail of comet 81P/Wild 2 and returned this cometary matter to Earth for laboratory analysis (Brownlee et al. 2006; Brownlee 2014);
- Interplanetary dust particles (IDPs) are small grains collected in the Earth stratosphere by NASA, that have sizes typically below 50 μm (Sect. 3.2). Among these particles, a subset, the chondritic-porous anhydrous IDPs, are thought to have a cometary origin.
- Ultracarbonaceous Antartic Micrometeorites (UCAMMs) recovered in Antarctic snow show compositions and stable isotope signatures also consistent with derivation from a cometary reservoir (Sect. 3.3).

Besides these materials for which a cometary origin is beyond doubt or reasonably well constrained, other cosmochemical materials on Earth have been proposed to be cometary in origin, such as some of the volatile-rich carbonaceous chondrites, as first suggested by Anders (1975), and others (e.g. McSween and Weissman 1989; Campins and Swindle 1998; Gounelle et al. 2006; Weisberg and Connolly 2008; Haack et al. 2011). Recently, a piece of carbon-rich material found in Egypt and nicknamed Hypatia has been interpreted as originating from a comet that exploded over Africa 35 Myr ago and made the Lybian glass strewnfield (Kramers et al. 2013).

3.1 Stardust

In 2004, NASA Discovery's Stardust mission succeeded in collecting grains from a Jupiter-family comet (JFC) named 81P/Wild 2, and in returning them to Earth two years later for laboratory analyses (Brownlee et al. 2006; Brownlee 2014). JFCs, a sub-population of the Kuiper belt object (KBO) family, are thought to represent primitive material consisting of silicates, organics and ice that was assembled in the outer solar system and stored beyond the orbit of Neptune since 4.5 Gyr. Before 1974, comet 81P/Wild 2 was on a 43-year orbit that ranged from Jupiter to beyond Uranus. Since then it is on a 6-year orbit crossing the Earth's region. Comet Wild 2 is a particularly active comet with up to 24 jets identified and a water loss estimated at 20000 tons per day. The spacecraft flew within 236 km of the comet's 4.5 km diameter nucleus. Cometary grains were collected during a flyby encounter by passive implantation into aerogel, a low-density silica foam, and by impacting onto the aluminum frame (Brownlee et al. 2003, 2004). On January 15th, 2006, the Stardust sample container returned safely to Earth, delivering the first cometary samples for laboratory analysis (Brownlee et al. 2006; Brownlee 2014). Due to the high differential velocity of 6.1 km/s between the spacecraft and the comet, cometary grains suffered mass loss and fragmentation during the collection process, to various extents that depended on the nature and density of the grains. While the finest material was mostly severely heated and mixed with the collecting silica aerogel during capture, some of the coarsest grains did not heat up to the melting point, and detailed mineralogical and geochemical analyses down to the nanometer scale revealed the nature of this cometary matter.

Contrary to some expectations, grains collected by Stardust are not dominated by interstellar grains coated with radiation-processed organics. Most of the analyzed materials are isotopically close to chondritic material, and were likely formed by the same set of high temperature processes that led to the formation of constituents of primitive meteorites. As for chondrites, Wild 2 grains contain chondrules, calcium-aluminum rich inclusions, Mg-rich olivines and pyroxenes. In comet Wild 2, such a high temperature material has been assembled with ice, demonstrating that large scale lateral mixing took place in the forming solar system.

There are nevertheless significant differences between comet Wild 2 and volatile-rich meteorites, e.g. carbonaceous chondrites. Organics in Wild 2 material are scarce, and there is little evidence for the occurrence of hydrated material, contrary to hydrous carbonaceous chondrites for which parent body aqueous alteration resulted in the formation of clay minerals and carbonates. It is not clear whether these differences are due to the high speed collection of grains or not. Wild 2 material appears to contain a mix of material more diverse than that found in chondrites, which may also point to a different processing history of cometary bodies compared to asteroids. Below is a summary of the isotopic compositions measured in Stardust material (McKeegan et al. 2006, and references therein).

- Bulk material presents moderate excesses of D with D/H ratios up to a factor of three higher than terrestrial (VSMOW). No hydrated minerals have been identified in Stardust samples, so it is likely that the measured D/H values are related to the carbonaceous phases, and are not representative of the D/H ratio of water in comet 81P/Wild 2. Variations in δD ($\delta D = [(D/H)/(D/H)_{VSMOW} - 1] \times 1000$) within a particle are also observed with some "hotspots" having a δD up to 2200 ± 900 permil. Because of the association with carbon and the lack of evidence for any hydrated minerals in Wild 2 grains, it is unlikely that the D/H values are related to cometary water.
- Overall, the $^{13}C/^{12}C$ ratio ($\delta^{13}C$ from -20 to -50 permil relative to terrestrial standard) shows moderate variations.
- Likewise, most $^{15}N/^{14}N$ ratios cluster around the terrestrial value and are also consistent with values found in most meteorites (Stadermann et al. 2008). A few $\delta^{15}N$ values range from $+100$ to $+500$ permil. So far, excesses of ^{15}N typical of cometary values (around $+800$ to $+1000$ per mil, Sect. 2.2) are not observed in bulk analysis. On a smaller scale, ^{15}N-rich hotspots with a maximum of $\delta^{15}N = 1300 \pm 400$ permil are observed, similar to the highest values found in refractory organic matter in interplanetary dust particles (IDPs, see Sect. 3.2) and in various types of carbonaceous chondrites (CR, CH-CB and

CM, Busemann et al. 2006; Briani et al. 2009). Several presolar grains have been identified showing much larger C and N isotope variations (from +60 to +964 permil for $\delta^{13}C$, and from −518 to −350 permil for $\delta^{15}N$), although these components appear to contribute little to the isotope variability due to their apparent scarcity (Floss et al. 2013). In summary, the Wild 2 grains display both low and high ^{15}N, indicating an unequilibrated mixture of a ^{15}N-poor material (perhaps protosolar N trapped in ice) component with a more refractory (high C/N) organic material and high $\delta^{15}N$ (McKeegan et al. 2006).

- The three oxygen isotopic compositions of silicate and oxide mineral grains from comet 81P/Wild 2 were measured using ion micro- and nanoprobe techniques. The average composition of olivine and pyroxene minerals, as well as of chondrule fragments, is consistent with average chondritic material (e.g. supplementary information in McKeegan et al. 2006; Nakamura et al. 2008; Bridges et al. 2012; Nakashima et al. 2012). In contrast, the O-isotope ratio of the refractory phase "Inti" exhibits a $\delta^{16}O$ enrichment plotting along the CCAM line with $\delta^{17}O = -62 \pm 20$ permil and $\delta^{18}O = -42 \pm 10$ permil (McKeegan et al. 2006, suppl. info). A detailed study of "Inti" subsequently performed by Simon et al. (2008) confirmed that Inti is indeed ^{16}O-rich, with $\delta^{18}O \sim \delta^{17}O \sim -40$ permil. The occurrence of silicates plotting close to the Terrestrial Fractionation (TF) line together with the ^{16}O-rich material (Ca-Al-rich inclusions) is in agreement with that observed in carbonaceous chondrites (McKeegan et al. 2006, suppl. info). These observations confirm that Wild 2 material contains oxides that were synthesized in the most inner regions of the solar system. Several presolar grains were identified in Stardust samples (at least 6). However, a large fraction of these tiny phases may have been destroyed during impact in the aerogel so that their initial abundance in Comet 81P/Wild 2 may have been higher than inferred from direct counting (Floss et al. 2013).
- Another interesting observation on Wild 2 material is the occurrence of CAIs as mentioned above. In the refractory minerals that were analyzed for Mg isotopes, no trace of the extinct ^{26}Al radioactivity ($T_{1/2} = 0.7$ Myr) could be found (Matzel et al. 2010; Ogliore et al. 2012; Nakashima et al. 2015), in contrast to the case of most of CAIs of primitive meteorites. The latter incorporated live ^{26}Al with an initial, "canonical" $^{26}Al/^{27}Al$ of 5×10^{-5}, indicating early and fast ($T < 1$ Myr) formation of these high temperature condensates. The lack of ^{26}Al in Wild 2 material could indicate that these CAIs formed before the injection of ^{26}Al in the solar nebula, or that they formed later than a few Myr when all the ^{26}Al was already transformed to ^{26}Mg. As these CAIs are not as refractory as those usually carrying ^{26}Al excesses in meteorites, the hypothesis of a late formation is favored. Late exchange of Mg isotopes through metamorphism or chemical alteration cannot either be excluded.
- Up to the Stardust sample return, noble gases had not been unambiguously identified in comets. This is unfortunate because these elements are chemically inert and have the potential to trace the physical conditions of comet formation such as the nature of ice, the formation, temperature, the composition of the ambient gas etc. Cometary light noble gases (He, Ne) have been found in Stardust aerogel (Marty et al. 2008). Gases are not located in the terminal grains implanted deeply in the foam (which were presumably dried up during the entry process). They appear to reside into the walls of the cavities formed when high velocity grains exploded during aerogel entry. The $^{3}He/^{4}He$ ratio (2.5×10^{-4}) is intermediate between that of the solar wind (post-D burning helium) and the protosolar nebula ratio, raising the possibility that cometary matter recorded the activity of the protoSun. By contrast, the neon isotopic composition is closer to that of Ne trapped in chondrites than to the solar composition, supporting the hypothesis of an asteroid-comet continuum and suggesting that this Ne component was trapped in dust and not in ice. Both

Fig. 3 Secondary electron micrograph of a chondritic porous (CP) anhydrous IDP. From Bradley (2005)

He and Ne from Stardust samples were released at high temperature indicating that their host is refractory (dust and/or organics) (Marty et al. 2008).

In summary, the collection of cometary grains from comet 81P/Wild 2 by the Stardust spacecraft and their analysis in the laboratory has demonstrated that a large fraction of dust trapped in this body is derived from precursors that share mineralogical and isotopic similarities with carbonaceous chondrites, that were thermally and chemically processed in the inner solar system, consistent with models advocating large scale radial mixing in the nascent solar system.

3.2 Interplanetary Dust Particles

Since 1981, NASA has been collecting interplanetary dust particles (IDPs) in the stratosphere, between 17–19 km altitude (Brownlee 1985). Due to their low densities, the fluffiest particles preferentially accumulate at such altitudes. Typical IDP sizes range from 1 to ~50 μm, most particles being in the ~2–15 μm size range. One class of IDPs, the chondritic porous anhydrous particles (CP-IDPs, Fig. 3) are thought to originate from comets. Such chondritic porous dust particles have also been identified from Antartic snow and ice collections (Noguchi et al. 2015).

CP-IDPs mainly consist of anhydrous crystalline phases (Mg-rich olivines, pyroxenes and low-Ni Fe sulfides Bradley 2005), as well as primitive glassy components called Glass Embedded with Metals and Sulfides (GEMS) that have a debated (solar vs. presolar) origin (Bradley 2013; Keller and Messenger 2013, and references therein). Early studies on the origin of IDPs used the helium and neon release temperature in IDPs to assess the temperature reached during atmospheric entry. This entry velocity may then be related to the asteroid vs. cometary origin (Nier and Schlutter 1993). This type of analyses first suggested a cometary origin for the CP-IDPs (Brownlee et al. 1995). They are usually rich in carbonaceous matter (from ~5 to ~45 wt%, (Keller et al. 1994), and their mineralogy is compatible with that of comets 1P/Halley and C/1995 O1 (Hale-Bopp) (e.g., Hanner and Zolensky 2010; Brunetto et al. 2011, and references therein). Some larger CP-IDPs are fragmented into several patches during impact in the collector, leading to the appellation of "cluster IDPs". It has been proposed that a large proportion (if not all) of CP- or cluster IDPs are of cometary origin (e.g., Bradley and Brownlee 1986; Ishii et al. 2008, and references therein). In addition, IDPs collected in April 2003 during a period predicted to show an increase in dust flux from comet 26P/Grigg-Skjellerup may originate from this Jupiter-family comet (Busemann et al. 2009).

The D/H ratios in CP- and cluster IDPs exhibit elevated values associated with the organic matter, typically ranging from D/H $\sim 10^{-4}$ to $\sim 8 \times 10^{-3}$, i.e. up to $\sim$ 50 times the terrestrial value in the case of one cluster IDP (Messenger 2000 and references therein). The D/H values of two IDPs possibly originating from the 26P/Grigg-Skjellerup collection (GSC-IDPs) also show high D/H values with ratios ranging from 5.4×10^{-4} to 1.5×10^{-3}.

In cluster IDPs, the nitrogen isotopes show substantial variations, with an average $^{14}N/^{15}N$ ratio around 236. The IDP organic matter exhibits ^{15}N-rich and ^{15}N-poor areas with $\delta^{15}N$ up to 1300 permil and $\delta^{15}N$ as low as -93 ± 4 permil (Messenger 2000; Floss et al. 2006). The two GSC-IDPs have bulk $^{14}N/^{15}N$ values varying from 238 ± 4 ($\delta^{15}N = 145 \pm 15$ permil) to 183 ± 6 ($\delta^{15}N = 485 \pm 35$ permil), with ^{15}N-rich hotspots showing $^{14}N/^{15}N$ values as low as 118 ± 18 ($\delta^{15}N = 1310 \pm 150$ permil) (Busemann et al. 2009).

The bulk carbon isotopes in normal and cluster IDPs are within the terrestrial range. In the GSC-IDPs, there are sub-regions showing carbon isotopic anomalies (^{12}C-rich, $\delta^{13}C = -280$ permil), suggesting the presence of primitive organic matter, or presolar grains (Busemann et al. 2009).

The bulk oxygen isotopic compositions of CP-IDPs are usually within the range of values observed for carbonaceous chondrites, although heavy oxygen isotopic compositions, with $\delta^{17}O$, $\delta^{18}O$ up to +80 to +200 permil, have been identified in one IDP (e.g., Starkey et al. 2014, and references therein). In the CP- and GSC-IDPs there are numerous localized isotopic anomalies suggesting an abundant content in presolar silicates, some of them possibly being GEMS (Messenger et al. 2003, 2005; Busemann et al. 2009; Keller and Messenger 2013, and references therein). The O isotopes are distinct in various solar system materials, and therefore provide a unique tool to trace genetic relationship between solar system reservoirs. The O isotopic composition of carbonaceous chondrites is related to that of chondritic porous IDPs, Stardust grains and UCAMMs (Sect. 3.3), and suggests a genetic continuum between carbonaceous chondrites and comets (Aléon 2010; Nakashima et al. 2012; Davidson et al. 2012; Starkey and Franchi 2013, and references therein).

3.3 Ultracarbonaceous Antarctic Micrometeorites

Micrometeorites dominated by organic matter were identified in the CONCORDIA collection (Duprat et al. 2010; Dobrică et al. 2011, 2012) and in a collection performed at Dome Fuji (Nakamura et al. 2005) (UCAMMs, for Ultracarbonaceous Antarctic Micrometeorites, Fig. 4). These particles contain more than 50 % of organic matter in volume, i.e. a carbon content up to ten times that of the most C-rich carbonaceous chondrites (e.g., Alexander et al. 2012). Such concentrations of carbonaceous matter are comparable with that of the most C-rich IDPs (Thomas et al. 1993; Keller et al. 2004), and compatible with CHON grains detected in comet Halley by the Giotto and VEGA space missions in 1986 (Kissel and Krueger 1987; Lawler and Brownlee 1992).

The hydrogen isotopic composition of UCAMMs exhibit extreme D enrichments over large areas (> 100 μm^2), with D/H ratios up to 10–30 times the VSMOW value (Duprat et al. 2010; Dartois et al. 2013; Slodzian et al. 2013; Bardin et al. 2014). This isotopic composition is a signature of a primitive organic matter from cold and dense regions of the protosolar cloud or protoplanetary disk. Remarkably, the extreme D excesses observed in UCAMMs are not limited to hotspots of 1–2 μm in size like usually observed in IDPs or meteoritic insoluble organic matter (e.g., Remusat et al. 2010, and references therein). The

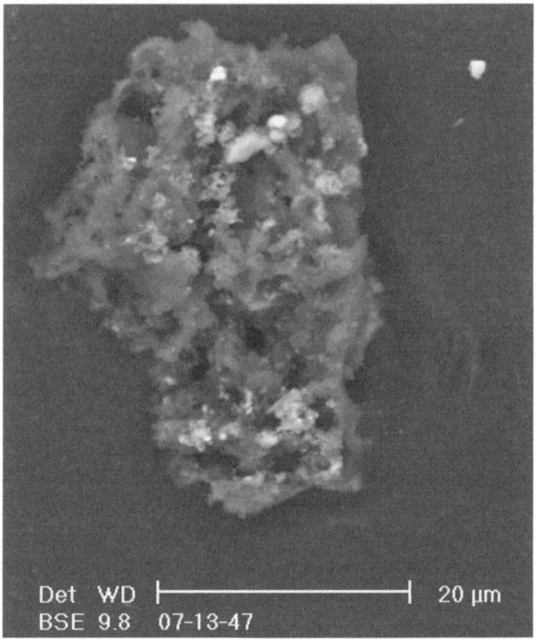

Fig. 4 Backscattered electron micrograph of a fragment of CONCORDIA Ultracarbonaceous Antarctic Micrometeorite (UCAMM). All *dark grey patches* are constituted of organic matter. *Light grey flakes* are silicates, and *bright spots* are Fe-Ni sulfides. From Duprat et al. (2010)

association in UCAMMs of primitive organic matter with mineral phases (including crystalline minerals) that show evidence of nebular processing (Dobrică et al. 2012), suggest that deuteration of the organic matter may have taken place in the outer regions of the protoplanetary disk rather than in the presolar molecular cloud core (Duprat et al. 2010). The organic matter itself is rich in N with concentrations of N/C ratios up to 0.15. The formation of such an N-rich organic matter can result from long term irradiation by non-attenuated galactic cosmic rays of N_2–CH_4 rich ices present at the surface of icy bodies in the external regions of the solar system (Dartois et al. 2013).

The nitrogen isotopic composition of UCAMMs is under investigation. It reveals a heterogeneous distribution of the $^{15}N/^{14}N$ ratios showing both ^{15}N-rich and ^{15}N-poor regions (Duprat et al. 2014).

The oxygen isotopic composition of UCAMMs is also currently under investigation. The bulk oxygen isotopic composition of two UCAMMs plots slightly above the terrestrial fractionation line with $\delta^{18}O \sim -5$ permil and $\delta^{17}O \sim 10$ permil (Kakazu et al. 2014).

3.4 Other Potential Cometary Samples on Earth

Some of the carbonaceous chondrites, especially those of the CI type (e.g., Orgueil), as well as CM or CR chondrites, may represent fragments of nuclei from extinct comets (Gounelle et al. 2006; Weisberg and Connolly 2008; Haack et al. 2011). Arguments for such an origin are: (i) ellipsoidal trajectories and high entry (18 km/s) velocities inferred from ground-based investigations, that point to a JFC origin; (ii) abundant organics (up to 5 % C) and hydrated minerals (equiv. 10–20 % water), leading to an estimated water/rock ratio of about 1, comparable to those of comets (e.g., comet 9P/Tempel 1); (iii) the similarity of oxygen isotope trends of Stardust oxides with those of carbonaceous chondrites (Sect. 3.1); (iv) elevated D/H or $^{15}N/^{14}N$ ratios; v) abundance of glassy phases. Their bulk D/H or N isotopic compositions are however not as elevated as those observed in comet water and CN,

HCN and NH_3,respectively. If these meteorites represent cometary nuclei, this difference may be due to loss of ice-bearing D and ^{15}N for CCs. Elevated D/H ratios in hydrated phases in Semarkona (unequilibrated LL3 chondrite) were however recently measured (up to 1.8×10^{-3}). These phases could result for the local alteration of silicates using D-rich water (ice grain?) that could be the source of part of the water emitted by Oort cloud comets (Piani et al. 2015).

Recently, Kramers et al. (2013) claimed that a peculiar cm-sized stone found in Egypt and nicknamed Hypatia could be a comet fragment that would have exploded in the atmosphere over Northern Africa 26 Myr ago, resulting in a glass strewnfield in the Lybian desert. Altogether, the absence of an impact crater large enough to account for the extent of the Lybian tektite field, physical considerations raised long ago (Urey 1957), and the peculiar composition of Hypatia (Kramers et al. 2013) are in agreement with such an origin. The Hypatia stone has a $^{40}Ar/^{36}Ar$ ratio lower than that of air, demonstrating an extraterrestrial origin. It is rich in carbon in the cubic form and hosts abundant nano- to microdiamonds. Hypatia's C isotope composition also clearly indicate an extraterrestrial origin and noble gas isotope signatures suggest the occurrence of unequilibrated material containing dust from AGB stars winds. Further elemental and isotopic measurements are underway to explore this exciting possibility.

4 Discussion

One viewpoint for the origin of cometary materials is that they are pristine remnants of the interstellar material that collapsed to form the solar nebula Greenberg (1982). Alternatively, cometary molecules may have formed in the cold outer regions of the solar nebula regions. The analogy between cometary and interstellar ice-composition indeed suggests that those materials formed by similar processes, i.e., ion-molecule and grain-surface reactions at low temperatures (Bockelée-Morvan et al. 2000). However, the presence of crystalline silicates and other high temperature condensates in comets indicates that comets incorporated also materials formed in the hot innermost regions of the disk (Brownlee et al. 2006; Bockelée-Morvan et al. 2002; Wooden 2008), or which underwent processing through, e.g., nebular shocks (Harker and Desch 2002), or in the surface layers during protosolar luminosity outbursts (Ábrahám et al. 2009). In the latter two cases, where a purely nebular origin may be indicated, mixing of material radially outwards or vertically downwards to the midplane, would also lead to cometary ices having a composition derived from a chemistry radically different from that of the interstellar medium.

Isotopic ratios in comets can provide constraints to the different scenarios. The ISM-comet connection can be tested by comparing isotopic ratios in both kinds of material. Tables 4 and 5 provide respectively hydrogen and nitrogen isotopic ratios measured in molecular clouds, dark cores and protostars.

4.1 Interstellar Isotopic Fractionation

Cosmic rays drive an ion-molecule chemistry in dense interstellar clouds. At the cold temperatures of these environments ($\sim$ 10 K) the zero-point energy difference between isotopically-substituted molecules can become important in exothermic ion-molecule reactions. For example, although D isotope-exchange reactions of the form

$$XH^+ + HD \rightleftharpoons XD^+ + H_2 + \Delta E \quad (1)$$

Table 3 Ion-molecule isotopic exchange reactions

No.	Reaction	$\Delta E/k$ (K)	Reference
R1	$H_3^+ + HD \rightleftharpoons H_2D^+ + H_2$	143.0	Giles et al. 1992
R2	$H_2D^+ + HD \rightleftharpoons HD_2^+ + H_2$	187.0	Giles et al. 1992
R3	$HD_2^+ + HD \rightleftharpoons D_3^+ + H_2$	234.0	Giles et al. 1992
R4	$CH_3^+ + HD \rightleftharpoons CH_2D^+ + H_2$	370.0	Millar et al. 1989
R5	$C_2H_2^+ + HD \rightleftharpoons C_2HD^+ + H_2$	550.0	Millar et al. 1989
R6	$^{13}C^+ + {}^{12}CO \rightleftharpoons {}^{12}C^+ + {}^{13}CO$	35.0	Watson et al. 1976
R7a	$^{14}N_2H^+ + {}^{14}N^{15}N \rightleftharpoons {}^{14}N^{15}NH^+ + {}^{14}N_2$	10.7	Adams and Smith 1981
R7b	$^{14}N_2H^+ + {}^{14}N^{15}N \rightleftharpoons {}^{15}N^{14}NH^+ + {}^{14}N_2$	2.3	Adams and Smith 1981
R8	$^{15}N^+ + {}^{14}N_2 \rightleftharpoons {}^{14}N^+ + {}^{14}N^{15}N$	28.3	Terzieva and Herbst 2000
R9	$^{15}N^+ + {}^{14}NO \rightleftharpoons {}^{14}N^+ + {}^{15}NO$	24.3	Terzieva and Herbst 2000
R10a	$^{15}N + {}^{14}N_2H^+ \rightleftharpoons {}^{14}N + {}^{14}N^{15}NH^+$	36.1	Terzieva and Herbst 2000
R10b	$^{15}N + {}^{14}N_2H^+ \rightleftharpoons {}^{14}N + {}^{15}N^{14}NH^+$	27.7	Terzieva and Herbst 2000
R11	$^{15}N + C^{14}NC^+ \rightleftharpoons {}^{14}N + C^{15}NC^+$	36.4	Terzieva and Herbst 2000
R12	$^{15}N + HC^{14}NH^+ \rightleftharpoons {}^{14}N + HC^{15}NH^+$	35.9	Terzieva and Herbst 2000

are exothermic in the forward direction (to the right) the higher abundance of H_2 relative to HD ($H_2/HD \sim 2 \times 10^{-5}$) means that the reverse reaction will dominate except at very low temperatures. Table 3 lists the major ion-molecule reactions leading to isotopic fractionation in low-temperature interstellar clouds. Although reaction (R1) has long been known to initiate D fractionation in molecular clouds (e.g., Millar et al. 1989), the enhanced depletion of CO onto dust and the low electron fractions found in high-density molecular cores both conspire to permit reactions (R2) and (R3) to play important roles (Roberts et al. 2003). The spin-state of the colliding H_2 molecules, whether *para* or the higher-energy *ortho* state, plays an important role in interstellar fractionation. The higher internal energy of *ortho*-H_2 can essentially quench ion-molecule deuteration when the ortho/para ratio (OPR) is high (Pagani et al. 2011). In such CO depletion cores, where protostars are about to form, gas-phase deuteration is enhanced and, as D_3^+ becomes the major ion, this leads to very high atomic D/H ratios. Atom addition reactions on dust, primarily starting from condensed CO, can lead to many organic compounds, such as formaldehyde, methanol, formic acid, acetaldehyde, isocyanic acid and formamide, molecules which are also identified in comets (e.g., Charnley and Rodgers 2008; Herbst and van Dishoeck 2009). High atomic D/H ratios in the condensing gas consequently leads to very high D/H ratios and the formation of many multiply-deuterated isotopologues for surface-formed molecules (Charnley et al. 1997); these latter species include D_2O, D_2CO, CHD_2OH, CD_3OH, NHD_2, ND_3 and D_2S (Parise et al. 2006; Ceccarelli et al. 2014). These characteristics are observed in the interstellar medium and Table 4 summarizes the range of D/H ratios found in cold dark clouds and in cores forming both low-mass and massive protostars. In the latter regions the highest D/H ratios are due to highly-fractionated ices being evaporated from warm dust grains (e.g., Charnley et al. 1997).

Reaction (R6) is the primary ion-molecule means of fractionating carbon in interstellar molecules. At low temperatures this has the effect that most ^{13}C nuclei become incorporated into ^{13}CO, leading to a relative depletion of ^{13}C in other interstellar molecules. This can have the secondary consequence that molecules formed on grains should have $^{12}C/^{13}C$ ratios similar to that of the accreted CO molecules in their carbonyl or carboxyl functional

Table 4 Interstellar deuterium fractionation ratios (%)

Molecule	Cold molecular clouds[1]	Low-mass protostars[2]	High-mass protostars[3]
HDO	–	0.04–8	0.03–2
		0.03–18	
NH_2D	5–13	7–10	0.7
DCN	1–7	0.4–7	0.05–2
DNC	1–13	2–6	0.06–12
CH_2DOH	10	37–65	0.06
CH_3DOD	1–3	1–5	0.05
HDCO	7–10	2–17	0.07–14
D_2CO	–	5–44	0.3
HDS	–	5–15	<2
DC_3N	2–13	–	1.5

[1]Ratios for L1544, TMC-1, L183: Howe et al. (1994), Turner (2001), Shah and Wootten (2001), Lis et al. (2008), Bizzocchi et al. (2014)

[2]Ratios from compilation of Taquet et al. (2014) for several protostars; exception is that DNC ratios are from Butner and Charnley (unpublished work). For HDO, values for the inner and outer regions of the protostellar envelope are given

[3]Ratios for various sub-components of Orion-KL (Hot Core, Compact Ridge, Extended Ridge), see for example Blake et al. (1987), Turner (1990), Minh et al. (1990), Mangum et al. (1991), Bergin et al. (2010), Esplugues et al. (2013), Neill et al. (2013). Entry for DNC is from the sample of Fontani et al. (2014)

groups, whereas other functional groups formed by addition of C atoms should be relatively depleted in ^{13}C (Charnley et al. 2004). Observations have confirmed this in the case of CO and methanol (Wirström et al. 2011). Isotope-selective photodissociation (Bally and Langer 1982) can also fractionate carbon in molecular clouds. In this case, the self-shielding of ^{12}CO molecules leads to regions where ^{13}CO is preferentially destroyed. However, most work has been focussed on understanding the concomitant fractionation resulting from isotope-selective photodissociation of $^{12}C^{18}O$ and $^{12}C^{17}O$ in the context of understanding the origin of the $^{16}O : {}^{17}O : {}^{18}O$ ratios measured in primitive meteoritic matter (Lee et al. 2008). Ion-molecule reactions involving oxygen isotopes produce negligible fractionation in interstellar clouds (Langer et al. 1984).

Significant isotopic enrichments in N-bearing interstellar molecules (relative to a putative $^{14}N/^{15}N$ elemental ratio of 440) can occur by the set of ion-molecule reactions listed in Table 3. Table 5 summarizes the $^{14}N/^{15}N$ ratios measured in molecular clouds and in the earliest phase of low-mass star formation ('Class 0').

Early chemical models of dense cores, in which nitrogen was assumed to be initially fully molecular, predicted that fractionation in amines is dependent on the level of CO depletion, as well as low temperatures. (Charnley and Rodgers 2002). Rodgers and Charnley (2008) showed that when N atoms are initially abundant, as one might expect since dense clouds form from diffuse gas, then reaction (R12) efficiently incorporates ^{15}N into $HC^{15}NH^+$. Electron dissociative recombination of $HC^{15}NH^+$ then produces $HC^{15}N$, $H^{15}NC$ and $C^{15}N$ and to ^{15}N enrichment in nitriles early in dense cloud evolution. These chemical models predict that nitriles should be more enriched than amines (Rodgers and Charnley 2008), as observed (Table 5). Thus, interstellar ^{15}N fractionation follows two distinct pathways, one rapid, involving N_2 and NH_3, and the other one much slower, involving atomic N, and responsible for the ^{15}N enrichment in HCN and other nitriles (Table 3). It is then possible that N_2 and

Table 5 Interstellar nitrogen isotope ratios

Source	Type	NH_3	N_2H^+	HCN	HNC	CN	Reference
L1544	dark core	>700	1000 ± 200	140–360	>27	500 ± 75	1, 2, 3, 4, 5
L1498	dark core	–	–	>75	>90	500 ± 75	4, 5
L1521E	dark core	–	–	151 ± 16	–	–	6
L1521F	dark core[§]	–	–	>51	24–31	–	4, 4
L183	dark core	$530\pm^{570}_{180}$	–	140–250	–	–	1, 3
NGC 1333-DCO$^+$	dark core	$360\pm^{260}_{110}$	–	–	–	–	1
NGC 1333-4A	Class 0 protostar	344 ± 173	–	–	–	–	7
Barnard 1	Class 0 protostar	$300\pm^{55}_{40}$	$400\pm^{100}_{65}$	$330\pm^{60}_{50}$	$225\pm^{75}_{45}$	$290\pm^{160}_{80}$	8
L1689N	Class 0 protostar	$810\pm^{600}_{250}$	–	–	–	–	4
Cha-MMS1	Class 0 protostar	–	–	–	135	–	9
IRAS 16293A	Class 0 protostar	–	–	163 ± 20	242 ± 32	–	10
R Cr A IRS7B	Class 0 protostar	–	–	287 ± 36	259 ± 34	–	10
OMC-3 MMS6	Class 0 protostar	–	–	366 ± 86	460 ± 65	–	10

References: (1) Gérin et al. (2009); (2) Bizzocchi et al. (2013); (3) Hily-Blant et al. (2013a); (4) Milam and Charnley (2012); (5) Hily-Blant et al. (2013b); (6) Ikeda et al. (2002); (7) Lis et al. (2010); (8) Daniel et al. (2013); (9) Tennekes et al. (2006); (10) Wampfler et al. (2014). For Barnard 1, the entry for the N_2H^+ ratios is that for $^{14}N^{15}NH^+$; a lower limit of 600 is obtained for the ratio derived from $^{15}N^{14}NH^+$

[§]L1521F does contain a very low luminosity object (VeLLO, Bourke et al. 2006)

NH_3 result from the separation of the nitrogen into at least two distinct reservoirs, with distinct ^{15}N enrichment, which never equilibrated. However, because of the small energy barrier for the first step of the ion-molecule reaction sequence

$$N^+ \xrightarrow{H_2} NH^+ \xrightarrow{H_2} NH_2^+ \xrightarrow{H_2} NH_3^+ \xrightarrow{H_2} NH_4^+ \xrightarrow{e^-} NH_3, NH_2 \qquad (2)$$

Ammonia formation also depends on the H_2 ortho/para spin ratio (OPR)—the higher energy ortho form can overcome the reaction barrier (Wirström et al. 2012). This can lead to reaction (R8) becoming important when the OPR is very small and to both enrichment and depletion of ^{15}N in NH_3 as the cloud core evolves.

The high $^{14}N_2/^{14}N^{15}N$ ratio measured in L1544 and Barnard 1 (Table 5) is a puzzle currently unresolved by existing models. The inefficiency of isotope-selective photodissociation of $^{14}N_2$ and $^{14}N^{15}N$ in molecular clouds (Heays et al. 2014) means that this process is excluded as a possible solution. The two existing measurements suggest a connection to the ammonia fractionation and measurements of these ratios in more sources is required.

The two existing measurements suggest a connection to the ammonia fractionation, and measurements of these ratios in more sources is now required. However, very recently, Roueff et al. (2015) published a study which re-assessed several of the ion-molecule fractionation pathways proposed by Terzieva and Herbst (2000), quantified some of the neutral-neutral reactions suggested by Rodgers and Charnley (2008), and considered new fractionation reactions for ^{13}C. Quantum-chemical calculations presented by Roueff et al. show that, in conflict with Terzieva and Herbst (2000), reactions (R9), (R10a), (R10b) and (R12) in Table 3 each possess activation energy barriers that will lead to negligible rates at around 10 K. Model calculations incorporating these new data indicate that, as expected, the calculated

$^{14}N/^{15}N$ ratios in HCN and HNC are much less enriched in ^{15}N than predicted in earlier models (e.g., Rodgers and Charnley 2008) and is observed (see Table 3). The calculations of Roueff et al. also predict that the ($^{14}N/^{15}N$ ratio in N_2H^+ should be close to elemental (≈ 400), in contrast to the depleted ratios observed in L1544 (~ 1000, Bizzocchi et al. 2013) and the enrichments predicted by previous theoretical models (e.g., Wirström et al. 2012). Thus, when compared to much of the published literature on interstellar ^{15}N fractionation, the work of Roueff et al. demonstrates that our current theoretical understanding of the underlying gaseous fractionation processes is far from complete.

4.2 Deuteration in the Solar Nebula

Cometary D/H values are lower than in most interstellar sources (Tables 1, 4). This low deuterium fractionation in comets can be interpreted by ion-molecule chemistry in gas phase at 25–35 K (Meier et al. 1998a, 1998b). Ion-molecule reactions in the outer solar nebula could also act to lower the interstellar fractionation (Aikawa and Herbst 1999). Alternatively, the low D-enrichment in H_2O might result from the mixing of D-rich water vapor originating from the pre-solar cloud or outer cold disk, and material reprocessed in the inner hot solar nebula (Hersant et al. 2001; Mousis et al. 2000; Kavelaars et al. 2011). Indeed, at high temperatures (>500 K), water molecules quickly exchange their deuterium with the main hydrogen reservoir H_2, so that the deuterium enrichment in water can not exceed the protosolar value of $0.21 \pm 0.04 \times 10^{-4}$ by more than a factor of 3 (Geiss and Gloeckler 1998) in high-T conditions. Transport mechanisms invoked so far include turbulent diffusion, which is also responsible for angular momentum transport in the solar nebula. Models are able to explain both the low, i.e. terrestrial, bulk deuterium enrichment measured in meteorites and the D/H ratio of $\sim 3 \times 10^{-4}$ measured in the water of Oort cloud comets, and Enceladus's plume (Hersant et al. 2001; Kavelaars et al. 2011; Waite et al. 2009). The higher deuterium enrichment in HCN measured in the Oort cloud comet Hale-Bopp (compared to the value in H_2O) might be explained by a lower efficiency of deuterium exchange between HCN and H_2 in the solar nebula (Mousis et al. 2000).

These models also predict that comets formed at higher distances from the Sun should exhibit higher D/H ratios as they incorporated less processed material from the inner solar nebula. It was generally agreed that Oort-cloud comets formed closer to the Sun than Jupiter-family comets (Dones et al. 2004, and references therein). However, in the framework of the Nice model, the Oort-cloud and the Scattered Disk, which is the source of the Jupiter-family comets, formed together as a result of late planetary migration which dispersed the trans-neptunian disk (Brasser and Morbidelli 2013). Therefore, in this picture, Jupiter-family comets are expected to exhibit similar or larger D/H than the Oort-cloud comets. The high deuterium enrichment measured in comet 67P/Churyumov-Gerasimenko is in line with these models (Altwegg et al. 2014), whereas the ocean-like D/H value found in the water of the Jupiter-family comets 103P/Hartley 2 and 45P/Honda-Mrkos-Pajdušáková is not. More recent calculations of the D/H distribution in the solar nebula showed that it is possible to obtain a non-monotonic gradient of the D/H ratio if fresh material from the infalling outer envelope fed the solar nebula continuously (Yang et al. 2013). Jacquet and Robert (2013) suggest that deuterium-poor ice grains condensed at the snow line were transported outwards in the comet-formation region, via turbulent diffusion. This scenario could also explain the distribution of D/H in CCs.

Another possible solution is that the Jupiter-family comets 103P/Hartley 2 and 45P/Honda-Mrkos-Pajdušáková formed closer to the Sun than other comets. According to the Grand Tack scenario, when the giant planets were still embedded in the nebular gas disk, there was a general radial mixing of the distribution of comets and asteroids (Walsh et al. 2011) so that the Kuiper Belt was populated by objects formed in a large range of heliocentric distance. Both the similarity of the D/H ratio in comets 103P/Hartley 2 and 45P/Honda-Mrkos-Pajdušáková with that found in carbonaceous chondrites, and the isotopic diversity observed in the Oort-cloud and Jupiter-family populations of comets are in agreement with such a scenario.

It was generally admitted that the extreme D-enrichments observed in primitive meteorites and IDPs are reflecting those acquired in the presolar cloud. The discovery of large D enrichments over large areas ($> 100\ \mu m^2$) of carbonaceous matter in UCAMMs would then require a large degree of preservation of such D-rich matter. Alternatively, cold chemistry in the outer regions of the protoplanetary disk could be responsible for such enrichments. The negative correlation between the D/H ratios of the CH bonds and their binding energy in the Insoluble Organic Matter of Orgueil carbonaceous meteorite also suggests that deuterium enrichments could have been acquired within the protoplanetary disk itself (Remusat et al. 2006). Whether this mechanism affected the D-fractionation observed in cometary HCN and H_2O has still to be investigated. Recent work by Cleeves et al. (2014) has demonstrated that, if cosmic-rays cannot penetrate the protosolar disk, then nebular chemistry alone cannot generate the D/H ratios measured in primitive materials such as comets and meteorites. In this case, all the D fractionation in the nebula must be directly inherited from the interstellar medium (Willacy et al. 2015).

4.3 Nitrogen Isotopes

In the solar system the $^{14}N/^{15}N$ ratio presents variations that remain difficult to interpret (Fig. 2). The analysis of Genesis solar wind samples (Marty et al. 2011), considered as representative of the protosolar nebula, provided a $^{14}N/^{15}N$ ratio of 441 ± 5. This value is in agreement with the in situ measurements made in the ammonia of Jupiter's atmosphere (Fouchet et al. 2004) which probably comes from primordial N_2 (Owen et al. 2001). These values differs from the terrestrial ratio measured in the atmospheric N_2 (272). The Earth's ratio is enriched in ^{15}N compared to the protosolar nebula and is similar to the bulk of ratios derived from the analysis of comet 81P/wild 2 grains (McKeegan et al. 2006). Nitrogen isotopic ratios have also been measured in Titan's atmosphere. This atmosphere is dominated by N_2 molecules and two in situ measurements, based on Cassini GCMS and INMS data (Niemann et al. 2010; Mandt et al. 2009) have been published: 167.7 ± 0.6 and 143. Because of the low abundance of primordial Ar observed by Cassini-Huygens, it is generally assumed that N_2 is of secondary origin in this atmosphere and was delivered in a less-volatile form, probably NH_3.

In comets, a similar ^{15}N enrichment is observed for both CN/HCN and ammonia (Fig. 2). This is in contrast with interstellar sources where molecules carrying the nitrile (–CN) functional group are more enriched in ^{15}N than the molecules carrying the amine (–NH) functional group (Table 5). As discussed in Sect. 4.1, the different studies of interstellar chemistry are able to explain the ^{15}N enrichment measured for cometary HCN but this is less clear for ammonia (Sect. 4.1).

Ion-molecule chemistry in dense interstellar and/or protostellar material could induce ^{15}N enrichment in the ammonia ice produced in such environments (Charnley and Rodgers 2002). Alternatively, the ^{15}N enrichment observed in cometary ices could result from isotopic fractionation processes that occurred in the late protosolar nebula (Aléon 2010).

A property of primitive solar system material, is that there is no strict correlation between deuterium and ^{15}N enrichments (Aléon 2010; Marty 2012, and references therein). Interestingly, Wirström et al. (2012) show that while the nitriles, HCN and HNC, have the greatest ^{15}N enrichment, this enrichment is not expected to correlate with extreme D enrichments.

Finally, models considering self-shielding of N_2 photodissociation can explain enrichment in ^{15}N for HCN, but for ammonia the enrichment in ^{15}N depends on the place in the nebula (Heays et al. 2014).

5 Prospects for the Future

5.1 Measurements from Rosetta

5.1.1 ROSINA

Mass spectrometry is a powerful tool to measure the isotopic ratios of the light volatiles in simple species of the coma. Mass spectrometers with high resolution (to resolve molecules on integer mass) and high dynamic range (to detect minor species) are required. On board the Rosetta orbiter is the Rosetta Orbiter Spectrometer for Ion and Neutral Analysis experiment (ROSINA). Its aim is to characterize *in situ* the volatiles in the coma of 67P/Churyumov-Gerasimenko (Balsiger et al. 2007). ROSINA consists of two complementary mass spectrometers, the Double Focusing Mass Spectrometer (DFMS) and the Reflectron Time of Flight mass spectrometer (RTOF), and a pressure sensor (COPS). DFMS has a mass per charge range of 12–150 amu/e and a mass resolution of 9000 at 50 % peak height at mass 28. It is optimized for very high mass resolution and large dynamic range and can resolve, e.g., CO and N_2 on mass/charge 28 amu/e, or ^{12}CH and ^{13}C on mass/charge 13 amu/e (Balsiger et al. 2007).

Since the first encounter at a comet by VEGA 1 on March 6, 1986, there have been several missions to comets, like Deep Space (19P/Borrelly), Stardust (81P/Wild 2), and Deep Impact (9P/Tempel 1). However, since Giotto none of them carried a mass spectrometer for *in situ* investigations of the gas coma. ROSINA will provide information about H_2O and other molecules in the coma (e.g., abundance of CO and N_2, Rubin et al. 2015, local abundances in the coma, extended sources, e.g., CN, etc.) and measure the isotopic composition of several species. In the following is a discussion of the capability of ROSINA to measure isotopic ratios.

For the light volatiles H, C, N, O, and S, DFMS is expected to measure at least the isotopic ratio in one molecule: for D/H in H_2O as it has been shown by Hässig et al. (2013), for $^{13}C/^{12}C$ in CH_4, CO_2, H_2CO, for $^{15}N/^{14}N$ directly in N, for $^{17}O/^{16}O$ in H_2O, for $^{18}O/^{16}O$ in H_2O, and for the sulfur isotopes directly in S and H_2S. There are three factors which influence whether an isotopologue can be detected or not:

- Mass resolution of DFMS, e.g., it is not possible to resolve the isotopologues of HCN.
- Abundance of main species and dynamic range: e.g., in order to determine the ratio of oxygen the peaks on mass 16 and 17 are needed since the signal for ^{17}O has to be separated from ^{16}OH, which might be much more abundant and thus hide the signal of ^{17}O. First results were derived already during the comet approach phase (see Sect. 2.1.1, Altwegg et al. 2014), and further results are expected in summer 2015 when 67P/Churyumov-Gerasimentko is at its perihelion.
- Some species are already present in the spacecraft background, e.g., CH_4 and HCN (for entire inventory see SchläPpi et al. 2010) and have to be accounted for.

An overview about the detection capabilities for isotopic ratios at comet 67P/Churyumov-Gerasimenko is given in Table 6.

Table 6 An overview on the detection capabilities of ROSINA for the light volatiles. The table is based on the current knowledge about abundances and isotopic ratios on comets (Bockelée-Morvan 2011; Jehin et al. 2009; Rousselot et al. 2014)

	Target species	Mass (amu)	Closest neighbour	Mass (amu)	Feasibility
D/H	HDO	19.0163	$H_2^{17}O$	19.0142	✓
	NH_2D	18.0323	$^{15}NH_3$	18.0230	(✓)
	CH_3D	17.0370	$^{13}CH_4$	17.0341	(✓)
	DCN	28.0166	CH_2N	28.0182	–
			$H^{13}CN$	28.0137	
$^{13}C/^{12}C$	$^{13}CH_3D$	18.0404	NDH_2	18.0323	(✓)
	^{13}C	13.0028	CH	13.0073	✓
	$^{13}CH_4$	17.0341	CH_3D	17.0370	✓
	$H^{13}CN$	28.0137	DCN	28.0166	–
			CH_2N	28.0182	
	^{13}CN	27.0059	HCN	27.0104	(✓)
	$^{13}CO_2$	44.9926	CHO_2	44.9971	✓
	$H_2^{13}CO$	31.0134	CH_3O	31.0178	✓
	^{13}CO	28.9977	CHO	29.0022	(✓)
$^{15}N/^{14}N$	$^{15}NH_3$	18.0230	NH_2D	18.0323	(✓)
	$HC^{15}N$	28.0074	N_2	28.0056	–
			$H^{13}CN$	28.0137	
	$C^{15}N$	26.9996	^{13}CN	27.0059	(✓)
	^{15}NN	29.0026	CO	27.9944	(✓)
			CHO	29.0022	(✓)
$^{17}O/^{16}O$	^{17}O	16.9986	OH	17.0022	?
	$H_2^{17}O$	19.0142	HDO	19.0163	✓
	$C^{17}O$	28.9986	^{13}CO	28.9977	-
	$CO^{17}O$	44.9935	$^{13}CO_2$	44.9926	(✓)
$^{18}O/^{16}O$	^{18}O	17.9986	H_2O	18.0100	✓
	$H_2^{18}O$	20.0143	HF	20.0057	✓
	$C^{18}O$	29.9986	NO	29.9974	?
	$CO^{18}O$	45.9935	NO_2	45.9924	?
$^{33}S/^{32}S$	^{33}S	32.9709	HO_2	32.9971	✓
	$H_2^{33}S$	34.9866	Cl	34.9683	✓
$^{34}S/^{32}S$	^{34}S	33.9673	H_2O_2	34.0049	✓
	$H_2^{34}S$	35.9830	HCl	35.9761	✓
			^{36}Ar	35.967	(✓)

✓: ROSINA will determine ratio, (✓): determination is dependent on abundance, ?: not enough data available to make predictions, (–): ROSINA will not be able to determine the isotopic ratio

5.1.2 Rosetta: COSAC and Ptolemy

The Philae Lander (Bibring et al. 2007b) has two Gas Chromatography (GC)—Mass Spectrometer instruments, COSAC (Goesmann et al. 2007) and Ptolemy (Wright et al. 2007).

The two instruments are complimentary in that COSAC aims to analyse the chemical composition of the organic compounds whereas Ptolemy aims to accurately measure the isotopic composition of the abundant chemical components, although there is some degree of overlap.

The Ptolemy mass spectrometer is an ion trap mass spectrometer (March and Todd 2005) with a mass range of 10 to 150 Da and unit mass resolution. With a low resolution mass spectrometer separation of isobaric interferences directly by mass spectrometry is not possible. However, Ptolemy uses various chemical and gas chromatographic methods to separate the molecular species of interest. Ptolemy includes a reference gas reservoir so that aliquots can be processed and analysed in the same procedure as the sample gas to give a direct comparison with known terrestrial values. Reference gases processed through the Ptolemy GC system have shown that the precision for $\delta^{13}C$, $\delta^{15}N$ and $\delta^{18}O$ of approximately 20 permil can be achieved.

Comet samples can be collected by the Philae Lander Sampler Drill and Distribution system (SD2), from depths of up to 30cm (Finzi et al. 2007). These samples will be transferred into one of either 10 High Temperature Ovens (HTO), which have a maximum temperature of $\sim 600\,^{\circ}C$, or 16 Medium Temperature Ovens (MTO), which have a maximum temperature of $180\,^{\circ}C$. The MTOs incorporate a window so that observations can be made of the sample by the CIVA microscopes (Bibring et al. 2007a) before and after heating. Samples are then heated in controlled temperature steps to evolve volatiles into the Ptolemy chemistry system. These gases will then be processed for isotopic analysis. One of the high temperature ovens contains a trapping material that passively collects coma gas when positioned in an exposed, low temperature position on the Lander balcony. The collected gas is then released into the Ptolemy system by heating the oven to $200\,^{\circ}C$. Evolved gas will then be processed and analyzed in a similar way to the drill samples. This will allow a direct comparison of the coma composition at the Lander site and the Orbiter.

Usually a small aliquot of evolved gas is analyzed directly to obtain a bulk compositional mass spectrum. The processing that the sample undergoes is dependent on the required isotopic analysis. The various sample processing schemes are:

1. Organic composition. Injection of the sample into GC column A (30 m CP-Sil 8CB) which separates organic molecules to obtain chemical composition of organic molecules.
2. Carbon and nitrogen isotopic analysis. Injection of the sample into GC column B (30 m CP-PoraPLOT Q) which separates CO and N_2 from CO_2. As the sample elutes from the column it passes through a Rhodium oxide reactor which oxidises the CO to CO_2. Isotopic exchange on the reactor gives the CO_2 a terrestrial oxygen isotopic signature thus allowing accurate carbon isotopic analysis (Santrock et al. 1985).
3. Oxygen and hydrogen isotopic analysis of water. The water is reduced to CO and H_2 by passing over a platinised carbon reactor which are then separated by GC column C (15 m 5 Angstroem molecular sieve MXT). Note that this is the only scheme which also gives a direct measurement of the $^{17}O/^{16}O$ ratio as well as the $^{18}O/^{16}O$ ratio.
4. Bulk hydrogen, carbon and nitrogen isotopic analysis. Combustion of the sample to H_2O, CO_2 and N_2. The carbon and nitrogen isotopes can then be analysed by either direct analysis or by using scheme (2) whilst water is analysed by scheme (3).
5. Noble gas analysis. Exposure of the sample to getters to remove all active gases to allow analysis of the noble gas concentrations and isotopic values.
6. Oxygen isotopic analysis of minerals. A reactor containing Asprey's salt (Asprey 1976) is heated to produce fluorine which is then used to liberate oxygen from the sample. Due to the reactive nature of this experiment this will be attempted near the end of the Lander mission.

Table 7 Ptolemy measurements of isotopic ratios

Species	Method	Isotope ratio	Requirements
Bulk C	Combustion of solid sample to CO_2	$^{13}C/^{12}C$	
Bulk N	Combustion of solid sample to N_2	$^{15}N/^{14}N$	
Bulk H	Combustion of solid sample to H_2O followed by reduction to CO and H_2	D/H	
H_2O	Reduction to CO and H_2 (channel C)	$^{18}O/^{16}O$	
		$^{17}O/^{16}O$	
		D/H	
CO_2	Direct analysis	$^{13}C/^{12}C$	(2)
		$^{18}O/^{16}O$	
	Channel B	$^{13}C/^{12}C$	
CO	Direct analysis	$^{13}C/^{12}C$	(1), (2)
		$^{18}O/^{16}O$	
	Channel B oxidation to CO_2	$^{13}C/^{12}C$	
CH_4	Removal CO and CO_2 by getters, combustion to CO_2, then direct analysis	$^{13}C/^{12}C$	
	Channel B, oxidation to CO_2	$^{13}C/^{12}C$	
N_2	Direct analysis		(2)
	Channel B, combustion to N_2	$^{15}N/^{14}N$	
NH_3	Channel B, combustion to N_2	$^{15}N/^{14}N$	
Noble gases	Removal of active gases by getters		
Minerals	Reaction with fluorine, direct analysis	$^{18}O/^{16}O$	

(1) $^{13}C/^{12}C$ measurement assumes $\delta^{17}O$—the terrestrial value within 200 permil

(2) Providing the CO/N_2 ratio is known

A summary of the various schemes and species targeted for isotopic analysis is shown in Table 7.

Ptolemy analysis of comet samples is a destructive technique that uses a relatively large amount of power and the target isotopic scheme has to be decided in advance and pre-programmed into the instrument. The planned First Science Sequence (FSS), which lasted from Lander touchdown until the non-rechargeable batteries were exhausted some 60 hours later, was to include the collection of 2 samples by SD2 from a depth of about 30cm from the comet surface. Both samples were to be delivered to HTOs, one for Ptolemy and the other for COSAC. The aim of the Ptolemy experiment was to heat a sample in a HTO in four temperature steps using the following schemes; −50 °C carbon and nitrogen isotopic analysis of CO and N_2, +100 °C oxygen and hydrogen isotopic analysis of water, +400 °C carbon and nitrogen isotopic analysis of CO, CO_2 and N_2, +600 °C bulk carbon and nitrogen isotopic analysis. This is then followed by a reference gas analysis.

Following a successful, but non-optimal landing, there was insufficient power for SD2 to collect a sample for Ptolemy. (SD2 did operate and attempted to collect a sample for COSAC, but it remains unknown whether or not a sample was delivered to the oven). The Ptolemy sequence was changed to a single isotopic analysis using the CASE oven in the hope that some material may have been collected during and after the landing. Although the sequence was successfully executed, the signal was too small to give meaningful isotope ratios.

The Long Term Science Sequence (LTS) was expected to start within a few weeks of the FSS where recharging of the Philae secondary batteries would allow intermittent operations for up to six months. In the current configuration it is hoped that the Lander will have sufficient power to begin operations in spring 2015. Ptolemy operation is severely restricted by power resources, however during the LTS the composition of the sample before analysis will be better characterized including ROLIS and CIVA imaging and ongoing observations from orbiter instruments e.g. ROSINA. The Ptolemy aim is to target particular abundant species for isotopic analysis, make coma observations using the adsorbent oven as the comet becomes more active and finally attempt a fluorination of the refractory residue from a High Temperature Oven.

5.1.3 MIRO

The main goals of the Microwave Instrument for the Rosetta Orbiter (MIRO) are to measure the thermal emission from nucleus sub-surface layers in two continuum channels (0.5 and 1.6 mm), and to characterize the outgassing of four fundamental molecules (H_2O, CO, NH_3, CH_3OH) from their lines near 0.5 mm (Gulkis et al. 2007). The fundamental rotational lines $1_{10} - 1_{01}$ of the three oxygen-related isotopologues of water ($H_2^{16}O$, $H_2^{18}O$, and $H_2^{17}O$) are observed simultaneously to determine the $^{18}O/^{16}O$ and $^{17}O/^{16}O$ ratios.

5.2 Measurements from Ground-Based and Space-Based Facilities

The Atacama Large Millimeter/submillimeter Array is currently under construction in northern Chile's Atacama desert at an altitude of 5000 meters above sea level. The ALMA interferometer will initially be composed of 66 antennas working at millimeter and submillimeter wavelengths (Wootten 2008). ALMA signifies the beginning of a new era for cometary science, in which routine measurements of the three-dimensional distributions of molecular species are now possible for almost all comets, with visual magnitudes of 8 or brighter. For the brightest comets, it will be possible to detect molecules which have strong lines in the microwave range, such as HDO, DCN, $H^{13}CN$, $HC^{15}N$, $H^{15}NC$, $H_2^{34}S$, and $C^{34}S$, isotopes of CH_3OH, etc. Hence, it will be possible to probe the isotopic diversity in the comet population. The prospects for ALMA observations of comets are discussed by Biver (2005) and Bockelée-Morvan (2008).

The upcoming of 40-m class optical telescopes will ease the characterization of deuterium fraction in water and methane in comets. At the European Extremely Large telescope (E-ELT) built by the European Southern Observatory, high-resolution spectroscopy in the L and M near-IR bands will be offered by one of the first generation instruments ELT-MIR.

5.3 On the Analyses of Cometary Samples on Earth

Samples of Comet 81P/Wild 2 returned by the Stardust mission are continuously being analyzed. Indigenous organic compounds recovered from this samples may include potential prebiotic compounds like glycine (Elsila et al. 2009). Further improvement in laboratory techniques may enable the measurement of the isotopic compositions of these compounds in order to ascertain their indigenous origin. In the coming year, with the forthcoming return of samples from primitive asteroids (Hayabusa 2 and Osiris-Rex missions), more constraints will be available on the relationship between primitive asteroids and carbonaceous chondrites (CI, CM, CR) that are suspected of having a cometary origin. The origin of GEMS in CP-IDPs is still debated, as some of these glassy inclusions (the largest ones) show isotopic anomalies possibly related to a presolar heritage, although their collection in silicon oil may bias the analyses (Bradley 2013; Keller and Messenger 2013). By contrast, UCAMMs are collected in water and do not suffer such bias. Analyzing the oxygen isotopic composition of GEMS in UCAMMs should provide more clues on the origin of these phases that seem ubiquitous in cometary material. UCAMMs offer an opportunity to directly study the association of organics and minerals in primitive dust. Measuring both the hydrogen and nitrogen isotopic compositions of organic matter in UCAMMs will shed more light on the formation and evolution of organic matter in the solar system (e.g., Floss et al. 2012; Haenecour et al. 2014; Bardin et al. 2014; Duprat et al. 2014). Since UCAMMs did not endure substantial alteration at atmospheric entry or during their stay in the snow, their content in presolar grains should also provide a valuable constraint on the interstellar heritage in cometary solids.

6 Summary

In this chapter, we reviewed the isotopic composition of cometary volatiles and grains, as well as that of potential cometary material available in collections on Earth. We discussed the isotopic ratios in the light of processes governing interstellar fractionation. Comparing isotopic ratios in comets and planets can shed light to the origin of volatiles in planets, as discussed in Mandt et al. (2015, this issue).

Important progresses in our knowledge of comet composition will benefit from current or future space missions to comets (e.g., the Rosetta mission of ESA) and from telescopic observations with yet available or new instrumentation (e.g., ALMA). Our understanding of comet composition in the context of the origin and evolution of our Solar System will greatly benefit from theoretical models of chemical and physical processes in the solar nebula, and from upcoming new results on the composition of protostellar sources and protoplanetary disks at different stages of evolution.

Acknowledgements This work was supported by NASA's Planetary Astronomy and Planetary Atmospheres Programs (SNM and SBC), and by the Swedish Space Board (ESW).

References

P. Ábrahám et al., Episodic formation of cometary material in the outburst of a young Sun-like star. Nature **459**, 224–226 (2009)

N.G. Adams, D. Smith, $^{14}N/^{15}N$ isotope fractionation in the reaction $N_2H^+ + N_2$—interstellar significance. Astrophys. J. **247**, L123–L125 (1981)

Y. Aikawa, E. Herbst, Deuterium fractionation in protoplanetary disks. Astrophys. J. **526**, 314–326 (1999)

J. Aléon, Multiple origins of nitrogen isotopic anomalies in meteorites and comets. Astrophys. J. **72**, 1342–1351 (2010)

C.M.O.D. Alexander, R. Bowden, M.L. Fogel, K.T. Howard, C.D.K. Herd, L.R. Nittler, The provenances of asteroids, and their contributions to the volatile inventories of the terrestrial planets. Science **337**(6095), 721–723 (2012)

K. Altwegg, Habilitationsschrift. University of Bern (1996)

K. Altwegg et al., 67P/Churyumov–Gerasimenko, a Jupiter family comet with a high D/H ratio. Science **347**, 1261952 (2014)

E. Anders, Do stony meteorites come from comets? Icarus **24**, 363–371 (1975)

E. Anders, N. Grevesse, Abundances of the elements—meteoritic and solar. Geochim. Cosmochim. Acta **53**, 197–214 (1989)

C. Arpigny, E. Jehin, J. Manfroid, D. Hutsemékers, R. Schulz, J.A. Stüwe, J.-M. Zucconi, I. Ilyin, Anomalous nitrogen isotope ratio in comets. Science **301**, 1522–1525 (2003)
L.B. Asprey, The preparation of very pure fluorine gas. J. Fluorine Chem. **7**, 359–361 (1976)
J. Bally, W.D. Langer, Isotope-selective photodestruction of carbon monoxide. Astrophys. J. **255**, 143–148 (1982)
H. Balsiger, K. Altwegg, J. Geiss, D/H and $^{18}O/^{16}O$ ratio in the hydronium ion and in neutral water from in situ ion measurements in comet P/Halley. J. Geophys. Res. **100**, 5827–5834 (1995)
H. Balsiger et al., ROSINA—Rosetta Orbiter Spectrometer for Ion and Neutral Analysis. Space Sci. Rev. **128**, 745–801 (2007)
N. Bardin, G. Slodzian, T.-D. Wu, D. Baklouti, E. Dartois, C. Engrand, R. Brunetto, J.-L. Guerquin-Kern, J. Duprat, D/H measurements in ultracarbonaceous antarctic micrometeorites using polyatomic ions with SIMS. Lunar Planet. Sci. Conf. Abstr. **45**, 2647 (2014)
E.A. Bergin et al., Herschel observations of EXtra-Ordinary Sources (HEXOS): the present and future of spectral surveys with Herschel/HIFI. Astron. Astrophys. **521**, L20–L27 (2010)
J.-P. Bibring, P. Lamy, Y. Langevin, A. Soufflot, M. Berthé, J. Borg, F. Poulet, S. Mottola, CIVA. Space Sci. Rev. **128**, 363–381 (2007a)
J.-P. Bibring, H. Rosenbauer, H. Boehnhardt, S. Ulamec, J. Biele, S. Espinasse, B. Feuerbacher, P. Gaudon, P. Hemmerich, P. Kletzkine, D. Moura, R. Mugnuolo, G. Nietner, B. Pätz, R. Roll, H. Scheuerle, K. Szegö, K. Wittmann (Philae Team), Rosetta lander ("Philae") investigations. Space Sci. Rev. **128**, 205–220 (2007b)
N. Biver, Comets with ALMA. ESA SP **577**, 151–156 (2005)
N. Biver, D. Bockelée-Morvan, J. Crovisier, D.C. Lis, R. Moreno, P. Colom, F. Henry, F. Herpin, G. Paubert, M. Womack, Radio wavelength molecular observations of comets C/1999 T1 (McNaught–Hartley), C/2001 A2 (LINEAR), C/2000 WM_1 (LINEAR) and 153P/Ikeya–Zhang. Astron. Astrophys. **449**, 1255–1270 (2006)
N. Biver, D. Bockelée-Morvan, J. Crovisier, A. Lecacheux, U. Frisk, Å. Hjalmarson, M. Olberg, H.-G. Florén, A. Sandqvist, S. Kwok, Submillimetre observations of comets with Odin: 2001–2005. Planet. Space Sci. **55**, 1058–1068 (2007)
N. Biver et al., *Composition and Outburst Follow-up Observations of Comet 17P/Holmes at the Nançay, IRAM and CSO Radio Observatories*. LPI Contributions, vol. 1405 (2008), p. 8146
N. Biver et al., Ammonia and other parent molecules in comet 10P/Tempel 2 from Herschel/HIFI and ground-based radio observations. Astron. Astrophys. **539**, 68 (2012)
L. Bizzocchi, P. Caselli, E. Leonardo, L. Dore, Detection of $^{15}NNH^+$ in L1544: non-LTE modelling of dyazenilium hyperfine line emission and accurate $^{14}N/^{15}N$ values. Astron. Astrophys. **555**, A109–A120 (2013)
L. Bizzocchi, P. Caselli, S. Spezzano, E. Leonardo, Deuterated methanol in the pre-stellar core L1544. Astron. Astrophys. **569**, A27–A34 (2014)
G.A. Blake, E.C. Sutton, C.R. Masson, T.G. Phillips, Molecular abundances in OMC-1—the chemical composition of interstellar molecular clouds and the influence of massive star formation. Astrophys. J. **315**, 621–645 (1987)
G.A. Blake, C. Qi, M.R. Hogerheijde, M.A. Gurwell, D.O. Muhleman, Sublimation from icy jets as a probe of the interstellar volatile content of comets. Nature **398**, 213–216 (1999)
D. Bockelée-Morvan, Cometary science with ALMA. Astrophys. Space Sci. **313**, 183–189 (2008)
D. Bockelée-Morvan, An overview of comet composition, in *The Molecular Universe*, ed. by J. Cernicharo, R. Bachiller. IAU Symposium, vol. 280 (Cambridge University press, Cambridge, 2011), pp. 261–274
D. Bockelée-Morvan et al., Deuterated water in comet C/1996 B2 (Hyakutake) and its implications for the origin of comets. Icarus **133**, 147–162 (1998)
D. Bockelée-Morvan et al., New molecules found in comet C/1995 O1 (Hale-Bopp). Investigating the link between cometary and interstellar material. Astron. Astrophys. **353**, 1101–1114 (2000)
D. Bockelée-Morvan, D. Gautier, F. Hersant, J.-M. Huré, F. Robert, Turbulent radial mixing in the solar nebula as the source of crystalline silicates in comets. Astron. Astrophys. **384**, 1107–1118 (2002)
D. Bockelée-Morvan, N. Biver, E. Jehin, A.L. Cochran, H. Wiesemeyer, J. Manfroid, D. Hutsemékers, C. Arpigny, J. Boissier, W. Cochran, P. Colom, J. Crovisier, N. Milutinovic, R. Moreno, J.X. Prochaska, I. Ramirez, R. Schulz, J.-M. Zucconi, Large excess of heavy nitrogen in both hydrogen cyanide and cyanogen from comet 17P/Holmes. Astrophys. J. **679**, L49–L52 (2008)
D. Bockelée-Morvan et al., Herschel measurements of the D/H and $^{16}O/^{18}O$ ratios in water in the Oort-cloud comet C/2009 P1 (Garradd). Astron. Astrophys. **544**, L15 (2012)
B.P. Bonev, M.J. Mumma, E.L. Gibb, M.A. DiSanti, G.L. Villanueva, K. Magee-Sauer, R.S. Ellis, Comet C/2004 Q2 (Machholz): parent volatiles, a search for deuterated methane, and constraint on the CH_4 spin temperature. Astrophys. J. **699**, 1563–1572 (2009)

T.L. Bourke, P.C. Myers, N.J. Evans II., M.M. Dunham, J. Kauffmann, Y.L. Shirley, A. Crapsi, C.H. Young, T.L. Huard, T.Y. Brooke, N. Chapman, L. Cieza, C.W. Lee, P. Teuben, Z. Wahhaj, The Spitzer c2d survey of nearby dense cores. II. Discovery of a low-luminosity object in the "Evolved Starless Core" L1521F. Astrophys. J. **649**, L37–L40 (2006)

J.P. Bradley, Interplanetary dust particles, in *Meteorites, Comets and Planets: Treatise on Geochemistry*, ed. by A.M. Davis, H.D. Holland, K.K. Turekian (Elsevier/Pergamon, Oxford, 2005), pp. 689–711

J.P. Bradley, How and where did gems form? Geochim. Cosmochim. Acta **107**(0), 336–340 (2013)

J.P. Bradley, D.E. Brownlee, Cometary particles—thin sectioning and electron beam analysis. Science **231**, 1542–1544 (1986)

R. Brasser, A. Morbidelli, Oort cloud and scattered disc formation during a late dynamical instability in the solar system. Icarus **225**, 40–49 (2013)

G. Briani, M. Gounelle, Y. Marrocchi, S. Mostefaoui, H. Leroux, E. Quirico, A. Meibom, Pristine extraterrestrial material with unprecedented nitrogen isotopic variation. Proc. Nat. Acad. Sci. **106**, 10522–10527 (2009)

J.C. Bridges, H.G. Changela, S. Nayakshin, N.A. Starkey, I.A. Franchi, Chondrule fragments from Comet Wild 2: evidence for high temperature processing in the outer solar system. Earth Planet. Sci. Lett. **341**, 186–194 (2012)

R.H. Brown, D.S. Lauretta, B. Schmidt, J. Moores, Experimental and theoretical simulations of ice sublimation with implications for the chemical, isotopic, and physical evolution of icy objects. Planet. Space Sci. **60**, 166–180 (2012)

D.E. Brownlee, Cosmic dust: collection and research. Annu. Rev. Earth Planet. Sci. **13**, 147–173 (1985)

D. Brownlee, The stardust mission: analyzing samples from the edge of the solar system. Annu. Rev. Earth Planet. Sci. **42**, 179–205 (2014)

D.E. Brownlee, D.J. Joswiak, D.J. Schlutter, R.O. Pepin, J.P. Bradley, S.G. Love, Identification of individual cometary IDPs by thermally stepped He release, in *Lunar Planet. Sci.*, vol. XXVI (1995), pp. 183–184

D.E. Brownlee et al., Stardust: comet and interstellar dust sample return mission. J. Geophys. Res., Planets **108**, 8111 (2003)

D.E. Brownlee et al., Surface of young Jupiter family comet 81 P/Wild 2: view from the Stardust Spacecraft. Science **304**, 1764–1769 (2004)

D. Brownlee et al., Comet 81P/Wild 2 under a microscope. Science **314**, 1711–1716 (2006)

R. Brunetto, J. Borg, E. Dartois, F.J.M. Rietmeijer, F. Grossemy, C. Sandt, L. Le Sergeant d'Hendecourt, A. Rotundi, P. Dumas, Z. Djouadi, F. Jamme, Mid-IR, Far-IR, Raman micro-spectroscopy, and FESEM-EDX study of IDP L2021C5: clues to its origin. Icarus **212**(2), 896–910 (2011)

H. Busemann, A.F. Young, C.M.O.D. Alexander et al., Interstellar chemistry recorded in organic matter from primitive meteorites. Science **312**, 727–730 (2006)

H. Busemann, A.N. Nguyen, G.D. Cody, P. Hoppe, A.L.D. Kilcoyne, R.M. Stroud, T.J. Zega, L.R. Nittler, Ultra-primitive interplanetary dust particles from the comet 26P/Grigg-Skjellerup dust stream collection. Earth Planet. Sci. Lett. **288**(1-2), 44–57 (2009)

H. Campins, T.D. Swindle, Expected characteristics of cometary meteorites. Meteorit. Planet. Sci. **33**, 1201–1211 (1998)

C. Ceccarelli, P. Caselli, D. Bockelée-Morvan, O. Mousis, S. Pizzarello, F. Robert, D. Semenov, Deuterium fractionation: the Ariadne's thread from the pre-collapse phase to meteorites and comets today, in *Protostars and Planets VI*, ed. by H. Beuther, R. Klessen, C. Dullemond, Th. Henning (University of Arizona Press, Tucson, 2014), pp. 859–882

S.B. Charnley, S.D. Rodgers, The end of interstellar chemistry as the origin of nitrogen in comets and meteorites. Astrophys. J. **569**, L133–L137 (2002)

S.B. Charnley, S.D. Rodgers, Interstellar reservoirs of cometary matter. Space Sci. Rev. **138**, 59–73 (2008)

S.B. Charnley, A.G.G.M. Tielens, S.D. Rodgers, Deuterated methanol in the Orion compact ridge. Astrophys. J. **482**, L203–L206 (1997)

S.B. Charnley, P. Ehrenfreund, T.J. Millar, A.C.A. Boogert, A.J. Markwick, H.M. Butner, R. Ruiterkamp, S.D. Rodgers, Observational tests for grain chemistry: posterior isotopic labeling. Mon. Not. R. Astron. Soc. **347**, 157–162 (2004)

L.I. Cleeves, E.A. Bergin, C.M.O.D. Alexander, F. Du, D. Graninger, K.I. Öberg, T.J. Harries, The ancient heritage of water ice in the solar system. Science **345**, 1590–1593 (2014)

A. Cochran, A.-C. Levasseur-Regourd, M. Cordiner, E. Hadamcik, J. Lasue, A. Gicquel, D.G. Schleicher, S.B. Charnley, M.J. Mumma, L. Paganini, D. Bockelée-Morvan, N. Biver, Y.-J. Kuan, The composition of comets: a pre-Rosetta look. Space Sci. Rev. (2015, this issue)

J. Crovisier, D. Bockelée-Morvan, P. Colom, N. Biver, D. Despois, D.C. Lis (The Team for Target-of-Opportunity Radio Observations of Comets), The composition of ices in comet C/1995 O1 (Hale-Bopp) from radio spectroscopy. Further results and upper limits on undetected species. Astron. Astrophys. **418**, 1141–1157 (2004)

F. Daniel, M. Gérin, E. Roueff, J. Cernicharo, N. Marcelino, F. Lique, D.C. Lis, D. Teyssier, N. Biver, D. Bockelée-Morvan, Nitrogen isotopic ratios in Barnard 1: a consistent study of the N_2H^+, NH_3, CN, HCN, and HNC isotopologues. Astron. Astrophys. **560**, AA3 (2013)

A.C. Danks, D.L. Lambert, C. Arpigny, The $^{12}C/^{13}C$ ratio in comet Kohoutek 1973f. Astrophys. J. **194**, 745–751 (1974)

E. Dartois, C. Engrand, R. Brunetto, J. Duprat, T. Pino, E. Quirico, L. Remusat, N. Bardin, G. Briani, S. Mostefaoui, G. Morinaud, B. Crane, N. Szwec, L. Delauche, F. Jamme, C. Sandt, P. Dumas, Ultracarbonaceous antarctic micrometeorites, probing the solar system beyond the nitrogen snow-line. Icarus **224**(1), 243–252 (2013)

J. Davidson, H. Busemann, I.A. Franchi, A NanoSIMS and Raman spectroscopic comparison of interplanetary dust particles from comet Grigg-Skjellerup and non-Grigg Skjellerup collections. Meteorit. Planet. Sci. **47**, 1748–1771 (2012)

A. Decock, E. Jehin, P. Rousselot, D. Hutsemékers, J. Manfroid, D. Cordier, in *International Comet Workshop*, April 1–3, 2014, Toulouse, France (2014)

E. Dobrică, C. Engrand, E. Quirico, G. Montagnac, J. Duprat, Raman characterization of carbonaceous matter in concordia antarctic micrometeorites. Meteorit. Planet. Sci. **46**(9), 1363–1375 (2011)

E. Dobrică, C. Engrand, H. Leroux, J.N. Rouzaud, J. Duprat, Transmission electron microscopy of concordia ultracarbonaceous antarctic micrometeorites (UCAMMs): mineralogical properties. Geochim. Cosmochim. Acta **76**, 68–82 (2012)

L. Dones, P.R. Weissman, H.F. Levison, M.J. Duncan, Oort cloud formation and dynamics, in *Comets II*, ed. by M.C. Festou et al., (University of Arizona Press, Tucson, 2004), pp. 153–174

K. Dressler, D.A. Ramsay, The electronic absorption spectra of NH_2 and ND_2. Philos. Trans. R. Soc. Lond. A **251**, 553–602 (1959)

J. Duprat, E. Dobrica, C. Engrand, J. Aléon, Y. Marrocchi, S. Mostefaoui, A. Meibom, H. Leroux, J.N. Rouzaud, M. Gounelle, F. Robert, Extreme deuterium excesses in ultracarbonaceous micrometeorites from central antarctic snow. Science **328**(5979), 742–745 (2010)

J. Duprat, N. Bardin, C. Engrand, D. Baklouti, R. Brunetto, E. Dartois, L. Delauche, M. Godard, G. Slodzian, T.-D. Wu, J.-L. Guerkin-Kern, Isotopic analysis of organic matter in ultra-carbonaceous micrometeorites. Meteorit. Planet. Sci. **49**(S1), A103 (2014)

P. Eberhardt, R. Meier, D. Krankowsky, R.R. Hodges, Methanol and hydrogen sulfide in comet P/Halley. Astron. Astrophys. **288**, 315–329 (1994)

P. Eberhardt, M. Reber, D. Krankowsky, R.R. Hodges, The D/H and $^{18}O/^{16}O$ ratios in water from comet P/Halley. Astron. Astrophys. **302**, 301–316 (1995)

J.E. Elsila, D.P. Glavin, J.P. Dworkin, Cometary glycine detected in samples returned by Stardust. Meteorit. Planet. Sci. **44**, 1323–1330 (2009)

G.B. Esplugues, J. Cernicharo, S. Viti, J.R. Goicoechea, B. Tercero, N. Marcelino, A. Palau, T.A. Bell, E.A. Bergin, N.R. Crockett, S. Wang, Combined IRAM and Herschel/HIFI study of cyano(di)acetylene in Orion KL: tentative detection of DC_3N. Astron. Astrophys. **559**, A51 (2013)

A.E. Finzi, F.B. Zazzera, C. Dainese, F. Malnati, P.G. Magnani, E. Re, P. Bologna, S. Espinasse, A. Olivieri, SD2—how to sample a comet. Space Sci. Rev. **128**, 281–299 (2007)

C. Floss, F.J. Stadermann, J.P. Bradley, Z.R. Dai, S. Bajt, G. Graham, A.S. Lea, Identification of isotopically primitive interplanetary dust particles: a NanoSIMS isotopic imaging study. Geochim. Cosmochim. Acta **70**, 2371–2399 (2006)

C. Floss, T. Noguchi, T. Yada, Ultracarbonaceous antarctic micrometeorites: origins and relationships to other primitive extraterrestrial materials, in *Lunar and Planetary Science Conference*, vol. 43 (2012), p. 1217

C. Floss, F.J. Stadermann, A.T. Kearsley, M.J. Burchell, W.J. Ong, The abundance of presolar grains in comet 81P/Wild 2. Astrophys. J. **763**(2), 140–151 (2013)

F. Fontani, T. Sakai, K. Furuya, N. Sakai, Y. Aikawa, S. Yamamoto, DNC/HNC and N_2D^+/N_2H^+ ratios in high-mass star-forming cores. Mon. Not. R. Astron. Soc. **440**, 448–456 (2014)

T. Fouchet, P.G.J. Irwin, P. Parrish, S.B. Calcutt, F.W. Taylor, C.A. Nixon, T. Owen, Search for spatial variation in the jovian $^{15}N/^{14}N$ ratio from Cassini/CIRS observations. Icarus **172**, 50–58 (2004)

J. Geiss, G. Gloeckler, Abundances of deuterium and Helium-3 in the protosolar cloud. Space Sci. Rev. **84**, 239–250 (1998)

M. Gérin, N. Marcelino, N. Biver, E. Roueff, L.H. Coudert, M. Elkeurti, D.C. Lis, D. Bockelé-Morvan, Detection of $^{15}NH_2D$ in dense cores: a new tool for measuring the $^{14}N/^{15}N$ ratio in the cold ISM. Astron. Astrophys. **498**, L9–L12 (2009)

E.L. Gibb, B.P. Bonev, G. Villanueva, M.A. DiSanti, M.J. Mumma, E. Sudholt, Y. Radeva, Chemical composition of comet C/2007 N3 (Lulin): another "Atypical" comet. Astrophys. J. **750**, 102–115 (2012)

A. Gicquel, S.N. Milam, G.L. Villanueva, A.J. Remijan, I.M. Coulson, Y.-L. Chuang, S.B. Charnley, M.A. Cordiner, Y.-J. Kuan, Ground-based multiwavelength observations of comet 103P/Hartley 2. Astrophys. J. **794**, 1–10 (2014)

K. Giles, N.G. Adams, D. Smith, A study of the reactions of H_3^+, H_2D^+, HD_2^+, and D_3^+ with H_2, HD, D_2 using a variable-temperature selected ion flow tube. J. Phys. Chem. **96**, 7645 (1992)

F. Goesmann, H. Rosenbauer, R. Roll, C. Szopa, F. Raulin, R. Sternberg, G. Israel, U. Meierhenrich, W. Thiemann, G. Munoz-Caro, COSAC, the COmetary SAmpling and Composition experiment on Philae. Space Sci. Rev. **128**, 257–280 (2007)

M. Gounelle, P. Spurny, P.A. Bland, The orbit and atmospheric trajectory of the orgueil meteorite from historical records. Meteorit. Planet. Sci. **41**, 135–150 (2006)

J.M. Greenberg, What are comets made of—a model based on interstellar dust, in *Comet Discoveries, Statistics, and Observational Selection*. IAU Colloq., vol. 61 (1982), pp. 131–163

S. Gulkis et al., Remote sensing of a comet at millimeter and submillimeter wavelengths from an orbiting spacecraft. Planet. Space Sci. **55**, 1050–1057 (2007)

H. Haack, R. Michelsen, G. Stober, D. Keuer, W. Singer, I. Williams, *CM Chondrites from Comets? New Constraints from the Orbit of the Maribo CM Chondrite Fall* LPI Contributions, vol. 1639 (2011), p. 9100

P. Haenecour, C. Floss, A. Wang, T. Yada, *Coordinated Analysis of Isotopic Anomalies in Antarctic Micrometeorites*. LPI Contributions, vol. 1783 (2014), p. 5017

M.S. Hanner, M.E. Zolensky, The mineralogy of cometary dust, in *Astromineralogy*, ed. by T. Henning (Springer, Berlin, 2010), pp. 203–226

D.E. Harker, S.J. Desch, Annealing of silicate dust by nebular shocks at 10 AU. Astrophys. J. **565**, L109–L112 (2002)

P. Hartogh et al., Ocean-like water in the Jupiter-family comet 103P/Hartley 2. Nature **478**, 218–220 (2011)

M. Hässig, K. Altwegg, H. Balsiger, J.J. Berthelier, U. Calmonte, M. Combi, J. De Keyser, B. Fiethe, S.A. Fuselier, M. Rubin, ROSINA/DFMS capabilities to measure isotopic ratios in water at comet 67P/Churyumov-Gerasimenko. Planet. Space Sci. **84**, 148–152 (2013)

A.N. Heays, R. Visser, R. Gredel, W. Ubachs, B.R. Lewis, S.T. Gibson, E.F. van Dishoeck, Isotope selective photodissociation of N_2 by the interstellar radiation field and cosmic rays. Astron. Astrophys. **562**, A61 (2014)

E. Herbst, E.F. van Dishoeck, Complex organic interstellar molecules. Annu. Rev. Astron. Astrophys. **47**, 427–480 (2009)

F. Hersant, D. Gautier, J.-M. Huré, A two-dimensional model for the primordial nebula constrained by D/H measurements in the solar system: implications for the formation of giant planets. Astrophys. J. **554**, 391–407 (2001)

P. Hily-Blant, L. Bonal, A. Faure, E. Quirico, The ^{15}N-enrichment in dark clouds and solar system objects. Icarus **223**, 582–590 (2013a)

P. Hily-Blant, G. Pineau des Forêts, A. Faure, R. Le Gal, M. Padovani, The CN/C^{15}N isotopic ratio towards dark clouds. Astron. Astrophys. **557**, A65 (2013b)

D.A. Howe, T.J. Millar, P. Schilke, C.M. Walmsley, Observations of deuterated cyanoacetylene in dark clouds. Mon. Not. R. Astron. Soc. **267**, 59–68 (1994)

W.F. Huebner, J.J. Keady, S.P. Lyon, Solar photorates for planetary atmospheres and atmospheric pollutants. Astrophys. Space Sci. **195**, 1–289 (1992)

D. Hutsemékers, J. Manfroid, E. Jehin, C. Arpigny, A. Cochran, R. Schulz, J.A. Stüwe, J.-M. Zucconi, Isotopic abundances of carbon and nitrogen in Jupiter-family and Oort-cloud comets. Astron. Astrophys. **440**, 21–24 (2005)

D. Hutsemékers, J. Manfroid, E. Jehin, J.-M. Zucconi, C. Arpigny, The ^{16}OH/^{18}OH and OD/OH isotope ratios in comet C/2002 T7 (LINEAR). Astron. Astrophys. **490**, L31–L34 (2008)

M. Ikeda, T. Hirota, S. Yamamoto, The H^{13}CN/HC^{15}N abundance ratio in dense cores: possible source-to-source variation of isotope abundances? Astrophys. J. **575**, 250–256 (2002)

H.A. Ishii, J.P. Bradley, Z.R. Dai, M. Chi, A.T. Kearsley, M.J. Burchell, N.D. Browning, F. Molster, Comparison of comet 81P/Wild 2 dust with interplanetary dust from comets. Science **319**, 447–450 (2008)

E. Jacquet, F. Robert, Water transport in protoplanetary disks and the hydrogen isotopic composition of chondrites. Icarus **223**, 722–732 (2013)

E. Jehin, J. Manfroid, A.L. Cochran, C. Arpigny, J.-M. Zucconi, D. Hutsemékers, W.D. Cochran, M. Endl, R. Schulz, The anomalous ^{14}N/^{15}N ratio in comets 122P/1995 S1 (de Vico) and 153P/2002 C1 (Ikeya-Zhang). Astrophys. J. **613**, 161–164 (2004)

E. Jehin, J. Manfroid, D. Hutsemékers, A.L. Cochran, C. Arpigny, W.M. Jackson, H. Rauer, R. Schulz, J.-M. Zucconi, Deep impact: high-resolution optical spectroscopy with the ESO VLT and the Keck I telescope. Astrophys. J. **641**, 145–148 (2006)

E. Jehin, J. Manfroid, D. Hutsemékers, C. Arpigny, J.-M. Zucconi, Isotopic ratios in comets: status and perspectives. Earth Moon Planets **105**, 167–180 (2009)

E. Jehin, D. Hutsemékers, J. Manfroid, A. Decock, M. Weiler, H. Kawakita, Y. Shinnaka, M. Hashimoto, D. Bockelée-Morvan, N. Biver, J. Crovisier, P. Hartogh, in *EPSC-DPS Joint Meeting 2011*, October 2–7, 2011 Nantes, France (2011), p. 1463
E. Jehin, J. Manfroid, H. Kawakita, D. Hutsemékers, M. Weiler, C. Arpigny, A. Cochran, O. Hainaut, H. Rauer, R. Schulz, J.-M. Zucconi, Optical spectroscopy of the B and C fragments of comet 73P/Schwassmann-Wachmann 3 at the ESO VLT, in *Asteroids, Comets, Meteors 2008*, July 14–18, 2008, Baltimore, Maryland. LPI Contributions, vol. 1405, p. 8319 (2008)
E.K. Jessberger, J. Kissel, Chemical properties of cometary dust and a note on carbon isotopes, in *Comets in the Post-Halley Era 167*, IAU Colloq., vol. 116 (1991), pp. 1075–1092
D. Jewitt, H.E. Matthews, T. Owen, R. Meier, Measurements of $^{12}C/^{13}C$, $^{14}N/^{15}N$, and $^{32}S/^{34}S$ ratios in comet Hale-Bopp (C/1995 O1). Science **278**, 90–93 (1997)
J.K. Jørgensen, F.L. Schöier, E.F. van Dishoeck, Molecular inventories and chemical evolution of low-mass protostellar envelopes. Astron. Astrophys. **416**, 603–622 (2004)
Y. Kakazu, C. Engrand, J. Duprat, G. Briani, N. Bardin, S. Mostefaoui, R. Duhamel, L. Remusat, Bulk oxygen isotopic composition of utracarbonaceous antarctic micrometeorites with the nanosims. Meteorit. Planet. Sci. **49**(S1), A195 (2014)
J.J. Kavelaars, O. Mousis, J.-M. Petit, H.A. Weaver, On the formation location of Uranus and Neptune as constrained by dynamical and chemical models of comets. Astrophys. J. **734**, LL30 (2011)
H. Kawakita, H. Kobayashi, Formation conditions of icy materials in comet C/2004 Q2 (Machholz). II. Diagnostics using nuclear spin temperatures and deuterium-to-hydrogen ratios in cometary molecules. Astrophys. J. **693**, 388–396 (2009)
H. Kawakita, J.-I. Watanabe, Fluorescence efficiencies of monodeuterio-methane in comets: toward the determination of the deuterium/hydrogen ratio in methane. Astrophys. J. **582**, 534–539 (2003)
H. Kawakita, J.-i. Watanabe, D. Kinoshita, M. Ishiguro, E. Nakamura, Saturated hydrocarbons in comet 153P/Ikeya-Zhang: ethane, methane, and monodeuterio-methane. Astrophys. J. **590**, 573–578 (2003)
H. Kawakita, J.-i. Watanabe, R. Furusho, T. Fuse, D.C. Boice, Nuclear spin temperature and deuterium-to-hydrogen ratio of methane in comet C/2001 Q4 (NEAT). Astrophys. J. Lett. **623**, L49–L52 (2005)
L.P. Keller, S. Messenger, On the origins of gems grains: a reply. Geochim. Cosmochim. Acta **107**, 341–344 (2013)
L.P. Keller, K.L. Thomas, D.S. McKay, Carbon in primitive interplanetary dust particles, in *Analysis of Interplanetary Dust*, ed. by M.E. Zolensky, T.L. Wilson, F.J.M. Rietmeijer, G.J. Flynn. AIP Conf. Proc. (Am. Inst. of Phys., New York, 1994), pp. 159–164
L.P. Keller, S. Messenger, G.J. Flynn, S. Clemett, S. Wirick, C. Jacobsen, The nature of molecular cloud material in interplanetary dust. Geochim. Cosmochim. Acta **68**, 2577–2589 (2004)
J. Kissel, F.R. Krueger, The organic component in dust from comet halley as measured by the puma mass spectrometer on board VEGA 1. Nature **326**, 755–760 (1987)
M. Kleine, S. Wyckoff, P.A. Wehinger, B.A. Peterson, The carbon isotope abundance ratio in comet Halley. Astrophys. J. **439**, 1021–1033 (1995)
J.D. Kramers, M.A.G. Andreoli, M. Atanasova, G.A. Belyanin, D.L. Block, C. Franklyn, C. Harris, M. Lekgoathi, C.S. Montross, T. Ntsoane, V. Pischedda, P. Segonyane, K.S. Viljoen, J.E. Westraadt, Unique chemistry of a diamond-bearing pebble from the Libyan desert glass strewnfield, SW Egypt: evidence for a shocked comet fragment. Earth Planet. Sci. Lett. **382**, 21–31 (2013)
D.L. Lambert, A.C. Danks, High-resolution spectra of C_2 Swan bands from comet West 1976 VI. Astrophys. J. **268**, 428–446 (1983)
W.D. Langer, T.E. Graedel, M.A. Frerking, P.B. Armentrout, Carbon and oxygen isotope fractionation in dense interstellar clouds. Astrophys. J. **27**, 581–604 (1984)
M.E. Lawler, D.E. Brownlee, CHON as a component of dust from comet Halley. Nature **359**, 810–812 (1992)
A. Lecacheux, N. Biver, J. Crovisier, D. Bockelée-Morvan, P. Baron, R.S. Booth, P. Encrenaz, H.-G. Florén, U. Frisk, Å. Hjalmarson, S. Kwok, K. Mattila, L. Nordh, M. Olberg, A.O.H. Olofsson, H. Rickman, A. Sandqvist, F. von Schèle, G. Serra, S. Torchinsky, K. Volk, A. Winnberg, Observations of water in comets with Odin. Astron. Astrophys. **402**, 55–58 (2003)
J.-E. Lee, E.A. Bergin, J.R. Lyons, Oxygen isotope anomalies of the Sun and the original environment of the solar system. Meteorit. Planet. Sci. **43**, 1351–1362 (2008)
D.C. Lis, D. Bockelée-Morvan, J. Boissier, J. Crovisier, N. Biver, S.B. Charnley, Hydrogen isocyanide in comet 73P/Schwassmann-Wachmann (Fragment B). Astrophys. J. **675**, 931–936 (2008)
D.C. Lis, A. Wootten, M. Gerin, E. Roueff, Nitrogen isotopic fractionation in interstellar ammonia. Astrophys. J. **710**, L49–L52 (2010)
D.C. Lis, N. Biver, D. Bockelée-Morvan, P. Hartogh, E.A. Bergin, G.A. Blake, J. Crovisier, M. de Val-Borro, E. Jehin, M. Küppers, J. Manfroid, R. Moreno, M. Rengel, S. Szutowicz, A Herschel study of D/H in water in the Jupiter-Family comet 45P/Honda-Mrkos-Pajdusakova and prospects for D/H measurements with CCAT. Astrophys. J. **774**, L3 (2013)

K.E. Mandt, J.H. Waite, W. Lewis et al., Isotopic evolution of the major constituents of Titan's atmosphere based on Cassini data. Planet. Space Sci. **57**, 1917–1930 (2009)

K. Mandt, O. Mousis, B. Marty, T. Cavalié, W. Harris, P. Hartogh, K. Willacy, Constraints from comets on the formation and volatile acquisition of the planets and satellites. Space Sci. Rev. (2015, this issue)

J. Manfroid, E. Jehin, D. Hutsemékers, A. Cochran, J.-M. Zucconi, C. Arpigny, R. Schulz, J.A. Stüwe, Isotopic abundance of nitrogen and carbon in distant comets. Astron. Astrophys. **432**, 5–8 (2005)

J. Manfroid, E. Jehin, D. Hutsemékers, A. Cochran, J.-M. Zucconi, C. Arpigny, R. Schulz, J.A. Stüwe, I. Ilyin, The CN isotopic ratios in comets. Astron. Astrophys. **503**, 613–624 (2009)

J.G. Mangum, R.L. Plambeck, A. Wootten, Fossil DCN in Orion-KL. Astrophys. J. **369**, 169–174 (1991)

R.E. March, J.F.J. Todd, *Quadrupole Ion Trap Mass Spectrometry* (Wiley, New York, 2005)

B. Marty, The origins and concentrations of water, carbon, nitrogen and noble gases on Earth. Earth Planet. Sci. Lett. **313**, 56–66 (2012)

B. Marty, R.L. Palma, R.O. Pepin, L. Zimmermann, D.J. Schlutter, P.G. Burnard, A.J. Westphal, C.J. Snead, S. Bajt, R.H. Becker, J.E. Simones, Helium and neon abundances and compositions in cometary matter. Science **319**(5859), 75–78 (2008)

B. Marty, M. Chaussidon, R.C. Wiens, A.J.G. Jurewicz, D.S. Burnett, A ^{15}N-poor isotopic composition for the solar system as shown by genesis solar wind samples. Science **332**, 1533–1536 (2011)

J.E.P. Matzel, H.A. Ishii, D. Joswiak, I.D. Hutcheon, J.P. Bradley, D. Brownlee, P.K. Weber, N. Teslich, G. Matrajt, K.D. McKeegan, G.J. MacPherson, Constraints on the formation age of cometary material from the NASA stardust mission. Science **328**(5977), 483–486 (2010)

K.D. McKeegan et al., Isotopic compositions of cometary matter returned by stardust. Science **314**, 1724–1728 (2006)

H.Y. McSween Jr., P.R. Weissman, Cosmochemical implications of the physical processing of cometary nuclei. Geochim. Cosmochim. Acta **53**, 3263–3271 (1989)

R. Meier, T.C. Owen, H.E. Matthews, D.C. Jewitt, D. Bockelée-Morvan, N. Biver, J. Crovisier, D. Gautier, A determination of the HDO/H_2O ratio in comet C/1995 O1 (Hale-Bopp). Science **279**, 842–844 (1998a)

R. Meier, T.C. Owen, D.C. Jewitt, H. Matthews, M. Senay, N. Biver, D. Bockelée-Morvan, J. Crovisier, D. Gautier, Deuterium in comet C/1995 O1 (Hale-Bopp): detection of DCN. Science **279**, 1707–1710 (1998b)

R. Meier, D. Wellnitz, S.J. Kim, M.F. A'Hearn, The NH and CH bands of comet C/1996 B2 (Hyakutake). Icarus **136**, 268–279 (1998c)

S. Messenger, Identification of molecular-cloud material in interplanetary dust particles. Nature **404**, 968–971 (2000)

S. Messenger, L.P. Keller, F.J. Stadermann, R.M. Walker, E. Zinner, Samples of stars beyond the solar system: silicate grains in interplanetary dust. Science **300**(5616), 105–108 (2003)

S. Messenger, L.P. Keller, D.S. Lauretta, Supernova olivine from cometary dust. Science **309**(5735), 737–741 (2005)

S.N. Milam, S.B. Charnley, Observations of nitrogen fractionation in prestellar cores: nitriles tracing interstellar chemistry. Technical report. Lunar and Planetary Institute Science Conference Abstracts vol. 43, p. 2618 (2012)

S.N. Milam, C. Savage, M.A. Brewster, L.M. Ziurys, S. Wyckoff, The $^{12}C/^{13}C$ isotope gradient derived from millimeter transitions of CN: the case for galactic chemical evolution. Astrophys. J. **634**, 1126–1132 (2005)

T.J. Millar, A. Bennett, E. Herbst, Deuterium fractionation in dense interstellar clouds. Astrophys. J. **340**, 906–920 (1989)

Y.C. Minh, W.M. Irvine, D. McGonagle, L.M. Ziurys, Observations of the H_2S toward OMC-1. Astrophys. J. **360**, 136–141 (1990)

O. Mousis, D. Gautier, D. Bockelée-Morvan, F. Robert, B. Dubrulle, A. Drouart, Constraints on the formation of comets from D/H ratios measured in H_2O and HCN. Icarus **148**, 513–525 (2000)

T. Nakamura, T. Noguchi, Y. Ozono, T. Osawa, K. Nagao, Mineralogy of ultracarbonaceous large micrometeorites. Meteorit. Planet. Sci. Suppl. **40**, 5046 (2005)

T. Nakamura, T. Noguchi, A. Tsuchiyama, T. Ushikubo, N.T. Kita, J.W. Valley, M.E. Zolensky, Y. Kakazu, K. Sakamoto, E. Mashio, K. Uesugi, T. Nakano, Chondrule-like objects in short-period comet 81P/Wild 2. Science **321**, 1664–1667 (2008)

D. Nakashima, T. Ushikubo, D.J. Joswiak, D.E. Brownlee, G. Matrajt, M.K. Weisberg, M.E. Zolensky, N.T. Kita, Oxygen isotopes in crystalline silicates of comet Wild 2: a comparison of oxygen isotope systematics between Wild 2 particles and chondritic materials. Earth Planet. Sci. Lett. **357**, 355–365 (2012)

D. Nakashima, T. Ushikubo, N.T. Kita, M.K. Weisberg, M.E. Zolensky, D.S. Ebel, Late formation of a comet Wild 2 crystalline silicate particle, Pyxie, inferred from Al-Mg chronology of plagioclase. Earth Planet. Sci. Lett. **410**, 54–61 (2015)

J. Neill, N.R. Crockett, E.A. Bergin, J.C. Pearson, L.-H. Xu, Deuterated molecules in Orion KL from Herschel/HIFI. Astrophys. J. **777**, 85–104 (2013)

H.B. Niemann, S.K. Atreya, J.E. Demick, D. Gautier, J.A. Haberman, D.N. Harpold, W.T. Kasprzak, J.I. Lunine, T.C. Owen, F. Raulin, Composition of Titan's lower atmosphere and simple surface volatiles as measured by the Cassini-Huygens probe gas chromatograph mass spectrometer experiment. J. Geophys. Res., Planets **115**, 12006 (2010)

A.O. Nier, D.J. Schlutter, The thermal history of interplanetry dust particles collected in the Earth's stratosphere. Meteoritics **28**, 675–681 (1993)

T. Noguchi, N. Ohashi, S. Tsujimoto, T. Mitsunari, J.P. Bradley, T. Nakamura, S. Toh, T. Stephan, N. Iwata, N. Imae, Cometary dust in Antarctic ice and snow: past and present chondritic porous micrometeorites preserved on the Earth's surface. Earth Planet. Sci. Lett. **410**, 1–11 (2015)

R.C. Ogliore, G.R. Huss, K. Nagashima, A.L. Butterworth, Z. Gainsforth, J. Stodolna, A.J. Westphal, D. Joswiak, T. Tyliszczak, Incorporation of a late-forming chondrule into Comet Wild 2. Astrophys. J. Lett. **745**, L19 (2012)

T. Owen, The isotope ratio $^{12}C/^{13}C$ in comet Tago-Sato (1969g). Astrophys. J. **184**, 33–44 (1973)

T. Owen, P.R. Mahaffy, H.B. Niemann, S. Atreya, M. Wong, Protosolar nitrogen. Astrophys. J. Lett. **553**, L77 (2001)

L. Pagani, E. Roueff, P. Lesaffre, Ortho-H_2 and the age of interstellar dark clouds. Astrophys. J. **739**, L35–L38 (2011)

B. Parise, C. Ceccarelli, A.G.G.M. Tielens, A. Castets, E. Caux, B. Lefloch, S. Maret, Testing grain surface chemistry: a survey of deuterated formaldehyde and methanol in low-mass class 0 protostars. Astron. Astrophys. **453**, 949–958 (2006)

L. Piani, F. Robert, L. Remusat, Micron-scale D/H heterogeneity in chondrite matrices: a signature of the pristine solar system water? Earth Planet. Sci. Lett. **415**, 154–164 (2015)

L. Remusat, F. Palhol, F. Robert, S. Derenne, C. France-Lanord, Enrichment of deuterium in insoluble organic matter from primitive meteorites: a solar system origin? Earth Planet. Sci. Lett. **243**, 15–25 (2006)

L. Remusat, Y. Guan, Y. Wang, J.M. Eiler, Accretion and preservation of D-rich organic particles in carbonaceous chondrites: evidence for important transport in the early solar system nebula. Astrophys. J. **713**(2), 1048–1058 (2010)

H. Roberts, G.A. Fuller, T.J. Millar, J. Hatchell, J.V. Buckle, Molecular D/H ratios in the dense gas surrounding low-mass protostars. Planet. Space Sci. **50**, 1173–1178 (2002)

H. Roberts, E. Herbst, T.J. Millar, Enhanced deuterium fractionation in dense interstellar cores resulting from multiply deuterated H_3^+. Astrophys. J. **591**, L41–L44 (2003)

S.D. Rodgers, S.B. Charnley, Nitrogen isotopic fractionation of interstellar nitriles. Astrophys. J. **689**, 1448–1455 (2008)

E. Roueff, J.C. Loison, K.M. Hickson, Isotopic fractionation of carbon, deuterium, and nitrogen: a full chemical study. Astron. Astrophys. **576**, A99 (2015), 18 pp.

P. Rousselot, E. Jehin, J. Manfroid, D. Hutsemékers, The $^{12}C_2/^{12}C^{13}C$ isotopic ratio in comets C/2001 Q4 (NEAT) and C/2002 T7 (LINEAR). Astron. Astrophys. **545**, A24–A30 (2012)

P. Rousselot et al., Toward a unique nitrogen isotopic ratio in cometary ices. Astrophys. J. **780**, L17–L21 (2014)

M. Rubin, K. Altwegg, H. Balsiger et al., Molecular nitrogen in comet 67P/Churyumov-Gerasimenko indicates a low formation temperature. Science **348**, 232–235 (2015)

J. Santrock, S.A. Studley, J.M. Hayes, Isotopic analyses based on the mass spectrum of carbon dioxide. Anal. Chem. **57**, 1444–1448 (1985)

B. SchläPpi et al., Influence of spacecraft outgassing on the exploration of tenuous atmospheres with in situ mass spectrometry. J. Geophys. Res. Space Phys. **115**, 12313 (2010)

R.Y. Shah, H. Wootten, Deuterated ammonia in galactic protostellar cores. Astrophys. J. **554**, 933–947 (2001)

Y. Shinnaka, H. Kawakita, H. Kobayashi, E. Jehin, J. Manfroid, D. Hutsemékers, C. Arpigny, Astrophys. J. **729**, 81–95 (2011)

Y. Shinnaka, H. Kawakita, M. Nagashima, K. Hitomi, A. Decock, E. Jehin, D.C. Boice, High-dispersion spectroscopic observations of comet C/2012 S1 (ISON) with the Subaru telescope. AAS/Division for Planetary Sciences Meeting Abstracts 46, #209.14 (2014a)

Y. Shinnaka, H. Kawakita, H. Kobayashi, M. Nagashima, D.C. Boice, $^{14}NH_2/^{15}NH_2$ ratio in comet C/2012 S1 (ISON) observed during its outburst in 2013 November. Astrophys. J. **782**, L16–L19 (2014b)

S.B. Simon, D.J. Joswiak, H.A. Ishii, J.P. Bradley, M. Chi, L. Grossman, J. Aléon, D.E. Brownlee, S. Fallon, I.D. Hutcheon, G. Matrajt, K.D. McKeegan, A refractory inclusion returned by Stardust from comet 81P/Wild 2. Meteorit. Planet. Sci. **43**, 1861–1877 (2008)

G. Slodzian, T.-D. Wu, N. Bardin, J. Duprat, C. Engrand, J.-L. Guerquin-Kern, Simultaneous hydrogen and heavier element isotopic ratio images with a scanning submicron ion probe and mass resolved polyatomic ions. Microsc. Microanal. **20**, 577–581 (2013)

F.J. Stadermann, P. Hoppe, C. Floss, P.R. Heck, F. Horz, J. Huth, A.T. Kearsley, J. Leitner, K.K. Marhas, K.D. McKeegan, T. Stephan, Stardust in stardust—the C, N, and O isotopic compositions of Wild 2 cometary matter in al foil impacts. Meteorit. Planet. Sci. **43**(1-2), 299–313 (2008)

N.A. Starkey, I.A. Franchi, Insight into the silicate and organic reservoirs of the comet forming region. Geochim. Cosmochim. Acta **105**, 73–91 (2013)

N.A. Starkey, I.A. Franchi, M.R. Lee, Isotopic diversity in interplanetary dust particles and preservation of extreme ^{16}O-depletion. Geochim. Cosmochim. Acta **142**, 115–131 (2014)

A. Stawikowski, J.L. Greenstein, The isotope ratio $^{12}C/^{13}C$ in a comet. Astrophys. J. **140**, 1280–1291 (1964)

V. Taquet, S.B. Charnley, O. Sipilä, Multilayer formation and evaporation of deuterated ices in prestellar and protostellar cores. Astrophys. J. **791**, 1–20 (2014)

P.P. Tennekes, J. Harju, M. Juvela, L.V. Tóth, HCN and HNC mapping of the protostellar core Chamaeleon-MMS1. Astron. Astrophys. **456**, 1037–1043 (2006)

R. Terzieva, E. Herbst, The possibility of nitrogen isotopic fractionation in interstellar clouds. Mon. Not. R. Astron. Soc. **317**, 563–568 (2000)

K.L. Thomas, G.E. Blanford, L.P. Keller, W. Klöck, D. McKay, Carbon abundance and silicate mineralogy of anhydrous interplanetary dust particles. Geochim. Cosmochim. Acta **57**, 1551–1566 (1993)

B.E. Turner, Detection of doubly deuterated interstellar formaldehyde (D_2CO)—an indicator of active grain surface chemistry. Astrophys. J. **362**, L29–L33 (1990)

B.E. Turner, Deuterated molecules in translucent and dark clouds. Astrophys. J. Suppl. Ser. **136**, 579–629 (2001)

H.C. Urey, Origin of tektites. Nature **179**, 556–557 (1957)

V. Vanysek, in *Comets, Asteroids, Meteorites: Interrelations, Evolution and Origins*, ed. by A.H. Delsemme. IAU Colloq., vol. 39 (1977), p. 499

G.L. Villanueva, M.J. Mumma, B.P. Bonev, M.A. Di Santi, E.L. Gibb, H. Böhnhardt, M. Lippi, A sensitive search for deuterated water in comet 8p/Tuttle. Astrophys. J. **690**, L5–L9 (2009)

J.H. Waite Jr. et al., Liquid water on Enceladus from observations of ammonia and ^{40}Ar in the plume. Nature **460**, 487–490 (2009)

K.J. Walsh, A. Morbidelli, S.N. Raymond, D.P. O'Brien, A.M. Mandell, A low mass for Mars from Jupiter's early gas-driven migration. Nature **475**, 206–209 (2011)

S.F. Wampfler, J.K. Jorgensen, M. Bizzarro, S.E. Bisschop, Observations of nitrogen isotope fractionation in deeply embedded protostars. Astron. Astrophys. **20**, 577–581 (2014)

W.D. Watson, V.G. Anicich, W.T. Huntress Jr., Measurement and significance of the equilibrium reaction $^{13}C^+ + ^{12}CO \rightleftharpoons ^{12}C^+ + ^{13}CO$ for alteration of the $^{13}C/^{12}C$ ratio in interstellar molecules. Astrophys. J. **205**, L165–L168 (1976)

H.A. Weaver, M.F. A'Hearn, C. Arpigny, M.R. Combi, P.D. Feldman, G.-P. Tozzi, N. Dello Russo, M.C. Festou, *Atomic Deuterium Emission and the D/H Ratio in Comets*. LPI Contributions, vol. 1405 (2008), p. 8216

M.K. Weisberg, H.C. Connolly, On the relationship between chondrites, comets, and asteroids, a petrologic perspective, in *Lunar Planet. Sci.*, vol. XXXIX (2008)

K. Willacy, A.C. Alexander, M. Ali-Dib, S.B. Ceccarelli Charnley, M. Doronin, Y. Ellinger, P. Gast, E. Gibb, S.N. Milam, O. Mousis, F. Pauzat, C. Tornow, E.S. Wirström, E. Zicler, Space Science Reviews (2015). this volume

E.S. Wirström, W.D. Geppert, Å. Hjalmarson, C.M. Persson, J.H. Black, P. Bergman, T.J. Millar, M. Hamberg, E. Vigren, Observational tests of interstellar methanol formation. Astron. Astrophys. **533**, 24–34 (2011)

E.S. Wirström, S.B. Charnley, M.A. Cordiner, S.N. Milam, Isotopic anomalies in primitive solar system matter: spin-state-dependent fractionation of nitrogen and deuterium in interstellar clouds. Astrophys. J. **757**, L11–L15 (2012)

D.H. Wooden, Cometary refractory grains: interstellar and nebular sources. Space Sci. Rev. **138**, 75–108 (2008)

P.M. Woods, K. Willacy, Carbon isotope fractionation in protoplanetary disks. Astrophys. J. **693**, 1360–1378 (2009)

A. Wootten, ALMA capabilities for observations of spectral line emission. Astrophys. Space Sci. **313**, 9–12 (2008)

I.P. Wright, S.J. Barber, G.H. Morgan, A.D. Morse, S. Sheridan, D.J. Andrews, J. Maynard, D. Yau, S.T. Evans, M.R. Leese, J.C. Zarnecki, B.J. Kent, N.R. Waltham, M.S. Whalley, S. Heys, D.L. Drummond, R.L. Edeson, E.C. Sawyer, R.F. Turner, C.T. Pillinger, Ptolemy—an instrument to measure stable isotopic ratios of key volatiles on a cometary nucleus. Space Sci. Rev. **128**, 363–381 (2007)

S. Wyckoff, P. Wehinger, Chemical abundance of comets, in *Reports of Planetary Astronomy* (1988), pp. 145–146

S. Wyckoff, E. Lindholm, P.A. Wehinger, B.A. Peterson, J.-M. Zucconi, M.C. Festou, The C-12/C-13 abundance ratio in comet Halley. Astrophys. J. **339**, 488–500 (1989)
S. Wyckoff, M. Kleine, B.A. Peterson, P.A. Wehinger, L.M. Ziurys, Carbon isotope abundances in comets. Astrophys. J. **535**, 991–999 (2000)
L. Yang, F.J. Ciesla, C.M.O. Alexander, The D/H ratio of water in the solar nebula during its formation and evolution. Icarus **226**, 256–267 (2013)
H. Yurimoto, K. Kuramoto, Molecular cloud origin for the oxygen isotope heterogeneity in the solar system. Science **305**, 1763–1766 (2004)
L.M. Ziurys, C. Savage, M.A. Brewster, A.J. Apponi, T.C. Pesch, S. Wyckoff, Cyanide chemistry in comet Hale-Bopp (C/1995 O1). Astrophys. J. **527**, L67–L71 (1999)

Space Sci Rev (2015) 197:85–99
DOI 10.1007/s11214-015-0160-0

The Subsurface Structure and Density of Cometary Nuclei

P.L. Lamy[1] · A. Herique[2] · I. Toth[1,3]

Received: 18 November 2014 / Accepted: 5 May 2015 / Published online: 22 May 2015

Abstract Little is known about the internal structure and density of cometary nuclei. Indirect evidences available so far are not compelling and these questions essentially remain a matter of speculation. It is therefore important to fully exploit the potential sources of information and this is particularly the case of radar observations which have the capability to probe the first few meters of cometary nuclei when they come sufficiently close to Earth. We review the available results and find that proper data are available for eight nuclei yielding their geometric radar albedo and the dielectric permittivity of their subsurface assuming that the scattering of the radar beam is predominantly specular. The range of permittivity is quite broad, extending from 1.7 to 3.1 although a more realistic interval is probably 2 to 3.1 implying pronounced diversity in the subsurface properties of cometary nuclei. A novel interpretation of these results is performed based on the calculation of the dielectric permittivity of various samples of three-phase mixtures of ice, dust and vacuum using two mixing formulas and on the introduction of ternary diagrams where the three axes correspond to the volumetric fraction of the three phases. The derived values of the permittivity supplemented by a general constraint on the dust-to-ice mass ratio define restricted regions in the ternary diagrams broadly imposing that the ice fraction lies in the range 0.1 to 0.2, the dust fraction in the range 0.2 to 0.5, and the porosity in the range 35 % to 75 %. The density of the subsurface of the considered eight nuclei is only constrained to the broad range 500 to 2000 $kg\,m^{-3}$ due to the poorly known density of the dust phase. However, the results unambiguously reveal considerable variation among cometary nuclei of the structure and properties of their subsurface layers.

✉ P.L. Lamy
philippe.lamy@lam.fr

1 Laboratoire d'Astrophysique de Marseille, UMR 7326, CNRS & Aix Marseille Université, Marseille, France

2 UJF-Grenoble 1 & CNRS-INSU, Institut de Planétologie et d'Astrophysique de Grenoble UMR 5274, Grenoble, France

3 Konkoly Observatory, MTA CSFK CSI, Budapest, Hungary

Keywords Comets: nucleus · Comets: nucleus surface · Radar: comets · Comets: individual: 2P/Encke, 8P/Tuttle, 26P/Grigg-Skjellerup, 67P/Churyumov-Gerasimenko, 103P/Hartley 2, C/1983 H1 (IRAS-Araki-Alcock), C/1983 J1 (Sugano-Saigusa-Fujikawa)

1 Introduction

Cometary nuclei are considered to be among the most primitive objects in the Solar System, and their properties are thought to preserve a record of the conditions in the early solar nebula when planetesimals formed, 4.5 billion years ago. Of particular interest is their internal structure which may give clues to the process of their accretion and their subsequent evolution in the solar system. This question has been a matter of debate dating back to the dispute between R. Lyttleton advocating the sand-bank model (Lyttleton 1948) and F. Whipple who opposed the "dirty snowball" model (Whipple 1950) which was finally confirmed by the radar observations of 2P/Encke and 26P/Grigg-Skjellerup, and later by the *VEGA* and Giotto missions. Whereas the existence of a monolithic body is now well established, Whipple's concept of "dirty snowball" has been challenged at least for some comets on the basis of their large refractory/volatile mass ratios. Based on their analysis of infrared observations of dust trails of eight short-period comets, Sykes and Walker (1992) found a mean ration of ≈ 3 and concluded that nuclei in general may be more like "frozen mudballs". A similar conclusion has been reached by Küppers et al. (2005) based on an analysis of the plume ejected from the nucleus of comet 9P/Tempel 1 by the Deep Impact experiment leading them to propose the term "icy dirtball". A large refractory/volatile mass ratio of 4 ± 1 has recently been determined for comet 67P/Churyumov-Gerasimenko (67P/C-G thereafter) based on the results of several instruments of the Rosetta mission (Rotundi et al. 2015).

Whatever the case, the detailed internal structure remains elusive and essentially a matter of speculation best illustrated by the well-known cartoon published by Weissman et al. (2004) which schematically illustrates four possible models: (i) Whipple's icy conglomerate model (Whipple 1950) as envisioned by Weissman and Kieffer (1981), (ii) the fractal aggregate model of Donn and Hughes (1985), (iii) the primordial rubble pile model of Weissman (1986), and (iv) the icy-glue model of Gombosi and Houpis (1986). A notable progress has been achieved by Belton et al. (2007) who proposed the "talps" or "layered pile" model prominently based on in situ imaging of the nucleus of comet 9P/Tempel 1 but also on those of comets 19P/Borrelly and 81P/Wild 2: on an ill-defined core, the growth of the nucleus proceeds by accumulation of successive layers or "talps" of increasing thickness and volume (see the cartoon in their Fig. 4). Analysis of the Rosetta/OSIRIS images do suggest ubiquitous layering on the surface of the nucleus of 67P/Churyumov-Gerasimenko with layers as thin as 3 m (Massironi et al. 2015).

Indirect evidences on their structure come from estimates of their bulk density. Lower limits can be obtained by considering their rotational periods and axial ratios, assuming that they are strengthless (i.e., that cometesimals are not physically bound) and that they are not rotating faster than the centrifugal limit for breakup. The results suggest that the fastest rotating nuclei are stable against centrifugal disruption if their bulk densities exceed $\sim 600\ \mathrm{kg\,m^{-3}}$ (Weissman et al. 2004; Lamy et al. 2004). Specific values can be calculated by separately estimating the mass of a nucleus (by modeling the non-gravitational force) and its volume; this led Rickman et al. (1987) to an average value $\lesssim 500\ \mathrm{kg\,m^{-3}}$ for about 30 ecliptic comets. The few cases of comets visited by space probes have not met early expectations and they generally led to loosely constrained intervals, for instance: 200–1500 $\mathrm{kg\,m^{-3}}$ for 1P/Halley (Rickman 1986, 1989; Sagdeev et al. 1988; Skorov and Rickman 1999),

100–830 kg m^{-3} for 19P/Borrelly (Farnham and Cochran 2002; Davidsson and Gutiérrez 2004), 600–800 kg m^{-3} for 81P/Wild 2 (Davidsson and Gutiérrez 2006), 200–700 kg m^{-3} for 9P/Tempel 1 (Davidsson et al. 2007), and 200–400 kg m^{-3} for 103P/Hartley 2 (Thomas et al. 2013). The Deep Impact experiment on comet 9P/Tempel 1 (A'Hearn 2008) unfortunately fell short of improving the picture. The investigators had hoped to track ballistically ejected boulders which eventually were not detected in the ejecta plume and they had to model the ejecta plume expansion as a "last-ditch" option to determine gravity resulting in a range of density of 200 to 1000 kg m^{-3} (Richardson et al. 2007). The recent in-situ observations of the nucleus of 67P/C-G cast a new light on this question. Combining its volume derived from a shape model constructed from images obtained with the OSIRIS narrow angle camera with its mass determined by the Radio Science Investigation (RSI) instrument, a bulk density of 470 ± 45 kg m^{-3} has been obtained (Sierks et al. 2015). This constitutes the very first robust determination of the density of a cometary nucleus. Whereas it excludes abnormally, not to say unrealistic low values (e.g., 100–300 kg m^{-3}) at least for this nucleus, it does imply significant porosity in the range 70–80 %.

Another line of evidence comes from the fragmentation of numerous nuclei and, in a few cases, of complete disintegration and disappearance (C/1999 S4 LINEAR, C/2011 W3 Lovejoy, and C/2012 S1 ISON for recent examples). In the specific case of comet D/1993 F1 Shoemaker-Levy 9 which tidally split under the gravitational influence of Jupiter, a bulk density of 250–700 kg m^{-3} was inferred as well as the size of some fragments (Asphaug and Benz 1996). But generally, this is not the case and we are left with only qualitative descriptions of the disruption (Boehnhardt 2004).

The CONSERT radio experiment (Kofman et al. 2007) on the Rosetta spacecraft is due to probe the interior of a nucleus for the first time and therefore raises considerable expectations, but it will leave open the question of whether the comets have the same internal structure. Any information that can be gathered on different nuclei are therefore of value to attempt characterizing the largest possible sample and thus detecting differences or similarities.

Following the pioneering work of Kamoun et al. (1982a,b) and Kamoun (1983), it has been realized that comets coming sufficiently close to Earth could produce detectable radar echoes from their nucleus (and also from their large coma grains). In addition to obtaining information simultaneously on its shape, size, and rotational state, the very fact that the radar beam penetrates inside the body allows characterizing the structure and density of its subsurface down to a few meters. The penetration depth very much depends upon the porosity of the soil being typically 10 wavelengths for a compact medium and exceeding 20 wavelengths for porosities of $\approx$65 % and larger (Campbell and Ulrichs 1969). For the Arecibo radar system at 2.38 GHz (12.6 cm) widely used for comet observations, this translates to a range of 1.3 to 2.6 m. Interpretation of the radar data in terms of subsurface properties has been traditionally limited to estimating the bulk density from the dielectric constant (or permittivity) and comparing it to "extreme" cases of dry snow and silicate powder (see Fig. 6 of Harmon et al. 2004). In a recent re-analysis of the 1982 radar observations of comet 67P/C-G, Kamoun et al. (2014) modeled the dielectric permittivity of porous mixtures of ice and silicate dust. Introducing a ternary diagram where the three axes correspond to the volumetric fractions of dust, ice, and vacuum, they showed that the observations delimit a narrow region of the diagram which in turn constrained the bulk density of the subsurface. Following a suggestion of the Reviewer of this article, we now extend this method of analysis to all cometary nuclei for which a radar cross-section has been measured so far.

This article is organized as follows. We briefly review the analysis of radar observations and present the results for cometary nuclei in Sect. 2. We next perform a complete analysis of the data and present our results in Sect. 3. We finally discuss the implications of our question of the density of the nuclei in Sect. 4 and then conclude.

2 Radar Observations of Cometary Nuclei

In a broad stroke, radar observations consist in transmitting a circularly polarized nearly monochromatic wave with a transmitted power P_t during a time approximately equal to the expected round-trip time of the radar signal and receiving the echo for nearly the same duration in the two (orthogonal) senses of circular polarization: opposite to that transmitted OC (for "opposite circular"), and same as transmitted SC (for "same circular"). The SC signal is not always available: first, the early observations had only OC detection and second, the SC signal may be too faint to be detected. The Doppler-only observations which are of interest here have been obtained with prominently the Arecibo S-band ($\lambda = 12.6$ cm) but also the Goldstone S-band ($\lambda = 12.9$ cm), and Goldstone X-band ($\lambda = 3.5$ cm) radar systems.

The data reduction and analysis procedures have been described in detail by Kamoun (1983) and summarized in Kamoun et al. (1982a, 1982b, see their Note No. 4) and in Harmon et al. (2004). The primary parameter derived from the radar echo is the radar cross-section σ of the target obtained by integrating the echo Doppler spectrum so as to retrieve the total received power P_r. These two quantities are linked by the radar equation:

$$P_r = \frac{P_t G^2 \lambda^2 \sigma}{(4\pi)^3 \Delta^4} \tag{1}$$

where G is the effective antenna gain, λ is the wavelength of the radar system, and Δ is the geocentric distance of the comet. Depending upon the case as mentioned above, either σ_{OC} or $\sigma = \sigma_{OC} + \sigma_{SC}$ are obtained. If the size of the nucleus is independently known, the geometric radar albedo is derived from the standard equation:

$$\hat{\sigma} = \sigma / A_p \tag{2}$$

where A_p is the apparent projected area of the nucleus. Since the radar beam penetrates the nucleus, the radar albedo is mainly sensitive to the bulk density of the top few meters

The analysis then considers the spectral distribution of the received echo in the different channels of the receiver for the two polarization states OC and SC. The bandwidth and shape of the spectrum are determined by the nucleus shape and orientation as well as by the angular scattering law of its surface. Generally, these parameters are unknown and the echo spectral shape is fitted to that of a sphere using a standard scattering law of the form $\cos^n \theta$ as a function of incidence angle θ. The echo spectrum is then simply expressed as $M\{1 - [2(f - f_0)/B]^2\}^{n/2}$, where M is the maximum amplitude of the echo, B is its total bandwidth, and f_0 represents the offset of the center of the spectrum from the prediction based on the ephemeris. A uniform scattering law corresponds to $n = 1$ and a Lambert law to $n = 2$.

If the scattering is predominantly specular, the square of the Fresnel coefficient at normal incidence ρ_0 is related to the radar albedo $\hat{\sigma}$ and the backscattering gain g via $\rho_0 \approx \hat{\sigma}/g$. For a $\cos^n \theta$ scattering law as already assumed in the geometrics optics approximation $g = (n + 2)/(n + 1)$. The real part of the dielectric permittivity ε is given by:

$$\varepsilon = \left(1 + \rho_0^{1/2}\right)^2 / \left(1 - \rho_0^{1/2}\right)^2. \tag{3}$$

Table 1 Radar observations of cometary nuclei

Comet	σ_{OC} (km^2)	σ	μ_c	R_{nuc} (km)	$\hat{\sigma}$	n	ε_1 ($n=1$)	ε_2 ($n=2$)	Ref.
2P/Encke (1980)	1.1 ± 0.7	–	–	2.4	0.06	–	2.2	2.4	[1, 2]
2P/Encke (2003)	–	1.02 ± 0.2	0.2	2.4	0.56	–	2.2	2.4	[3]
8P/Tuttle	–	4.5	–	3.1	0.15	0.6	–	–	[4]
26P/G-S	0.5 ± 0.13	–	<0.3	1.3	0.09	–	2.7	2.9	[5]
67P/C-G	≤ 0.7	–	–	2.1	≤ 0.05	1	≤ 2.1	–	[4, 6]
103P/Hartley 2	–	0.061	0.37	–	0.05	–	2.1	2.2	[7]
P/2005 JQ5	–	0.042 ± 0.009	0.49	0.7	0.027	1	1.7	–	[8]
C/IAA	2.14 ± 0.4	–	0.105	3.5	0.06	–	2.2	2.4	[9]
C/SSF	0.034 ± 0.008	–	0.23	0.37	0.10	–	2.9	3.1	[10]

Notes:
26P/G-S: 67P/Grigg-Skjellerup;
67P/C-G: 67P/Churyumov-Gerasimenko;
P/2005 JQ5: P/2005 JQ5 (Catalina);
C/IAA: C/1983 H1 (IRAS-Araki-Alcock);
C/SSF: C/1983 J1 (Sugano-Saigusa-Fujikawa);
σ_{OC}: geometric radar albedo for opposite circular polarization;
σ: total radar cross section;
μ_c: circular polarization ratio;
R_{nuc}: effective radius of the cometary nucleus;
$\hat{\sigma}$: geometric radar albedo;
n: power exponent of the scattering law;
ε_1, ε_2: permittivity respectively corresponding to $n = 1$ and to $n = 2$.

References:
1 = Kamoun et al. (1982b);
2 = Kamoun (1983);
3 = Harmon and Nolan (2005);
4 = Harmon et al. (2010);
5 = Kamoun et al. (1999);
6 = Kamoun et al. (2014);
7 = Harmon et al. (2011);
8 = Harmon et al. (2006);
9 = Harmon et al. (1989);
10 = Campbell et al. (1983).

Polarization properties of the radar echo are useful for characterizing the surface of the nucleus. From the OC echo, one mostly learns on the quasi-specular scattering properties of the surface since a single reflection reverses the sense of circular polarization. From the SC echo, one learns on the depolarization by wavelength-scale roughness or multiple scattering properties. When the two echos are available, they lead to two cross-sections σ_{OC} and σ_{SC} enabling to define the circular polarization ratio $\mu_c = \sigma_{SC}/\sigma_{OC}$ According to Harmon et al. (2004), μ_c lies in the range 0.1 to 0.5 for cometary nuclei (see also Table 1). At the low end, say for $\mu_c \leq 0.2$, the OC echo is significantly larger than the SC and the quasi-specular scattering predominates thus justifying the above derivation of the permittivity. However, when μ_c reaches values up to 0.5, depolarization becomes significant and so do non-specular scattering processes thus questioning the validity of the above derivation. There is no simple formalism to describe this situation so we retain the above expression of the permittivity keeping in mind that it may introduce some error.

Table 1 lists the radar albedos obtained so far and circular polarization ratios when available. Note that a few other nuclei have been detected through their radar echo but their unknown size precludes the derivation of their radar albedo. We briefly comment these results below.

- 2P/Encke is the only comet observed at two apparitions, 1980 (OC-only mode) and 2003 (dual mode). According to Harmon and Nolan (2005), their mean $\sigma_{OC} = 0.84 \pm 0.17$ km^2 obtained in 2003 is not significantly different from the 1980 value but may be partly explained by the different viewing geometries of an elongated body. Whatever the case, the impact on the permittivity is nil.
- 8P/Tuttle was further imaged by the delay-Doppler method enabling the direct determination of the projected area of its nucleus. It has by far the largest radar albedo of all listed nuclei. Note also the value of $n = 0.6$ indicating that the nucleus is limb-brightened, "a condition diagnostic of considerable blocky roughness at super-wavelength scales" according to Harmon et al. (2010).
- 26P/Grigg-Skjellerup was the first comet to be observed in dual-polarization reception but the SC echo was not detected.
- 67P/Churyumov-Gerasimenko was not detected at its 1982 apparition but the recent reanalysis of the data has led Kamoun et al. (2014) to constrain its permittivity to $\approx$2.
- 103P/Hartley 2 was also imaged by the delay-Doppler method but we favor the cross-sections based on the EPOXI flyby imaging (Thomas et al. 2013) of this highly elongated nucleus (henceforth, a radius is not quoted in Table 1). Then the radar albedo ranges from 0.045 to 0.054 and we retain the mean value of 0.05.
- P/2005 JQ5 (Catalina) was the first comet to be imaged by the delay-Doppler method enabling the direct determination of the projected area of its nucleus. Note the very large value of $\mu_c = 0.49$, by far the largest of the listed nuclei whereas the radar albedo has the smallest value of 0.027.
- C/1983 H1 (IRAS-Araki-Alcock) was also observed with the Goldstone S-band radar ($\lambda = 12.9$ cm) giving $\sigma_{OC} = 2.25$ km^2 in excellent agreement with the Arecibo result reported in Table 1, and with the Goldstone X-band radar ($\lambda = 3.5$ cm) giving $\sigma_{OC} = 4.44$ km^2 and $\mu_c = 0.25$ (Goldstein et al. 1984). This cross-section, twice the S-band value has been explained by the varying cross-section of an elongated nucleus possibly combined with wavelength dependence of the cometary material. We adopt an effective radius of 3.5 km that Groussin et al. (2010) obtained from their in-depth analysis and interpretation of all visible, infrared and radio observations.
- C/Sugano-Saigusa-Fujikawa (1983J1) has the smallest nucleus among those listed in Table 1 but properties (radar albedo, permittivity) similar to those of 26P/Grigg-Skjellerup, a surprising result since they probably come from different reservoirs.

C/1996 B2 (Hyakutake) was detected with the Goldstone X-band radar (Harmon et al. 1996; Harmon et al. 1997) giving $\sigma_{OC} = 0.13 \pm 0.03$ km^2 and an extremely large $\mu_c = 0.49 \pm 0.10$ but similar to that of P/2005 JQ5 (Catalina). The size of its nucleus poses a problem with conflicting results between the infrared observations giving a radius in the range 2.1 to 2.4 km (Sarmecanic et al. 1997; Fernández et al. 1996; Lisse et al. 1999) and the radio continuum observations giving an upper limit of 1.05 km (Altenhoff et al. 1999). The former result leads to an incredibly small radar albedo in the range 0.01 to 0.015 whereas the latter result implies a realistic lower limit of 0.06. In view of this situation, we have left this nucleus out of the list of Table 1.

Unless the exponent n is directly obtained from the observations, two values of the permittivity ε_1 and ε_2 are calculated corresponding to $n = 1$ (uniform scattering law) and $n = 2$

(Lambert law). The nucleus of P/2005 JQ5 (Catalina) has by far the lowest permittivity (1.7) but it may possibly be biased by its very large polarization ratio (0.49). Otherwise, the permittivity values lie in the range 2.1 to 3.1 which is quite broad and imply pronounced diversity in the subsurface properties of cometary nuclei.

3 Subsurface Structure of Cometary Nuclei

The analysis of the observational data and their interpretation in terms of the subsurface properties of cometary nuclei follow the method developed for the case of comet 67P/C-G (Kamoun et al. 2014). In a first approach, we separately consider the two main phases of cometary material, water-ice and refractory material. Laboratory measurements of dry snow (defined as a mixture of ice particles and air voids and henceforth assimilated to porous ice) by Hallikainen et al. (1986) and their compilation of previous works, and later by Matzler (1996) show that the real part of its dielectric permittivity is independent of temperature and frequency in the microwave region and that it is a function of snow density only. Figure 4 of Hallikainen et al. (1986) indicates that the range of permittivity of 1.7 to 3.1 that we found for the eight cometary nuclei corresponds to a range of density of $\approx$350 $kg\,m^{-3}$ to 900 $kg\,m^{-3}$, hence a range of porosity of $\approx$60 % to 0. Campbell and Ulrichs (1969) measured the dielectric properties of a large variety of rocks (36 samples) and mineral powders (25 samples) at two wavelengths, 0.45 and 35 GHz. Granting that the former is close to that of the radar observation and that the values of the permittivity vary only slightly between the above two wavelengths, their results at 0.45 GHz are adequate to our analysis. The permittivity of *porous* rocks such as pumice and volcanic ash shale exceeds 2.4 whereas it exceed 5.0 for all *solid* rocks which are therefore ruled out for the observed nuclei. For powdered rocks, their measurements indicate that the permittivity is prominently a function of powder density with a weak dependence on temperature and no dependence on particle size distribution or compressive history of the powder. For instance, at a density of 1000 $kg\,m^{-3}$, the permittivity of their 25 powders falls in the range 1.9–2.1. This density already implies a large porosity in the range 65 to 70 % based on the bulk densities of minerals expected to be present in cometary nuclei (see discussion in the next section). Another interesting result concerns the penetration depth: according to their Fig. 14, it exceeds 20 wavelengths at 0.45 GHz for most of their samples (in particular olivine powders) at a density of 1000 $kg\,m^{-3}$. This confirms that the Arecibo radar beam probably penetrated the nuclei to depths of the order of 2.5 m.

Whereas a layer of pure porous ice is excluded at the surface of cometary nuclei based on all observational evidences, the case of a $\approx$2.5 m thick layer of regolith is conceivable (and possible for dead comets) but unlikely for active comets as ice is expected to be closer to the surface on both observational and physical grounds. Indeed, the Deep Impact observations of 9P/Tempel 1 suggest that a surface layer of thickness less than 1 m is depleted in ice (A'Hearn 2008) and the model of cometry dust mantle formation of Rosenberg and Prialnik (2009) indicates that its thickness varies from 1 cm to about 10 cm, much less than the expected penetration depth of the radar beam. We therefore assume that the top 2.5 m thick layer is an homogeneous and porous three-phase mixture of various proportions of ice, dust and vacuum whose effective (or average) dielectric permittivity is calculated using two mixing formulas: the classical Maxwell-Garnet mixing law and the more general Looyenga mixing law [see the review of Sihvola (2004) for detail]. We consider all compositions of the surface material bracketed by the above two laws to limit model sensitivity. The ice permittivity is taken equal to 3.2 corresponding to hexagonal water ice, the only form of ice which can be present at or close to the surface for thermodynamical reasons. A first dust permittivity of ≈5.8 is obtained from the laboratory measurements of two meteoritic samples: LEW 85320 L5-type-meteorite and MET 01260 LL5-type-meteorite (Heggy et al. 2012). A second, somewhat larger permittivity of 7.0 corresponds to the equivalent permittivity of a mixture of 40 % serpentine, 41.5 % montmorillonite, 5.8 % olivine, 8.8 % magnetite and 3.9 % pyrrhotite, representative of the mineralogical composition of the Orgueil C-type meteorite and calculated by Beck and Buttarazzi (private communication). These two values of 5.8 and 7.0 are taken as mean values for the whole surface/subsurface averaging possible thermal effects.

The results are presented on ternary diagrams (Figs. 1, 2, 3) where the three axes corresponds to the volumetric fractions of dust, ice, and vacuum. Rather than considering the specific values of the permittivity listed in Table 1 and in order to be more general to help interpreting future observations, we present a set of diagrams corresponding to values ranging from 1.5 to 4.0 with a step of 0.5. Visual interpolation between two diagrams for intermediate values is ample for the present characterization, especially in view of the various approximations of our procedure. On a given diagram, the two continuous lines bound a first region imposed by the selected permittivity using a dust permittivity of 5.8; the upper line corresponds to the Maxwell-Garnet law and the lower one to the Looyenga law. The two broken lines bound a second region with the same conditions as above except that it corresponds to a dust permittivity of 7.0. We can see that this latter parameter has very little influence (<5 %), less than the mixing laws, in the case of large porosities, but that their influences become comparable as the porosity decreases.

An additional limitation on the ternary diagrams results from the constraint on the dust-to-ice mass ratio coming from our present understanding of the composition of cometary nuclei. Based on infrared and millimeter observations that are able to detect the largest dust grains, on the analysis of the excavated material by the impact on the nucleus of comet 9P/Tempel 1 to a depth of about 15 m, and the recent determination for comet 67P/C-G, the dust-to-ice ratio cannot be lower than 2.5 to 3.5 consistent with the concept of "frozen mudball" and "icy dirtball" presented in the introduction. Adopting a realistic range for the density of the dust phase (see next section), the above condition simply imposes that the dust-to-ice volumetric ratio be larger than ≈1 and eliminates half of the ternary diagrams (light gray regions) bounded by the vertical dashed-dotted lines. The allowed region (where the calculated permittivity is equal to the selected value) consequently shrinks to the dark gray quadrilateral in the case of a dust permittivity equal to 5.8; the corresponding quadrilateral for a dust permittivity equal to 7.0 is not highlighted but can be easily visualized as bounded by the two dashed lines. Varying the dust-to-gas mass ratio would cause the vertical dashed-dotted line to rotate around the top summit of the ternary diagram but in a very restricted angular range unless extreme values are considered. Therefore the allowed region is not very sensitive to this parameter.

To summarize the implications of the permittivity for the porosity and ice and dust fractions of the nuclei (respectively f_{ice} and f_{dust}), we regroup them in five different classes and consider the "barycenter" of the allowed regions to obtain typical values of the fractions and of the porosity. With the lowest permittivity of 1.7, the nucleus of P/2005 JQ5 (Catalina) has consequently a very large porosity, typically 75 % and ice and dust fractions of respectively 0.1 and 0.2. However, it also has the largest polarization ratio (0.49) which may bias these values as noted above. We next have a group of nuclei with permittivity of 2.1 to 2.2 for which a typical solution corresponds to a porosity of 60 %, $f_{\text{ice}} = 0.1$ and $f_{\text{dust}} = 0.3$. Nuclei with a permittivity of 2.4 typically have a porosity of 50 %, $f_{\text{ice}} = 0.15$ and $f_{\text{dust}} = 0.35$. A permittivity of 2.7 leads to a typically porosity of 40 %, $f_{\text{ice}} = 0.15$ and $f_{\text{dust}} = 0.45$. Finally, nuclei with permittivity of 2.9 to 3.1 may be typically characterized by a porosity of

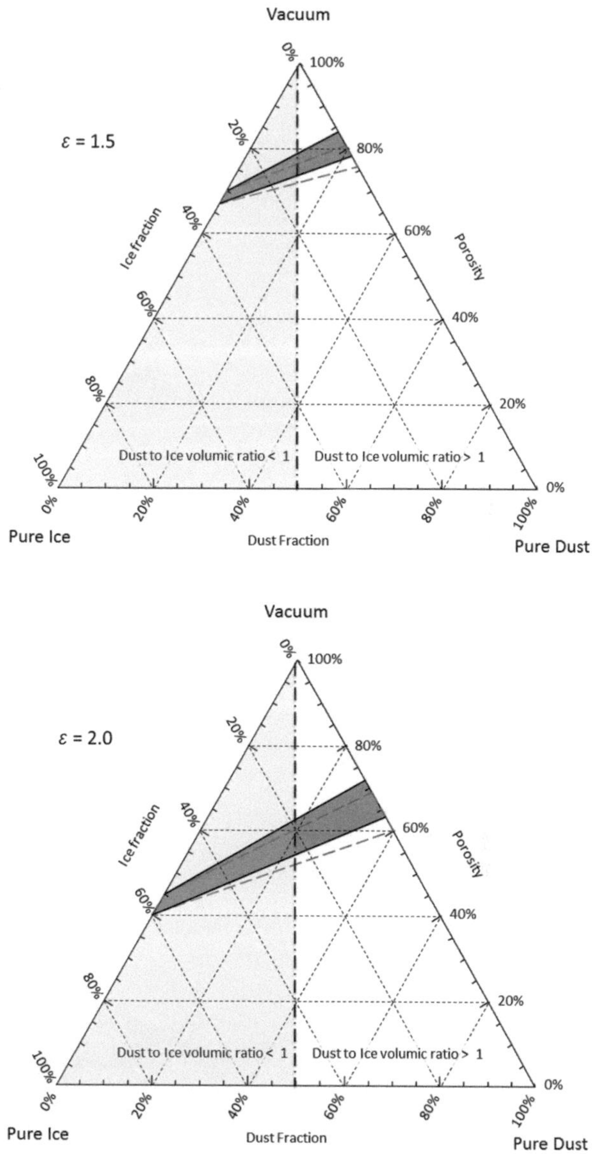

Fig. 1 Ternary diagrams for two values of the dielectric permittivity $\varepsilon = 1.5$ (*top*) and $\varepsilon = 2.0$ (*bottom*)

35 % and $f_{\text{ice}} = 0.2$ and $f_{\text{dust}} = 0.5$. Altogether, the results available so far imply a considerable range of porosity, broadly 35 % to 75 %, ice fractions in the narrow range of 0.1 to 0.2 and dust fractions in the range of 0.2 to 0.5. Note that all above fractions are volumetric.

4 Implications for the Density of the Nuclei

We can estimate the bulk density of the nucleus subsurface layer probed by the radar beam from our results coming from the ternary diagrams. This requires specifying the densities of the two phases, ice and dust. Hexagonal water ice has a bulk density of 920 $\text{kg}\,\text{m}^{-3}$

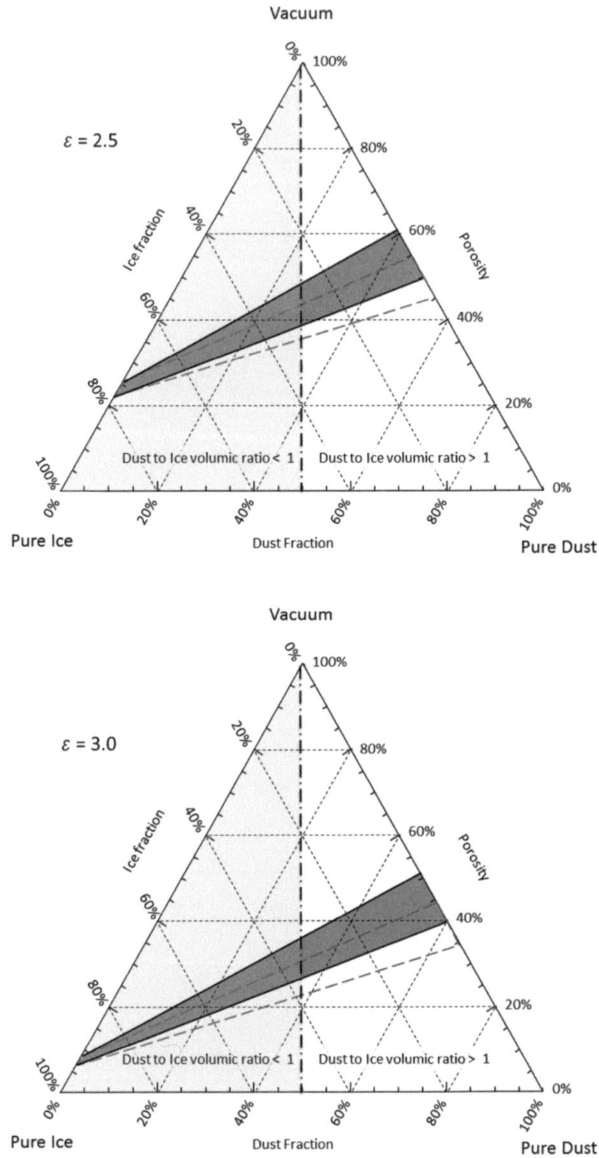

Fig. 2 Ternary diagrams for two values of the dielectric permittivity $\varepsilon = 2.5$ (*top*) and $\varepsilon = 3.0$ (*bottom*)

but the question is more complex for the dust phase and has been discussed in detail by Kamoun et al. (2014) when they analyzed the case of 67P/C-G. They considered various interplanetary matters which can potentially be associated to comets and whose density can be measured or reliably inferred:

1. Primitive meteorites of petrologic type 3: ordinary chondrites and CV carbonaceous chondrites.
2. Chemically primitive meteorites based on the abundance of the non-volatile elements similar to the Sun: CI carbonaceous chondrites.
3. Meteoroids with JFC-like orbits and with Halley-type orbits.

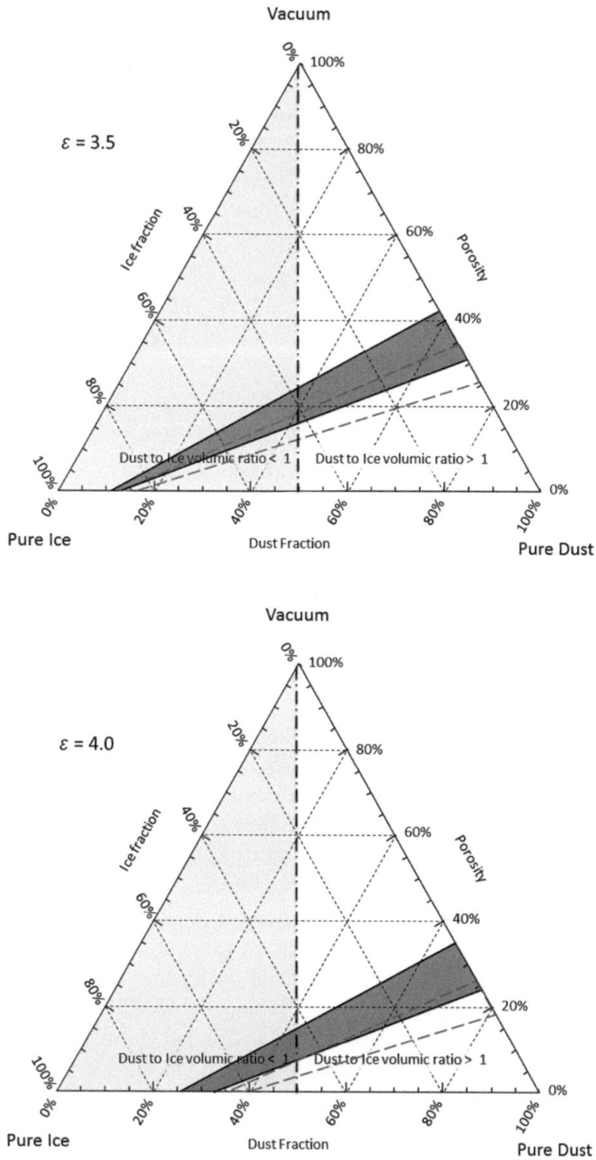

Fig. 3 Ternary diagrams for two values of the dielectric permittivity $\varepsilon = 3.5$ (*top*) and $\varepsilon = 4.0$ (*bottom*)

They concluded that a range of grain density from 2000 to 3500 kg m^{-3} widely encompasses the different possibilities summarized above and they accordingly considered four values, 2000, 2500, 3000 and 3500 kg m^{-3}. It is to be noted that the grain density is the relevant quantity since the porosity is inherently part of our modeling as evidenced by the ternary diagram.

The corresponding bulk densities for the five classes of permittivity defined in the previous section and the above four values of the grain density are given in Table 2 where values have been rounded for simplicity. Altogether, they range from ≈500 to ≈1900 kg m^{-3}, that is a factor of almost 4, but part of this variation stems from the grain density. For example, at a constant density of 2000 kg m^{-3}, the range is only ≈500 to ≈1200 kg m^{-3}. If we further

Table 2 Density of the subsurface of cometary nuclei for different values of their permittivity and for different grain densities (see text for detail)

Permittivity	1.7	2.1–2.2	2.4	2.7	2.0–3.1
f_{ice}	0.1	0.1	0.15	0.15	0.2
f_{dust}	0.2	0.3	0.35	0.45	0.5
Porosity (%)	75	60	50	40	35
Density ($\rho_{dust} = 2000$)	500	700	800	1000	1200
Density ($\rho_{dust} = 2500$)	600	800	1000	1300	1400
Density ($\rho_{dust} = 3000$)	700	1000	1200	1500	1700
Density ($\rho_{dust} = 3500$)	800	1100	1400	1700	1900

f_{ice}, f_{dust}: volumetric fractions of ice and dust.

ρ_{dust}: density of the dust phase ($kg\,m^{-3}$).

Density $= f_{ice} \times 920 + f_{dust} \times \rho_{dust}$ ($kg\,m^{-3}$).

consider that the lowest permittivity of 1.7 is biased by the large circular polarization ratio, the above range shrinks to $\approx$700 to $\approx$1200 $kg\,m^{-3}$. In any case, these results show considerable variation of the structure and properties of the subsurface layers among cometary nuclei. We note that they are consistent with the upper limit of 1600 $kg\,m^{-3}$ for the surface layers of the nucleus of 9P/Tempel 1 derived by A'Hearn (2008) if we except the two extreme cases.

Whether these results obtained for the top $\approx$2.5 m layer may be extended to the overall nuclei raises the question of their internal structure and homogeneity, a matter still widely open and subject to intense debates and speculations. In a broad stroke, our density values fall in the range of the present determinations but this is not very compelling since these ranges are quite large; they however do not support densities as low as 100 to 400 $kg\,m^{-3}$. In their review of the structure and density of cometary nuclei for the Comets II book, Weissman et al. (2004) found a range of 500 to 1200 $kg\,m^{-3}$—well compatible with our result—"with a perhaps best current value of 0.6 $g\,cm^{-3}$" (600 $kg\,m^{-3}$).

Let us consider the case of the nucleus of 67P/C-G for which a robust determination of 470 ± 45 $kg\,m^{-3}$ now exists (Sierks et al. 2015). In their re-analysis of the 1982 radar observations, Kamoun et al. (2014) concluded that the density of the top $\approx$2.5 m layer lies in the range of $\approx$600 to $\approx$1000 $kg\,m^{-3}$. On the one hand, there is sufficient margin on the parameters to consider that there is an agreement between the bulk density and the lowest value of the density of the subsurface layer; for instance, a density of the dust phase of 1250 $kg\,m^{-3}$ (instead of the considered minimum value of 2000 $kg\,m^{-3}$) would bring the two densities in agreement. On the other hand, it is entirely possible that the difference in densities reflects a true physical property of the subsurface layer. In the framework of the "talps" or "layered pile" model of Belton et al. (2007) and depending upon several parameters but prominently the dust-to-ice ratio in the different layers, it is conceivable that the density decreases from the surface inward so that the resulting bulk density could be lower than the subsurface density.

Another crucial contribution of the OSIRIS images is what probably constitutes the first direct glimpse into the interior of a cometary nucleus. The 900 m-high Hathor "cliff" most likely reveals a cut through the nucleus. It is pervaded by regular layering with lineaments and terraces intimately associated. Perpendicular lineaments are also conspicuous and interpreted as fractures (Massironi et al. 2015). Visual inspection of the best resolved images do not indicate any major inhomogeneity. This is also the case of the ≈200 m deep cylindrical pits as seen from their walls. They however reveal the presence of ≈3 m size pebbles whose origin, perhaps primordial, is presently unclear. Finally, Sierks et al. (2015)'s remark that "the agreement between the axis of rotation and the axis of maximum moment of inertia reinforces the concept of weakly bonded icy dust aggregates, with porosity at small scales relative to the size of the nucleus (rather than large voids)" provides some ground to our model.

5 Conclusion

We reconsidered and re-interpreted past radar observations of eight cometary nuclei for which the size were known, to constrain the dielectric permittivity of their subsurface layer down to typically 2.5 m. We found a broad range of values, from 1.7 to 3.1 with the lowest value possibly biased by non-specular scattering so that a more robust interval is probably 2 to 3.1 implying a large diversity of the subsurface properties of cometary nuclei. Based on laboratory measurements and our modeling of porous mixtures of ice and dust, the permittivity allows constraining the volumetric fractions in the mixtures. The results imply a large range of porosity, broadly 35 % to 75 %, ice fractions in the narrow range of 0.1 to 0.2 and dust fractions in the range of 0.2 to 0.5. The density of the subsurface of the considered eight nuclei is only constrained to the broad range 500 to 2000 $\mathrm{kg\,m^{-3}}$ due to the poorly known density of the dust phase. However, the results unambiguously reveal considerable variation among cometary nuclei of the structure and properties of their subsurface layers. The case of the nucleus of 67P/C-G was examined in the light of the recent determination of its bulk density from in-situ measurements of its volume and its mass. We concluded that an homogeneous nucleus is possible if the density of the dust phase is of the order of 1250 $\mathrm{kg\,m^{-3}}$. Alternatively and for larger density of the dust phase, different porosities and bulk densities of the subsurface layer and the interior are conceivable.

Acknowledgements We are grateful to P. Beck and I. Buttarazzi (IPAG/UJF Grenoble) for their mineralogical and dielectric modeling of the Orgueil meteorite.

References

M.F. A'Hearn, Deep impact and the origin and evolution of cometary nuclei. Space Sci. Rev. **138**, 237–246 (2008)

W.J. Altenhoff, J.H. Bieging, B. Butler, H.M. Butner, R. Chini et al., Coordinated radio continuum observations of comets Hyakutake and Hale-Bopp from 22 to 860 GHz. Astron. Astrophys. **348**, 1020–1034 (1999)

E. Asphaug, W. Benz, Size, density, and structure of Comet Shoemaker-Levy 9 inferred from the physics of tidal breakup. Icarus **121**, 225–248 (1996)

M.J.S. Belton, P. Thomas, J. Veverka, P. Schultz, M.F. A'Hearn et al., The internal structure of Jupiter family cometary nuclei from deep impact observations: The "talps" or "layered pile" model. Icarus **187**, 332–344 (2007)

H. Boehnhardt, Split comets, in *Comets II*, ed. by M.C. Festou, H.U. Keller, H.A. Weaver (University of Arizona Press, Tucson, 2004), pp. 301–316

M.J. Campbell, J. Ulrichs, Electrical properties of rocks and their significance for lunar radar observations. J. Geophys. Res. B, Solid Earth Planets **74**, 5867–5881 (1969)

D.B. Campbell, J.K. Harmon, A.A. Hine, I.I. Shapiro, B.G. Marsden, G.H. Pettengill, Arecibo radar observations of comets IRAS-Araki-Alcock and Sugano-Saigusa-Fujikawa (Invited). Bull. Am. Astron. Soc. **15**, 800 (1983)

B.J.R. Davidsson, P.J. Gutiérrez, Estimating the nucleus density of Comet 19P/Borrelly. Icarus **168**, 392–408 (2004)
B.J.R. Davidsson, P.J. Gutiérrez, Non-gravitational force modeling of Comet 81P/Wild 2. I. A nucleus bulk density estimate. Icarus **180**, 224–242 (2006)
B.J.R. Davidsson, P.J. Gutiérrez, H. Rickman, Nucleus properties of Comet 9P/Tempel 1 estimated from non-gravitational force modeling. Icarus **191**, 547–561 (2007)
B. Donn, D.W. Hughes, Characteristics and implications of a comet nucleus formed by random grain accretion. Bull. Am. Astron. Soc. **17**, 689 (1985)
T.L. Farnham, A.L. Cochran, A McDonald observatory study of Comet 19P/Borrelly: Placing the deep space 1 observations into a broader context. Icarus **160**, 398–418 (2002)
Y.R. Fernández, C.M. Lisse, A. Kundu, M.F. A'Hearn, W.F. Hoffmann, A. Dayal, The nucleus of Comet Hyakutake (C/1996 B2). Bull. Am. Astron. Soc. **28**, 1088 (1996)
R.M. Goldstein, R.F. Jurgens, Z. Sekanina, A radar study of Comet IRAS-Araki-Alcock 1983d. Astron. J. **89**, 1745–1754 (1984)
T.I. Gombosi, H.L.F. Houpis, An icy-glue model of cometary nuclei. Nature **324**, 43–44 (1986)
O. Groussin, P.L. Lamy, L. Jorda, The nucleus of Comet C/1983 H1 IRAS-Araki-Alcock. Planet. Space Sci. **58**, 904–912 (2010)
M.T. Hallikainen, F.T. Ulaby, M. Abdelrazik, Dielectric properties of snow in the 3 to 37 GHz range. IEEE Trans. Antennas Propag. **AP–34**, 1329 (1986)
J.K. Harmon, M.C. Nolan, Radar observations of Comet 2P/Encke during the 2003 apparition. Icarus **176**, 175–183 (2005)
J.K. Harmon, D.B. Campbell, A.A. Hine, I.I. Shapiro, B.G. Marsden, Radar observations of Comet IRAS-Araki-Alcock 1983d. Astrophys. J. **338**, 1071–1093 (1989)
J.K. Harmon, S.J. Ostro, K.D. Rosema, R.F. Jurgens, R. Winkler et al., Radar echoes from Comet Hyakutake (C/1996 B2). Bull. Am. Astron. Soc. **28**, 1195 (1996)
J.K. Harmon, S.J. Ostro, L.A.M. Benner, K.D. Rosema, R.F. Jurgens et al., Radar detection of the nucleus and coma of Comet Hyakutake (C/1996 B2). Science **278**, 1921–1924 (1997)
J.K. Harmon, M.C. Nolan, S.J. Ostro, D.B. Campbell, Radar studies of comet nuclei and grain comae, in *Comets II*, ed. by M.C. Festou, H.U. Keller, H.A. Weaver (University of Arizona Press, Tucson, 2004), pp. 265–279
J.K. Harmon, M.C. Nolan, J.-L. Margot, D.B. Campbell, L.A.M. Benner, J.D. Giorgini, Radar observations of Comet P/2005 JQ5 (Catalina). Icarus **184**, 285–288 (2006)
J.K. Harmon, M.C. Nolan, J.D. Giorgini, E.S. Ellen, Radar observations of 8P/Tuttle: A contact-binary comet. Icarus **207**, 499–502 (2010)
J.K. Harmon, M.C. Nolan, E.S. Howell, J.D. Giorgini, P.A. Taylor, Radar observations of Comet 103P/Hartley 2. Astrophys. J. Lett. **734**, L2, 4 pp. (2011)
E. Heggy, E.M. Palmer, W. Kofman, S.M. Clifford, K. Righter, A. Herique, Radar properties of comets: Parametric dielectric modeling of Comet 67P/Churyumov-Gerasimenko. Icarus **221**, 925–939 (2012)
P.G. Kamoun, Radar observations of cometary nuclei. Ph.D. thesis, Massachusetts Institute of Technology (MIT) (1983)
P.G. Kamoun, D.B. Campbell, S.J. Ostro, G.H. Pettengill, I.I. Shapiro, Comet Encke—Radar detection of nucleus. Science **216**, 293–295 (1982a)
P.G. Kamoun, G.H. Pettengill, I.I. Shapiro, Radar detectability of comets, in *Comets*, ed. by L.L. Wilkening (University of Arizona Press, Tucson, 1982b), pp. 288–296
P. Kamoun, D. Campbell, G. Pettengill, I. Shapiro, Radar observations of three comets and detection of echoes from one: P/Grigg-Skjellerup. Planet. Space Sci. **47**, 23–28 (1999)
P. Kamoun, P.L. Lamy, I. Toth, A. Herique, Constraints on the subsurface structure and density of the nucleus of Comet 67P/Churyumov-Gerasimenko from Arecibo radar observations. Astron. Astrophys. **568**, A21, 8 pp. (2014)
W. Kofman, A. Herique, J.-P. Goutail, T. Hagfors, I.P. Williams et al., The Comet Nucleus Sounding Experiment by Radiowave Transmission (CONSERT): A short description of the instrument and of the commissioning stages. Space Sci. Rev. **128**, 413–432 (2007)
M. Küppers, I. Bertini, S. Fornasier, P.J. Gutierrez, S.F. Hviid et al., A large dust/ice ratio in the nucleus of Comet 9P/Tempel 1. Nature **437**, 987–990 (2005)
P.L. Lamy, I. Toth, Y.R. Fernández, H.A. Weaver, The sizes, shapes, albedos, and colors of cometary nuclei, in *Comets II*, ed. by M.C. Festou, H.U. Keller, H.A. Weaver (University of Arizona Press, Tucson, 2004), pp. 223–224
C.M. Lisse, Y.R. Fernández, A. Kundu, M.F. A'Hearn, A. Dayal, L.K. Deutch, G.G. Fazio, J.L. Hora, W.F. Hoffmann, The nucleus of Comet Hyakutake (C/1996 B2). Icarus **140**, 189–204 (1999)
R.A. Lyttleton, On the origin of comets. Mon. Not. R. Astron. Soc. **108**, 465–475 (1948)

M. Massironi, E. Simioni, F. Marzari, G. Cremonese, L. Giacomini et al., The two independent and primitive envelopes of the bilobate nucleus of Comet 67P/Churyumov-Gerasimenko. Nature (2015, in press)
C. Matzler, Microwave permittivity of dry snow. IEEE Trans. Geosci. Remote Sens. **34**, 573–581 (1996)
J.E. Richardson, H. Melosh, C.M. Lisse, C. Brian, A ballistics analysis of the deep impact ejecta plume: Determining Comet Tempel 1's gravity, mass, and density. Icarus **190**, 357–390 (2007)
H. Rickman, Masses and densities of comets Halley and Kopff, in *Proc. of the Workshop on the Comet Nucleus Sample Return Mission*, ed. by O. Meilta. ESA SP-249, Noordwijk, The Netherlands (1986), pp. 195–205
H. Rickman, The nucleus of Comet Halley—Surface structure, mean density, gas and dust production. Adv. Space Res. **9**(3), 59–71 (1989)
H. Rickman, L. Kamél, M.C. Festou, Cl. Froeschlé, Estimates of masses, volumes and densities of short-period comet nuclei, in *ESA Proceedings of the International Symposium on the Diversity and Similarity of Comets*, ed. by E.J. Rolfe, B. Battrick. ESA SP-278, Noordwijk, The Netherlands (1987), pp. 471–481
E.D. Rosenberg, D. Prialnik, Fully 3-dimensional calculations of dust mantle formation for a model of Comet 67P/Churyumov-Gerasimenko. Icarus **201**, 740–749 (2009)
A. Rotundi, H. Sierks, V. Della Corte, M. Fulle, P. Gutiérrez, L. Lara et al., Dust measurements in the coma of Comet 67P/Churyumov-Gerasimenko inbound to the Sun. Science **347**(6220), Special edition: Rosetta, id aaa-3905 6 pp. (2015)
R.Z. Sagdeev, P.E. Elyasberg, V.I. Moroz, Is the nucleus of Comet Halley a low density body? Nature **331**, 240–242 (1988)
J. Sarmecanic, M. Fomenkova, B. Jones, T. Lavezzi, Constraints on the nucleus and dust properties from mid-infrared imaging of Comet Hyakutake. Astrophys. J. **483**, L69–L72 (1997)
H. Sierks, C. Barbieri, P.L. Lamy, R. Rodrigo, D. Koschny et al., On the nucleus structure and activity of Comet 67P/Churyumov-Gerasimenko. Science **347**(6220), Special edition: Rosetta, id aaa-1044, 5 pp. (2015)
A. Sihvola, *Electromagnetic Mixing Formulas and Applications* (The Institution of Electrical Engineers, London, 2004), p. 242
Y.V. Skorov, H. Rickman, Gas flow and dust acceleration in a cometary Knudsen layer. Planet. Space Sci. **47**, 935–949 (1999)
M.V. Sykes, R.G. Walker, The nature of comet nuclei, in *In Lunar and Planetary Inst., Asteroids, Comets, Meteors 1991* (1992), pp. 587–591
P.C. Thomas, M.F. A'Hearn, J. Veverka, M.J.S. Belton, J. Kissel et al., Shape, density, and geology of the nucleus of Comet 103P/Hartley 2. Icarus **222**, 550–558 (2013)
P.R. Weissman, Are cometary nuclei primordial rubble piles? Nature **320**, 242–244 (1986)
P.R. Weissman, H.H. Kieffer, Thermal modeling of cometary nuclei. Icarus **47**, 302–311 (1981)
P.R. Weissman, E. Asphaug, S.C. Lowry, Structure and density of cometary nuclei, in *Comets II*, ed. by M.C. Festou, H.U. Keller, H.A. Weaver (University of Arizona Press, Tucson, 2004), pp. 337–357
F.L. Whipple, A comet model, I. The acceleration of Comet Encke. Astrophys. J. **111**, 375–394 (1950)

Space Sci Rev (2015) 197:101–150
DOI 10.1007/s11214-015-0192-5

Laboratory Studies Towards Understanding Comets

Murthy S. Gudipati[1,2] · Ninette Abou Mrad[3] · Jürgen Blum[4] · Steven B. Charnley[5] · Thierry Chiavassa[3] · Martin A. Cordiner[5] · Olivier Mousis[6] · Grégoire Danger[3] · Fabrice Duvernay[3] · Bastian Gundlach[4] · Paul Hartogh[7] · Ulysse Marboeuf[8] · Irakli Simonia[9] · Tsitsino Simonia[9] · Patrice Theulé[3] · Rui Yang[10,11]

Received: 14 November 2014 / Accepted: 27 July 2015 / Published online: 1 September 2015

Abstract This review presents some of the recent advancements in our understanding of comets facilitated by laboratory studies, need for new laboratory simulations, and predictions for future explorations. With the spacecraft Rosetta at the comet 67P/Churyumov–Gerasimenko, a large volume of science data is expected to follow early results that have been published recently. The most surprising of them being hard ice shell that bounced the lander Philae a couple of times before settling on the comet. Long evaded molecular nitrogen has now been detected in the comet 67P/CG. The observed density of 470 $kg\,m^{-3}$ is in line with other comet observations, whereas the nature and composition of hydrocarbons

✉ M.S. Gudipati
gudipati@jpl.nasa.gov

1 Science Division, Jet Propulsion Laboratory, California Institute of Technology, 4800 Oak Grove Drive, Pasadena, CA 91109, USA

2 IPST, University of Maryland at College Park, College Park, MD 20742, USA

3 Aix-Marseille Université, PIIM UMR-CNRS 7345, 13397, Marseille, France

4 Institut für Geophysik und extraterrestrische Physik, Technische Universität Braunschweig, Mendelssohnstr. 3, 38106 Braunschweig, Germany

5 Astrochemistry Laboratory, Code 691, NASA Goddard Space Flight Center, Greenbelt, MD 20771, USA

6 Aix-Marseille Université, CNRS, LAM (Laboratoire d'Astrophysique de Marseille) UMR 7326, 13388, Marseille, France

7 Max Planck Institute for Solar System Research, Justus-von-Liebig-Weg 3, 37077 Göttingen, Germany

8 Physics Institute and Center for Space and Habitability, University of Bern, Sidlerstrasse 5, CH-3012 Bern, Switzerland

9 Laboratory for Interdisciplinary Investigation in Cosmochemistry and Astronomy at the School of Natural Sciences and Engineering, Ilia State University, Cholokashvili str., 3/5, Tbilisi 0162, Georgia

10 Former Postdoctoral Fellow at the University of Maryland, College Park, MD 20742, USA

11 Present address: Chemistry Department, Fudan University, 20 Handan Road, Shanghai 200433, China

detected on the surface are still a puzzle. Observation of D/H ratio that deviates significantly from Earth's water D/H ratio brings back to the table the long-standing question whether or not water on Earth was delivered by comet impacts. Our review summarizes some of the critical laboratory work that helps improve our understanding of cometary interior (whether amorphous or crystalline or containing clathrates), cometary surface (rich of complex organics), cometary coma and tail (D/H ratio, negative ions, and photoluminescence). Outstanding questions are also discussed.

Keywords Comets · Laboratory · Experimental · Ice · Volatiles · Clathrates · Origin of life · Organics · Review · Isotope ratios · Negative ions · Coma · Outgassing · Origin of solar systems

1 Introduction

There are excellent reviews (A'Hearn 2011; Bockelée-Morvan 2010; Lisse et al. 2013; Mumma and Charnley 2011) and books (Festou et al. 2004) or book chapters on comets, their anatomy, physics, chemistry, and dynamics. More popular books on comets are also emerging (Eicher 2013) to reach out to the curiosity of majority of the humans, who inevitably would have seen or will see a comet in their lifetime. In brief, comets are icy bodies with all sizes (kilometers diameter/length) and shapes, highly porous, mostly with dark surface and have highly eccentric orbit around the Sun. When comets get closer to the Sun, their surface temperature increases resulting in outgassing of volatiles, flow of jets containing molecules, ions, and dust/ice particles, all of which become a comet's coma and subsequently tail of a comet that spans for hundreds of thousands of kilometers. Comets originate from either the Kuiper Belt or Oort cloud. In either case, comets are supposed to contain and preserve the most primitive (primordial) ice, dust, and organics from interstellar medium that would have gravitationally collapsed into a dense molecular cloud and subsequently to a protostar, whose ice-line kept the KBOs and Oort cloud objects unaltered in their interior (Irvine et al. 1996; Wooden 2008). Thus, the connection between comets and interstellar ice grains is critical to understand. Also important is the role of comets (and asteroids) in the origin of life on Earth, as the earliest organisms seem to have emerged about 3.8 Gyr ago (Fedo et al. 2006), soon after the late heavy bombardment (Bottke et al. 2012; Broz et al. 2013; Gomes et al. 2005) around 4.1–3.8 Gyr ago.

Molecular and atomic composition of cometary coma/tail is relatively well studied and understood (Lisse et al. 2005; Moreno et al. 2003; Mumma and Charnley 2011), but their origin—whether outgassing or photochemical or both still needs clarification. Nature of charged species in cometary coma needs a better understanding as well. Surfaces of short-period Jupiter family comets do not show permanent/significant water-ice presence, whereas refrosting could occur on some parts and can be seen at large incidence angles at the morning side as patches of water-ice that is not intimately mixed with the surface dust (Groussin et al. 2013). Cometary surface is thus expected to be composed of non-volatile refractory organics and silicate grains. Thickness of this surface crust is an unknown still and it is not clear how deep is water-ice buried (centimeters to meters). Several models discuss the cometary interior composition and thermal properties (Davidsson et al. 2013; Prialnik 2002), some involving amorphous ice grains and others ignoring the water-ice phase (Gortsas et al. 2011; Kuhrt 1999). There is a general conscience that comets are highly porous (porosity >50 %), low density (~ 0.3–$0.6\ \mathrm{g/cm^3}$), and poor thermal conductors with low thermal inertia of ~ 10–$50\ \mathrm{J\,K^{-1}\,m^{-2}\,s^{-0.5}}$ (Gulkis et al. 2015). It is not clear whether the interior of a comet is composed of small grains (micron-size) or a wide distribution of grains, rubble, and blocks. It is not clear whether the interstellar ice grain composition (Greenberg 1963), containing large amount of volatiles trapped in amorphous ice, is retained and preserved in a comet's nucleus. Outgassing of a comet typically begins close to 3 au (where the daytime temperature starts raising beyond 180 K). Carbon monoxide is seen early in the formation of a cometary coma—so called "lighting up the comet" during a comet's journey towards perihelion. Is this CO from the subsurface, trapped within amorphous ice or trapped in the grain boundaries of loosely bound crystalline ice grains, or photochemically made from the surface refractory organics? What is the chemical composition of organics in cometary tail—charged and neutral? How this composition connected to surface and interior composition of a comet? These are only a few samples of a suit of questions for which laboratory studies (in collaboration with modeling and observational studies) would help find answers. In this review article we try to bring as many of these questions/answers/predictions/postulates as possible by focusing on the laboratory studies to address the four parts of a comet: interior, surface, outgassing, and coma. The review is also structured into the same four parts. First we discuss briefly the outstanding questions for which laboratory studies are urgently needed and then we present some of the studies that helped resolve ongoing and past observations.

2 Motivation for Laboratory Studies

The pool of solar system icy bodies (Kuiper Belt Objects and Oort Cloud), where comets are known to originate from, are thought to be the least processed presolar material left behind. Comets are exposed to extreme variations in temperature and solar radiation; as a consequence, a comets surface, coma, and tail are intimately connected to its interior make up, which is a challenge even today to understand. A comparison of interstellar ice grain composition (in the solid phase) and a comet's volatile coma composition (outgassing) is shown in Fig. 1. The data is normalized to water abundance. This bar plot helps visualize the fact that many species in cometary outgassing are very similar in abundance as compared to the interstellar ices, indicating that the composition of volatiles is preserved. Cometary coma shows far more oxidized species that have not (yet) been identified or detected in interstellar ice grains. We should keep in mind that the only spectroscopic method so far used to study interstellar ice grains is the infrared absorption in the mid-IR region, where many species could have overlapping IR absorptions (unresolved) or their absorption could be weak and buried under strongly absorbing molecular species. On the other hand, in cometary coma/tail, each molecule is observed through a variety of spectroscopic techniques (infrared, microwave/far infrared, ultraviolet, etc.), and in the gas-phase these molecules show much sharper absorption profiles, avoiding overlapping bands. At this juncture it is not clear whether the interstellar ice grains lack the diversity of oxidation products seen in cometary outgassing or whether it is merely a limitation of the observational methods. Laboratory studies presented in Sect. 3 deal with complex chemistry that could occur in presolar conditions and properties of volatiles trapped in amorphous ice vs. crystalline ice containing clathrates of volatiles in a comet's interior.

Thanks to recent missions to comets, we have unprecedented volume of data on the surface topography and geology of comets. While most of the cometary surface material is suggested to be organic crust and dust mix—resulting in a low albedo material, the exact composition of a cometary surface, its thickness, its thermal conductivity, its gas permeability, and its photochemical activity—are still far from well understood. Microwave instrument for the Rosetta orbiter (MIRO) of Rosetta mission has shown that the thermal inertia of

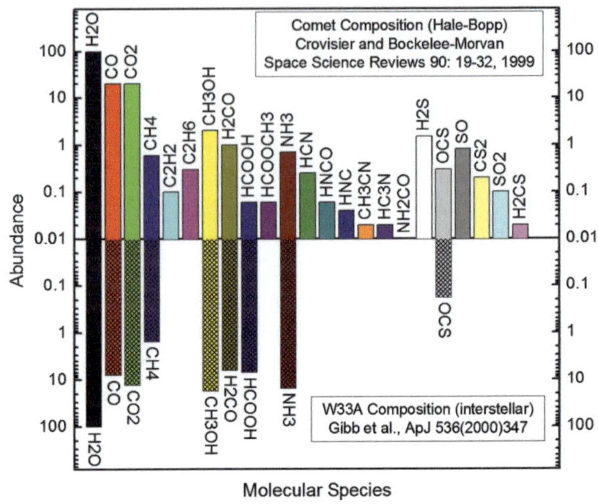

Fig. 1 Molecular abundances of comet outgassing (Hale-Bopp) compared with interstellar ice composition (W33A). A striking similarity among the major volatile species can be seen, indicating that interstellar ice grain composition is preserved to a greater extent in comets

comet 67P/CG is very low and within a few centimeters the temperature could significantly fall several tens of K (Gulkis et al. 2015), consistent with a thermally insulating powdered surface. Laboratory experiments are gaining momentum in simulating the surfaces, how the surface reacts to extreme temperature changes, radiation, and volatiles from the interior are some of the few examples of what needs to be addressed in the lab. Along these lines, recent studies (Lignell and Gudipati 2015) subjecting cometary ice analogs to thermal cycles revealed that phase separation occurs between amorphous ice that crystallizes, expelling the impurity organics that form the outer crust. These laboratory studies are in agreement with recent Rosetta observations that hard crust (likely crystalline ice) is covered with organic dust. Some of the laboratory studies may have to look for observations/modeling to find answers. Understanding the physical properties of cometary analogues provides the possibility to better constrain the conditions under which comets have formed. For example, the water-driven dust activity of comets can only be explained if comets were to be formed by gravitational instability (see Sect. 4.1). Rosetta's VIRTIS (visible and infrared thermal imaging spectrometer) has clearly shown that a comets surface is rich with organics. We do not yet exactly know what kind of organics these are and whether they are intimately mixed with silicate grains and what ratio these two materials have at different heliocentric distances of a comet, these are all unknowns.

Section 4 deals with radiation-induced chemical evolution of simple organic volatiles shown in Fig. 1. Production of complex non-volatile organics as well as potential biologically relevant prebiotic organics is observed under laboratory conditions. In the laboratory, irradiation by UV photons or charged particles followed by thermal heating of astrophysical ice analogues that contain simple molecules such as H_2O, CH_3OH, NH_3 and CO_2 have produced a suite of complex organic molecules similar to those found in small bodies of our Solar system. When UV photons penetrate into ices, they break bonds, producing radicals and activated species, which during warm up lead to the formation of photoproducts. Diffusion and reaction barriers control this radiation-induced chemistry in cometary and astrophysical ices. As a result a suit of complex organics are synthesized on cometary surface and near surface, of which some of them volatile and others refractory (organic compounds with a larger desorption energy than water). Refractory residue analyses have shown an important molecular diversity with interesting similarities with the soluble or insoluble organic matter found in carbonaceous meteorites. Investigation of the formation mechanisms of the most abundant molecules present in refractory residues is an important challenge that can drive the search for these species in cometary environments. Accordingly a mechanistic approach of cometary reactivity has been developed on polyoxymethylene (POM) and hexamethyltetramine (HMT) two molecular compounds detected in the organic residue that will be searched for by the Rosetta instruments. These experiments give constraint on cometary environment in which they can be search for as well as on the possibility to detect them by instruments on-board of the Rosetta mission. Characterization of the volatile species is also important, since they certainly take a part in the subsequent formation of these refractory residues, and give a screening of molecules desorbing during a comet heating.

Section 5 is devoted to volatiles outgassed during perihelion approach of a comet, which are the most studied due to their access for ground and space based observatories. As a comet approaches its perihelion, its surface temperature increases resulting in sublimation of atoms and molecules from its surface (non directional). This sublimation forms the coma of a comet and subsequently evolves into comet's tail. As the temperature increases with decreasing distance to the Sun, jets containing both gas and dust/ice grains are routinely observed. Very few laboratory studies are available that focus on the jets of a comet (Laufer et al. 2013), where as extensive work is focused on understanding the coma and tail—through the outgassing process.

One of these studies involves D/H ratios of cometary water in order to connect it with presolar ice grains and whether or not comets delivered water on Earth during the early stages of our solar system. The formation of water on dust grains in the interstellar medium at very low temperatures favors heavier isotopes. Very high deuterium enrichments in protostellar envelopes with D/H ratios of about 0.1 % and 1 % were reported by Butner et al. (2007) and Ceccarelli et al. (2005) respectively. This is a factor 50 to 500 higher than the D/H ratio of protosolar hydrogen (21 ppm). Isotopic exchange processes in the inner protoplanetary disk under temperatures between 600 and 1300 K and pressures between 10 μbar and 1 mbar (see e.g. Yang et al. 2013) would lead to a defractionation of highly deuterium-enriched water to the hydrogen D/H ratio of 21 ppm ("equilibrated water"). With increasing heliocentric distance the efficiency of the defractionation process decreases on the one hand, and mixing processes with highly deuterated water vapor lead to an increase of the D/H ratio on the other hand. The degree of the D/H ratio increase with heliocentric distance in models (e.g. Drouart et al. 1999; Mousis et al. 2000; Horner et al. 2007 and Kavelaars et al. 2011) can be constrained by observations of the D/H ratio in comets. In the "classical" view, the Oort cloud comets (OCCs) have formed in the vicinity of the gas giants while Jupiter family comets (JFCs) at larger heliocentric distances in the Kuiper belt and scattered disk. Therefore the models predict higher D/H ratios for JFCs than for OCCs. Observations in OCCs (Balsiger et al. 1995; Eberhardt et al. 1995; Biver et al. 2006; Bockelée-Morvan et al. 1998 and 2012; Meier et al. 1998; Hutsemékers et al. 2008; Villanueva et al. 2009) reveal D/H ratios between 206 and 409 ppm, while recent observations in JFCs show values between 161 ppm (Hartogh et al. 2011) and below 200 ppm (Lis et al. 2013), i.e. the opposite of the model predictions. Based on these new constraints, recent models addressing the topic of the D/H distribution in the solar system suggest a possible overlap of the OCC and JFC formation region (Albertsson et al. 2014) or that the defractionation process happened already while the disk was still forming. In the latter case transport of defractionated water from the inner disk outwards would compensate angular momentum from still infalling material (Yang et al. 2013). The discrepancy between prediction and observation may also be interpreted in terms of the Grand Tack scenario (Walsh et al. 2011) along with the migration of the gas giants and related mixing of material over large heliocentric distances or the interpretation of Hogerheijde et al. (2011)

based on measurements of the ortho-to-para ratio of water in the nearest classical T-Tauri stellar system that comets contain heterogeneous ice mixtures collected across the entire solar nebula during the early stages of planetary birth. If true one should expect a large range of D/H ratios in each single comet and depending of its formation history related to processes of the disk formation and evolution a more or less distinct variability of the D/H ratio in cometary comas during their presence in the inner solar system. Unfortunately all existing D/H observations in comet have been performed only during a single orbital phase so that no constraints about this hypothesis exist yet. This situation will change with the on-going Rosetta mission where water D/H ratios in the JFC 67P/Churyumov-Gerasimenko will be determined continuously over heliocentric distances of at least between 1.3 to 3.7 AU. First results from ROSINA (*Rosetta* Orbiter Spectrometer for Ion and Neutral Analysis) instrument measurements on 67P/CG outgassing put the earlier theories that water on Earth has similar D/H ratios as the cometary bodies to question (Altwegg et al. 2015). As discussed in Sect. 5, more systematic and rigorous laboratory studies are needed that help understand the evolution of D/H ratios with thermal and photochemical processing connecting gas-phase D/H ratios with solid-phase composition.

Section 6 is devoted to the understanding of molecular composition of cometary coma and tail. Ionic species that form jets in a comet gas tail, especially the negative ions, that is directed away from the Sun, are somewhat less understood (Meech et al. 2005; Mumma and Charnley 2011) compared to the neutral gas and dust in a cometary tail. Negative ions could help understand the production mechanism of photoelectrons, their energies, and atomic and molecular composition of the comet itself. Finally, some of the outgassing atoms and molecules are highly fluorescent such as the polycyclic aromatic hydrocarbons (PAHs), some of them observed in cometary tails. Molecular fluorescence excited by solar radiation could contribute to understand the composition of comet coma and tail, connecting this composition back to the surface and interior composition of a comet.

3 Comet Interior

The nucleus of a comet is the invisible elephant in the room. Results from the Comet Nucleus Sounding Experiment by Radiowave Transmission (CONSERT) instrument of Rosetta mission to comet 67P/CG would soon shed some new light on the interior or a comet (Hegler et al. 2013). We have a reasonable constrains on the density of a comet (0.2–0.6 $\mathrm{g\,cm^{-3}}$), meaning it is a highly porous water-ice and organic-volatile rich body mixed with silicate/mineral dust grains. Rosetta mission has determined (Sierks et al. 2015) the density of the comet 67P/CG to be 470 $\mathrm{kg\,cm^{-3}}$, well within the above-mentioned range. However, we have very little data on the interior structure, whether it is uniform throughout or whether it consists of rock-like high-density boulders or fresh-snow like ultra-low-density regions or a mix of gradients in between. We explore in this section some of the laboratory work pertinent to a comet's nucleus. We presently have reasonable understanding of ices and volatile mixes, but studies are only emerging on the inclusion of dust (silicate/mineral grains) in the mixtures. One of the outstanding issues of cometary nucleus is the phase of water ice in the core—amorphous or crystalline (Sect. 3.1)? It is likely that the crust of a comet would have crystalline ice due to its exposure to higher temperatures than $\sim$130 K for a prolonged period of time. In the crystalline ice regions of a comet clathrate hydrates become important (Sect. 3.2). One should not confuse crystalline clathrates with amorphous ices with trapped volatiles, which are called "gas-laden amorphous ices".

3.1 Is Present-Day Comet's Interior Made of Amorphous or Crystalline Ice?

While interstellar ice grains are predominantly made of amorphous ice (Öberg et al. 2011), most of the solar system icy surfaces are detected to be made of crystalline ice, including some of the KBOs (Brown et al. 2012a; Licandro et al. 2006). One of the key questions pertinent to comets is: What is the ice composition of a comet's nucleus interior: amorphous or crystalline? If amorphous, then why most of the KBOs have crystalline ice surface? If crystalline (interior), when did the phase-transition from amorphous to crystalline occur—during the protostar phase, or during the past 4.5 Gyr evolution? Can volatiles be trapped in the crystalline ice grains—such as clathrates—to the same extent as the amorphous ices, which are known to trap large amount of volatiles (Laufer et al. 2013; Laufer et al. 1987)? We will discuss here pros and cons for a cometary interior to be amorphous or crystalline by comparing the existing experimental and observational evidence. Eventually only a cryogenic (<50 K) comet subsurface sample analysis and return (CCSSAR—pronounced as Caesar) would answer the ultimate question: Do the comets carry the primordial interstellar ice grains in their pristine form?

Gudipati and coworkers recently developed a 2-color laser ablation and laser ionization mass spectrometry (2C-LAIMS) method, whereby ice was resonantly excited in the OH vibrations in ice at 2.94 microns to efficiently ablate the ice into the gas-phase, ionize the gas-phase atoms, molecules and clusters using a UV laser, and analyze the species with a time-of-flight mass spectrometer, in order to determine the chemical composition of the ice at any given temperature without having to heat the sample (Gudipati and Yang 2012; Henderson and Gudipati 2014, 2015; Yang and Gudipati 2014). This method, based on the pioneering work by Zare and Clemett on laser desorption mass spectrometry (Clemett et al. 1991), complements the other in situ spectroscopic techniques such as infrared, Raman, UV, and fluorescence (Allodi et al. 2013). As shown in Fig. 2, these studies clearly showed that under radiation (UV or electron) organic impurities in ices undergo complex chemistry even under interstellar temperatures ($\sim$5 K), which means it is very likely that the interstellar ice grains may host far more diverse molecular species than so far observed either in the dense molecular cloud phase or during the protostar stage, without annealing the ices to crystalline phase. These complex organics could have been preserved further during the formation of KBO and Oort Cloud icy bodies. Similar work is being pursued by several other researchers using laser-desorption mass spectrometry methods (Bennett et al. 2011; Maity et al. 2015).

One may quickly conclude: because the interstellar ice grains are made of amorphous ices, because the amorphous ices can accommodate large amount of impurities, and because cometary outgassing volatiles are strikingly similar to interstellar ice grain composition, cometary nucleus should consist of amorphous ice grains and the primordial interstellar composition should have been preserved in cometary nucleus. However, there are some modeling studies that do not consider amorphous form to be necessary (Gortsas et al. 2011; Kuhrt 1999) and comet interior consisting of volatiles trapped in crystalline ice clathrates are proposed (Marboeuf et al. 2012a, 2012b; Mousis et al. 2015) challenging the amorphous ice interior hypothesis.

Most of the comet nucleus thermal models do not ask the question whether the interior ice is amorphous or crystalline, rather these models use thermal conductivity as a proxy for porosity—whether amorphous or crystalline in nature. Thermal conductivity of amorphous ice (with or without impurities) is not constrained in the laboratory (Andersson and Inaba 2005; Andersson and Suga 2002; Bar-Nun and Laufer 2003; English et al. 2010; Handa et al. 1986; Kouchi et al. 1992; Waluyo et al. 2008), whereas these values are well constrained for

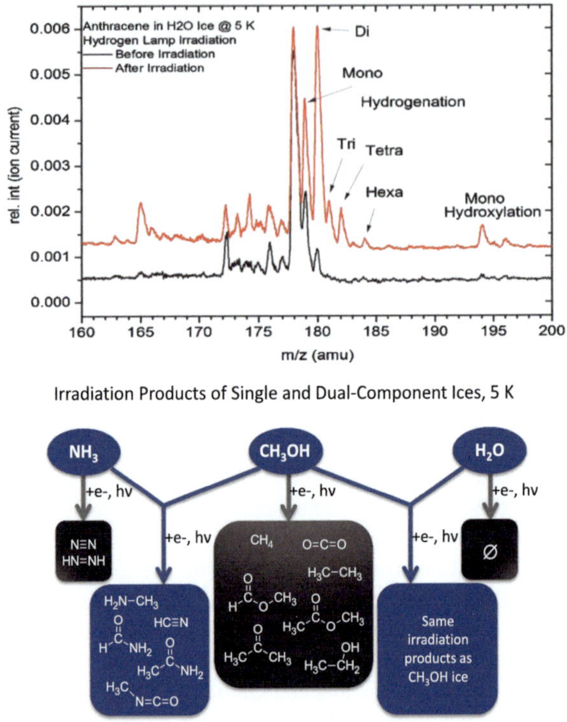

Fig. 2 *Top*: Two-color, two-step laser and ablation (laser) ionization mass spectra (2S-LAIMS) of a polycyclic aromatic hydrocarbon (PAH) anthracene ($C_{14}H_{10}$) in H_2O ice at 5 K before (*black curve*) and after (*red curve*) Lyman-α UV radiation processing at 5 K. Mass peaks due to hydrogenation (up to 8) and hydroxylation (one) can clearly be seen in the mass spectrum measured after radiation processing at 5K indicating complex radiation chemistry involving organics and water matrix can occur even in the coldest parts of the Universe, including interstellar ice grains that eventually transform to outer solar system and cometary nucleus. Reproduced partly from (Gudipati and Yang 2012). *Bottom*: Complex organics, including some key prebiotic molecules produced from radiation processed interstellar and cometary ice analogs investigated using the 2S-LAIMS methodology. Reproduced from Henderson and Gudipati (2015)

crystalline ice at cryogenic temperatures (Andersson and Inaba 2005; Hansen and McCord 2004; Kim et al. 2003; Smith et al. 2003). Thermal conductivity, porosity, grain size, dust (silicate) grain composition, and volatile fraction—all these parameters are correlated to each other in modeling a cometary interior thermal inertia. One of these models proposes that crystalline ice properties could be used to compute thermal properties of a comet as well as its outgassing behavior (Gortsas et al. 2011), which puts a fly in the ointment. Is this possible that cometary nucleus could be composed of crystalline ice and still behave very similar in each orbit from aphelion to perihelion and back—with respect to outgassing volatiles (with similar number density)? Or is it only that in the upper crust of a comet (a few cm to a few meters—to be constrained) ice becomes crystalline, leaving the interior to be amorphous? A short period comet's aphelion is ~5 au (Jupiter), leaving the minimum surface temperature to be ~120 K in the daytime. Could this be sufficient to crystallize the interior ice of a ~1 km diameter comet? These are the questions for which we still lack credible experimental data and theoretical models (that rely heavily on experimental data).

Yet another important aspect of amorphous or crystalline ice is the laboratory work on pure water-ice may not always represent water-ice heavily loaded with trapped species. In fact, both in comets and interstellar ice grains though water alone is the most abundant molecule, when the abundances of all the organic impurities are added together, then water to total impurity ratio is almost 1:1 (see Fig. 1). So, cometary interior must have a mechanism to trap extremely high quantity of organic volatile impurity, and still maintain a high porosity to get overall density to be ~ 0.2–$0.4\ \mathrm{g\,cm^{-3}}$ (Thomas et al. 2013). Laboratory studies show that while pure amorphous water ice undergoes exothermic (releasing heat) crystallization at higher temperatures, crystallization of amorphous ice with impurities trapped is an endothermic (takes heat) process (Kouchi and Sirono 2001). Though quantitative data on heat of crystallization of impurity-laden amorphous ices still need to be obtained in the laboratory, if true and if the magnitude of endothermic heat consumption to be significant, spontaneous break-up of comets can no longer be attributed to runaway crystallization process. Further, it is necessary that at one or other point in the history of a comet's life, there should have been sufficient heat deposited into the interior with very poor thermal conductivity, in order to make the amorphous to crystalline transition—adding another barrier to spontaneous crystallization that has so far been assumed in the literature. Implications of this endothermic crystallization of amorphous ices would include retention of primordial interstellar ice composition in the interior and much closer to the surface than so far thought.

More laboratory work on physical properties such as thermal conductivity, heat of crystallization, and thermal and photochemical processes in amorphous ices containing impurities is critical to constrain our understanding of a comet's interior. At the same time more laboratory data is needed on crystalline ice generated from amorphous ice annealing or directly made at higher temperatures, its volatiles trapping abilities. Another mechanism for trapping volatiles is in the form of clathrates and hydrates. Some work is done in the laboratory on high-pressure generated clathrates and their relevance to astrophysical and cometary ices (Dartois et al. 2013; Dartois and Deboffle 2008). Laboratory work on low-temperature and low-pressure clathrates (molecules trapped in crystalline ice) would lead us to a better understanding of the role of crystalline ices under low-gravity, low-pressure, and low-temperature conditions, as discussed in the next section.

3.2 Role of Clathrates: Formation, Equilibrium Conditions, and Composition of Clathrate

Clathrate hydrates (hereafter clathrates) are ice-like crystalline solids forming from water and non-stochiometric amounts of small non-polar molecules under appropriate conditions of temperature and pressure. Water molecules form cages or cavities, which may, or may not, contain guest molecules. To be stable, clathrate structures require that a minimum amount of guest molecules be trapped into the cages without forming any covalent or hydrogen bonds to the water molecules. The most common guest molecules encaged in terrestrial clathrates are of organic aliphatic nature like methane, ethane or propane. However other small inorganic molecules like carbon dioxide and nitrogen can also be encaged in clathrates. Clathrates are considered today as one of the most important reservoirs of fossil energy on Earth (Sloan and Koh 2008). The two most common clathrate structures known in natural environments are "structure I" and "structure II". In structure I, the unit cell consists of 46 water molecules and can trap up to 8 guest molecules. In structure II, the unit cell is composed of 136 water molecules and can incorporate 24 guest molecules at most.

Of particular interest is the clathrate structure that can incorporate volatile molecules from the surrounding gas phase in the water ice structure during the course of the cooling of the protosolar nebula. This structure of water ice can cause chemical fractionation of the

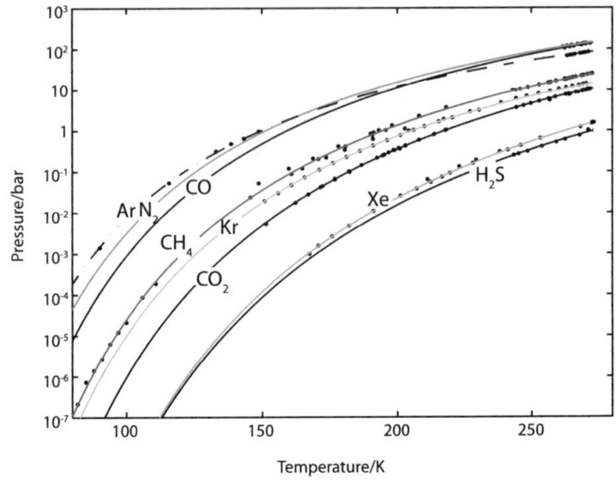

Fig. 3 Equilibrium pressure of single guest clathrates of CO, CO_2, CH_4, H_2S, N_2, Xe, Kr, and Ar. Dots are experimental data and *solid lines* are fits using Eq. (1). See Fray et al. (2010), Marboeuf et al. (2012a), Mohammadi and Anderson (2010, 2011, 2013) for further details on the experimental data

trace gases, depending on various parameters such as ambient temperature, pressure, and chemistry. Here, we discuss the formation of clathrates at low pressure ($\ll$1 bar) and low temperature ($\ll$270 K) under laboratory conditions. We also review the accessible data representing the equilibrium conditions of species that have significant relevance for cometary environments. Experiments showing the fractionation of gaseous mixtures once trapped in clathrates are also depicted. Because of the scarce number of experimental data regarding the equilibrium and composition of clathrates formed at low temperature and pressure conditions, we finally discuss the assumptions made in models computing the formation conditions and composition of icy planetesimals condensed in the protosolar nebula.

3.2.1 Review of Existing Experiments

Clathrates are composed of a cubic structure of water ice that forms cages within which guest molecules, initially in the gas phase, are trapped by Van der Waals forces. The formation/dissociation of clathrates at the surface of icy grains in the protosolar nebula or in comets occurs by gas-water ice interaction. The equilibrium pressure of guest molecules defines the stability of this structure. At a given temperature, the cages are stable as long as the pressure of the gas phase is equal to or greater than the dissociation pressure of the clathrate structure. Below the equilibrium gas pressure of the guest molecule, the cages become unstable and dissociate, leaving only the water ice and volatile molecules in the gas phase. Numerous experimental studies determined the equilibrium pressure of single guest clathrate hydrates for several molecules (see Fig. 3). The clathrate equilibrium conditions are usually depicted via a Arrhenius-type law:

$$\ln\bigl(P^{cl}(x)\bigr) = A_0(x) + A_1(x)/T, \tag{1}$$

where $P^{cl}(x)$ is the equilibrium pressure of the guest molecule x expressed in bar and T in K. $A_0(x)$ and $A_1(x)$ are constant parameters that depend on the nature of the guest molecule x (see values in Hersant et al. 2004; Fray et al. 2010; Marboeuf et al. 2012a and references therein).

Among the large list of clathrate forming volatiles, a dozen of guest species are of particular interest in planetary science: CO, CO_2, CH_4, CH_3OH, N_2, NH_3, H_2S, Xe, Kr, and Ar. Except N_2 and Xe, Kr and Ar, these molecules are the most abundant volatile species

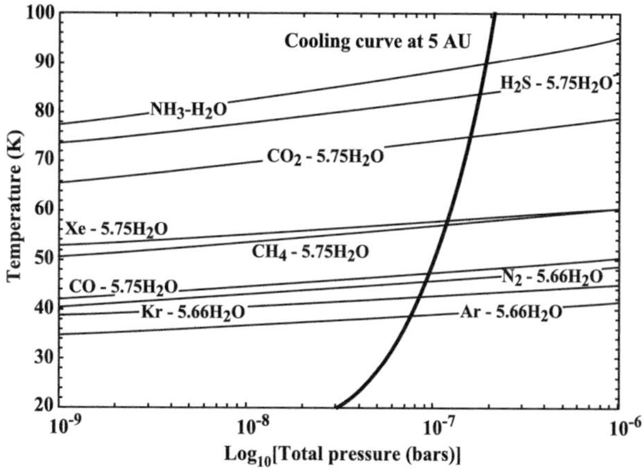

Fig. 4 Equilibrium curves of clathrates (*horizontal lines*) together with an evolutionary track of the protosolar nebula at the heliocentric distance of 5 AU (adapted from Mousis and Alibert 2006). This evolutionary track corresponds to the thermodynamic pathway followed by the disk during its cooling at the considered distance (the evolution of the disk proceeds from high to low temperatures). Abundances of various elements are solar

observed in comets and in ISM (Gibb et al. 2000, 2004; Bockelée-Morvan et al. 2004; van Dishoeck 2004; Mumma and Charnley 2011). The equilibriums of single guest clathrates of CH_4, CO_2, H_2S, Xe and Kr are fairly well constrained at low temperature (mainly in the 150–270 K range) and low pressure (>1 mbar most of the time; see data in Fray et al. 2010; Marboeuf et al. 2012a, and Mohammadi and Richon 2013). However, for the highly volatile species CO, N_2, and Ar, the experimental data are very scarce and mainly at high temperature (>260 K) (Holder et al. 1980; Mohammadi and Anderson 2010, 2011), expected for the single guest clathrate of Ar for which data exist below 150 K (Barrer and Edge 1967). Meanwhile, as discussed in Fray et al. (2010) and Marboeuf et al. (2012a), experimental data for Xe and Kr clathrates from Barrer and Edge (1967) present a progressive deviation compared to other measurements, indicating that their data are not reliable. Hence, clathrates of highly volatile species CO, N_2 and Ar are not well constrained in thermodynamic conditions close to those met in comets and in the protosolar nebula (see Fig. 4). Yet, such data are very important to understand the fractionation of these species during the formation and evolution of ices in the protosolar nebula (Iro et al. 2003; Mousis et al. 2009b, 2012a, 2012b, 2014), comets, icy moons and dwarf and terrestrial planets (Marboeuf et al. 2012b; Mousis et al. 2002, 2009a, 2011, 2013a, 2013b; Thomas et al. 2007, 2008). It is important to note that no experimental data concerning the stability curve of the CH_3OH clathrate has been reported in the literature and the conditions under which it forms stoichiometric hydrates or clathrate (Blake et al. 1991; Notesco and Bar-Nun 2000) are still unclear. Moreover, NH_3 is known to form rather stoichiometric hemihydrates ($2NH_3$–H_2O) and/or monohydrates (NH_3–H_2O) under low-pressure conditions (Lewis 1972; Bertie and Refaat Shehata 1984; Lunine and Stevenson 1987; Moore et al. 2007). This molecule has also been assigned a role of anti-freeze for water ice and clathrate formation, modifying the stability region of the solid ice and methane clathrate hydrate phases as a thermodynamic inhibitor (Choukroun et al. 2006, 2010; Fortes and Choukroun 2010; Shin et al. 2012). However, one study has shown that ammonia can form clathrates and participate synergistically in the formation of the cages in the presence of methane gas at low temperature (Shin et al. 2012), but the conditions under which clathrates are formed remain unclear and no experimental data allowing us to determine the conditions of formation of the NH_3 clathrate has been reported in the literature.

Of particular interest is the fractionation of volatile species during the formation of clathrates in presence of a gas mixture composed of several different volatile molecules. The presence of several volatile species in the gas phase can generate the formation of a multiple

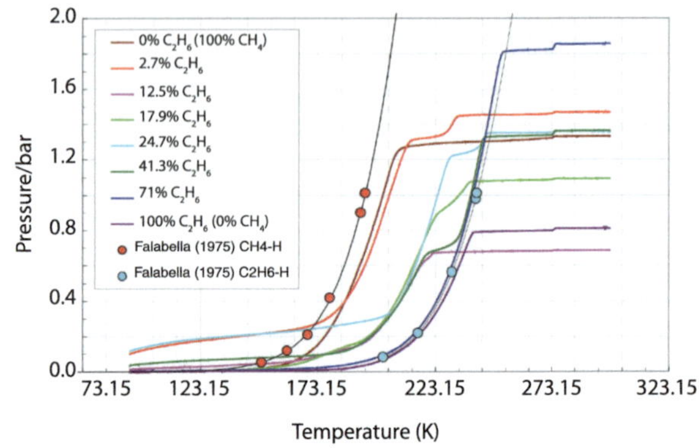

Fig. 5 Equilibrium pressure of $CH_4 + C_2H_6$ hydrate for several different abundances of species in the gas phase (adapted from Hachikubo et al. 2008)

guest clathrate whose equilibrium pressure varies as a function of the gas phase composition and the temperature (see Fig. 5). The trapping efficiencies of gases in clathrate in equilibrium with a gas mixture are very scarce. They are currently limited for temperatures greater than 200 K (Kang et al. 2001; Rydzy et al. 2007; Hachikubo et al. 2007, 2008), for a very small number of dataset, and for molecules which are not necessarily of great interest in planetary science ($C_3H_8 + Ar$, $C_2H_6 + Ar$, $CO_2 + N_2$, $CO_2 + N_2 + C_4H_8O$, $CH_4 + CO_2$, $CH_4 + C_2H_6 + C_3H_8$). No other experimental data, to our current knowledge, give the fractionation of volatile molecules in clathrates in equilibrium with a gas mixture at lower temperature (<200 K). Other studies giving the equilibrium pressure of multiple guest clathrates exist for different dataset and volatile species, but these studies do not give the abundance of volatile species both in the gas phase and in clathrate, and are limited most of the time to temperatures close to 270 K or above the triple point (Maekawa 2001; Linga et al. 2007; Mohammadi and Anderson 2009; Yasuda et al. 2010). Yet, both informations on the mole fractions of species in the clathrate structure and in the surrounding gas phase are needed to characterize the fractionation process during volatile trapping. In particular, the investigation of the composition of clathrates formed from gas mixtures of $CO + N_2$, $CO + CO_2$, $CO + CH_4$, and mixtures of noble gas ($Xe + Kr$, $Xe + Ar$, $Kr + Ar$, and $Xe + Kr + Ar$) are fundamental for studies of comets. While H_2O is the predominant molecule and least volatile other than the refractory larger organics, CO is the most abundant volatile specie in comets, and CO_2 and CH_4 the two other abundant carbon-bearing species (Bockelée-Morvan et al. 2004). N_2 is currently not observed in comets and it could not be trapped in the clathrate structure, explaining its apparent under-abundance in comets (Iro et al. 2003; Mousis et al. 2012a). The presence of noble gases in comets is still under debate.

Of great interest is also the fractionation of isotopic ratios between gas and clathrates. Isotopic ratios measurements are the classical methods for investigating the origin of solar system materials. But these measurements are very scarce in clathrates. Only a few data on the isotopic fractionation of carbon in molecules C_2H_6 and CH_4 have been obtained, but for temperatures greater than 250 K (Hachikubo et al. 2007).

Another issue remains the kinetics of formation and dissociation of clathrates from/to pure crystalline water ice and gas interaction. Two main stages of the clathrate formation process are distinguished (Kuhs et al. 2006). The first stage, relatively rapid, corresponds to the formation of a clathrate film over the pure crystalline ice surface (stage I), and the second, which dominates when the whole ice grain is covered by a clathrate layer, is the growth of the shell of clathrate phase around ice grains (stage II), with gas and water mass transports (diffusion) through the clathrate layer. For the first stage, a few kinetics experimental data have been determined only for pure clathrates of CH_4 and CO_2, mainly in the 185–270 K range (Schmitt et al. 1989; Staykova et al. 2003; Wang et al. 2002; Kuhs et al. 2006; Falenty et al. 2011). The authors have assumed that the value of the kinetics parameter of formation/dissociation of clathrates follows an Arrhenius-type function of temperature. However, the physical parameters that may affect the value of kinetics are multiple: temperature, activity of the ice surface (mobility of water molecules), type of gas molecule, thickness of clathrate formed, thermal history, etc. The kinetics conditions of the formation of the clathrate structure from pure crystalline water ice and gas species are currently poorly constrained at low temperature and the kinetic feasibility of the formation of clathrates in the cooling disk or in icy objects is still under debate, even if appropriate thermodynamics conditions are encountered for clathrate formation (Marboeuf et al. 2012b, 2014, Marboeuf and Schmitt 2014, Mousis et al. 2008, 2009a, 2009b, 2012a, 2014).

3.2.2 Prediction of Clathrates Composition from Thermodynamic Models

Because of the scarcity of experimental data at thermodynamic conditions representative of those encountered in the protosolar nebula, pressure-temperature equilibrium curves are usually extrapolated via the fit of existing data down to ~20 K (Gautier et al. 2001; Hersant et al. 2004, 2008; Mousis et al. 2009b, 2012a). For example, the equilibrium curve of N_2 clathrate is experimentally known at pressures higher than ~60 bars (Lunine and Stevenson 1985), implying that it has to be extrapolated over more than 13 orders of pressure magnitudes (given that solar N_2/H_2 is $\sim 10^{-5}$; Lodders et al. 2009) to derive its formation temperature at nebular conditions. The case of CO is very similar since its equilibrium curve is derived from experimental data obtained at pressures above 135 bars (Mohammadi et al. 2005). The extrapolated equilibrium curves are used to infer the temperature and pressure corresponding to the formation conditions of clathrates in the protosolar nebula via the use of protoplanetary disks models (see Fig. 4).

These calculations allow providing estimates of the composition of icy planetesimals condensed in the protosolar nebula (Marboeuf et al. 2008; Mousis et al. 2008) but their reliability depends upon the quality of the fits of the experimental data, which remain unconstrained given their large deficiency at low-pressure conditions. One also has to mention that the kinetics of clathrate formation is not considered in these calculations, due again to the absence of experimental data at low-pressure range. Kinetics is an important factor because a low formation rate would impede the agglomeration of large amounts of clathrates in the building blocks of icy bodies of the outer solar system. On the other hand, turbulent mixing in the protosolar nebula may ease the formation of clathrates since the collision between grains would expose fresh surfaces of ice to the gas and then favor its entrapment. In any case, because the question of clathrate formation kinetics in the protosolar nebula remains open, all existing models assume equilibrium formation conditions. Another restriction of the clathration efficiency in the protosolar nebula is the availability of crystalline water, which usually condenses at ~150 K. A full clathration of volatiles present in the gas phase of the protosolar nebula requires that the abundance of crystalline water is ~2 times higher than that of protosolar oxygen (Gautier et al. 2001; Mousis et al. 2014). If the water abundance is assumed to be protosolar, then a significant fraction of volatiles cannot be trapped in clathrates (Mousis et al. 2009b) and forms pure condensate at lower temperature in the 20–30 K range if the nebula effectively cooled down to these temperatures.

When the amount of available water is lower than the one required for the full clathration of the volatiles present in the gas phase, it is preferable to use of a statistical

Fig. 6 Composition of the volatile phase incorporated in planetesimals formed beyond the snow line in the protosolar nebula as a function of their formation temperature (adapted from Mousis et al. 2014). The water abundance is assumed to be sufficiently high to allow full trapping of volatiles in the clathrate phase. Computations have been made with a statistical model derived from the approach of Van der Waal and Platteeuw (1959)

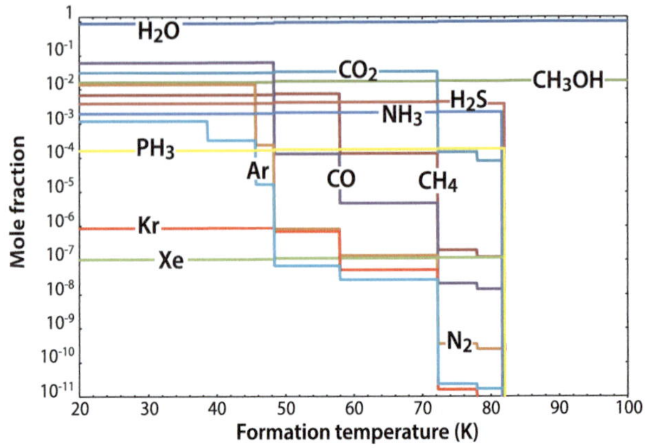

thermodynamic model (Van der Waal and Platteeuw 1959; Parrish and Prausnitz 1972; Lunine and Stevenson 1985; Thomas et al. 2007, 2008; Mousis et al. 2013a, 2013b) for the computation of the planetesimals composition because the different molecules do not have the same propensities for trapping. Statistical thermodynamic models are based on a very simple expression of the interaction energy between the guest and the surrounding water molecules. These models are commonly used in industry and science because they save substantial experimental effort for the determination of (i) the equilibrium pressure of a clathrate formed from various mixtures; and (ii) the mole fraction of the different species trapped in clathrate from a given gas or liquid phase. Their accuracy depends on the reliability of the parameters used to calculate the water-molecules interactions and also on the type of molecules considered. These models are fairly well constrained near the triple point of ice but their validity remains unverified in the 100–200 K temperature range because there is no experimental data against which the individual interaction potential can be fitted. Figure 6 represents the composition of planetesimals agglomerated from clathrates formed in the outer part of the protosolar nebula and calculated from a statistical thermodynamic approach (Mousis et al. 2014). It shows that all species under consideration can be trapped in clathrates at temperatures at high as $\sim$80 K in the nebula but with very different mole fractions. The mole fractions of volatiles trapped in clathrates vary as a function of the disk's temperature. The final composition of planetesimals will depend on the temperature at which the protosolar nebula has cooled down before its dissipation.

4 Comet Surface

4.1 Thermal Properties of Cometary Surfaces

Laboratory experiments were carried out in order to investigate the heat and mass transfer in cometary surface layers (Gundlach et al. 2011a; Gundlach and Blum 2012). Therefore, an experimental setup was constructed to measure the gas production rate and the heat conductivity of cylindrical cometary analogue samples (diameter: 2.5 cm; typical height: 3 cm) at low pressures ($< 10^{-5}$ mbar) and low temperatures (between 120 K and 300 K). The cometary analogue samples were composed of hexagonal water ice (produced in a freezer) covered by a dust layer, which was composed of SiO_2 grains with different radii (20 ± 8 μm, 89 ± 30 μm, 186 ± 63 μm, 254 ± 42 μm and 463 ± 32 μm). The outgassing rate of the water

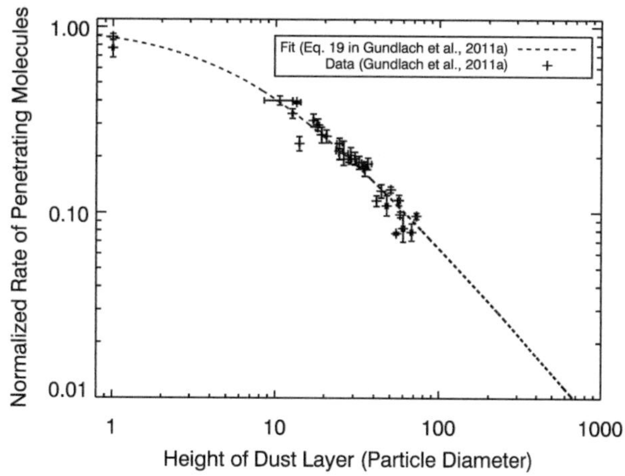

Fig. 7 Normalized rate of penetrating molecules through a dust layer with respect to the height of the dust layer (adapted from Fig. 8 in Gundlach et al. 2011a). The data (plusses) are obtained by covering a sublimating ice surface by dust layers with variable thickness and a volume filling factor of 0.59. The *dashed curve* shows the fit to the data, which is motivated by the fact that the dust layer acts like a resistor to the gas flux (Eq. 19 in Gundlach et al. 2011a)

ice through the dust layer was measured for different temperatures of the water ice surface and for different heights of the covering dust layer. These experiments showed that the presence of a dust layer could sufficiently decrease the outgassing rate of water ice in vacuum (see Fig. 7). For dust layers possessing thicknesses larger than approximately ten particle diameters, the outgassing rate is inversely proportional to the thickness of the covering dust layer. This behavior can be explained by the idea that the dust layer acts like a resistor to the gas flux.

Additionally, the experimental setup was used systematically determine the temperature dependency of the so-called sublimation coefficient (temperature range: $\sim$160 K to $\sim$220 K), introduced by Kossacki et al. (1999). This coefficient describes the deviation of the sublimation rate measured in the laboratory from the sublimation rate calculated by the Hertz-Knudsen formula (see Eqs. 2 and 3 in Gundlach et al. 2011a). These experiments revealed that the sublimation coefficient is unity at temperatures below $\sim$194 K (see Eq. 20 in Gundlach et al. 2011a) and decreases steeply (over a temperature range of $\sim$20 K) to a value of 0.15 at higher temperatures. This implies that the Hertz-Knudsen formula overestimates the gas production rate of water ice by approximately one order of magnitude at temperatures above $\sim$194 K.

Thermal conductivity of the surface layer of a comet is an important physical parameter that needs to be constrained. While this is a mix of dust, organics, and ice, all these may play a critical role. Thermal conductivity of comets is known to be very low, e.g., $\sim 1 \times 10^{-3}$ $\mathrm{WK^{-1}\,m^{-1}}$ (Groussin et al. 2007) in the case of 9P/Tempel 1, estimated from thermal inertia measurements (50 $\mathrm{J\,m^{-2}\,K^{-1}\,s^{-0.5}}$). Unfortunately, the thermal inertia estimation strongly depends on the surface roughness (Davidsson et al. 2013), which is not well known in the case of Tempel 1. Heat conductivity of porous, amorphous ice is reported to be 3×10^{-2} $\mathrm{WK^{-1}\,m^{-1}}$ (Bar-Nun and Laufer 2003), which is larger than that comet Tempel 1. In the laboratory (KOSI experiments), the heat conductivity of porous cometary analogue materials, composed of different non-volatile and volatile materials, was measured to be much higher at $\sim 3 \times 10^{-1}$ W/Km (Spohn and Benkhoff 1990). This value is lower than the bulk heat conductivity of water ice but much higher than the heat conductivity estimated for the surface material of Tempel 1.

The heat conductivity of dust layers in vacuum was measured by using the same experimental setup (Gundlach and Blum 2012). Therefore, dust layers of different heights were illuminated by an external light source with intensities ranging from 0.3 to 5.3 Solar constants. During these experiments, the bottom and the surface temperature of the dust layers were continuously measured by a temperature sensor and an infrared camera, respectively. The temperature data were then used to compute the heat conductivity of the samples, which depends on the particle size (between 40 μm and 1 mm) and on the temperature of the dust layers. These measurements showed that the heat conductivity of dust layers in vacuum is relatively low (between $10^{-1}\ \mathrm{W\,m^{-1}\,K^{-1}}$ and $10^{-2}\ \mathrm{W\,m^{-1}\,K^{-1}}$; see Figs. 10 and 11 in Gundlach and Blum 2012), and that the heat is transported by different mechanisms: (1) by heat conduction through the solid network of particles and (2) by radiation within the pores of the dust layers. The heat flow through the solid network of particles dominates for small particle sizes and at low temperatures (e.g., for particle sizes smaller than $\sim$ 15 μm and for temperatures below $\sim$300 K). In this case, the heat conductivity of the bulk material determines the temperature dependency of the heat conductivity of the dust layer. With increasing particle radius, R, the heat conductivity decreases as $\lambda \sim R^{-1/3}$, where λ is the heat conductivity of the dust layer (see Eq. 12 in Gundlach and Blum 2012). The heat transport is dominated by radiation, if the dust layers are composed of larger particles, or at higher temperatures (e.g., for particle sizes larger than $\sim$ 15 μm and temperatures above $\sim$300 K). In this case, the heat conductivity strongly depends on temperature ($\lambda \sim T^3$, where T is the temperature of the dust layer) and increases linearly with increasing particle radius ($\lambda \sim R$; see Eqs. 5 and 13 in Gundlach and Blum 2012).

Additionally, the heat conductivity of aggregate layers (dust layers composed of aggregates, which consist of micrometer-sized particles; see Fig. 13 in Gundlach and Blum 2012) was measured by Krause et al. (2011). The heat conductivity of aggregate layers is approximately an order of magnitude lower (between $10^{-2}\ \mathrm{W\,m^{-1}\,K^{-1}}$ and $10^{-3}\ \mathrm{W\,m^{-1}\,K^{-1}}$; see their Fig. 2) than the heat conductivity of dust layers composed of single solid particles. However, radiation can be an important heat transport mechanism in the case of aggregate layers, which can effectively increase the heat conductivity (for aggregate sizes larger than $\sim$ 0.1 mm; see Fig. 8). For example, an aggregate layer consisting of 1 cm-sized aggregates at a temperature of 300 K possess a heat conductivity of $1 \times 10^{-1}\ \mathrm{W\,m^{-1}\,K^{-1}}$, which is approximately two orders of magnitude higher than the heat conductivity without radiative heat transport.

If cometary surface layers are composed of aggregates, the strong temperature dependency of the radiative heat transport ($\lambda \sim T^3$) implies diurnal variations of the heat conductivity on the order of one to three orders of magnitude. Thus, future measurements should aim at investigating the diurnal variations of the heat conductivity (or, the thermal inertia) of cometary surface materials in order to derive the structure (i.e., aggregate size and volume filling factor) of the surface material.

Further experiments were carried out in order to investigate the tensile strength of the cometary surface material (Blum et al. 2014). Therefore, dust aggregates of two different sizes (0.66 ± 0.14 mm and 1.29 ± 0.29 mm) were poured carefully onto a grating inside a glass tube to form aggregate layers with several centimeters in height. With the help of a vacuum pump, different pressure gradients were applied to the aggregate layers until a break-up of the samples was observed. From the pressure difference at the moment of the breakup, the tensile strength of the aggregate layers was derived and yielded 1.3 ± 0.9 Pa in the case of the 0.66 mm-sized aggregates and 0.8 ± 0.7 Pa in the case of the 1.29 mm-sized aggregates, respectively. This ultra-low tensile strength is in a good agreement with model calculations, which predict a decreasing tensile strength with increasing aggregate size (see Skorov and Blum 2012, for details).

In addition to the laboratory investigations mentioned previously, μm-sized water-ice particles were produced and used in the laboratory in order to generate a more realistic cometary

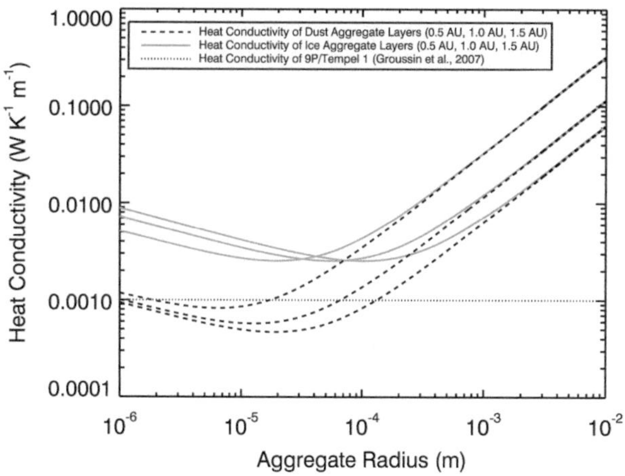

Fig. 8 Heat conductivity of aggregate layers with different aggregate sizes covering a comet nucleus. The heat conductivity of the aggregate layers is derived by the heat conductivity model developed by Gundlach and Blum (2012). The calculations were performed for ice aggregates (*gray, solid curves*) and for dust aggregates (*black, solid curves*) with a Bond albedo of 0.04 and at different heliocentric distances: 0.5 AU, 1.0 AU and 1.5 AU (from *left* to *right*, respectively). This model is capable to predict the heat conductivity for aggregate size smaller than the diurnal skin depth, which is typically in the cm-range. Additionally, the heat conductivity of the surface material of Comet 9P/Tempel 1 (Groussin et al. 2007) is shown by the dotted line. Here, we would like to remind the reader that the heat conductivity of the surface material is derived from the thermal inertia estimation, which strongly depends on the surface roughness of the nucleus

analog material (Gundlach et al. 2011b). The μm-sized water-ice particles were produced by spraying with an inhalator directly into liquid nitrogen, or into a cold ($T < 200$ K) sedimentation chamber. The particles produced by these two methods are visibly spherical (see Fig. 3 in Jost et al. 2013) and possess a mean radius of $1.47^{+0.96}_{-0.58}$ μm, with particle radii ranging from ~ 0.2 μm to ~ 7 μm (Jost et al. 2013).

First experiments with this novel sample material revealed a relatively high surface energy of water ice (0.19 J m^{-2}; Gundlach et al. 2011b) in comparison to silica particles (0.02 J m^{-2}; Poppe et al. 2000), which implies that water-ice particles can stick at higher collision velocities than silica particles. The specific surface energy of the μm-sized water ice particles was measured by observing the gravitational restructuring of ice aggregates composed of μm-sized water-ice particles with a long-distance microscope (Gundlach et al. 2011b).

Furthermore, the μm-sized water-ice particles, produced by spraying the water droplets into the cold sedimentation chamber, were used to measure their sticking threshold (the velocity until the particles stick in collisions; Gundlach and Blum 2015). Therefore, the μm-sized water-ice particles were extracted by a vacuum system from the cold sedimentation chamber into a vacuum chamber. Inside the vacuum chamber, the μm-sized water-ice particles formed an ice-particle jet, and collided with a cryogenically cooled target. These experiments showed that the sticking threshold of the μm-sized water-ice particles is at ~ 10 m s^{-1}, which is approximately ten times higher than the sticking threshold estimated for micrometer-sized silica particles (~ 1 m s^{-1}; Poppe et al. 2000).

In conclusion, experimental studies presented in this section aim at supporting models on cometary activity and to help preparing future comet sample-return mission. So far, experiments have been focused on investigating outgassing properties of water ice under cometary conditions (i.e., illuminated, at low temperatures, under low pressure, and covered by dust layers), which have shown that dust layers have a major influence on the water ice activity. Our studies (Gundlach et al. 2011a) confirmed the unexpected behavior of water ice to decrease its activity around ~200 K by almost one order of magnitude, proposed by Kossacki et al. (1999). It was found that porosity of the material has a strong influence on the effective heat conductivity (Gundlach and Blum 2012). Cometary surface material has been simulated by layers composed of dust aggregates (which themselves consist of micrometer-sized dust particles) and we found that these dust aggregate layers possess ultra-low tensile strength (Blum et al. 2014), confirming earlier predictions by Skorov and Blum (2012). Using micrometer-sized water ice particles, it was shown that water ice particles can stick at velocities up to $\sim 10\ \mathrm{m\,s^{-1}}$ (Gundlach and Blum 2015), which is approximately one order of magnitude higher compared to the sticking threshold velocity of micrometer-sized silica particles at $1.2\ \mathrm{m\,s^{-1}}$ (Poppe et al. 2000). Further studies are needed to determine the tensile properties of organic dust, crystalline and amorphous ice aggregates, and differences in the physical properties of volatiles trapped in crystalline and amorphous ices, including clathrates.

4.2 Refractive Organic Matter on Comets—Laboratory Studies

Many experimental devices have been developed to investigate the formation of organic molecules in cometary environments. In the laboratory, irradiation by UV photons or charged particles followed by thermal heating of astrophysical ice analogues that contain simple molecules such as H_2O, CH_3OH, NH_3 and CO_2 have produced a suite of complex organic molecules similar to those found in small bodies of our Solar system, including amino acids and nucleobases (Gerakines et al. 1996; Muñoz-Caro et al. 2002; Bernstein et al. 1995, 2002). When UV photons are penetrating into astrophysical ices, they break bonds, producing radicals and activated species. It has recently been shown by Gudipati and coworkers (Gudipati and Yang 2012; Henderson and Gudipati 2014; Yang and Gudipati 2014) that even at very low temperatures (5 K) UV-irradiation of water-ice containing organics (PAHs) lead to hydrogenation and oxygenation/hydroxylation. While these primary chemical processes between organic radicals/ions with radicals and ions produced from water matrix (H, OH, O, etc.), UV-irradiation by itself may not produce large complex organic compounds. When the irradiated ices are subjected to annealing to higher temperatures the photoproducts become mobile and could react more readily forming an non-volatile complex organic residue. During this warming, a large part of volatile compounds desorb with the most abundant molecule, water. The characterization of these volatile species is important, since they certainly take a part in the subsequent formation of this refractory residue. Here, the refractory organic residue refers to all compounds having a higher temperature desorption than that of water, which is the most abundant molecule of a cometary ice analogue.

4.2.1 Characterization of the Whole Residues

As already described, after the sublimation of volatile organic compounds (VOC), a residue is present at the surface of sample holders. These residues are supposed to be the first step toward organic matter found in comets, asteroids or meteorites. That is why they are considered as potential analogues of cometary organic matter. Different analytical methods have been tested to determine the molecular constitution of such residues and have shown important concordance between these analogues and the soluble organic matter of meteorites.

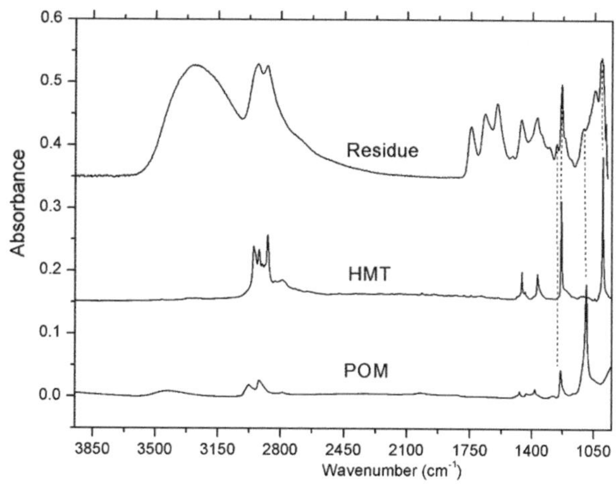

Fig. 9 Infrared spectrum of an organic residue obtained after UV photolysis (9 h) and warming at 300 K of an ice mixture with $H_2O : CH_3OH : NH_3$ composition compared with infrared spectra of pure HMT recorded at 300 K and pure POM recorded at 300 K. The residue was produced by Pierre de Marcellus using the MICMOC experiment at IAS, France (Danger et al. 2013)

A method usually used for characterizing these residues is the infrared spectroscopy, because it provides an in situ residue analysis without sample degradation, and a direct comparison to observational data. Infrared spectra of these residues have shown the presence of polar compounds such as amines, carboxylic acid or amides functions (Allamandola et al. 1988; Bernstein et al. 1995; Muñoz-Caro and Schutte 2003). Some specific compounds have also been identified, such as the hexamethylenetetramine (HMT), which seems to represent a large part of the organics constituting these residues. In addition to these compounds, some polymers have been characterized, such as the polyoxymethylene (POM) as shown in Fig. 9. Infrared spectroscopy was also used for estimating the impact of the source of energetic processes applied to the ice before its warming. It was shown that infrared spectroscopy of residue spectra presenting rather similar characteristics are obtained between an ice initially irradiated by VUV photons or by ions (Islam et al. 2014; Muñoz-Caro et al. 2014), even if some abundance variations can be observed. However, if infrared spectroscopy can give first clues on the most abundant molecules present in such residues as well as on their formation mechanisms (see Sect. 4.3), a drawback of this analytical method is its low sensitivity and the difficulties to obtain a clear identification of molecular species in a complex mixture due to the overlapping of broad solid-state bands. Consequently, numerous other analytical techniques have been used for determining the molecular constitution of these residues.

Mass spectrometry is another analytical method that can provide the characterization of the entire residue without the need of sample derivatization or degradation. Laser desorption mass spectrometry was used to identify polyaromatic hydrocarbons (PAH) (Dworkin et al. 2004; Spencer et al. 2008). Nevertheless, depending on the resolution used, mass spectrometry cannot be sufficient for the analysis of the molecular complexity present in refractory residues. New mass spectroscopy systems presenting high resolution mass spectrometry (HRMS) such as FT-ICR (Fourier Transform Ion Cyclotron Resonance) or orbitrap, present enough resolution for that sort of characterization. These techniques were used for the analysis of the Murchison meteorite (Schmitt-kopplin et al. 2010) and the analysis of biomolecules coming from the hydrolysis of tholins (Neish et al. 2010). Recently, the orbitrap was used for the analysis of residues formed by the VUV irradiation and the warming of ice analogues containing $CH_3OH : NH_3 : H_2O$ with $^{12}CH_3OH$ or $^{13}CH_3OH$ (Danger et al. 2013). In the protocol used, residues are dissolved in methanol and directly injected in the orbitrap without any degradation. Molecules are ionized trough an electrospray source (ESI),

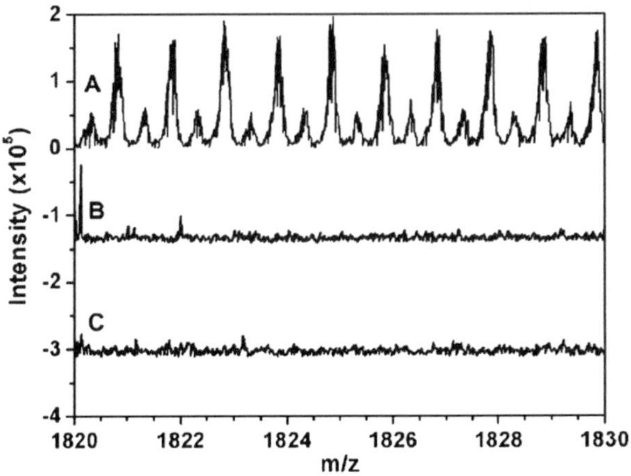

Fig. 10 FT-orbitrap SIM spectra obtained in ESI negative mode for the m/z range 1820–1830 after recovery in methanol of (**A**) a refractory residue formed from an ice mixture containing $^{12}CH_3OH : NH_3 : H_2O$ (1:1:3), (**B**) a blank coming from an ice mixture deposited on the MgF_2 window and warmed without VUV irradiation, and (**C**) a blank coming from a VUV irradiation of the MgF_2 window, without gas mixture deposit. Between blanks and sample formation, all the different steps are exactly the same. Reprinted from Danger et al. (2013) with permission from Elsevier

which provides smooth molecule ionization and the molecular ion detection. The mass spectra present an intensity that progressively decreases when m/z increases. Furthermore, the HRMS provides a zoom in this distribution, which emphasizes the diversity of molecules constituting this refractory residue. While the started ice is only composed of three simple molecules H_2O, CH_3OH and NH_3, HRMS analyses have shown that more than ten of thousand of ions (i.e. structures) are detected with masses up to 4.000 Da. These results confirmed a previous report of the high m/z ratios up to 2600 detected by MALDI-TOF technique in similar organic residues (Modica et al. 2013). An example of a mass spectrum obtained with orbitrap in negative ESI mode is displayed in Fig. 10A into the range m/z 1820 to 1830 for a residue analysis coming from an ice containing $^{12}CH_3OH : NH_3 : H_2O$ in ratio 1:1:3. If this profile is compared to two different blanks (Fig. 10 traces B and C), ions are only detected in the residue, confirming that molecular species detected come from the photochemistry and the warming of the initial ice. Therefore, the composition of residues is extremely rich and complex. It has to be noted that a warming alone cannot lead to the molecular diversity observed (Fig. 10B). This molecular diversity is only obtained if VUV irradiations are associated with a warming. This reactivity aspect will be further discussed in Sect. 4.3. Furthermore, next to this global residue analysis, the HRMS can also provide in a same run the analysis of specific molecules. HRMS has for example confirmed the presence of HMT already observed by infrared spectroscopy in such residues. Finally, this whole residue analysis has allowed obtaining an estimation of the residue elemental composition.

4.2.2 Search for Specific Molecules in Residues

Even if HRMS analyses can provide in some cases the search for specific molecules, more accurate analytical techniques are used for this aim. GC-MS or LC-MS were extensively used. They provide the identification of various amino acids (Bernstein et al. 2002; Muñoz-Caro et al. 2002). Interestingly, amino acids are detected before residue hydrolysis, their yield increasing after hydrolysis (Meinert et al. 2012). These results suggest different amino acid formation coming from different pathways (Elsila et al. 2007; Danger et al. 2011a; Danger et al. 2011b; Bossa et al. 2010). GCxGC-MS has provided the identification of 30 amino acids (Meinert et al. 2012). Even if the initial ices are different as well as the physical source of energy (photons, electron or ions) applied to these ices, in each case amino acids are detected. Interesting experiments using VUV circularly polarized light instead of a linearly polarized light commonly used, have shown that enantiomeric excess up to 2.54 % in the L form can be obtained on amino acids by this process, the sign of these enantiomeric excesses depending on the helicity and the energy of the CPL (de Marcellus et al. 2011b; Modica et al. 2014). The use of UV-CPL is a scenario that could explain the detection of enantiomeric excess in some meteorites. The enantiomeric excess formed by this process could then be enhanced in parent bodies of meteorites. Other analyses also made without a preliminary residue hydrolysis provide the identification of various linear or cyclic compounds containing nitrogen and/or oxygen chemical functions (Meierhenrich et al. 2005; Nuevo et al. 2010; de Marcellus et al. 2011a; Materese et al. 2014), such as urea, hydantoin, glycolic acid, glycerol or hexahydro-1, 3, 5-triazine (TMT). All these molecules are considered as possible sources of matter for the development of a prebiotic chemistry in a specific environment (Danger and Le Sergeant d'Hendecourt 2012). Other analyses provide the confirmation of the HMT presence; the influence of the initial ice mixture composition on its formation being investigated using GC-MS (Cottin et al. 2001; Muñoz-Caro et al. 2004). HPLC analyses also provide the detection of amphiphilic molecules (Dworkin et al. 2001). Some ices were doped with pyridine derivatives and their analyses have shown the presence of nucleic bases such as uracil, cytosine and thymine (Nuevo et al. 2012; Nuevo et al. 2014).

Residue characterizations show an important molecular diversity (Danger et al. 2013). Interesting similarity can be found between these residues and the soluble organic matter (SOM) of meteorites. The whole characterization described a similar molecular diversity, and specific analyses demonstrate the presence of common compounds. Even if these residues cannot be considered as the SOM of astrophysical objects, they can be considered as analogues of the first step that will then lead to the SOM formation.

These analyses also demonstrate a necessity of an energetic process at the beginning of the ice deposit. If this energetic process is not present no reactivity or only a low reactivity occurs depending on the initial ice composition. However, the reactivity leading to refractory residues is much more complex than an only VUV irradiation or ion bombardment. The ice heating after irradiation or bombardment is an essential process that will then lead to the residue formation (Theule et al. 2013). As it will be discussed in the next section, physical processes (VUV irradiations or ion bombardment) have to be associated to a thermal reactivity for explaining the molecular diversity observed. Based on these observations, we can thus propose that the organic matter detected in interplanetary objects comes from the association of high energetic processes and thermal heating.

4.3 Precursors for Refractory Organics

4.3.1 Water Ice Impact on Chemical Reactivity

In cometary environments, water is the most abundant molecule. As for desorption of volatiles, water will play an important role in the chemical composition (Martin-Domenech et al. 2014) and reactivity (Jenniskens and Blake 1994). Indeed, at low temperature, the chemical reactivity is dominated by the kinetics of diffusion of the molecules in the

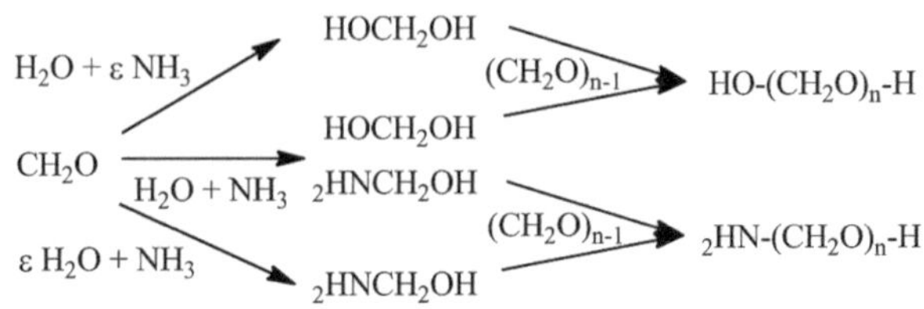

Fig. 11 Formaldehyde (CH_2O) reactivity in ices containing various H_2O/NH_3 ratios

amorphous solid water (ASW, also known as amorphous ice). Barrier-less reactions involving photo-produced radicals (Schutte 1995) or low-barrier reactions involving neutrals (Theule et al. 2013) are strongly limited by the ability of reactants to diffuse and meet each other. The diffusion-limited reactivity mechanisms are poorly known. The diffusion of species in the ice is itself poorly known. While bulk and surface diffusion coefficients have been measured in crystalline ice (Livingston et al. 2002; Varotsos and Zellner 2010), little measurements exist for diffusion in ASW (Mispelaer et al. 2013; Daschbach et al. 2004). Measuring reliable diffusion coefficients in water ice at low-temperature is a current challenge. Diffusion of volatiles in ASW seems to be related to the water ice substrate crystallization (Mispelaer et al. 2013). Understanding the implication of this seemingly crystallization-driven diffusion on reactivity is central to quantify the kinetics of complex organic matter formation in ice.

However, a water environment have others impacts on reactivity than diffusion. Experiments have shown that water can act as a catalyst for the formation of complex organic molecules. For instance, the formation of POM initiated by HCN (NC-(CH_2-O)-H) can only occur in presence of water, since water dissociates HCN forming $^{-}$CN that can then react on H_2CO (Danger et al. 2014). Furthermore, as discussed previously, molecules are trapped in the water matrix during the ice formation on grains. This trapping also plays a fundamental role in reactivity. A study on the reactivity between acetone $(CH_3)_2CO$ and ammonia NH_3 at low temperature and low pressure (20–300 K, 10^{-9} mbar) for the formation of $_2$HN-C$(CH_3)_2$-OH, an amino acid precursor, have shown that without water, the reaction does not occur, because reactants desorb before reacting (Fresneau et al. 2014). However, when acetone/NH_3 is mixed with a water ice, reactants are trapped in the water matrix up to the water desorption temperature, at which sufficient energy is available for acetone and NH_3 to react before their desorption. These examples demonstrate that the role of water in reactivity is fundamental for the formation of complex molecules in cometary environment. We also note that HCN polymers are also proposed to exist on comets and contribute to the potential prebiotic molecules that could have lead to the origin of life on Earth (Matthews and Minard 2006).

4.3.2 Formaldehyde Chemistry in Cometary Ices: POM Formation

The presence of POM (HO-(CH_2O)-H) on grains of the comet 1P/Halley was first claimed in 1987 by Huebner (Huebner 1987), because mass spectra recorded by the PICCA instrument on the *Giotto* mission showed a regular pattern of peaks consistent with a POM-like structure. However, it was shown afterward that this regular pattern can also be due to the presence of complex molecules containing C, H, O and N atoms, including POM-like structures (Mitchell et al. 1992). Although the presence of POM in comets has not yet been confirmed, some experimental results support its presence in cometary grains. POM and POM-like polymers can indeed be formed from the heating of interstellar/cometary ice analogues (Schutte et al. 1993a, 1993b; Danger et al. 2012, 2014; Duvernay et al. 2014a;

Noble et al. 2012). POM has also been detected in laboratory analogues of interstellar/cometary grains by the thermal processing of ices containing H_2O, H_2CO and NH_3 (Vinogradoff et al. 2011). Thermally induced formaldehyde reactions at low temperature in $H_2O : NH_3 : H_2CO$ ices have also shown that two other organic compounds than POM are distinguished by infrared spectroscopy: methyleneglycol ($HOCH_2OH$) (Duvernay et al. 2014a, 2014b) and aminomethanol (NH_2CH_2OH) (Bossa et al. 2009b), which are the reaction products of H_2CO with H_2O or NH_3, respectively (Fig. 11). The formation of such compounds in cometary environments is likely and therefore should be searched for by space missions dedicated to cometary analysis such as the *Rosetta* mission.

4.3.3 Formaldehyde Chemistry in Cometary Ices: HMT Formation

Formaldehyde is therefore a precursor for the formation of oxygen bearing compounds, but as we will discuss here, it is also a precursor for the formation of compounds bearing only nitrogen as a heteroatom. As previously mentioned, among the complex organic molecules present in the organic residues produced in the laboratory, hexamethylenetetramine (HMT, $C_6H_{12}N_4$) has been observed by numerous authors. It can reach up to 50 % of residue mass (Danger et al. 2013). HMT was initially proposed as a good tracer of photochemical processing in the pre-cometary phase (Bernstein et al. 1995) because it was supposed to be formed through the photolysis of the interstellar ice. However, HMT can be also formed either after proton irradiation of $H_2O : CO : CH_3OH : NH_3$ ices (Cottin et al. 2001) or by only heating at room temperature, without any energetic processing (photons or particles), of $H_2CO : NH_3 : HCOOH$ ices (Vinogradoff et al. 2011, 2012, 2013a, 2013b). Therefore, HMT cannot be considered anymore only as a tracer of photolysis. Furthermore, the thermal formation suggests that its formation cannot be explained only by UV photolysis processes. In fact, an association between UV and thermal processes should occur during the processing of a canonical $H_2O : CO : CH_3OH : NH_3$ ice mixture (Vinogradoff et al. 2013b). Indeed, this complex organic molecule has never been detected at low temperature after irradiation of interstellar/cometary ice analogues. It is only observed after the warming of a photolyzed-ice until 300 K. The mechanistic investigation of the HMT formation in such conditions allows elucidating the role played by irradiation and thermal processes. This mechanism is divided in two parts. The first part of this reaction occurs during the photolysis of the ice, where several photo-products are formed such as H_2CO and HCOOH. As described in Sect. 3.2.2, depending on the initial ice ratio, H_2CO can react with ammonia when the photolyzed ice is warmed, leading to the formation of NH_2CH_2OH. This molecule is then a bridge between POM-like structures $H_2N\text{-}(CH_2O)_n\text{-}H$ and molecules bearing only nitrogen as an heteroatom. Since HCOOH is formed during the UV irradiation step, during the warming HCOOH catalyzes the H_2NCH_2OH dehydration leading to the formation of methanimine CH_2NH. This imine formation opens a way through the formation of various linear or cyclic compounds bearing only C, H and N atoms. During this process, a specific cyclic intermediate can be formed, trimethylenetriamine (TMT), which is the last step for the HMT formation at around 300 K in laboratory conditions. The whole mechanism of HMT formation is displayed in Fig. 12.

Considering that the reaction has been performed in conditions that mimic those encountered in cometary environments, the HMT formation in such environments is likely, which thus justifies the fact that it will be searched for by the forthcoming *Rosetta* mission.

4.3.4 POM vs. HMT in Astrophysical Ices

POM and HMT are two very different complex molecules. POM is an oxygen-bearing polymer while HMT is nitrogen bearing cyclic molecule. However, as previously discussed,

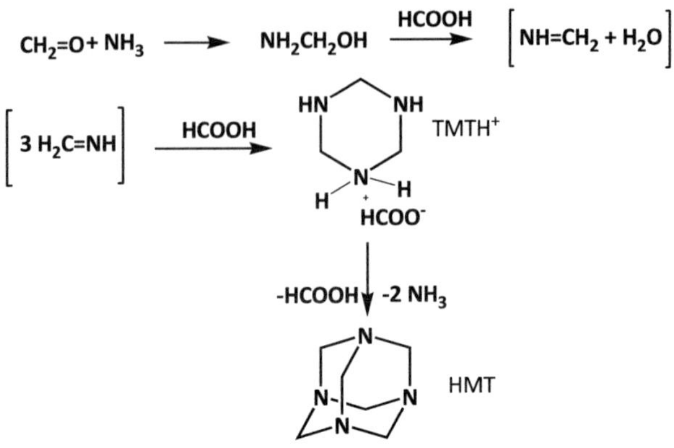

Fig. 12 Formation mechanism of HMT in photolyzed cometary ices

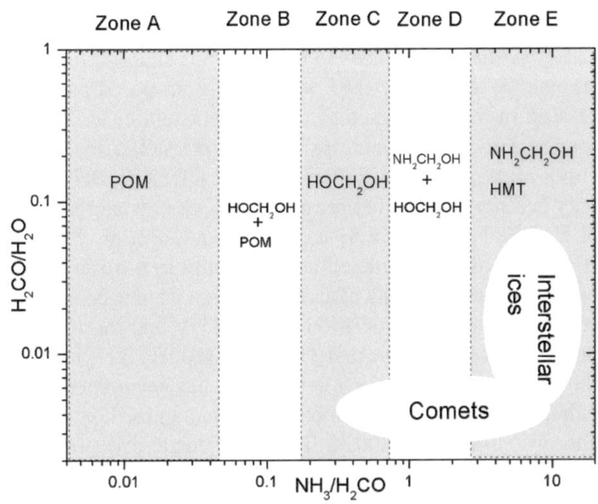

Fig. 13 The different major products formed form the warming of $H_2O : NH_3 : H_2CO$ ice mixtures as a function of the $NH_3 : H_2CO$ concentration and H2CO:H2O concentration ratios. *Horizontal* and *vertical ellipses* represent the typical ice compositions detected in cometary environments and in star-forming regions. Reproduced from Duvernay et al. (2014a)

they both have the same precursor (Figs. 11 and 12): the formaldehyde H_2CO and their formation depends strongly on the initial ice composition and especially on the NH_3/H_2CO ratio (see Fig. 13). The POM is only the major product when the NH_3/H_2CO ratio is lower than 4×10^{-2} (zone A in Fig. 13), whereas the NH_2CH_2OH (and thus the HMT since NH_2CH_2OH is one of its precursors) will be the dominant product for an NH_3/H_2CO ratio higher than 3 (zone E in Fig. 13). Therefore, the warming of interstellar/cometary ices may form NH_2CH_2OH (and thus HMT if HCOOH is present) or POM, depending on the initial ice composition. An excess of ammonia in regard to formaldehyde would favor the NH_2CH_2OH formation, while an excess of formaldehyde would favor POM formation. Shown in Fig. 13 (horizontal and vertical ellipses) are the NH_3/H_2CO ratios and H_2CO/H_2O ratios for several star-forming regions obtained from Spitzer data (Zasowski et al. 2009) (Öberg et al. 2011) as well as for cometary ices (Mumma and Charnley 2011). The typical ice composition of star-forming regions falls in the NH_2CH_2OH domain (zone E in Fig. 13) while the typical cometary ice composition falls in both NH_2CH_2OH and $HOCH_2OH$ domains (zones C–E in Fig. 13). This means that cometary ices that are supposed to be formed from the accretion of primitive ices at the early stage of solar system formation may preferentially contain $HOCH_2OH$ and NH_2CH_2OH (and HMT). However, POM formation in cometary ices cannot be definitively ruled out since the NH_3 column densities reported for star-forming regions in the literature are usually upper limits. Regions belonging to star-forming environments, where the NH_3 column density is low enough (compared to H_2CO) might exist and accordingly POM formation is plausible.

Thus the detection of both HMT and POM in cometary environments is not unlikely since cometary nuclei may have accreted matter coming from different disk regions with various ice compositions. In fact, the detection of POM and HMT (and its precursors) by the Rosetta mission will be another proof of the heterogeneity of the cometary nucleus composition. Since the ratio between these three products (POM, $HOCH_2OH$ and NH_2CH_2OH) depends strongly on the initial ice composition (Fig. 13). Consequently, their detection could give information on the pre-cometary/interstellar ice composition.

5 Comet Outgassing

5.1 Isotopic Abundances and the Origins

It is customary to use isotopic abundances to date the materials (Aléon and Robert 2004; Alexander et al. 2010; Duprat et al. 2010; Lecluse et al. 1996; Tomescu et al. 2009) or to compare origins of different bodies such as planets, comets, asteroids, meteorites etc. (Feuchtgruber et al. 2013; Robert et al. 2000). Isotopes of trapped, unaltered materials that do not interact with their surrounding over astronomical time scales (Myrs to Gyrs) may carry information on their origin (Marty et al. 2008). However, when the material is continuously subjected to radiation and/or thermal processing (including sublimation) and exchange processes with its surroundings, the measured isotopic ratios may not be representative of the initial isotopic ratios of the materials. Then a detailed knowledge or estimation of the history of the body is necessary to extrapolate present day observed isotope ratios to early solar system conditions. Almost exclusively the D/H ratios observed for water have been made in the gas-phase (HDO/H_2O line intensities), which may or may not represent the ice D/H ratios (vide infra).

Brown et al. (2012b) have shown from their laboratory work that when ice sublimation is much slower than molecular diffusion within the ice, enrichment of heavy isotope (D) occurs in the solid ice and lighter isotope (H) in the sublimed water—say in the case of a comet surface vs. coma/tail—hence the observations need to be carefully interpreted in the light of these laboratory data. In addition to thermally simulated altering the original isotope ratios, it is also possible to enrich or deplete heavier isotopes through radiation-induced processes. Photochemical enrichment of ^{13}C isotope in CO_2 solid films upon Lyman-α irradiation (10.2 eV) due to kinetic isotope effect has been reported by Yuan and Yates (2013).

Recently Yang and Gudipati investigated hydrogenated polycyclic aromatic hydrocarbons (PAHs) embedded in D_2O ice (Yang and Gudipati 2014). In the absence of any photons or electrons, no D/H exchange between the hydrogen of PAHs and Deuterium of the ice host could be observed (Fig. 14). As soon as UV or electrons were bombarded on the ice, an exchange between the H atoms of PAHs and D atoms of D_2O ice occurred, leading to almost complete deuteration of the PAH molecule. The same scenario can be reverted—when deuterium enriched organic molecule trapped in H_2O ice is subjected to radiation process, D of the organics will be exchanged with H of ice host, altering the D/H ratio of both the organics and the ice. Similar D/H exchange has been observed (Bernstein et al. 1999) when a room temperature residue was analyzed using laser ablation mass spectrometry of coronene ($C_{24}H_{12}$) trapped in D_2O ice, irradiated with UV and subsequently warmed

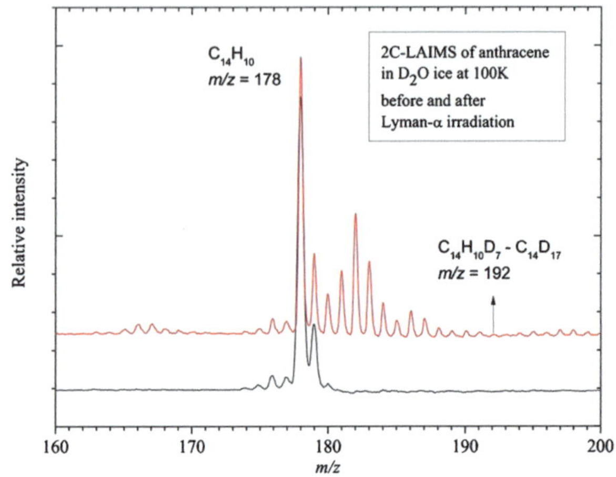

Fig. 14 Hydrogen enriched anthracene ($C_{14}H_{10}$) trapped in D_2O ice. No D/H exchange before UV irradiation (*black curve*). Upon Lyman-a irradiation both deuteration and D/H exchange occur resulting in a wide range of molecules with fractional deuteration and D/H exchange. Compete D/H exchange gives $C_{14}D_{10}$ ($m/z = 188$), complete deuteration + D/H exchange gives $C_{14}D_{20}$ ($m/z = 198$). Taken from Yang and Gudipati (2014)

up to room temperature—subliming the ice and leaving the non-volatile residue behind. Very recent modeling studies concluded negligible contribution from ionization-mediated chemistry towards deuterium enrichment in ices of a protostar (Cleeves et al. 2014).

Even in the absence of radiation or thermal processes exemplified above, under dark and cold environment, deuterium and hydrogen exchange can still occur through quantum chemical "tunneling" processes (Andersson et al. 2011; Hama and Watanabe 2013). Under these conditions reaction barrier for a D/H exchange through normal chemical potential is insurmountable, but because hydrogen atom acts as both a particle and a wave, there is a certain probability of hydrogen to move from one side of this energy barrier to the other side through the quantum chemical tunneling. Tunneling is temperature-independent. Hence tunneling could occur in the interstellar medium at 10 K or in KBOs or in comets, or in ices on Earth at 250 K with the same efficiency for a given system. Quantum chemical tunneling could thus lead to an exchange and mixing of D/H isotopes in the bulk ice even though it might have started with isotope enrichment at one location (such as in the case of a comet's interior or on its surface). Based on the above recent experimental and theoretical data, it is necessary for the community to ensure that the observations made today can hold be compared to the times of their origin a few billion years ago.

5.1.1 D/H Ratios Measured on Comet Outgassing (Coma/Tail)

In spite of several observational data on the D/H ratios in interstellar and solar system water (mostly observed in the gas-phase), there is a critical need to get more observational data to improve the statistics and error bars, particularly for highly active solar system bodies such as comets. While the lower limit of deuterium abundance seems to be set by galactic and solar nebula D/H ratio of $\sim 2 \times 10^{-5}$, the upper limit observed so far in water seems to be $\sim 2 \times 10^{-2}$ in cold interstellar clouds. Water on Earth and comets (outgassing) fall in between at $\sim$ 1–3 $\times 10^{-4}$. Recent Rosetta ROSINA data for the D/H ratio of water from the comet 67P/CG ($5.3 \pm 0.7 \times 10^{-4}$) puts the hypothesis—water on Earth could have come from comet impacts—to question. Data on the bulk ice D/H ratio in a comet's nucleus is not available so far, as this can only be achieved through *in situ* sample analysis or cryogenic comet sample return. Once the planetesimals are formed there would be less direct radiation effect on the D/H ratios, but before their formation, in addition to the other processes mentioned above, radiation induced fractionation of D/H on a ice grain should also be taken into account. The role of quantum chemical tunneling—that is temperature independent and can occur at the coldest, darkest parts of icy bodies with the same efficiency (Hama et al. 2014; Trixler 2013)—needs to be evaluated in the laboratory through experimental simulations.

5.2 Molecular Nitrogen in Comets

Contrary to CO, non-polar volatiles such as N_2 and O_2 have evaded direct observation through remote sensing due to their transparency in the infrared spectral region (Wyckoff et al. 1991). All these three molecules have similar volatility, suggesting N_2 and O_2 to be present in cometary nucleus, coma, and tail. Molecular oxygen is yet to be detected unambiguously, whereas Rosetta mission with its ROSINA instrument has successfully detected N_2 recently (Rubin et al. 2015), depleted by about a factor of 25 compared to protostellar N_2/CO ratio. A formation temperature of $\sim$30 K is inferred from these observations, which could be explained by either amorphous ice with trapped N_2 and CO or crystalline ice with N_2 and CO clathrates. Though the observation of N_2 by Rosetta resolved the long-standing issue of "where is N_2", the question of amorphous or crystalline interior is still open. Laboratory studies are necessary to address the question of D/H ratios as well as the N_2 vs. CO trapped in amorphous vs. crystalline ices.

5.3 Volatile Organic Compounds (VOCs) from Cometary Ice Analogs

5.3.1 Volatile Desorption from Water Matrix

The desorption of volatiles in comets is a complex phenomenon (Schaller and Brown 2007). Due to the size and mass of the comet, volatiles sublime and possibly form a vapor pressure controlled atmosphere, with a dynamical accretion-desorption phenomenon. Depending on the mass and temperature of the body, atmospheric escape can occur through a variety of processes, such as the Jean or the hydrodynamics escapes. On the microscopic scale, volatiles need to be accessible to the surface to desorb. They need to diffuse in the bulk of the cometary ice before reaching the surface. Their desorption processes are thus intimately linked to the water ice substrate (Jenniskens and Blake 1994). Amorphous water ice crystallization induces a fast release of the volatiles (Smith et al. 1997), the so-called molecular volcano process. Volatiles, which are not released through the diffusion-desorption process or the molecular volcano process are trapped in crystalline ice and will co-desorb along with the water ice mantle. The mechanisms for the entrapment of volatile gases trapped within amorphous solid water (ASW) have been studied by many authors (Notesco et al. 2003; Collings et al. 2003, 2004). This trapping of volatile gas in ice is of fundamental importance for chemical reactions and for the outgassing of interstellar ice in star forming regions and in comets (Marboeuf et al. 2012a, 2012b; Bar-Nun et al. 2013; Bar-Nun and Laufer 2013). The fraction of trapped volatiles within the ice and their release are depending on the thickness of the ice and on the heating rate (May et al. 2013a, 2013b). When extrapolating May et al.'s results (May et al. 2013b) to thick ice and to complex molecules, we can reasonably assume that co-desorption with the ice mantle dominates the volatile desorption. Another pathway to desorb volatiles from water ice is through photodissociation caused by vacuum ultraviolet (VUV) photons resulting in smaller atoms and molecules that escape the ice even at lower temperatures corresponding to larger heliocentric distances of a comet. For example VUV (63–310 nm) photochemistry of $H_2O : CO_2 : NH_3$ (1:1:1) ice produces H_2, NH_2, OH, O_2, and CO as volatiles exiting the ice (Chen et al. 2011).

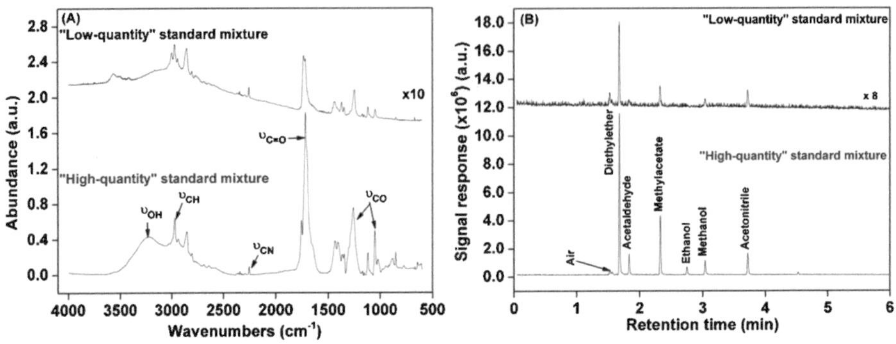

Fig. 15 Recovery of volatile organic compound (VOC) through the VAHIIA interface after their deposition on a cold finger (20 K) inside the vacuum chamber (10 mbar). Infrared spectra of ices (**A**) and GC-MS after the VOC recovery during the ice warming (**B**). Standard mixture used: diethylether, acetaldehyde, methanol, ethanol, methylacetate and acetonitrile. Reprinted with permission from Abou Mrad et al. (2014). Copyright 2014 American Chemical Society

5.3.2 Volatile Analyses from Ice Analogues

Some volatile compound analyses have been done on ice analogues, often on simple chemical systems, since their analyses with mass spectrometers (usually low mass resolution QMS) allow the confirmation of their identification previously obtained with infrared spectroscopy. As examples, it can be cited the glycine salt formation from a first thermal condensation of methylamine NH_2CH_3 and carbon dioxide CO_2 leading to the corresponding carbamate, followed by its UV irradiation forming this glycine salt (Bossa et al. 2009a); or the formation of aminoacetonitrile through UV irradiations of an acetonitrile and ammonia ice mixture (Danger et al. 2011b).

Volatile analyses have also been performed to understand products coming from thermal or photo-degradation of specific compounds such as polyoxymethylene (POM) or hexamethylenetetramine (HMT), that could explain the distribution of some species in comets such as formaldehyde or hydrogen cyanide (Cottin et al. 2000, 2002). These analyses were sometimes obtained by a direct mass spectrometer or by GC-MS after the condensation of volatile species on a cold finger and their subsequent warming (Cottin et al. 2000). Other authors used mass spectrometry for the study of gas trapping in water at low temperature (Notesco et al. 2003). However, no such studies were done for the characterization of volatile species coming from the heating of complex ice analogues, which are relevant to interstellar or cometary ices.

Recently, an analytical system devoted to the analysis of volatile organic compounds (VOC) has been developed (Abou Mrad et al. 2014). It consists in coupling a GC-MS system through a specific interface (the VAHIIA interface) to the vacuum chamber for obtaining an online analysis of these VOCs. The VAHIIA interface allows the recovery of VOC coming from the heating of cometary ice analogues and their subsequent transfer to the GC-MS for VOC separation and characterization. Using a standard mixture, this system was validated and has shown its ability to analyze and identify compounds deposited on the cold finger inside the vacuum chamber and subsequently warmed (Fig. 15). The forthcoming results will thus give important information on the reactivity that leads to residue formation. Furthermore, the GC-MS being a technique used and developed for space missions such as for the COSAC instrument on the Lander Philae of the *Rosetta* mission (Szopa et al. 1999,

2002a, 2002b, 2003), the VAHIIA experiment should provide interesting information for the future data analyses coming from this mission.

6 Comet Coma and Tail

6.1 Negative Ions in Cometary Coma

An active positive-ion (cation) chemistry has long been known to occur in the collisional inner coma of comets (e.g. Rodgers et al. 2004). However, it is only recently that the possibility of an extensive negative-ion (anion) chemistry has been considered in theoretical models. Although several studies have been made, more experimental data on anionic processes involving cometary molecules are required. An important and unexpected discovery made by the RPA-1 electron analyzer on the *Giotto* spacecraft was the detection of negative ions (i.e., *anions*) in the inner coma of comet 1P/Halley: O^-, OH^-, C^-, CH^- and CN^- (Chaizy et al. 1991). Additional unidentified anions were also detected within the 22–65 and 85–110 amu (m/z) mass regions, leading Chiazy et al. to conclude that these were the negatively-charged versions of the complex 'CHON' organics detected in Halley's coma (e.g., Altwegg et al. 1999). Prior to this, Wekhof (1981) had speculated on the possible presence of negative ions in cometary comae but, nevertheless, the Halley results came as a surprise. Subsequently, the presence of C_2^- has been claimed in optical spectra of comets Scorichenko-George (1990 VI) and C/2000 WM1 (LINEAR) (Churyumov et al. 1993, 2002).

Dalgarno and McCray (1973) and Herbst (1981) suggested that negative ions could be present in dense interstellar clouds and detailed predictions were made for specific species (e.g. Millar et al. 2000). These predictions were validated by the recent detection of anions in interstellar clouds and circumstellar ouflows. The anions C_4H^-, C_6H^-, and C_8H^- have been detected in the envelope of the C-rich AGB star IRC+10216 (McCarthy et al. 2006; Cernicharo et al. 2007; Remijan et al. 2007); C_4H^- and C_6H^- are widespread in molecular clouds (Cordiner et al. 2013). Particularly relevant for comet chemistry is the fact that the emission line profiles in IRC+10216 indicate that the anions are forming in the region of the outflow where the interstellar UV radiation field is driving the chemistry. Bare carbon chains and various carbene (such as $H_2C=C=C$:) and cumulene (such as $H_2C=C=CH_2$) chains are detected directly in almost all comets (e.g., C_2 and C_3), are present as fragmentation products in the Halley data (e.g., C_4H), are suggested by the *Stardust* data, and are expected on the basis of interstellar chemistry (Mitchell et al. 1992; Altwegg et al. 1999; Geiss et al. 1999; Schleicher and Farnham 2004; Sandford et al. 2006). Thus, we should expect that a similar anionic photochemistry would also occur in cometary plasmas.

Amongst the various anion formation processes, three are most relevant for cometary comae:

(1) The elementary process of *radiative electron attachment*, REA,

$$e^- + \mathrm{CN} -> \mathrm{CN}^- + h\nu$$

is favored by a deep potential well for electron attachment (i.e., a large *electron affinity*, EA, of the neutral) and the presence of a large number of rotational and vibrational degrees of freedom for energy dispersal and IR photon emission. Competition with electron *autodetachment* from the intermediate activated complex leads to slow rates for radiative electron attachment to atoms and small molecules, $\sim 10^{-14}$ cm^3 s^{-1}. Organic radicals with significant EA values, and many internal degrees of freedom, can have

REA rates approaching the collisional rate of $\sim 10^{-7}$–10^{-8} cm^3 s^{-1} (Herbst and Osamura 2008), making them prime candidates for anion formation by electron radiative attachment in dense, UV-irradiated astrophysical sources (Millar et al. 2007).

(2) *Dissociative electron attachment*, DEA

$$e^- + \text{HCN} -> \text{CN}^- + \text{H}$$

Experimental studies of DEA to acetylene and formamide have been performed (Szymanska and Mason 2011, Szymanska et al. 2014a, 2014b). In the former case product channels for both C_2^- and C_2H^- exist, suggesting that these anions may be present in cometary comae. Quantum-chemical calculations for rates of DEA and REA to long carbon-chain molecules that may be relevant to comets (polar and non-polar polyynes) have also been computed (Carelli et al. 2013a, 2013b).

(3) *Threshold ion-pair production* (TIPP; also called polar photodissociation),

$$h\nu + \text{HCN} -> \text{CN}^- + \text{H}^+$$

should also be important in forming negative ions (Berkowitz et al. 1969; Massey 1976; Berkowitz 1979; Suits and Hepburn 2006; Hu et al. 2006). For example, Solar UV photons in the range $\approx$15.2–19 eV could produce CN^- from HCN.

The major loss route for cometary anions is *photodetachment*

$$\text{CN}^- + h\nu -> \text{CN} + e^-$$

The photodetachment cross-sections for small ions such as OH^- are large and yield rapid photodetachment rates in the unattenuated Solar UV field at 1 AU (Wekhof 1981). Complex molecules with higher EA values may have much lower cross-sections (e.g., Smith et al. 1978). The lifetimes of anions in the inner coma can be much longer than those calculated for the unattenuated Solar UV flux. Differential optical depth effects in the inner coma may tend to inhibit photodetachment, which destroys anions, but not substantially affect the TIPP processes forming them. Specifically, the high-energy photons, which drive TIPP, can penetrate deeper, producing anions in the dense, inner regions of the coma. The lower-energy photons, which cause photodetachment, however, will be absorbed in the outer coma.

These and other negative-ion processes were incorporated into the first model of cometary anion chemistry by Cordiner and Charnley (2014). Calculations showed that CN^-, C_4H^-, C_6H^-, and C_8H^- could all be expected to form in the coma and thus could provide candidate identifications for several of the peaks in the *Giotto* mass spectra at 1P/Halley. Anions may also be detected by the Rosetta mission to comet 67P/Churyumov-Gerasimenko. Part of the Rosetta Plasma Consortium, the ion and electron sensor (IES) contains two electrostatic plasma analyzers (one each for ions and electrons), operating over the energy/charge range 1 eV/e to 22 keV/e (Burch et al. 2007). Similar to the Giotto electron electrostatic analyser (EESA) that detected various negatively-charged molecules in the coma of comet Halley (Chaizy et al. 1991), the IES should have the capability of detecting negatively-charged particles in the coma of comet 67P, including molecular anions and negative nanograins (Goldstein et al. 2012). In contrast to Giotto, which traversed Halley's coma at about 68 km/s and was sensitive to negative ions with masses 1–830 amu, Rosetta's orbital speed with respect to the comet 67P/Churyumov-Gerasimenko will be small relative to the 1 km/s outflow velocity of the coma gases. It is therefore expected that the Rosetta IES will be sensitive to singly-charged negative ions in the coma with a higher mass range (190–4.2×10^6 amu), covering large molecular anions and organic macromolecules with tens to hundreds of thousands of carbon atoms.

Thus, it is likely that any anionic molecular species detected by Rosetta would be derived from large organic compounds. Possible candidates are polycyclic aromatic hydrocarbons (PAHs), polymers of aliphatic and aromatic hydrocarbons, or polymers with highly polar/electronegative functional groups. PAHs are ubiquitous in many astrophysical environments and have previously been considered as components of cometary dust and ices (see review by Li 2009). Direct detection of PAHs in cometary comae is difficult and the tentative identifications of phenanthrene ($C_{14}H_{10}$) and pyrene ($C_{16}H_{10}$) in UV spectra of 1P/Halley, obtained by the Vega-2 spacecraft (Moreels et al. 1994; Clairemidi et al. 2004), remain inconclusive. Nevertheless, PAHs are clearly evident in the Stardust samples returned from comet 81P/Wild 2 (Sandford et al. 2006) and so they and their anions may indeed be found at 67P/Churyumov-Gerasimenko.

Development of coma chemistry models for PAH anion formation requires laboratory data and quantum-chemical calculations of the rates of anion processes. Some relevant work is the calculation of REA rates for phenanthrene, perylene ($C_{20}H_{12}$) and coronene ($C_{24}H_{12}$) by Carelli et al. (2013c). Experiments on DEA to PAHs have also been reported (Szymanska and Mason 2011). Future models of cometary anions will require more laboratory data and theoretical calculations on absolute cross sections and rates for anion chemistry involving abundant coma molecules. These will be particularly important if Rosetta detects anions in coma of comet 67P/Churyumov-Gerasimenko.

6.2 Fluorescence from Frozen Hydrocarbon Particles (FHPs) in the Coma and on the Surface

6.2.1 Emission Spectra of Active Comets

One of the most effective methods of cometary investigation is the spectroscopic observation of these bodies with subsequent analysis of the obtained data. Cometary spectra consists of two principal parts—emission lines and bands radiated by gaseous compounds of cometary comas, and faint solar continuum scattered by dusty and icy particles of the cometary halos. Gaseous comas are cool and rarefied shells around cometary nuclei and emit fluoresce on the specific wavelengths in UV, visible, and IR ranges of the electromagnetic spectrum. Daughter molecules as the result of photo degradation of parent species are main compounds of gaseous cometary comas. The following molecules and ions were already detected in comets: C_2, C_3, CH, CN, CO_2, NH, NH_2, OH, CH II, CO II (Arpigny 1995; Cochran et al. 2012). This list is incomplete, surely, being the subject of permanent growth. Parent molecule or their degradation products were detected as well in the cometary bodies, particularly, H_2O, HCN, C_2H_2, CH_3OH, C_2H_6, H_2CO and others (Mumma and Charnley 2011). The mineral and the icy halos of comets are clouds of fine dispersed grains responsible for the scattering of solar radiation. Gaseous comas and grains of halos form the cometary atmospheres as complex of gas and dust around icy nuclei. The cometary spectra are rich also with the series of narrow emission lines of unknown nature. These so-called unidentified emissions were registered in the spectra of many comets (Brown et al. 1996; Cochran and Cochran 2002; Cremonese et al. 2007; Kobayashi and Kawakita 2009; Dello Russo et al. 2013; A'Hearn et al. 2014). Many observers assigned these emissions to multiple ionized molecules (Wyckoff et al. 1999; Cochran and Cochran 2002; Kawakita and Watanabe 2002). New theory suggests luminescence nature of unidentified emissions in the form of photoluminescence by frozen hydrocarbon particles (FHPs) of cometary atmospheres (Simonia 2004, 2007, 2011a, 2011b, 2013). Approximately 10 % of the previously unidentified emissions were identified as photoluminescence of cometary

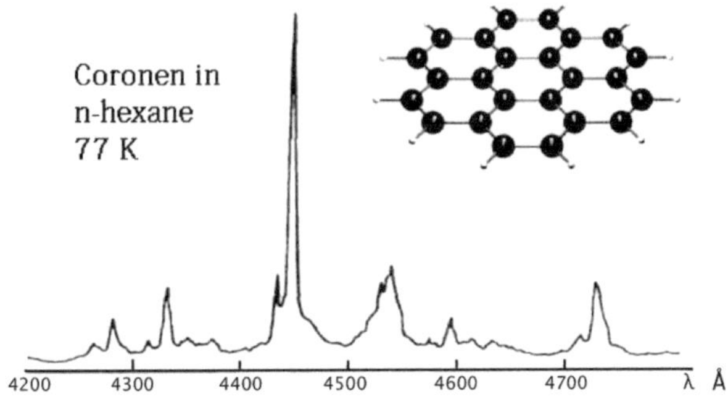

Fig. 16 Quasilinear photoluminescence spectrum of Coronene in n-hexane at 77 K (Shpolskii 1960)

FHPs, though 90 % of the mentioned emissions remained in the status of unidentified and more laboratory work is needed to assign these emission bands. The peculiarities of cometary atmospheres provide the complicated character of cometary spectra in the form of various features superposition. The cometary nuclei are rich with complex organic substance (Ehrenfreund and Charnley 2000; Gudipati and Allamandola 2003; Clairemidi et al. 2004; Crovisier and Bockelée-Morvan 2007; Li 2009; Kobayashi and Kawakita 2009; Zellner et al. 2011).

6.2.2 Frozen Hydrocarbon Particles (FHP) in Cometary Coma

When the comets approach the Sun, processes of thermo destruction and sublimation of cometary ices are started. Micro fragments of nucleus ices form the spherical halo with various diameters. The sources of cometary hydrocarbons provide organic compounds for icy cometary halos as shells of frozen hydrocarbon particles (FHPs). Each FHP is characterized with inherent size, shape, and color. FHPs with substitutional solid solution structures have nano sizes, but molecularly dispersed FHPs micro sizes. Shapes of individual FHPs depend on structural peculiarities of icy particles. FHPs with substitutional structures may be flat, but molecular dispersed particles—bulky and elliptic.

The solar ultraviolet radiation excites FHPs, which produce photoluminescence cometary halos in range of 4000–8000 Å. Chemical compositions of cometary FHPs and physical peculiarities of Shpolskii microcrystals have similar properties of their photoluminescence spectra including positions and profiles of emission lines. Photoluminescence spectra of the cometary FHPs at low temperature ($T < 80$ K) have the standard quasilinear forms—series of multiple narrow emission lines with average line width $\Delta\lambda \leq 1$ Å. Laboratory analogs of such spectra are called quasilinear photoluminescence spectra of Shpolskii matrix (PAH in n-alkanes) (Fig. 16).

We also note that PAHs trapped and isolated in water-ice grains such as in ice particles in cometary tails also show strong photoluminescence properties, particularly their emission spectra very similar to gas-phase spectra but much broader due to inhomogeneous broadening (Allodi et al. 2013).

Photoluminescence spectra of the cometary frozen hydrocarbons could demonstrate also wide featureless bands (case of resonance dimmers). Such featureless bands may be observed as certain quasicontinuum. Similar phenomenon was reported by Churyumov and Kleshchenok (1999), Churyumov et al. (2013). On the other hand the width of emission lines and general character of photoluminescence spectra of Shpolskii matrix depends also on temperature of polycrystalline substance. According to Shpolskii (1959) acyclic matrix provides quasilinear character of frozen PAHs luminescence spectra. In case of glass-similar matrix the same PAHs demonstrates featureless photoluminescence emissions. For the cometary case it means probably that quasi-continual luminescence emissions belong to amorphous mixtures of hydrocarbons. Narrow luminescence emissions indicate PAHs in n-alkanes matrix—crystalline FHPs. It might be the qualitative tool for remote determination of hydrocarbons state in the cometary matter.

Frozen hydrocarbon particles of the cometary halos interact also with gaseous compounds of comas. Physically such interactions are probably nonelastic in the form of absorption and adsorption of comas molecules by the solid substance of FHPs. Main result of the mentioned processes is enrichment and encapsulation of FHP substance with the neutral and ionic inorganic molecules of the cometary origin. Photoluminescence spectra of enriched organic probably differ from the original in positions and profiles of emission lines.

Solar UV and visible radiations probably excites also fluorescence of cometary solid organic in infrared range within 2–13 μm. FHPs of the cometary halos in this case as complex of fine dispersed icy grains are subject of the so-called infrared fluorescence phenomenon. Surfaces of cometary nuclei, including ice and mineral components are irradiated permanently by solar electromagnetic radiation and fluxes of the charged particles. Photo-processing and electrons bombardment of nuclei surfaces provides transformation of physical and chemical characteristics of surface substance. Within time scale of comets evolution the substance of nuclei surfaces is transformed completely due to external factors and forces. In other words, surface layers of the periodic comets nuclei contain new substance—the result of evolution. The substance of inner layers of the cometary nuclei including hydrocarbons, which have never been processed by the solar radiation, has primordial properties. Such a substance is called—the cometary relicts. The main properties of the cometary relicts are probably following: (a) extremely low temperature of the substance; (b) complex character of the substance with primordial structural peculiarities; (c) possible presence of prebiotic compounds.

6.2.3 *Frozen Hydrocarbon Particles in Cometary Jets and Flares*

In time of active cometary processes including flares and outbursts the certain fragments of relict substance (hydrocarbons mixtures) might be delivered in to the cometary comas. The solar radiation could excite photoluminescence and cathodoluminescence of newly delivered relicts. Depending on the chemical composition and the physical properties of the specific relicts such luminescence may be in the form of featureless emissions or highly structured quasilinear spectra. Duration of luminescence of the relicts may be rather short. The intense luminescence emissions of the relicts may be quenched by the solar infrared radiation. On the other hand, complex molecules of the same relict substance may be dissociated on daughter formations in the field of the solar hard ultraviolet radiation. Taking in to account high quantum yield of hydrocarbons luminescence we have proposed that ground based observations for registration of the relict substance luminescence must be made in time of outbursts or immediately after this phenomena. The cometary relicts in the form of clouds of icy micrograins will be processed by the solar radiation. Therefore, within 6–8 hours after outburst luminescence properties of relict substance will be lost or changed. The cometary outbursts or flashes spectra have character of featureless emissions in range 3900–5500 Å (Cochran et al. 1980). Frozen organic mixtures, in certain cases, demonstrate photoluminescence spectra in the form of single bump in almost the same spectral range. Registration of photoluminescence of the cometary relicts including complex organic substance could become an effective tool for investigation of pristine matter of the Solar System including prebiotic compounds. This new conceptual hypothesis has been developed and presented by Simonia and Simonia (2013).

Table 1 Photoluminescence emissions in the spectra of Hyakutake comet

Cometary emissions (Å)	LabPhotolum lines (Å)	PAHs in n-alkanes	Reference
5102.235	5101.8	1-amino, 9, 10-anthraquinone + n-heptane	Nakhimovsky et al. (1989)
5110.064	5110	Naphtacene + n-nonane	Nakhimovsky et al. (1989)
5314.689	5315.4	9, 10-anthraquinone + n-heptane	Nakhimovsky et al. (1989)
5318.079	5318.1	2-methyl, 1, 4-naphtoquinone + n-hexane	Nakhimovsky et al. (1989)
5332.924	5332.5	1, 2-benzopyrene + tetrahydrofuran	Kirkbright and Lima (1974)
5909.68	5910	Tetraphene + n-hexane	Teplitskaiya et al. (1978)

6.2.4 Photoluminescence of Complex Organics of C/1996 B2 (Hyakutake) and C/1995 O1 (Hale-Bopp) Comets

The successful identification of previously unidentified cometary emissions is the important task for cometary organic investigations. Here two comets, namely C/1996 B2 Hyakutake and C/1995 O1 Hale-Bopp, are selected for identification of previously unidentified or unknown emissions. Morrison et al. (1997) observed Hyakutake comet with a Fiber-fed echelle spectrograph attached to the 1-m telescope of Ritter observatory at March 23 and 27, 1996; receiver—CCD; resolving power of spectrograph $R = 25800$; average signal-to-noise ratio was equal 200. Hyakutake comet was observed on the heliocentric distances $r = 1.08$; 1.00 AU. The mentioned authors found in the spectrum of Hyakutake comet the standard emissions of several molecules and ions. They revealed also 18 emission lines of unknown nature, which acquired the status of the unidentified emissions and used these lines for the comparative analysis.

Zhang et al. (2001) observed Hale-Bopp comet with a Cude Echelle spectrograph mounted on the 2.16 m telescope of Beijing astronomical observatory (Xinglong, China) at March 26, 28 and April 22, 1997; receiver—CCD; resolving power of spectrograph $R = 44000$; 37000 for the blue and red path; average signal-to-noise ratio was equal 180. Hale-Bopp comet was observed on the heliocentric distances $r = 0.920$; 0.918; 0.991 AU. The mentioned authors found in the spectra of Hale-Bopp comet the standard emissions of several molecules and ions. They revealed also 73 emission lines with unknown nature. For the comparative analysis works of Kirkbright and Lima (1974), Teplitskaiya et al. (1978), and Nakhimovsky et al. (1989) are referred to as the sources of laboratory data. These sources have been selected on the bases of supposition that substance of frozen hydrocarbon particles of the cometary atmospheres has the same structure as Shpolskii matrix—aromatic molecules trapped in n-alkane ices. The comparative analysis was made with the following strategy: (a) coincidence of the spectral positions of the observational and laboratory emission lines with accuracy of $+/-1$ Å; (b) profiles similarities of observational and laboratory emission lines.

All specific circumstances, including differences in wavelength scales, were taken in to consideration. The results of the comparative analysis for both comets are presented in Tables 1, 2.

In the spectrum of Hyakutake comet 18 emission lines were tabulated as unidentified (Morrison et al. 1997). We have identified 6 emissions as photoluminescence of PAHs fixed in n-alkanes. In other words 33.3 % of previously unidentified emissions represent photoluminescence of frozen hydrocarbon particles of icy halo of Hyakutake comet. In the spectrum of Hale-Bopp comet 73 emission lines were tabulated as unidentified (Zhang et al. 2001).

Table 2 Photoluminescence emissions in the spectrum of Hale-Bopp comet

Cometary emissions (Å)	LabPhotolum lines (Å)	PAHs in n-alkanes	Reference
4838.27	4838.2	Perylene + n-hexane	Nakhimovsky et al. (1989)
4850.66	4850	2-methylperylene + n-hexane	Nakhimovsky et al. (1989)
5318.19	5318.1	2-methyl, 1, 4-naphtoquinone	Nakhimovsky et al. (1989)
5333.01	5332.5	1, 2-benzopyrene + tetrahydrofuran	Kirkbright and Lima (1974)
5351.438	5352	Chrysene + n-hexane	Teplitskaiya et al. (1978)
5428.19	5427.5	1, 2-benzopyrene + tetrahydrofuran	Kirkbright and Lima (1974)
5909.64	5910	Tetraphene + n-hexane	Teplitskaiya et al. (1978)

We have identified 7 emissions as photoluminescence of PAHs in n-alkanes. 9.5 % of the total quantity of previously unidentified emission lines in the spectrum of Hale-Bopp comet represents photoluminescence of frozen hydrocarbon particles of icy halo of this comet (Table 2).

Figure 17 represents certain fragments of spectral profiles from Morrison et al. (1997), Kirkbright and Lima (1974), and Zhang et al. (2001), respectively. Photoluminescence emission line $\lambda = 5332.5$ Å of 1, 2-benzopyrene (Kirkbright and Lima 1974), coincides in spectral position and profile with previously unidentified emissions in the spectra of Hyakutake and Hale-Bopp comets. Other emission lines of these comets spectra also coincides in spectral positions and profiles with laboratory photoluminescence emission lines of PAHs in n-alkanes. In fact the substance of Hyakutake and Hale-Bopp comets is rich in complex organic in the form of frozen mixtures of aromatic and acyclic hydrocarbons. The inner layers of Hyakutake and Hale-Bopp comets nuclei contain the bulk organic zones—sources of hydrocarbons. Organic relicts are delivered from cometary nuclei in to atmosphere on the short heliocentric distances as the natural markers of the Solar System past.

In other words, Schpolskii matrix in the form of frozen polycrystalline mixture (prebiotics in n-alkanes) luminesces as series of very narrow emission lines. In this case luminescence of such mixtures is excited by ultraviolet radiation. It means that series of very narrow emission lines of the cometary spectra (early unidentified emissions) might belong to luminescence of FHPs with prebiotic compounds. Results of the comparative analisys and this hypothesis has been obtained, developed, and presented by Simonia and Simonia (2013, 2014).

6.2.5 Implications to Detection of Aromatic Molecules on Comets

The frozen hydrocarbons are components of cometary ice. The cometary substance is probably rich in the complex aromatics and acyclics. These hydrocarbons are comprised unified structures including solid solutions, polycrystalline mixtures, and substitutional matrices. The potential cometary prebiotics or biotics are related with the complex organic. Therefore the significance of remote revealing of complex organic of comets is high. The intense solar ultraviolet radiation, the fluxes of charged particles, and the structure of cometary organic substance provide bright luminescence phenomena of cometary organic ice with possibilities of its detection. We have suggested that Shpolskii matrix in the form of polycrystalline mixtures (PAHs in n-alkanes) might be the main unit of cometary complex organic substance.

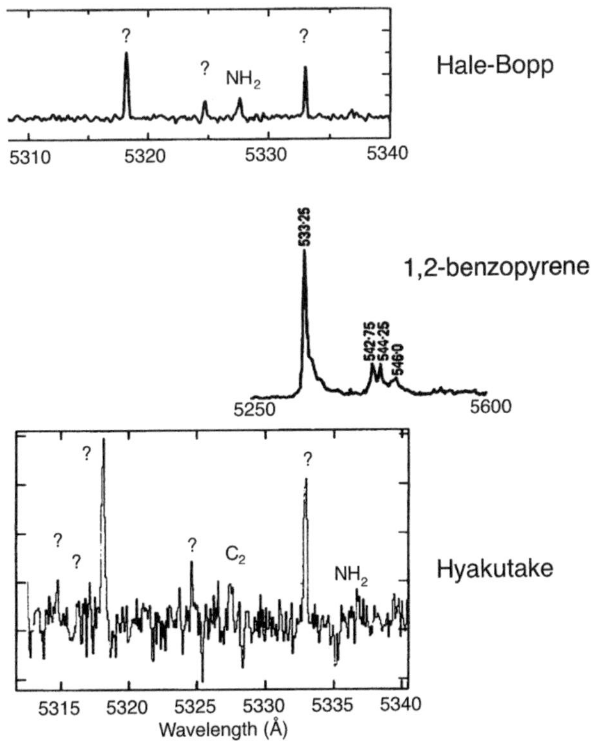

Fig. 17 Unidentified emissions in spectra of Hale-Bopp (5333.01 Å) and Hyakutake (5332.92 Å) comets; Luminescence of 1, 2-benzopyrene (5332.5 Å)

Frozen hydrocarbon particulate (FHP) matter of the cometary nuclei will be the main source of fine dispersed organic icy particles of the cometary halos. In time of the cometary outbursts or flares the certain portions of cometary relict matter penetrates in to cometary comas with subsequent photo-excitation of this matter. The de-excitation of previously excited molecules (relict molecules) will be accompanied by bright luminescence emission. Optical spectra of the cometary bursts could be rich in unknown emission lines or featureless bands especially in the blue range. Photoluminescence spectra of polycyclic aromatic hydrocarbons at temperatures $T > 100$ K has featureless character in the form of extended blue emission (4000–4900 Å). In our opinion for the first time this extended blue emission of the cometary spectra was registered by Bobrovnikoff (1927a, 1927b). In terminology of Bobrovnikoff "violet type spectra" were registered for the several comets. Archived spectroscopic data of the cometary outbursts or flares must be reanalyzed in light of this problem.

We have also described the cometary FHPs with molecularly dispersed structure and substitutional solid solution structure. Both molecular systems have binary nature. The first class of the icy particles consists of aromatic molecules (guest) fixed in whole acyclic matrix (host). The second class of the icy particles consists of aromatics (or carbon rings) substituted in acyclic chains. Such solids belong to the so-called Shpolskii matrix, which is the source of bright luminescence in the field of UV radiation. We have studied the spectra of Hyakutake and Hale-Bopp comets in search of FHPs photoluminescence emissions. The comparative analysis of the optical spectra of the mentioned comets with photoluminescence spectra of FHPs laboratory analogs was performed. For Hyakutake comet 6 previously unidentified lines were identified as cometary FHPs photoluminescence. For Hale-Bopp comet—7 emissions were assigned to photoluminescence of frozen hydrocarbon particles of icy halo.

The main properties of analogs of frozen hydrocarbon matter of comets discussed here show the possibilities of revealing of prebiotics as pristine matter of the Solar System. Several unidentified emission bands from comets could be associated with laboratory spectra of PAHs trapped in Shpolskii organic matrix, and new suppositions were presented. Photoluminescence has been shown to be an important molecular property to investigate and identify cometary organic matter. At the same time we are rather far from the complete understanding of cometary organics' nature and its interactions with external forces. More laboratory photoluminescence data is needed on realistic analogs of cometary ices interacting with organics, minerals, and radiation and UV wavelengths that simulate solar spectrum also into the vacuum ultraviolet region ($\sim$ 100–300 nm).

7 Conclusions

In this review we have shown that though our understanding of comets has tremendously improved in the past decades, several key questions remain unanswered, for which serious answers can only be obtained with cryogenic comet subsurface sample return (CCSSR) missions. Until then the composition and geology of a comet nucleus remains a puzzle, including whether water-ice exists in amorphous or crystalline phase, whether clathrates can exist or even dominate the nucleus, etc. Crystalline ice alone does not trap large quantities of impurities (volatiles) within the lattice, resulting in the volatiles to be expelled into ice grain boundaries. How such a scenario would fit into observation of outgassing of short period comets needs to be explored both in the laboratory and through modeling studies. Water and volatiles condensed at very low temperatures forming high-volatile-content amorphous ice grains (similar to interstellar ice grains) explains readily many observed phenomena of comets. Crystalline water-ice can not be ruled out, but to explain large quantities of trapped volatiles it is necessary to invoke clathrates that are formed at higher temperature and pressure or condensation of volatiles along with crystalline ice grains at very low temperatures ($\sim$30 K). Laboratory measurements on the thermal properties of realistic cometary analogs have begun and need to be intensified in order to constrain the same on comets. Thanks to extensive efforts made in the laboratory, we have now a better understanding of the breadth of volatile and non-volatile organics that can be formed on the surface and near surface. While PAHs may be potential candidates to be looked at for both their emission as FHP or as negatively charged ions in cometary coma, cometary surface and crust could host a suite of organic matter from abiotic to prebiotic to biogenic. Many of these molecules have been detected and analyzed in the laboratory on cometary ice analogs. A few observations on isotope enrichment (D/H ratio) from comets seem to fall in a range (within a factor of 2–5) of terrestrial D/H ratio. While the observations are robust, how to interpret them in connection with the origin of our solar system seems to be heavily debated. More laboratory work is warranted in this connection, especially taking into account potential channels of equilibration, exchange, and defractionation.

Since the beginning of 2015 science results from the Rosetta mission are becoming public. In addition to being the first ever landing on a comet surface (unlike impacting a comet during the Deep Impact mission), Philae revealed unexpected hardness of comet 67P/CG shell, which could be understood as being due to hard crystalline ice beneath the expelled organics during warm-up of its surface as the comet's surface gets warmer. Long-standing hypothesis that comet's dark low-albedo surface is organic-rich, is proven by Rosetta's VIRTIS observations. However, it still remains unresolved what is the molecular composition of these organics, for which more observations and perhaps future comet surface sample return (touch-and-go) could help. Laboratory studies are also critical to evaluate a wide range of conditions and compositions possible to interpret the observed spectra. For a better understanding of the interior, we still need to wait for final results from the CONSERT results to be published. Microwave sounder instrument (MIRO) on Rosetta measured low thermal inertia of 67P/CG, in accordance with the expected low-thermal conductivity and high porosity of cometary surface dust. Most surprising results also came from the ROSINA mass spectrometer—that water from 67P/CG is highly enriched with heavier isotope D, putting these observations to the highest deuterium enrichment observed so far indicating extremely cold interior preserving high-deuterium content from interstellar ices. This observation is in line with the detection of N_2 and the evaluation of N_2/CO ratio leading to $\sim$30 K as the temperature of cometary core during its formation. Each of these observations raise more questions, for which we also future comet missions, particularly that perform in-situ analysis of subsurface composition of a comet at least a meter or two deep and bring back these cryogenic comet samples at $\sim$30 K leading us in to the next-generation comet science.

Acknowledgements Murthy Gudipati thanks partial funding from the following NASA programs: Planetary Atmospheres, Cassini Data Analysis Programs, Spitzer Science Center, and Astrobiology Institute Node Early Habitable Environments (NASA Ames). JPL's DRDF and R&TD funding for infrastructure of the "ice spectroscopy laboratory" is also gratefully acknowledged. This research was carried out at the Jet Propulsion Laboratory, California Institute of Technology, under a contract with the National Aeronautics and Space Administration. Olivier Mousis thanks the support of the A*MIDEX project (no. ANR-11-IDEX-0001-02) funded by the "Investissements d'Avenir" French Government program, managed by the French National Research Agency (ANR). Grégoire Danger and collaborators thank funding by the French national programs "Physique Chimie du Milieu Interstellaire" (P.C.M.I, INSU), "Programme National de Planétologie" (P.N.P, INSU), "Environnements Planetaires et Origines de la Vie" (E.P.O.V, CNRS) and the "Centre National d'Etudes Spatiales" (C.N.E.S) from its exobiology program as well as the ANR project VAHIIA (Grant ANR-12-JS08-0001-01) of the French Agence Nationale de la Recherche. Ulysse Marboeuf thanks funding from the Swiss National Science Foundation. The work of Steven B. Charnley and Martin A. Cordiner was supported by NASA's Planetary Atmospheres Program and the Goddard Center for Astrobiology.

References

M.F. A'Hearn, Comets as building block, in *Annual Review of Astronomy and Astrophysics*, vol. 49, ed. by S.M. Faber, E. VanDishoeck (2011), pp. 281–299

M.F. A'Hearn, D.D. Wellnitz, R. Meier, λ4430 emission by comet Hyakutake. The diffuse interstellar bands, in *Proceedings of the International Astronomical Union*. IAU Symposium, vol. 297 (2014), pp. 216–218

T. Albertsson, D. Semenov, T. Henning, Chemodynamical deuterium fractionation in the early solar nebular: the origin of water on earth and in asteroids and comets. Astrophys. J. **784**, L39 (2014)

J. Aléon, F. Robert, Interstellar chemistry recorded by nitrogen isotopes in solar system organic matter. Icarus **167**(2), 424–430 (2004)

C.M.O. Alexander, S.D. Newsome, M.L. Fogel, L.R. Nittler, H. Busemann, G.D. Cody, Deuterium enrichments in chondritic macromolecular material-implications for the origin and evolution of organics, water and asteroids. Geochim. Cosmochim. Acta **74**(15), 4417–4437 (2010)

L.J. Allamandola, S.A. Sandford, G.J. Valero, Photochemical and thermal evolution of interstellar/precometary ice analogs. Icarus **76**, 225 (1988)

M.A. Allodi, R.A. Baragiola, G.A. Baratta, M.A. Barucci, G.A. Blake, P. Boduch, J.R. Brucato, C. Contreras, S.H. Cuylle, D. Fulvio, M.S. Gudipati, S. Ioppolo, Z. Kanuchova, A. Lignell, H. Linnartz, M.E. Palumbo, U. Raut, H. Rothard, F. Salama, E.V. Savchenko, E. Sciamma-O'Brien, G. Strazzulla, Complementary and emerging techniques for astrophysical ices processed in the laboratory. Space Sci. Rev. **180**(1–4), 101–175 (2013)

K. Altwegg, H. Balsiger, J. Geiss, Composition of the volatile material in Halley's coma from in situ measurements. Space Sci. Rev. **90**, 3–18 (1999)

K. Altwegg, H. Balsiger, A. Bar-Nun, J.J. Berthelier, A. Bieler, P. Bochsler, C. Briois, U. Calmonte, M. Combi, J. De Keyser, P. Eberhardt, B. Fiethe, S. Fuselier, S. Gasc, T.I. Gombosi, K.C. Hansen, M. Hassig, A. Jackel, E. Kopp, A. Korth, L. Leroy, U. Mall, B. Marty, O. Mousis, E. Neefs, T. Owen, H. Reme, M. Rubin, T. Semon, C.Y. Tzou, H. Waite, P. Wurz, 67p/Churyumov-Gerasimenko, a Jupiter family comet with a high d/h ratio. Science **347**, 1261952 (2015)

O. Andersson, A. Inaba, Thermal conductivity of crystalline and amorphous ices and its implications on amorphization and glassy water. Phys. Chem. Chem. Phys. **7**(7), 1441–1449 (2005)
O. Andersson, H. Suga, Thermal conductivity of amorphous ices. Phys. Rev. B **65**(14), 4 (2002)
S. Andersson, T.P.M. Goumans, A. Arnaldsson, Tunneling in hydrogen and deuterium atom addition to CO at low temperatures. Chem. Phys. Lett. **513**(1–3), 31–36 (2011)
C. Arpigny, Spectra of comets: ultraviolet and optical regions, in *Workshop on Laboratory and Astronomical High Resolution Spectra, Proceedings of ASP Conference no. 81 held in Brussels*. Astronomical Society of the Pacific Conference Series, vol. 81 (1995), pp. 362–382
H. Balsiger, K. Altwegg, J. Geiss, D/H and O-18/O-16 Ratio in the hydronium ion and in neutral water from in-situ ion measurements in comet Halley. J. Geophys. Res. Space Phys. **100**, 5827–5834 (1995)
A. Bar-Nun, D. Laufer, First experimental studies of large samples of gas-laden amorphous "cometary" ices. Icarus **161**(1), 157–163 (2003)
A. Bar-Nun, D. Laufer, CO2 as the driving force of comet Hartley 2's activity—an experimental study. Planet. Space Sci. **86**, 160 (2013)
A. Bar-Nun, D. Laufer, O. Rebolledo, S. Malyk, H. Reisler, C. Wittig, Gas trapping in ice and its release upon warming, in *The Science of Solar System Ices* (2013), p. 487
R.M. Barrer, A.V.J. Edge, Gas hydrates containing argon, krypton and xenon: kinetics and energetics of formation and equilibria. Proc. R. Soc. Lond. A **300**, 1–24 (1967)
C.J. Bennett, T. Hama, Y.S. Kim, M. Kawasaki, R.I. Kaiser, Laboratory studies on the formation of formic acid (HCOOH) in interstellar and cometary ices. Astrophys. J. **727**(1), 27 (2011). doi:10.1088/0004-637x/727/1/27
J. Berkowitz, *Photoabsorption, Photoionization, and Photoelectron Spectroscopy* (Academic Press, San Diego, 1979)
J. Berkowitz, W.A. Chupka, T.A. Walter, Photoionization of HCN: the electron affinity and heat of formation of CN. J. Chem. Phys. **50**, 1497–1500 (1969)
M.P. Bernstein, S.A. Sandford, L.J. Allamandola, S. Chang, M.A. Scharberg, Organic compounds produced by photolysis of realistic interstellar and cometary ice analogs containing methanol. Astrophys. J. **454**, 327–344 (1995)
M.P. Bernstein, S.A. Sandford, L.J. Allamandola, J.S. Gillette, S.J. Clemett, R.N. Zare, UV irradiation of polycyclic aromatic hydrocarbons in ices: production of alcohols, quinones, and ethers. Science **283**(5405), 1135–1138 (1999)
M.P. Bernstein, J.P. Dworkin, S.A. Sandford, G.W. Cooper, L.J. Allamandola, Racemic amino acids from the ultraviolet photolysis of interstellar ice analogues. Nature **416**, 401 (2002)
J.E. Bertie, M. Refaat Shehata, Ammonia dihydrate: preparation, x-ray powder diffraction pattern and infrared spectrum of NH3 · 2H2O at 100 K. J. Chem. Phys. **81**, 27–30 (1984)
N. Biver, D. Bockelée-Morvan, J. Crovisier, D.C. Lis, R. Moreno, P. Colom, F. Henry, F. Herpin, G. Paubert, M. Womack, Radio wavelengths observations of comets C/1999 T1 (Mc Naught-Hartley), C/2001 A2 (LINEAR), C/2000 WM1 (LINEAR) and 153P/Ikeya-Zhang. Astron. Astrophys. **449**, 1255–1270 (2006)
D. Blake, L.J. Allamandola, S.A. Sandford, D. Hudgins, F. Freund, Clathrate hydrate formation in amorphous cometary ice analogs in vacuo. Science **254**, 548–551 (1991)
J. Blum, B. Gundlach, S. Mühle, J.M. Trigo-Rodriguez, Comets formed in solar-nebula instabilities! An experimental and modeling attempt to relate the activity of comets to their formation process. Icarus **235**, 156–169 (2014)
N.T. Bobrovnikoff, On the spectra of comets. Astrophys. J. **66**, 439–464 (1927a)
N.T. Bobrovnikoff, On the spectrum of Halley's comet. Astrophys. J. **66**, 145–169 (1927b)
D. Bockelée-Morvan, Comets, tracers of the early solar nebula, in *Physics and Astrophysics of Planetary Systems*, vol. 41, ed. by T. Montmerle, D. Ehrenreich, A.M. Lagrange (2010), pp. 313–324
D. Bockelée-Morvan, D. Gautier, D.C. Lis, K. Young, J. Keene, T. Phillips, T. Owen, J. Crovisier, P.F. Goldsmith, E.A. Bergin, D. Despois, A. Wootten, Deuterated water in comet C/1996 B2 (Hyakutake) and its implications for the origin of comets. Icarus **133**, 147–162 (1998)
D. Bockelée-Morvan, J. Crovisier, M.J. Mumma, H.A. Weaver, The composition of cometary volatiles, in *Comets II* (2004), pp. 391–423
D. Bockelée-Morvan, N. Biver, B. Swinyard, M. de Val-Borro, J. Crovisier, P. Hartogh, D.C. Lis, R. Moreno, S. Szutowicz, E. Lellouch, E. Emprechtinger, G.A. Blake, R. Courtin, C. Jarchow, M. Kidger, M. Küppers, M. Rengel, G.R. Davis, T. Fulton, D. Naylor, S. Sidher, H. Walker, Herschel measurements of the D/H and $^{16}O/^{18}O$ ratios in water in the Oort-cloud comet C/2009 P1 (Garradd). Astron. Astrophys. **544**, L15 (2012)
J.B. Bossa, F. Duvernay, P. Theule, F. Borget, L. d'Hendecourt, T. Chiavassa, Methylammonium methylcarbamate thermal formation in interstellar ice analogs: a glycine salt precursor in protostellar environments. Astron. Astrophys. **506**, 601–608 (2009a)

J.B. Bossa, P. Theule, F. Duvernay, T. Chiavassa, NH2CH2OH thermal formation in interstellar ices contribution to the 5–8 mu m region toward embedded protostars. Astrophys. J. **707**, 1524 (2009b)
J.B. Bossa, F. Borget, F. Duvernay, P. Theulé, T. Chiavassa, How a usual carbamate can become an unusual intermediate: a new chemical pathway to form glycinate in the interstellar medium. J. Phys. Org. Chem. **23**, 333–339 (2010)
W.F. Bottke, D. Vokrouhlicky, D. Minton, D. Nesvorny, A. Morbidelli, R. Brasser, B. Simonson, H.F. Levison, An archaean heavy bombardment from a destabilized extension of the asteroid belt. Nature **485**(7396), 78–81 (2012)
M.E. Brown, A.H. Boucher, H. Spinrad, E.M. Jons-Krull, A high-resolution catalog of cometary emission lines. Astron. J. **112**, 1197–1225 (1996)
R.H. Brown, D.S. Lauretta, B. Schmidt, J. Moores, Experimental and theoretical simulations of ice sublimation with implications for the chemical, isotopic, and physical evolution of icy objects. Planet. Space Sci. **60**(1), 166–180 (2012a). doi:10.1016/j.pss.2011.07.023
M.E. Brown, E.L. Schaller, W.C. Fraser, Water ice in the Kuiper Belt. Astron. J. **143**(6) (2012b). doi:10.1088/0004-6256/143/6/146
M. Broz, A. Morbidelli, W.F. Bottke, J. Rozehnal, D. Vokrouhlicky, D. Nesvorny, Constraining the cometary flux through the asteroid belt during the late heavy bombardment. Astron. Astrophys. **551** (2013). doi:10.1051/0004-6361/201219296
J.L. Burch, R. Goldstein, T.E. Cravens, W.C. Gibson, R.N. Lundin, C.J. Pollock, J.D. Winningham, D.T. Young, RPC-IES: the ion and electron sensor of the Rosetta Plasma Consortium. Space Sci. Rev. **128**(1–4), 697–712 (2007). doi:10.1007/s11214-006-9002-4
H.M. Butner, S.B. Charnley, C. Ceccarelli, S.D. Rodgers, J.R. Pardo, B. Parise, J. Cernicharo, G.R. Davis, Discovery of interstellar heavy water. Astrophys. J. **659**, L137–L140 (2007)
F. Carelli, M. Satta, T. Grassi, F.A. Gianturco, Carbon-rich molecular chains in protoplanetary and planetary atmospheres: quantum mechanisms and electron attachment rates for anion formation. Astrophys. J. **774**(2), 97 (2013a). doi:10.1088/0004-637X/774/2/97
F. Carelli, T. Grassi, F. Sebastianelli, F.A. Gianturco, Electron-attachment rates for carbon-rich molecules in protoplanetary atmospheres: the role of chemical differences. Mon. Not. R. Astron. Soc. **428**(2), 1181–1184 (2013b). doi:10.1093/mnras/sts100
F. Carelli, T. Grassi, F.A. Gianturco, Electron attachment rates for PAH anions in the ISM and dark molecular clouds: dependence on their chemical properties. Astron. Astrophys. **549**, A103 (2013c). doi:10.1051/0004-6361/201219990
C. Ceccarelli, C. Dominik, E. Caux, B. Lefloch, P. Caselli, Discovery of deuterated water in a young protoplanetary disk. Astrophys. J. **631**, L81–L84 (2005)
J. Cernicharo, M. Guélin, M. Agündez, K. Kawaguchi, M. McCarthy, P. Thaddeus, Astronomical detection of C4H-, the second interstellar anion. Astron. Astrophys. **467**(2), L37–L40 (2007). doi:10.1051/0004-6361:20077415
P. Chaizy, H. Reme, J.A. Sauvaud, C. D'Uston, R.P. Lin, D.E. Larson, D.L. Mitchell, K.A. Anderson, C.W. Carlson, A. Korth, D.A. Mendis, Negative ions in the coma of comet Halley. Nature **349**, 393–396 (1991). doi:10.1038/349393a0
Y.J. Chen, M. Nuevo, C.C. Chu, Y.G. Fan, T.S. Yih, W.H. Ip, H.S. Fung, C.Y.R. Wu, Photo-desorbed species produced by the UV/euv irradiation of an $H_2O : CO_2 : NH_3$ ice mixture. Adv. Space Res. **47**(9), 1633–1644 (2011)
M. Choukroun, G. Tobie, O. Grasset, Ammonia, a methane hydrate inhibitor—implications for Titan's atmospheric methane, in *37th Annual Lunar and Planetary Science Conference*, vol. 37 (2006). Abstract 1640
M. Choukroun, O. Grasset, G. Tobie, C. Sotin, Stability of methane clathrate hydrates under pressure: influence on outgassing processes of methane on Titan. Icarus **205**(2), 581–593 (2010). doi:10.1016/j.icarus.2009.08.011
K.I. Churyumov, V.V. Kleshchenok, Parameters of luminescence cometary continuum in spectra of comets Schaumasse (24P), Scoritchenko-George (C/1989 Y1), and Hale-Bopp (C/1995 O1), in *AAS, DPS Meeting*, vol. 31 (1999), p. 17.13
K.I. Churyumov, G.F. Chernyi, A.I. Shapovalova, Discovery of emission from the negative carbon ion C2 in comet Scorichenko-George (1990 VI). Astron. Lett. **19**(5), 332–334 (1993)
K.I. Churyumov, I.V. Luk'yanyk, A.A. Berezhnoi, V.H. Chavushyan, L.S. Sandoval, A.A. Palma, Optical spectroscopy of comet C/2000 WM1 (LINEAR) at the Guillermo Harro Astrophysical Observatory in Mexico. Earth Moon Planets **90**(1), 361–368 (2002)
K.I. Churyumov, V.A. Ponomarenko, V.V. Kleshchonok, A.R. Baranski, Luminescent continuum in spectra of the comet C/2007 N3 (Lulin). Astron. Circ. **1603**, 1–4 (2013)
J. Clairemidi, P. Brechignac, G. Moreels, D. Pautet, Tentative identification of pyrene as a polycyclic aromatic molecule in UV spectra of comet P/Halley: an emission from 368 to 384 nm. Planet. Space Sci. **57**, 761–772 (2004)

L.I. Cleeves, E.A. Bergin, C.M.O.D. Alexander, F. Du, D. Graninger, K.I. Öberg, T.J. Harries, The ancient heritage of water ice in the solar system. Science **345**(6204), 1590–1593 (2014). doi:10.1126/science.1258055

S.J. Clemett, L.J. Kovalenko, C.R. Maechling, R.N. Zare, 2-step laser mass-spectrometry—analysis at high spatial-resolution of cosmochemical samples. Meteoritics **26**(4), 328–329 (1991)

A.L. Cochran, W.D. Cochran, A high spectral resolution atlas of comet 122P/de Vico. Icarus **157**, 297–308 (2002)

A. Cochran, E.S. Barker, W. Cochran, Spectrophotometric observations of P/Schwassmann-Wachmann 1 during outburst. Astron. J. **85**, 474–477 (1980)

A.L. Cochran, E.S. Barker, C.L. Gray, Thirty years of cometary spectroscopy from McDonald Observatory. Icarus **218**, 144–168 (2012)

M.P. Collings, J.W. Dever, H.J. Fraser, M.R.S. McCoustra, D.A. Williams, Carbon monoxide entrapment in interstellar ice analogs. Astrophys. J. **583**, 1058 (2003)

M.P. Collings, M.A. Anderson, R. Chen, J.W. Dever, S. Viti, D.A. Williams, M.R.S. McCoustra, A laboratory survey of the thermal desorption of astrophysically relevant molecules. Mon. Not. R. Astron. Soc. **354**, 1133 (2004)

M.A. Cordiner, S.B. Charnley, Negative ion chemistry in the coma of comet 1P/Halley. Meteorit. Planet. Sci. **49**, 21–27 (2014)

M.A. Cordiner, J.V. Buckle, E.S. Wirström, A.O.H. Olofsson, S.B. Charnley, On the ubiquity of negative ions in the dense interstellar medium. Astrophys. J. **770**, 48–57 (2013)

H. Cottin, M.C. Gazeau, J.F. Doussin, F. Raulin, An experimental study of the degradation of polyoxymethylene at 122, 147, and 193 nm. Photochem. Photobiol. **135**, 53 (2000)

H. Cottin, C. Szopa, M.H. Moore, Production of hexamethylenetetramine in photolyzed and irradiated interstellar cometary ice analogs. Astrophys. J. Lett. **561**, 139 (2001)

H. Cottin, B. Souleyman, F. Raulin, M.C. Gazeau, Photodegradation of hexamethylenetetramine by VUV and its relevance for CN and HCN extended sources in comets. Adv. Space Res. **30**, 1481 (2002)

G. Cremonese, M.T. Capria, M.C. De Sanctis, Catalog of the emission lines in the visible spectrum of comet 153P/Ikeya-Zhang. Astron. Astrophys. **461**, 789–792 (2007)

J. Crovisier, D. Bockelée-Morvan, A search for PAHs in the ISO spectra of comet Hale-Bopp, in *Molecules in Space and Laboratory, Meeting in Paris, France, May 14–18*, vol. 19, ed. by J.L. Lemaire, F. Combes (2007)

A. Dalgarno, R.A. McCray, The formation of interstellar molecules from negative ions. Astrophys. J. **181**, 95–100 (1973)

G. Danger, L. Le Sergeant d'Hendecourt, From astrochemistry to prebiotic chemistry: an hypothetical approach toward astrobiology, in *SF2A-2012 Proc. Annu. Meet. French Soc. Astron. Astrophys.*, ed. by S. Boissier (2012), pp. 263–267

G. Danger, F. Borget, M. Chomat, F. Duvernay, P. Theulé, J.C. Guillemin, S. Le Sergeant d'Hendecourt, T. Chiavassa, Experimental investigation of aminoacetonitrile formation through the Strecker synthesis in astrophysical-like conditions: reactivity of methanimine (CH2NH), ammonia (NH3), and hydrogen cyanide (HCN). Astron. Astrophys. **535**, A47 (2011a)

G. Danger, J.B. Bossa, P. de Marcellus, F. Borget, F. Duvernay, P. Theule, T. Chiavassa, L. D'Hendecourt, Experimental investigation of nitrile formation from VUV photochemistry of interstellar ices analogs: acetonitrile and amino acetonitrile. Astron. Astrophys. **525**, A30 (2011b)

G. Danger, F. Duvernay, P. Theule, F. Borget, T. Chiavassa, Hydroxyacetonitrile (HOCH2CN) Formation in astrophysical conditions. Competition with the aminomethanol, a glycine precursor. Astrophys. J. **756**, 11 (2012)

G. Danger, F. Orthous-Daunay, P. de Marcellus, P. Modica, V. Vuitton, F. Duvernay, L. Flandinet, L. Le Sergeant d'Hendecourt, R. Thissen, T. Chiavassa, Characterization of laboratory analogs of interstellar/cometary organic residues using very high resolution mass spectrometry. Geochim. Cosmochim. Acta **118**, 184–201 (2013)

G. Danger, A. Rimola, N. Abou Mrad, F. Duvernay, G. Roussin, P. Theule, T. Chiavassa, Formation of hydroxyacetonitrile (HOCH2CN) and polyoxymethylene (POM)-derivatives in comets from formaldehyde (CH2O) and hydrogen cyanide (HCN) activated by water. Phys. Chem. Chem. Phys. **16**, 3360 (2014)

E. Dartois, D. Deboffle, Methane clathrate hydrate FTIR spectrum implications for its cometary and planetary detection. Astron. Astrophys. **490**(3), L19–L22 (2008)

E. Dartois, M. Bouzit, B. Schmitt, Clathrate hydrates: FTIR spectroscopy for astrophysical remote detection, in *ECLA: European Conference on Laboratory Astrophysics*, vol. 58, ed. by C. Stehle, C. Joblin, L. Dhendecourt (2013), pp. 219–224

J.L. Daschbach, G.K. Schenter, P. Ayotte, R.S. Smith, B.D. Kay, Helium diffusion through H2O and D2O amorphous ice: observation of a lattice inverse isotope effect. Phys. Rev. Lett. **92**, 198306 (2004)

B.J.R. Davidsson, P.J. Gutierrez, O. Groussin, M.F. A'Hearn, T. Farnham, L.M. Feaga, M.S. Kelley, K.P. Klaasen, F. Merlin, S. Protopapa, H. Rickman, J.M. Sunshine, P.C. Thomas, Thermal inertia and surface roughness of comet 9p/Tempel 1. Icarus **224**(1), 154–171 (2013)

P. de Marcellus, M. Bertrand, M. Nuevo, F. Westall, L. Le Sergeant d'Hendecourt, Prebiotic significance of extraterrestrial ice photochemistry: detection of hydantoin in organic residues. Astrobiology **11**, 847 (2011a)

P. de Marcellus, C. Meinert, M. Nuevo, J.-J. Filippi, G. Danger, D. Deboffle, L. Nahon, L. Le Sergeant d'Hendecourt, U.J. Meierhenrich, Non-racemic amino acid production by ultraviolet irradiation of achiral interstellar ice analogs with circularly polarized light. Astrophys. J. **727**, L27 (2011b)

N. Dello Russo, R.J. Vervack Jr., H.A. Weaver et al., A high-resolution infrared spectral survey of 103P/Hartley 2 on the night of the EPOXI closest approach. Icarus **222**, 707–722 (2013)

A. Drouart, B. Dubrulle, D. Gautier, F. Robert, Structures and transport in the solar nebula from constraints on deuterium enrichment and giant planets formation. Icarus **140**(1), 129–155 (1999)

J. Duprat, E. Dobrica, C. Engrand, J. Aléon, Y. Marrocchi, S. Mostefaoui, A. Meibom, H. Leroux, J.N. Rouzaud, M. Gounelle, F. Robert, Extreme deuterium excesses in ultracarbonaceous micrometeorites from central antarctic snow. Science **328**(5979), 742–745 (2010)

F. Duvernay, G. Danger, P. Theulé, T. Chiavassa, A. Rimola, Formaldehyde chemistry in cometary ices: on the prospective detection of NH2CH2OH, HOCH2OH, and POM by the on-board Rosina instrument of the Rosetta mission. Astrophys. J. **791**, 75 (2014a)

F. Duvernay, A. Rimola, P. Theule, G. Danger, T. Chiavassa, Formaldehyde chemistry in cometary ices: the case of HOCH2OH formation. Phys. Chem. Chem. Phys. **16**, 24200 (2014b)

J.P. Dworkin, D.W. Deamer, S.A. Sandford, L.J. Allamandola, Self-assembling amphiphilic molecules: synthesis in simulated interstellar/precometary ices. Proc. Natl. Acad. Sci. USA **98**, 815 (2001)

J.P. Dworkin, J.S. Gillette, M.P. Bernstein, S.A. Sandford, L.J. Allamandola, J.E. Elsila, D.R. McGlothlin, R.N. Zare, An evolutionary connection between interstellar ices and IDPs? Clues from mass spectroscopy measurements of laboratory simulations. Adv. Space Res. **33**, 67 (2004)

P. Eberhardt, M. Reber et al., The D/H and O-18/O-16 ratios in water from comet P/Halley. Astron. Astrophys. **302**(1), 301–316 (1995)

P. Ehrenfreund, S.B. Charnley, Organic molecules in the interstellar medium, comets, and meteorites: a voyage from dark clouds to the Early Earth. Annu. Rev. Astron. Astrophys. **38**, 427–483 (2000)

D.J. Eicher, *Comets! Visitors from Deep Space* (2013)

J.E. Elsila, J.P. Dworkin, M.P. Bernstein, M.P. Martin, S.A. Sandford, Mechanisms of amino acid formation in interstellar ice analogs. Astrophys. J. **660**, 911 (2007)

N.J. English, J.S. Tse, R. Gallagher, Thermal conductivity in amorphous ices from molecular dynamics. Phys. Rev. B **82**(9), 4 (2010)

A. Falenty, G. Genov, T.C. Hansen, W.H. Kuhs, A.N. Salamatin, Kinetics of CO 2 hydrate formation from water frost at low temperatures: experimental results and theoretical model. J. Phys. Chem. C **115**, 4022–4032 (2011)

C.M. Fedo, M.J. Whitehouse, B.S. Kamber, Geological constraints on detecting the earliest life on earth: a perspective from the early archaean (older than 3.7 gyr) of Southwest Greenland. Philos. Trans. R. Soc. Lond. B, Biol. Sci. **361**(1470), 851–867 (2006)

M.C. Festou, U. Keller, H.A. Weaver (eds.), *Comets II*. Space Science Series (University of Arizona Press, Tucson, 2004)

H. Feuchtgruber, E. Lellouch, G. Orton, T. de Graauw, B. Vandenbussche, B. Swinyard, R. Moreno, C. Jarchow, F. Billebaud, T. Cavalie, S. Sidher, P. Hartogh, The d/h ratio in the atmospheres of uranus and neptune from Herschel-PACS observations. Astron. Astrophys. **551** (2013)

A.D. Fortes, M. Choukroun, Phase behavior of ices and hydrates. Space Sci. Rev. **153**, 185 (2010)

N. Fray, U. Marboeuf, O. Brissaud, B. Schmitt, Equilibrium data of methane, carbon dioxide, and xenon clathrate hydrates below the freezing point of water. Applications to astrophysical environments. J. Chem. Eng. Data **55**, 5101–5108 (2010)

A. Fresneau, G. Danger, A. Rimola, P. Theule, F. Duvernay, T. Chiavassa, Trapping in water—an important prerequisite for complex reactivity in astrophysical ices: the case of acetone (CH3)2C = O and ammonia NH3. Mon. Not. R. Astron. Soc. **443**, 2991 (2014)

D. Gautier, F. Hersant, O. Mousis, J.I. Lunine, Enrichments in volatiles in Jupiter: a new interpretation of the Galileo measurements. Astrophys. J. **550**, L227–L230 (2001)

J. Geiss et al., Rare atoms, molecules and radicals in the coma of 1P/Halley. Space Sci. Rev. **90**, 253–268 (1999)

P.A. Gerakines, W.A. Schutte, P. Ehrenfreund, Ultraviolet processing of interstellar ice analogs. Astron. Astrophys. **312**, 289 (1996)

L. Gibb et al., An inventory of interstellar ices toward the embedded protostar W33A. Astrophys. J. **536**, 347–356 (2000)

E.L. Gibb, D.C.B. Whittet, A.C.A. Boogert, A.G.G.M. Tielens, Interstellar ice: the infrared space observatory legacy. Astrophys. J. Suppl. Ser. **151**, 35–73 (2004)
R. Goldstein, D.C. Boice, J.L. Burch, T.W. Hill, Are charged nanograins at Enceladus a model for 67P/Churyumov-Gerasimenko? AGU abstract #P43C-1935 (2012)
R. Gomes, H.F. Levison, K. Tsiganis, A. Morbidelli, Origin of the cataclysmic late heavy bombardment period of the terrestrial planets. Nature **435**(7041), 466–469 (2005)
N. Gortsas, E. Kuhrt, U. Motschmann, H.U. Keller, Thermal model of water and CO activity of comet c/1995 o1 (Hale-Bopp). Icarus **212**(2), 858–866 (2011)
J.M. Greenberg, Interstellar grains. Annu. Rev. Astron. Astrophys. **1**, 267 (1963)
O. Groussin, M.F. A'Hearn, J.Y. Li, P.C. Thomas, J.M. Sunshine, C.M. Lisse, K.J. Meech, T.L. Farnham, L.M. Feaga, W.A. Delamere, Surface temperature of the nucleus of comet 9p/Tempel 1. Icarus **187**(1), 16–25 (2007)
O. Groussin, J.M. Sunshine, L.M. Feaga, L. Jorda, P.C. Thomas, J.Y. Li, M.F.A. ÄôHearn, M.J.S. Belton, S. Besse, B. Carcich, T.L. Farnham, D. Hampton, K. Klaasen, C. Lisse, F. Merlin et al., The temperature, thermal inertia, roughness and color of the nuclei of comets 103p/Hartley 2 and 9p/Tempel 1. Icarus **222**(2), 580–594 (2013)
M.S. Gudipati, L.J. Allamandola, Facile generation and storage of polycyclic aromatic hydrocarbon ions in astrophysical ices. Astrophys. J. **596**(5), L195–L198 (2003)
M.S. Gudipati, R. Yang, In-situ probing of radiation-induced processing of organics in astrophysical ice analogs: novel laser desorption laser ionization time-of-flight mass spectroscopic studies. Astrophys. J. Lett. **756**(1), L24 (2012), 25 pp.
S. Gulkis, M. Allen, P. von Allmen, G. Beaudin, N. Biver, D. Bockelée-Morvan, M. Choukroun, J. Crovisier, B.J.R. Davidsson, P. Encrenaz, T. Encrenaz, M. Frerking, P. Hartogh, M. Hofstadter, W.H. Ip, M. Janssen, C. Jarchow, S. Keihm, S. Lee, E. Lellouch, C. Leyrat, L. Rezac, F.P. Schloerb, T. Spilker, Subsurface properties and early activity of comet 67p/Churyumov-Gerasimenko. Science **347**, 6220 (2015)
B. Gundlach, J. Blum, Outgassing of icy bodies in the solar system—II. Heat transport in dry, porous surface dust layers. Icarus **219**, 618–629 (2012)
B. Gundlach, J. Blum, The stickiness of micrometer-sized water-ice particles. Astrophys. J. **798**, 34 (2015). doi:10.1088/0004-637X/798/1/34
B. Gundlach, Y.V. Skorov, J. Blum, Outgassing of icy bodies in the solar system—I. The sublimation of hexagonal water ice through dust layers. Icarus **213**, 710–719 (2011a)
B. Gundlach, S. Kilias, E. Beitz, J. Blum, Micrometer-sized ice particles for planetary-science experiments—I. Preparation, critical rolling friction force, and specific surface energy. Icarus **214**, 717–723 (2011b)
A. Hachikubo, T. Kosaka, M. Kida, A. Krylov, H. Sakagami, H. Minami, N. Takahashi, H. Shoji, Isotopic fractionation of methane and ethane hydrates between gas and hydrate phases. Geophys. Res. Lett. **34**, L21502–L21506 (2007)
A. Hachikubo, R. Nakagawa, D. Kubota, H. Sakagami, N. Takahashi, H. Shoji, Dissociation heat of mixed-gas hydrate composed of methane and ethane, in *Proceedings of the 6th International Conference on Gas Hydrates (ICGH-2008)* (2008)
T. Hama, N. Watanabe, Surface processes on interstellar amorphous solid water: adsorption, diffusion, tunneling reactions, and nuclear-spin conversion. Chem. Rev. **113**(12), 8783–8839 (2013)
T. Hama, H. Ueta, A. Kouchi, N. Watanabe, H. Tachikawa, Quantum tunneling hydrogenation of solid benzene and its control via surface structure. J. Phys. Chem. Lett. **5**(21), 3843–3848 (2014)
Y.P. Handa, O. Mishima, E. Whalley, High-density amorphous ice. 3. Thermal-properties. J. Chem. Phys. **84**(5), 2766–2770 (1986)
G.B. Hansen, T.B. McCord, Amorphous and crystalline ice on the galilean satellites: a balance between thermal and radiolytic processes. J. Geophys. Res. **109**, E01012-01019 (2004)
P. Hartogh, D.C. Lis, D. Bockelée-Morvan, M. de Val-Borro, N. Biver, M. Küppers, M. Emprechtinger, E.A. Bergin, J. Crovisier, M. Rengel, R. Moreno, S. Szutowicz, G.A. Blake, Ocean-like water in the Jupiter-family comet 103P/Hartley 2. Nature **478**(7368), 218–220 (2011)
S. Hegler, C. Statz, R. Hahnel, D. Plettemeier, A. Herique, W. Kofman, Operation of consert aboard Rosetta during the descent of philae. Planet. Space Sci. **89151**, 158 (2013)
B.L. Henderson, M.S. Gudipati, Plume composition and evolution in multicomponent ices using resonant two-step laser ablation and ionization mass spectrometry. J. Phys. Chem. A **118**(29), 5454–5463 (2014)
B.L. Henderson, M.S. Gudipati, Direct detection of complex organic products in ultraviolet (Ly alpha) and electron-irradiated astrophysical and cometary ice analogs using two-step laser ablation and ionization mass spectrometry. Astrophys. J. **800**(1), 66 (2015)
E. Herbst, Can negative molecular ions be detected in dense interstellar clouds? Nature **289**, 656–657 (1981)
E. Herbst, Y. Osamura, Calculations on the formation rates and mechanisms for CnH anions in interstellar and circumstellar media. Astrophys. J. **679**, 1670–1679 (2008)

F. Hersant, D. Gautier, J.I. Lunine, Enrichment in volatiles in the giant planets of the Solar System. Planet. Space Sci. **52**, 623–641 (2004)
F. Hersant, D. Gautier, G. Tobie, J.I. Lunine, Interpretation of the carbon abundance in Saturn measured by Cassini. Planet. Space Sci. **56**, 1103–1111 (2008)
M.R. Hogerheijde, E.A. Bergin, C. Brinch, L.I. Cleeves, J.K.J. Fogel, G.A. Blake, C. Dominik, D.C. Lis, G. Melnick, D. Neufeld, O. Panic, J.C. Pearson, L. Kristensen, U.A. Yildiz, E.F. van Dishoeck, Detection of the water reservoir in a forming planetary system. Science **334**(6054), 338–340 (2011)
H.D. Holder, G. Corbin, K.D. Papadopoulos, Thermodynamic and molecular properties of gas hydrates from mixtures containing methane, argon, and krypton. Ind. Eng. Chem. Fundam. **19**, 280–286 (1980)
J. Horner, O. Mousis et al., Constraints on the formation regions of comets from their D:H ratios. Earth Moon Planets **100**(1–2), 43–56 (2007)
Q.J. Hu, Q. Zhang, J.W. Hepburn, Threshold ion-pair production spectroscopy of HCN. J. Chem. Phys. **124**, 074310 (2006)
W.B. Huebner, First polymer in space identified in comet Halley. Science **237**, 628 (1987)
D. Hutsemékers, J. Manfroid, E. Jehin, J.-M. Zucconi, C. Arpigny, The 16OH/18OH and OD/OH isotope ratios in comet C/2002 T7 (LINEAR). Astron. Astrophys. **490**(3), L31–L34 (2008)
N. Iro, D. Gautier, F. Hersant, D. Bockelée-Morvan, J.I. Lunine, An interpretation of the nitrogen deficiency in comets. Icarus **161**, 511–532 (2003)
W.M. Irvine, D. Bockelée-Morvan, D.C. Lis, H.E. Matthews, N. Biver, J. Crosivier, J.K. Davies, W.R.F. Dent, D. Gaultier, P.D. Godfrey, J. Keene, A.J. Lovell, T.C. Owen, T.G. Phillips, H. Rauer, F.P. Schloerb, M. Senay, K. Young, Spectroscopic evidence for interstellar ices in comet Hyakutake. Nature **383**, 418–420 (1996)
F.G.A. Islam, G.A. Baratta, M.E. Palumbo, Simultaneous UV- and ion processing of astrophysically relevant ices. The case of CH3OH:N2 solid mixtures. Astron. Astrophys. **561**, A73 (2014)
P. Jenniskens, D.F. Blake, Structural transitions in amorphous water ice and astrophysical implications. Science **265**, 753 (1994)
B. Jost, B. Gundlach, A. Pommerol, J. Oesert, S.N. Gorb, J. Blüm, N. Thomas, Micrometer-sized ice particles for planetary-science experiments—II. Bidirectional reflectance. Icarus **225**, 352–366 (2013)
S.-P. Kang, H. Lee, B.-J. Ryu, Enthalpies of dissociation of clathrate hydrates of carbon dioxide, nitrogen, (carbon dioxide + nitrogen), and (carbon dioxide + nitrogen + tetrahydrofuran). J. Chem. Thermodyn. **33**, 513–521 (2001)
J.J. Kavelaars, O. Mousis et al., On the formation location of Uranus and Neptune as constrained by dynamical and chemical models of comets. Astrophys. J. Lett. **734**(2), L30 (2011)
H. Kawakita, J. Watanabe, Unidentified bands in comet Ikeya-Zhang (C/2002 C1): the correlation between unidentified bands and H2O+. Astrophys. J. **574**, L183–L185 (2002)
K.J. Kim, S.H. Lee, J. Ulrich, Experimental thermal conductivity of ice crystalline layer in layer melt crystallization. J. Ind. Eng. Chem. **9**, 111–116 (2003)
G.F. Kirkbright, C.G. Lima, The detection and determination of polynuclear aromatic hydrocarbons by luminescence spectrometry utilizing the Shpol'skii effect at 77 K. Analyst **99**, 338–354 (1974)
H. Kobayashi, H. Kawakita, Formation conditions of icy materials in comet C/2004 Q2 (Machholz). I. Mixing ratios of organic volatiles. Astrophys. J. **703**, 121–130 (2009)
K.J. Kossacki, W.J. Markiewicz, Y.V. Skorov, N.I. Kömle, Sublimation coefficient of water ice under simulated cometary-like conditions. Planet. Space Sci. **47**, 1521–1530 (1999)
A. Kouchi, S. Sirono, Crystallization heat of impure amorphous H(2)O ice. Geophys. Res. Lett. **28**(5), 827–830 (2001)
A. Kouchi, J.M. Greenberg, T. Yamamoto, T. Mukai, Z.F. Xing, Extremely low thermal conductivity of amorphous ice—relevance to comet evolution. Astrophys. J. **388**(2), L73–L76 (1992)
M. Krause, J. Blum, Y.V. Skorov, M. Trieloff, Thermal conductivity measurements of porous dust aggregates: I. Technique, model and first results. Icarus **214**, 286–296 (2011)
E. Kuhrt, H2O-activity of comet Hale-Bopp. Space Sci. Rev. **90**(1–2), 75–82 (1999)
W.F. Kuhs, D.K. Staykova, A.N. Salamatin, Formation of methane hydrate from polydisperse ice powders. J. Phys. Chem. B **110**, 13283–13295 (2006)
D. Laufer, E. Kochavi, A. Bar-Nun, Structure and dynamics of amorphous water ice. Phys. Rev. B **36**(17), 9219–9227 (1987)
D. Laufer, A. Bar-Nun, I. Pat-El, R. Jacovi, Experimental studies of ice grain ejection by massive gas flow from ice and implications to comets, Triton and Mars. Icarus **222**(1), 73–80 (2013)
C. Lecluse, F. Robert, D. Gautier, M. Guirand, Deuterium enrichment in giant planets. Planet. Space Sci. **44**(12), 1579–1592 (1996)
J.S. Lewis, Low temperature condensation from the solar nebula. Icarus **16**, 241–252 (1972)
A. Li, PAHs in comets: an overview, in *Deep Impact as a World Observatory Event: Synergies in Space, Time, and Wavelength*. ESO Astrophysics Symposia (Springer, Berlin, 2009), pp. 161–175

J. Licandro, N. Pinilla-Alonso, M. Pedani, E. Oliva, G.P. Tozzi, W.M. Grundy, The methane ice rich surface of large TNO 2005 FY9: a Pluto-twin in the trans-Neptunian belt? Astron. Astrophys. **445**(3), L35–L38 (2006)

A. Lignell, M.S. Gudipati, Mixing of the immiscible: hydrocarbons in water-ice near the ice crystallization temperature. J. Phys. Chem. A **119**, 2607–2613 (2015). doi:10.1021/jp509513s

P. Linga, R. Kumar, P. Englezos, Gas hydrate formation from hydrogen/carbon dioxide and nitrogen/carbon dioxide gas mixtures. Chem. Eng. Sci. **62**, 4268–4276 (2007)

D.C. Lis, N. Biver, D. Bockelée-Morvan, P. Hartogh, E.A. Bergin, G.A. Blake, J. Crovisier, M. de Val-Borro, E. Jehin, M. Küppers, J. Manfroid, R. Moreno, M. Rengel, S. Szutowicz, A Herschel study of D/H in water in the Jupiter-family comet 45P/Honda-Mrkos-Pajdusáková and prospects for D/H measurements with CCAT. Astrophys. J. Lett. **774**(1), L3 (2013)

C.M. Lisse, M.F. A'Hearn, T.L. Farnham, O. Groussin, K.J. Meech, U. Fink, D.G. Schleicher, The coma of comet 9P/Tempel 1. Space Sci. Rev. **117**(1–2), 161–192 (2005)

C.M. Lisse, A. Bar-Nun, D. Laufer, M. Belton, W. Harris, H. Hsieh, D. Jewitt, Cometary ices, in *The Science of Solar System Ices*, vol. 356, ed. by M.S. Gudipati, J. Castillo-Rogez (Springer, New York, 2013), pp. 455–485

F.E. Livingston, J.A. Smith, S.M. George, General trends for bulk diffusion measurments in ice films using infrared laser resonant desorption. J. Phys. Chem. A **106**, 6309 (2002)

K. Lodders, H. Palme, H.-P. Gail, *Abundances of the Elements in the Solar System*. Landolt Börnstein, vol. 44 (2009)

J.I. Lunine, D.J. Stevenson, Thermodynamics of clathrate hydrate at low and high pressures with application to the outer solar system. Astrophys. J. Suppl. Ser. **58**, 493–531 (1985)

J.I. Lunine, D.J. Stevenson, Clathrate and ammonia hydrates at high pressure—application to the origin of methane on Titan. Icarus **70**, 61–77 (1987)

T. Maekawa, Equilibrium conditions for gas hydrates of methane and ethane mixtures in pure water and sodium chloride solution. Geochem. J. **35**, 59–66 (2001)

S. Maity, R.I. Kaiser, B.M. Jones, Formation of complex organic molecules in methanol and methanol-carbon monoxide ices exposed to ionizing radiation—a combined FTIR and reflectron time-of-flight mass spectrometry study. Phys. Chem. Chem. Phys. **17**(5), 3081–3114 (2015)

U. Marboeuf, B. Schmitt, How to link the relative abundances of gas species in coma of comets to their initial chemical composition? Icarus **242**, 225–248 (2014)

U. Marboeuf, O. Mousis, D. Ehrenreich, Y. Alibert, A. Cassan, V. Wakelam, J.-P. Beaulieu, Composition of ices in low-mass extrasolar planets. Astrophys. J. **681**, 1624–1630 (2008)

U. Marboeuf, N. Fray, O. Brissaud, B. Schmitt, D. Bockelée-Morvan, D. Gautier, Equilibrium pressure of ethane, acetylene, and krypton clathrate hydrates below the freezing point of water. J. Chem. Eng. Data **57**, 3408–3415 (2012a)

U. Marboeuf, B. Schmitt, J.-M. Petit, O. Mousis, N. Fray, A cometary nucleus model taking into account all phase changes of water ice: amorphous, crystalline, and clathrate. Astron. Astrophys. **542**, A82–A105 (2012b)

U. Marboeuf, A. Thiabaud, N. Cabral, Y. Alibert, W. Benz, From stellar nebula to planetesimals. Astron. Astrophys. **570**, A35–A55 (2014)

R. Martin-Domenech, G.M. Muñoz-Caro, J. Bueno, F. Goesmann, Thermal desorption of circumstellar and cometary ice analogs. Astron. Astrophys. **564**, A8 (2014)

B. Marty, R.L. Palma, R.O. Pepin, L. Zimmermann, D.J. Schlutter, P.G. Burnard, A.J. Westphal, C.J. Snead, S. Bajt, R.H. Becker, J.E. Simones, Helium and neon abundances and compositions in cometary matter. Science **319**(5859), 75–78 (2008)

H. Massey, *Negative Ions* (Cambridge University Press, Cambridge, 1976)

C.K. Materese, D.P. Cruikshank, S.A. Sandford, H. Imanaka, M. Nuevo, D.W. Douglas, Ice chemistry on outer solar system bodies: carboxylic acids, nitriles, and urea detected in refractory residues produced from the UV photolysis of N2:CH4:CO-containing ices. Astrophys. J. **788**, 111–121 (2014)

C.N. Matthews, R.D. Minard, Hydrogen cyanide polymers, comets and the origin of life. Faraday Discuss. **133**, 393–401 (2006)

A.R. May, S.R. Smith, B.D. Kay, The release of trapped gases from amorphous solid water films. II. "Bottom-up" induced desorption pathways. J. Chem. Phys. **138**, 104502 (2013a)

R.A. May, R.S. Smith, B.D. Kay, The release of trapped gases from amorphous solid water films. I. "Top-down" crystallization-induced crack propagation probed using the molecular volcano. J. Phys. Chem. **138**, 104501 (2013b)

M.C. McCarthy, C.A. Gottlieb, H. Gupta, P. Thaddeus, Laboratory and astronomical identification of the negative molecular Ion C6H-. Astrophys. J. **652**, L141–L144 (2006)

K.J. Meech et al., Deep impact: observations from a worldwide earth-based campaign. Science **310**(5746), 265–269 (2005)

R. Meier, T.C. Owen, D.C. Jewitt, H.E. Matthews, M. Senay, N. Biver, D. Bockelée-Morvan, J. Crovisier, D. Gautier, Science **279**(6054), 942 (1998)

U.J.L. Meierhenrich, G.M. Muñoz-Caro, W.A. Schutte, W.H.-P.P. Thiemann, B. Barbier, A. Brack, G.M. Muñoz-Caro, Precursors of biological cofactors from ultraviolet irradiation of circumstellar/interstellar ice analogues. Eur. J. Chem. **11**, 4895–4900 (2005)

C. Meinert, J.J. Filippi et al., N-(2-Aminoethyl)glycine and amino acids from interstellar ice analogues. Chem. Plus Chem. **77**, 186–191 (2012)

T.J. Millar, E. Herbst, R.P.A. Bettens, Large molecules in the envelope surrounding IRC+10216. Mon. Not. R. Astron. Soc. **316**, 195–203 (2000)

T.J. Millar, C. Walsh, M.A. Cordiner, R. Ni Chuimin, E. Herbst, Hydrocarbon anions in interstellar clouds and circumstellar envelopes. Astrophys. J. **662**, L87–L90 (2007)

F. Mispelaer, P. Theule, H. Aouididi, J. Noble, F. Duvernay, G. Danger, P. Roubin, O. Morata, T. Hasegawa, T. Chiavassa, Diffusion measurments of CO, HNCO, H2CO and NH3 in amorphous water ice. Astron. Astrophys. **555**, A13 (2013)

D.L. Mitchell, R.P. Lin, C.W. Carlson, A. Korth, H. Reme, D.A. Mendis, The origin of complex organic ions in the coma of comet Halley. Icarus **98**, 125–133 (1992)

P. Modica, P. de Marcellus, D. Baklouti, R. Brunetto, M. Noun, S. Della Negra, L. Le Sergeant d'Hendecourt, Organic residues from ultraviolet irradiation of interstellar ice analogs. EAS Publ. Ser. **58**, 343 (2013)

P. Modica, C. Meinert, P. de Marcellus, L. Nahon, U.J. Meierhenrich, L. Le Sergeant d'Hendecourt, Enantiomeric excesses induced in amino acids by ultraviolet circularly polarized light irradiation of extraterrestrial ice analogs: a possible source of asymmetry for prebiotic chemistry. Astrophys. J. **788**, 79 (2014)

A.H. Mohammadi, R. Anderson, Phase equilibria of clathrate hydrates of tetrahydrofuran + hydrogen sulfide and tetrahydrofuran + methane. Ind. Eng. Chem. Res. **48**, 7838–7841 (2009)

A.H. Mohammadi, R. Anderson, Ice—Clathrate hydrate—Gas phase equilibria for air, oxygen, nitrogen, carbon monoxide, methane, or ethane + water system. Ind. Eng. Chem. Res. **49**, 3976–3979 (2010)

A.H. Mohammadi, R. Anderson, Ice—Clathrate hydrate—Gas phase equilibria for argon + water and carbon dioxide + water systems. Ind. Eng. Chem. Res. **50**, 1452–1454 (2011)

A.H. Mohammadi, D. Richon, Clathrate hydrate phase equilibria for carbonyl sulfide, hydrogen sulfide, ethylene or propane + water system below the freezing point. Chem. Eng. Commun. **200**(12), 1635–1644 (2013)

A.H. Mohammadi, R. Anderson, B. Tohidi, Carbon monoxide clathrate hydrates: equilibrium data and thermodynamic modeling. AIChE J. **51**(10), 2825–2833 (2005)

M.H. Moore, R.F. Ferrante, R.L. Hudson, J.N. Stone, Ammonia water ice laboratory studies relevant to outer Solar System surfaces. Icarus **190**, 260–273 (2007)

G. Moreels, J. Clairemidi, P. Hermine, P. Bréchignac, P. Rousselot, Detection of a polycyclic aromatic molecule in comet P/Halley. Astron. Astrophys. **282**, 643–656 (1994)

F. Moreno, L.M. Lara, J.J. Lopez-Moreno, O. Munoz, A. Molina, J. Licandro, The dust tail of comet c/1999 t1 Mcnaught-Hartley. Astron. Astrophys. **399**(2), 789–794 (2003)

N.D. Morrison, D.C. Knauth, C.L. Mulliss, W. Lee, High-Resolution Optical Spectra of the Head of the Comet C/1996 B2 (Hyakutake). Publ. Astron. Soc. Pac. **109**, 676–681 (1997)

O. Mousis, Y. Alibert, Modeling the Jovian subnebula. II. Composition of regular satellite ices. Astron. Astrophys. **448**, 771–778 (2006)

O. Mousis, D. Gautier, D. Bockelée-Morvan, F. Robert, B. Dubrulle, A. Drouart, Constraints on the formation of comets from D/H ratios measured in H2O and HCN. Icarus **148**(2), 513–525 (2000)

O. Mousis, D. Gautier, D. Bockelée-Morvan, An evolutionary turbulent model of Saturn's subnebula: implications for the origin of the atmosphere of Titan. Icarus **156**, 162–175 (2002)

O. Mousis, Y. Alibert, D. Hestroffer, U. Marboeuf, C. Dumas, B. Carry, J. Horner, F. Selsis, Origin of volatiles in the main belt. Mon. Not. R. Astron. Soc. **383**, 1269–1280 (2008)

O. Mousis, J.I. Lunine, C. Thomas, M. Pasek, U. Marboeuf, Y. Alibert, V. Ballenegger, D. Cordier, Y. Ellinger, F. Pauzat, S. Picaud, Clathration of volatiles in the solar nebula and implications for the origin of Titan's atmosphere. Astrophys. J. **691**, 1780–1786 (2009a)

O. Mousis, U. Marboeuf, J.I. Lunine, Y. Alibert, L.N. Fletcher, G.S. Orton, F. Pauzat, Y. Ellinger, Determination of the minimum masses of heavy elements in the envelopes of Jupiter and Saturn. Astrophys. J. **696**, 1348–1354 (2009b)

O. Mousis, J.I. Lunine, S. Picaud, D. Cordier, J.H. Waite, K.E. Mandt, Removal of Titan's atmospheric noble gases by their sequestration in surface clathrates. Astrophys. J. **740**, L9 (2011)

O. Mousis, J.I. Lunine, N. Madhusudhan, T.V. Johnson, Nebular water depletion as the cause of Jupiter's low oxygen abundance. Astrophys. J. **751**, L7 (2012a)

O. Mousis, J.I. Lunine, E. Chassefière, F. Montmessin, A. Lakhlifi, S. Picaud, J.-M. Petit, D. Cordier, Mars cryosphere: a potential reservoir for heavy noble gases? Icarus **218**, 80–87 (2012b)

O. Mousis, J.I. Lunine, K.E. Mandt, E. Schindhelm, H.A. Weaver, S.A. Stern, J.H. Waite, R. Gladstone, A. Moudens, On the possible noble gas deficiency of Pluto's atmosphere. Icarus **225**, 856–861 (2013a)
O. Mousis, A. Lakhlifi, S. Picaud, M. Pasek, E. Chassefière, On the abundances of noble and biologically relevant gases in Lake Vostok. Antarct. Astrobiol. **13**, 380–390 (2013b)
O. Mousis, J.I. Lunine, L.N. Fletcher, K.E. Mandt, M. Ali-Dib, D. Gautier, S. Atreya, New insights on Saturn's formation from its nitrogen isotopic composition. Astrophys. J. Lett. **796**(2), L28 (2014)
O. Mousis, E. Chassefiere, N.G. Holm, A. Bouquet, J.H. Waite, W.D. Geppert, S. Picaud, Y. Aikawa, M. Ali-Dib, J.-L. Charlou, P. Rousselot, Methane clathrates in the solar system. Astrobiology **15**(4), 308–326 (2015)
N. Abou Mrad, F. Duvernay, P. Theulé, T. Chiavassa, G. Danger, Development and optimization of an analytical system for volatile organic compound analysis coming from the heating of interstellar/cometary ice analogues. Anal. Chem. **86**, 8391 (2014)
M.J. Mumma, S.B. Charnley, The chemical composition of comets-emerging taxonomies and natal heritage, in *Annual Review of Astronomy and Astrophysics, Palo Alto*, vol. 49, ed. by S.M. Faber, E. Van Dishoeck (2011), pp. 471–524
G.M. Muñoz-Caro, U.J. Meierhenrich, W.A. Schutte, B. Barbier, A. Arcones Segovia, H. Rosenbauer, W.H.-P. Thiemann, A. Brack, J.M. Greenberg, Amino acids from ultraviolet irradiation of interstellar ice analogues. Nature **416**, 403–406 (2002)
G.M. Muñoz-Caro, E. Dartois, P. Boduch, H. Rothard, A. Domaracka, A. Jiménez-Escobar, Comparison of UV and high-energy ion irradiation of methanol: ammonia ice. Astron. Astrophys. **566**, A93 (2014)
G.M. Muñoz-Caro, W.A. Schutte, UV-photoprocessing of interstellar ice analogs: new infrared spectroscopic results. Astron. Astrophys. **412**, 121–132 (2003)
G.M. Muñoz-Caro, U. Meierhenrich, W.A. Schutte, W.H.-P. Thiemann, J.M. Greenberg, UV-photoprocessing of interstellar ice analogs: detection of hexamethylenetetramine-based species. Astron. Astrophys. **413**, 209–216 (2004)
L.A. Nakhimovsky, M. Lamotte, J. Joussot-Dubien, *Handbook of low temperature electronic spectra of polycyclic aromatic*. Hydrocarbons. Phys. Sci. Data, vol. 40 (Elsevier, Amsterdam, 1989)
C.D. Neish, Á. Somogyi, M.A. Smith, Titan's primordial soup: formation of amino acids via low-temperature hydrolysis of Tholins. Astrobiology **10**, 337–347 (2010)
J.A. Noble, P. Theule, F. Mispelaer, F. Duvernay, G. Danger, E. Congiu, F. Dulieu, T. Chiavassa, The desorption of H2CO from interstellar grains analogues. Astron. Astrophys. **543**, A5 (2012)
G. Notesco, A. Bar-Nun, The effect of methanol clathrate–hydrate formation and other gas-trapping mechanisms on the structure and dynamics of cometary ices. Icarus **148**, 456–463 (2000)
G. Notesco, A. Bar-Nun, T. Owen, Gas trapping in water ice at very low deposition rates and implications for comets. Icarus **162**, 183–189 (2003)
M. Nuevo, J.H. Bredehoft, U. Meierhenrich, L. D'Hendecourt, W.H.P. Thiemann, J. Bredehöft Urea, Glycolic acid, and glycerol in an organic residue produced by ultraviolet irradiation of interstellar = pre-cometary ice analogs. Astrobiology **10**, 245–256 (2010)
M. Nuevo, S. Milam, S. Sandford, Nucleobases and prebiotic molecules in organic residues produced from the ultraviolet photo-irradiation of pyrimidine in NH3 and H2O + NH3 ices. Astrobiology **12**, 192 (2012)
M. Nuevo, Y.-J. Chen, W.-J. Hu, J.-M. Qiu, S.-R. Wu, H.-S. Fung, C.-C. Chu, T.-S. Yih, W.-H. Ip, C.-Y.R. Wu, Irradiation of pyrimidine in pure H2O ice with high-energy ultraviolet photons. Astrobiology **14**, 119 (2014)
K.I. Öberg, A.C.A. Boogert, K.M. Pontoppidan, S. van den Broek, E.F. van Dishoeck, S. Bottinelli, G.A. Blake, N.J. Evans, The spitzer ice legacy: ice evolution from cores to protostars. Astrophys. J. **740**(2), 109 (2011)
W.R. Parrish, J.M. Prausnitz, Dissociation pressures of gas hydrates formed by gas mixtures. Ind. Eng. Chem. Process Des. Dev. **11**, 26–35 (1972)
T. Poppe, J. Blum, T. Henning, Analogous experiments on the stickiness of micronsized preplanetary dust. Astrophys. J. **533**, 454–471 (2000)
D. Prialnik, Modeling the comet nucleus interior: application to comet c/1995 o1 Hale-Bopp. Earth Moon Planets **89**(1–4), 27–52 (2002)
A.J. Remijan, J.M. Hollis, F.J. Lovas, M.A. Cordiner, T.J. Millar, A.J. Markwick-Kemper, P.R. Jewell, Detection of C8H- and comparison with C8H toward IRC+10216. Astrophys. J. **664**, L47–L50 (2007)
F. Robert, D. Gautier, B. Dubrulle, The solar system d/h ratio: observations and theories. Space Sci. Rev. **92**(1–2), 201–224 (2000)
S.D. Rodgers, S.B. Charnley, W.F. Huebner, D.C. Boice, Physical processes and chemical reactions in cometary comae, in *Comets II*, ed. by M. Festou, H.U. Keller, H.A. Weaver (University of Arizona Press, Tucson, 2004), pp. 505–522
M. Rubin, K. Altwegg, H. Balsiger, A. Bar-Nun, J.J. Berthelier, A. Bieler, P. Bochsler, C. Briois, U. Calmonte, M. Combi, J. De Keyser, F. Dhooghe, P. Eberhardt, B. Fiethe, S.A. Fuselier, S. Gasc, T.I. Gombosi, K.C.

Hansen, M. Haessig, A. Jaeckel, E. Kopp, A. Korth, L. Le Roy, U. Mall, B. Marty, O. Mousis, T. Owen, H. Reme, T. Semon, C.Y. Tzou, J.H. Waite, P. Wurz, Molecular nitrogen in comet 67p/Churyumov-Gerasimenko indicates a low formation temperature. Science **348**(6231), 232–235 (2015)

M.B. Rydzy, J.M. Schicks, R. Naumann, J. Erzinger, Dissociation enthalpies of synthesized multicomponent gas hydrates with respect to the guest composition and cage occupancy. J. Phys. Chem. B **111**, 9539–9545 (2007)

S.A. Sandfordet et al., Organics captured from comet 81P/Wild 2 by the Stardust spacecraft. Science **314**, 1720–1724 (2006)

E.L. Schaller, M.E. Brown, Volatile loss and retention on Kuiper belt objects. Astrophys. J. **659**, L61 (2007)

D.G. Schleicher, T.L. Farnham, Photometry and imaging of the coma with narrowband filters, in *Comets II*, ed. by M. Festou, H.U. Keller, H.A. Weaver (University of Arizona Press, Tucson, 2004), pp. 449–469

B. Schmitt, S. Espinasse, R.J.A. Grim, J.M. Greenberg, J. Klinger, Laboratory studies of cometary ice analogues, in *Physics and Mechanics of Cometary Materials*. ESA Spec. Publ., vol. 302 (1989), pp. 65–69

P. Schmitt-kopplin, Z. Gabelica, R.D. Gougeon, A. Fekete, B. Kanawati, M. Harir, I. Gebefuegi, G. Eckel, N. Hertkorn, High molecular diversity of extraterrestrial organic matter in Murchison meteorite revealed 40 years after its fall. Proc. Natl. Acad. Sci. USA **107**, 2763 (2010)

W.A. Schutte, The formation of organic molecules in astronomical ices. Adv. Space Res. **16**, 53 (1995)

W. Schutte, L. Allamandola, S. Sandford, An experimental study of the organic molecules produced in cometary and interstellar ice analogs by thermal formaldehyde reactions. Icarus **104**, 118–137 (1993a)

W.A. Schutte, L.J. Allamandola, S.A. Sandford, Formaldehyde and organic molecule production in astrophysical ices at cryogenic temperatures. Science **259**, 1143–1145 (1993b)

K. Shin, R. Kumar, K.A. Udachin, S. Alavi, J.A. Ripmeester, Ammonia clathrate hydrates as new solid phases for Titan, Enceladus, and other planetary systems. Proc. Natl. Acad. Sci. **109**, 14785–14790 (2012)

E. Shpolskii, Emission spectroscopic analysis of organic compounds. Usp. Fiz. Nauk **68**(1), 51–69 (1959)

E. Shpolskii, Line fluorescence spectra of organic compounds and their applications. Usp. Fiz. Nauk **71**(2), 215–242 (1960)

H. Sierks, C. Barbieri, P.L. Lamy, R. Rodrigo, D. Koschny, H. Rickman et al., On the nucleus structure and activity of comet 67P/Churyumov-Gerasimenko. Science **347**, 6220 (2015)

I.A. Simonia, Unidentified cometary emission lines as the photoluminescence of frozen hydrocarbon particles. Astron. Lett. **30**, 863–868 (2004)

I.A. Simonia, On the possible luminescence nature of unidentified cometary emissions. Astrophys. Space Sci. **312**, 27–33 (2007)

I. Simonia, Organic component of cometary ice. Astrophys. Space Sci. **332**(1), 91–98 (2011a)

I. Simonia, Frozen hydrocarbons in comets. Astron. J. **141**, 56–61 (2011b)

I. Simonia, Organic molecules of cometary substance. AIP Conf. Proc. **1543**, 99–114 (2013)

I. Simonia, Ts. Simonia, Relict substance of comets, in *European Planetary Science Congress* University College London, September 8–13 (2013). EPSC 2013-108

I. Simonia, Ts. Simonia, Complex organic and possible prebiotics in the cometary matter in *International Cometary Workshop* Toulouse, France, April 1–3 (2014)

Y.V. Skorov, J. Blum, Dust release and tensile strength of the non-volatile layer of cometary nuclei. Icarus **221**, 1–11 (2012)

E.D. Sloan, C.A. Koh, *Clathrate Hydrates of Natural Gases*, 3rd edn. (2008)

G.P. Smith, L.C. Lee, P.C. Cosby, J.R. Peterson, J.T. Moseley, Photodissociation and photodetachment of molecular negative ions. V. Atmospheric ions from 7000 to 8400 Å. J. Chem. Phys. **68**, 3818–3822 (1978)

R.S. Smith, C. Huang, E.K.L. Wong, B.D. Kay, The molecular volcano: abrupt CCl 4 desorption driven by the crystallization of amorphous solid water. Phys. Rev. Lett. **79**, 909 (1997)

J.A. Smith, F.E. Livingston, S.M. George, Isothermal desorption kinetics of crystalline H2O, (H2O)-O-18, and d2o ice multilayers. J. Phys. Chem. B **107**(16), 3871–3877 (2003)

M.K. Spencer, M.R. Hammond, R.N. Zare, Laser mass spectrometric detection of extraterrestrial aromatic molecules: mini-review and examination of pulsed heating effects. Proc. Natl. Acad. Sci. USA **105**, 18096 (2008)

T. Spohn, J. Benkhoff, Thermal history models for kosi sublimation experiments. Icarus **87**(2), 358–371 (1990)

D.K. Staykova, W.F. Kuhs, A.N. Salamatin, T. Hansen, Formation of porous gas hydrates from ice powders: diffraction experiments and multistage model. J. Phys. Chem. B **107**, 10299–10311 (2003)

A.G. Suits, J.W. Hepburn, Ion pair dissociation: spectroscopy and dynamics. Annu. Rev. Phys. Chem. **57**, 431–465 (2006)

C. Szopa, R. Sternberg, D. Coscia, H. Cottin, F. Raulin, F. Goesmann, H. Rosenbauer, Gas chromatography for in situ analysis of a cometary nucleus: characterization and optimization of diphenyl/dimethylpolysiloxane stationary phases. J. Chromatogr. A **863**, 157 (1999)

C. Szopa, U.J. Meierhenrich, D. Coscia, L. Janin, F. Goesmann, R. Sternberg, Gas chromatography for in situ analysis of a cometary nucleus IV. Study of capillary column robustness for space application. J. Chromatogr. A **982**, 303 (2002a)

C. Szopa, R. Sternberg, D. Coscia, F. Raulin, C. Vidal-Madjar, H. Rosenbauer, Gas chromatography for in situ analysis of a cometary nucleus III. Multi-capillary column system for the cometray sampling and the composition experiment of the Rosetta lander probe. J. Chromatogr. A **953**, 165 (2002b)

C. Szopa, R. Sternberg, F. Raulin, H. Rosenbauer, What can we expect from the in situ chemical investigation of a cometary nucleus by gas chromatography: first results from laboratory studies. Planet. Space Sci. **51**, 863–877 (2003)

E. Szymanska, N.J. Mason, Dissociative electron attachment to hydrocarbons: a laboratory study. Paper #343 presented at The Molecular Universe, IAUS 280, Toledo, Spain (2011)

E. Szymanska, B.G. Nair, N.J. Mason, E. Krishnakumar, Dissociative electron attachment to formamide. XXVII International Conference on photonic, electronic and atomic collisions. J. Phys. Conf. Ser. **388**, 052085 (2014a)

E. Szymanska, I. Cadez, E. Krishnakumar, N.J. Mason, Electron impact induced anion production in acetylene. Phys. Chem. Chem. Phys. **16**, 3425–3432 (2014b)

T.A. Teplitskaiya, T.A. Alekseeva, M.M. Valdman, *Atlas of Quasilinear Luminescence Spectra of Aromatic Molecules* (Moscow University, Moscow, 1978)

P. Theule, F. Duvernay, G. Danger, F. Borget, J.B. Bossa, V. Vinogradoff, F. Mispelaer, T. Chiavassa, Thermal reactions in interstellar ice: a step towards molecular complexity in the interstellar medium. Adv. Space Res. **52**, 1567–1579 (2013)

C. Thomas, O. Mousis, V. Ballenegger, S. Picaud, Clathrate hydrates as a sink of noble gases in Titan's atmosphere. Astron. Astrophys. **474**, L17–L20 (2007)

C. Thomas, S. Picaud, O. Mousis, V. Ballenegger, A theoretical investigation into the trapping of noble gases by clathrates on Titan. Planet. Space Sci. **56**, 1607–1617 (2008)

P.C. Thomas, M.F. A'Hearn, J. Veverka, M.J.S. Belton, J. Kissel, K.P. Klaasen, L.A. McFadden, H.J. Melosh, P.H. Schultz, S. Besse, B.T. Carcich, T.L. Farnham, O. Groussin, B. Hermalyn, J.Y. Li, D.J. Lindler, C.M. Lisse, K. Meech, J.E. Richardson, Shape, density, and geology of the nucleus of comet 103p/Hartley 2. Icarus **222**(2), 550–558 (2013)

A.M.F. Tomescu, L.M. Pratt, G.W. Rothwell, P.K. Strother, G.C. Nadon, Carbon isotopes support the presence of extensive land floras pre-dating the origin of vascular plants. Palaeogeogr. Palaeoclimatol. Palaeoecol. **283**(1–2), 46–59 (2009)

F. Trixler, Quantum tunnelling to the origin and evolution of life. Curr. Org. Chem. **17**(16), 1758–1770 (2013)

J.H. Van der Waal, J.C. Platteeuw, Clathrate solutions, in *Advances in Chemical Physics*, vol. 2 (Interscience, New York, 1959), pp. 1–57

E.F. van Dishoeck, ISO spectroscopy of gas and dust: from molecular clouds to protoplanetary disks. Annu. Rev. Astron. Astrophys. **42**, 119–167 (2004)

C.A. Varotsos, R. Zellner, A new modeling tool for the diffusion of gases in ice or amorphous binary mixture in the polar stratosphere and the upper troposphere. Atmos. Chem. Phys. **10**, 3099 (2010)

G.L. Villanueva, M.J. Mumma, B.P. Bonev, M.A. Di Santi, E.L. Gibb, H. Böhnhardt, M. Lippi, A sensitive search for deuterated water in comet 8p/Tuttle. Astrophys. J. Lett. **690**(1), L3–L9 (2009)

V. Vinogradoff, F. Duvernay, G. Danger, P. Theule, T. Chiavassa, New insight into the formation of hexamethylenetetramine (HMT) in interstellar and cometary ice analogs. Astron. Astrophys. **530**, A128 (2011)

V. Vinogradoff, A. Rimola, F. Duvernay, G. Danger, P. Theule, T. Chiavassa, The mechanism of hexamethylenetetramine (HMT) formation in the solid state at low temperature. Phys. Chem. Chem. Phys. **14**, 12309–12320 (2012)

V. Vinogradoff, F. Duvernay, G. Danger, P. Theulé, T. Chiavassa, Thermal evolution of interstellar ice analogues containing formaldehyde. EAS Publ. Ser. **58**, 337 (2013a)

V. Vinogradoff, N. Fray, F. Duvernay, G. Briani, G. Danger, H. Cottin, P. Theule, T. Chiavassa, Importance of thermal reactivity for hexamethylenetetramine formation from simulated interstellar ices. Astron. Astrophys. **551**, A128 (2013b)

K.H. Walsh, A. Morbidelli, S.N. Raymond, D.P. O'Brien, A.M. Mandell, A low mass for Mars from Jupiter's early gas-driven migration. Nature **475**(7355), 206–209 (2011)

I. Waluyo, D. Nordlund, L.A. Naslund, H. Ogasawara, L.G.M. Pettersson, A. Nilsson, Spectroscopic evidence for the formation of 3-d crystallites during isothermal heating of amorphous ice on pt(111). Surf. Sci. **602**(11), 2004–2008 (2008)

X. Wang, A.J. Schultz, Y. Halpern, Kinetics of methane hydrate formation from polycrystalline deuterated ice. J. Phys. Chem. A **106**, 7304–7309 (2002)

A. Wekhof, Negative ions in comets. Moon Planets **24**, 157–173 (1981)

D.H. Wooden, Cometary refractory grains: interstellar and nebular sources. Space Sci. Rev. **138**(1–4), 75–108 (2008)

S. Wyckoff, S.C. Tegler, L. Engel, Nitrogen abundance in comet Halley. Astrophys. J. **367**(2), 641–648 (1991)

S. Wyckoff, R.S. Heyd, R. Fox, Unidentified molecular bands in the plasma tail of comet Hyakutake (C/1996 B2). Astrophys. J. **512**(L73), L76 (1999)

R. Yang, M.S. Gudipati, Novel two-step laser ablation and ionization mass spectrometry (2S-LAIMS) of actor-spectator ice layers: Probing chemical composition of d2o ice beneath a H2O ice layer. J. Chem. Phys. **140**(10), 202–207 (2014)

L. Yang, F.J. Ciesla, C.M.O.D. Alexander, The D/H ratio of water in the solar nebula during its formation and evolution. Icarus **226**(1), 256–267 (2013)

K. Yasuda, H. Ogawa, R. Ohmura, Phase equilibrium for clathrate hydrates formed with methane + ethane + propane mixtures at temperatures below the freezing point of water. J. Chem. Eng. Data **55**, 358–361 (2010)

C.Q. Yuan, J.T. Yates, Isotope effect in the photochemical decomposition of CO2 (ice) by lyman-alpha radiation. J. Chem. Phys. **138**(15) (2013)

G. Zasowski, F. Kemper, D.M. Watson, E. Furlan, C.J. Bohac, C. Hull, J.D. Green, Spitzer infrared spectrograph observations of class I/Ii objects in Taurus: composition and thermal history of the circumstellar ices. Astrophys. J. **694**, 459–478 (2009)

N.E.B. Zellner, V.P. McCaffrey, E. Bennett, M. Gudipati, Chemistry and astrochemistry of simple sugars: implications for asteroid, meteorite, or comet delivery, in *42nd Lunar and Planetary Science Conference, held March 7–11, 2011 at the Woodlands, Texas*, vol. 1608 (2011), p. 1586

H.W. Zhang, G. Zhao, J.Y. Hu, A catalogue of emission lines in spectra of Comet C/1995 O1 (Hale-Bopp). Astron. Astrophys. **367**, 1049–1055 (2001)

Space Sci Rev (2015) 197:151–190
DOI 10.1007/s11214-015-0167-6

The Composition of the Protosolar Disk and the Formation Conditions for Comets

K. Willacy[1] · C. Alexander[2] · M. Ali-Dib[3] · C. Ceccarelli[4,5] · S.B. Charnley[6] · M. Doronin[7,8] · Y. Ellinger[7] · P. Gast[9] · E. Gibb[10] · S.N. Milam[6] · O. Mousis[11] · F. Pauzat[7] · C. Tornow[9] · E.S. Wirström[12] · E. Zicler[7]

Received: 6 November 2014 / Accepted: 22 May 2015 / Published online: 26 June 2015

Abstract Conditions in the protosolar nebula have left their mark in the composition of cometary volatiles, thought to be some of the most pristine material in the solar system. Cometary compositions represent the end point of processing that began in the parent molecular cloud core and continued through the collapse of that core to form the protosun and the solar nebula, and finally during the evolution of the solar nebula itself as the cometary bodies were accreting. Disentangling the effects of the various epochs on the final composition of a comet is complicated. But comets are not the only source of information about the so-

✉ K. Willacy
karen.willacy@jpl.nasa.gov

[1] MS 169-506, NASA Jet Propulsion Laboratory, California Institute of Technology, 4800 Oak Grove Drive, Pasadena, CA 91109, USA

[2] MS 321-590, NASA Jet Propulsion Laboratory, California Institute of Technology, 4800 Oak Grove Drive, Pasadena, CA 91109, USA

[3] Institut UTINAM, CNRS/INSU, UMR 6213, Université de Franche-Comté, 41 bis Avenue de l'Observatoire, BP1615, 25010 Besançon, France

[4] IPAG, Univ. Grenoble Alpes, 38000 Grenoble, France

[5] IPAG, CNRS, 38000 Grenoble, France

[6] Astrochemistry Laboratory, Mail Code 691, NASA Goddard Space Flight Center, Greenbelt, MD 20771, USA

[7] Laboratoire de Chimie Théorique (LCT), Sorbonne Université, UPMC Univ Paris 06, CNRS UMR, 7616, 75252 Paris CEDEX 05, France

[8] Laboratoire de Physique Moléculaire pour l'Atmosphère et l'Astrophysique (LERMA/LPMAA), Sorbonne Université, UPMC Univ Paris 06, CNRS UMR, 7092, 75252 Paris CEDEX 05, France

[9] Institute of Planetary Research (DLR), Rutherfordstraße 2, 12489 Berlin, Germany

[10] Department of Physics and Astronomy, University of Missouri – St. Louis, St. Louis, MO 63121, USA

[11] LAM (Laboratoire d'Astrophysique de Marseille) UMR 736, Aix-Marseille Université, CNRS, 13388, Marseille, France

[12] Onsala Space Observatory, Department of Earth and Space Sciences, Chalmers University of Technology, 439 92, Onsala, Sweden

lar nebula. Protostellar disks around young stars similar to the protosun provide a way of investigating the evolution of disks similar to the solar nebula while they are in the process of evolving to form their own solar systems. In this way we can learn about the physical and chemical conditions under which comets formed, and about the types of dynamical processing that shaped the solar system we see today.

This paper summarizes some recent contributions to our understanding of both cometary volatiles and the composition, structure and evolution of protostellar disks.

Keywords Protostellar disks · Solar nebula · Comets · Chemistry

Abbreviations

AGB	Asymptotic giant branch
ALMA	Atacama Large Millimeter Array
CAI	Calcium-rich aluminium-rich inclusion
FUN	Fractionated and unknown nuclear isotopic effects
ISM	Interstellar medium
ISRF	Interstellar radiation field
IRAM	Institut de Radioastronomie Millimétrique telescope
MRI	Magneto-rotational instability
MHD	Magnetohydrodynamical
OPR	Ortho-to-para ratio
PSN	Protosolar nebula
SLR	Short-lived radionuclide
Gaseous inner disk	Inside the iceline
Inner disk	Inside 35 AU
Outer disk	Outside 35 AU

1 Introduction

Cometary volatiles are some of the most pristine of solar system materials, having remained relatively unprocessed since the comets formed in the inner solar system some 4.2 billion years ago. Consequently they retain in their composition signatures of the chemical and physical conditions under which they formed. Observations of cometary comae provide the mixing ratios or relative abundances of various species and have demonstrated that these can vary greatly between comets, e.g., A'Hearn et al. (2012) and Mumma and Charnley (2011) and references therein, with little correlation with dynamical family. The origin of these compositional variations is not well understood but must lie in the physical and chemical processes that were active at the time and location at which the cometary volatiles were formed. To understand these processes requires not only a study of the comets themselves but also of protostellar disks similar to the one from which our solar system formed. Observations of protostellar disks allow for the important disk processes to be investigated while they are occurring and can test models of the evolution of the early solar system.

Some key questions that can potentially be answered by interdisciplinary investigations involving both protostellar disk modeling and observations, and the observations of comets are:

1. Can synergy between protoplanetary disk modeling and compositional studies of comets inform us about the chemical history and processes in the early solar disk?
2. What is the origin of the diversity of observed cometary compositions?

3. Where did comets form in the solar nebula?
4. Are D/H ratios in volatile molecules and spin temperatures cosmogonic indicators for comets? If so, what can they tell us about the formation conditions in the early solar system?
5. What is the range of D/H ratios in cometary water and can we infer the degree to which comets contributed volatiles to the early terrestrial planets?
6. Can cometary volatile abundances be used to constrain the extent of radial and vertical mixing in the early protoplanetary disk?

This paper summarizes some recent developments in our understanding of cometary compositions and their relationship to the conditions under which the comets formed. The paper is arranged as follows. Section 2 covers some observational and modeling studies of protostellar disks. Section 3 discusses fractionation processes in protostellar disks including using deuterium fractionation as a tracer of the links between comets, interstellar chemistry and nebular chemistry. Section 4 considers the origin of the nitrogen deficiency in comets and Sect. 5 discusses spin temperature as a cosmogenic indicator in comets. Section 6 presents a possible explanation for the observations of neutral sodium cometary tails. Finally, Sect. 7 summarizes this paper and briefly discusses the potential impact of Rosetta on our understanding of the formation conditions of comets.

2 Overview of the Structure and Composition of Protostellar Disks

Protostellar disks form during the star formation process. A molecular cloud core collapses under gravity forming a protostar in the center. As material continues to fall inwards conservation of angular momentum results in the formation of a disk around the central protostar. In the early stages, the disks are gas-dominated with masses $\sim 1\ \%$ of the mass of the protostar. They rotate about the protostar with a Keplerian velocity profile. Disks contain a wide range of physical conditions, ranging from cold ($\sim$ 10–20 K in the outer disk midplane) to hot (with gas temperatures up to several thousand K in the surface layers). Irradiation conditions also vary, with the surface layers experiencing the full force of the stellar UV field, but with little radiation reaching the midplane. Consequently there are a wide variety of chemical conditions depending on location within the disk. The star and disk formation process, and the chemistry and dynamics of disks are discussed in more detail in recent reviews by Dutrey et al. (2014); Pontoppidan et al. (2014) and Turner et al. (2014).

Disks can be very turbulent places. Turbulence is difficult to detect directly but evidence for its effects can be seen in the presence of small dust grains high above the disk midplane. Without turbulence these grains would quickly be removed from these high altitudes by settling (Dullemond and Dominik 2004). The dynamical motions also drive collisions between grains, which can lead to both grain growth, eventually forming planetesimals, and to creation of new small grains or dust. With grain growth, the grains will eventually reach a size at which their motions decouple from those of the gas and they begin to settle under the influence of gravity to the midplane, where they can undergo further collisions and growth to form planetesimals. The removal of grains from the surface layers allows the stellar UV to penetrate deeper into the disk, changing its composition. Turbulence can also drive chemical changes by bringing together material processed in different parts of the disk and conversely chemistry can affect the dynamics by controlling the ionization levels. Consequently dynamics and chemistry strongly influence each other (see Sect. 2.3.2).

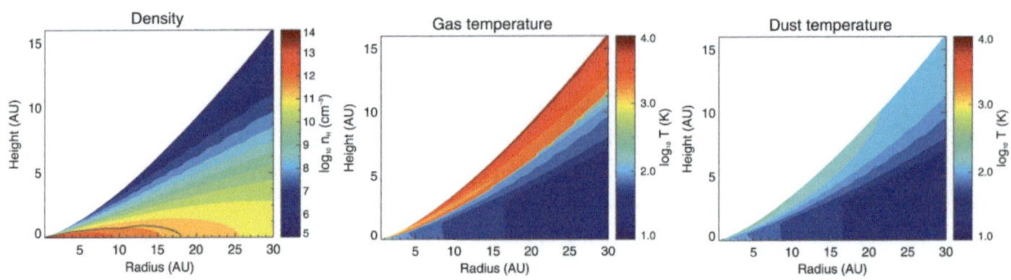

Fig. 1 The physical structure of a typical T Tauri disk. Shown are the density (*left*), gas temperature (*center*) and dust temperature (*right*). The density and dust temperatures are calculated by a hydrodynamical model of Paola D'Alessio with a similar mass and surface density profile to the minimum mass solar nebula. The gas temperature is determined via thermal balance and is coupled to the dust temperature throughout most of the disk, but is much hotter in the surface layers due to the photoelectric effect. The *black contour line* on the density plot indicates the likely location of the dead zone, determined by the location where the ionization level $x(e) < 10^{-12}$. In this region the turbulent motions driven by MRI will be severely inhibited

In recent years protostellar disks have been an area of active research, with their molecular content, dust properties, and physical conditions being studied in detail. The observational work has been coupled with simulations that can interpret the observations and make predictions about those regions of the disks which are not amenable to direct observational study. This synergy of the observations and modeling work has revealed a wealth of information about the chemistry and physics of disks, and in turn has informed our understanding of the early history of our own solar system. This section reviews some of the recent observational and modeling work related to protostellar disks, concentrating on the early stages of evolution when the disk is still gas-dominated (the T Tauri phase).

2.1 The Structure of a Protostellar Disk

Protostellar disks are characterized by strong radial and vertical gradients in temperature and density (Fig. 1). The stellar UV field controls the temperature and hence the chemistry throughout much of the disk. The penetration of the UV into the disk is governed by the opacity which is mainly provided by dust absorption. In the surface layers, where the UV field is unattenuated, the photoelectric effect can efficiently heat the gas to much higher temperatures than the dust. Close to the midplane, where the UV field is reduced, the dust and grain temperatures are coupled.

These gradients in physical conditions lead to a wide variety of chemical environments (Fig. 2). Models have found that outside of the water ice line the disk can be divided vertically into three chemically distinct layers. At the surface, the disk is dominated by the effects of UV. Ices are desorbed (either thermally or by photodesorption) and efficient photodissociation means that the gas is mainly composed of atoms and their ions. In the midplane most molecules are frozen out onto the surfaces of dust grains. Between these two layers is a region where the grains are warm enough to allow at least some molecules to thermally desorb and which is shielded enough that photodissociation is inefficient, allowing the molecules to survive once in the gas. It is this warm molecular layer that is detected in most gas phase observations of the outer regions of protostellar disks (see Sect. 2.2).

As the distance to the protostar decreases the disk warms up. In the midplane, the ice mantles will begin to desorb, with different molecules being released at different radii, depending on their binding energies—the most volatile species being released at larger radii than the less volatile ones. This results in a series of ice lines, which mark the transition

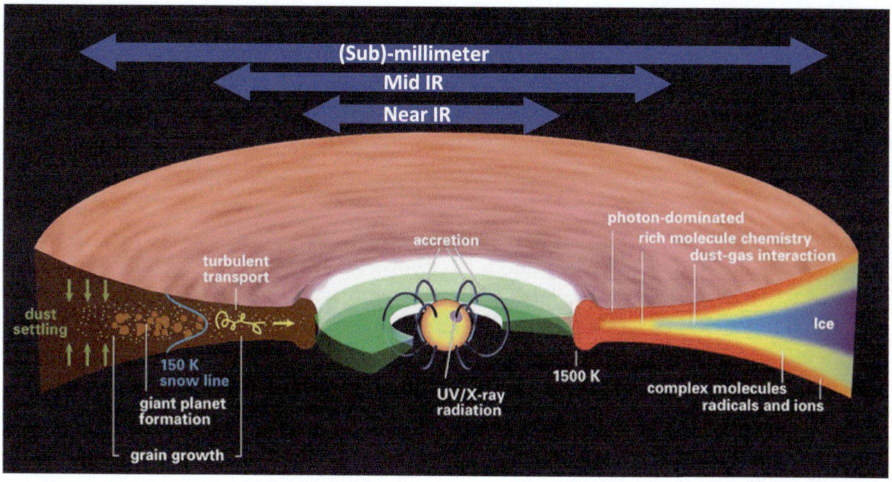

Fig. 2 The physical and chemical environment of a 1–5 Myr old protostellar disk around a Sun-like star. Reprinted from Henning and Semenov (2013) with permission. ©(2013) American Chemical Society

from ice to vapor for each molecule (see Sect. 2.4). The icelines are curved, and in the outer disk are almost horizontal with the molecule existing as ice below this line and as gas above it (see Pontoppidan et al. 2014). Once inside the water iceline (within a few AU of the protostar) the molecules are all in the gas phase.

Figure 2 also illustrates some of the main physical processes that are active in disks. In particular, turbulence is a crucial process, enabling dust coagulation and driving chemical changes by mixing material processed under different physical conditions within the disk. What drives this turbulence is still a matter of debate but one of the main candidate processes is the magneto-rotational instability (MRI; e.g., Balbus and Hawley 1991; Hawley and Balbus 1991; Hawley et al. 1996) which can efficiently transport angular momentum in ionized accretion disks. A minimum level of ionization is required (in the minimum mass solar nebula the ionization fraction, $x(e)$, must be $> 10^{-12}$ at 1 AU; Inutsuka and Sano 2005). This can be difficult to achieve in the shielded midplane regions of a protostellar disk, where the high column densities prevent UV and cosmic rays from penetrating. Hence there are regions in the disk where the MRI is not active. These are known as "dead zones" and they are predicted to occur throughout much of the midplane in the planet-forming region. The approximate location of the dead zone in the disk shown in Fig. 1 is indicated by the black contour line in the density plot.

The dependence of MRI-driven turbulence on the ionization means that there is a close link between the chemistry and the dynamics. The chemistry determines the ionization level and therefore the strength of the turbulence, but the dynamical motions themselves can drive chemical changes through mixing. This in turn affects the ionization and consequently the dynamics (e.g., Inutsuka and Sano 2005; Ilgner and Nelson 2006, 2008; Turner et al. 2007). Hence the evolution of chemistry and dynamics are inextricable linked.

2.2 Observations of Protostellar Disks

Protostellar disks have been observed at wavelengths from the UV to the millimeter. Each wavelength can probe a different region of the disk. The majority of information about the molecular composition of disks comes from the infrared and the (sub)-millimeter. The disk is mostly optically thick in the infrared, therefore these observations are restricted to the warm gaseous region inside of the iceline or to the surface layers. At these wavelengths gaseous molecules can be observed in both absorption and emission. Ices can also be observed in absorption if illuminated by the central star in an edge-on disk, e.g., Pontoppidan et al. (2005), but the spectra can be difficult to interpret due to contributions from material outside of the disk. Near- and mid-infrared observations are most sensitive to the inner regions of disks, while the far-infrared can probe the surface layers at larger radii (see Henning and Semenov 2013).

Spitzer, operating in the mid-infrared, has shown that inside of the iceline the disk is chemically rich, with the terrestrial planet-forming region containing organics such as C_2H_2, HCN, as well as CO, CO_2, water and OH (e.g., Carr and Najita 2008; Salyk et al. 2008; Pontoppidan et al. 2010). The detected molecules are the result of a rich gas phase molecular chemistry, in a region where temperatures range from a few hundred kelvin up to 1000 K, and densities are high ($> 10^8\ \mathrm{cm}^{-3}$).

Spitzer observations of the gaseous inner disk have also been used to make inferences about the conditions beyond the iceline. For example, Najita et al. (2013) found the HCN/H_2O ratio inside of the iceline to be correlated with disk mass, suggesting that this ratio may reflect changes to the C/O ratio induced by the formation of icy planetesimals outside of the iceline. Outside of the iceline much of the oxygen is tied up in water ice, whereas the carbon is mainly in the gas phase in the form of more volatile molecules such as CO. Planetesimals outside of the iceline will form from grains coated with water ice. Since the planetesimals do not move about the disk as easily the small grains and the gas, this will trap the oxygen outside of the iceline and the material that moves inwards to the inside of the iceline will be enriched in carbon. Consequently the gaseous inner disk will be enriched in organics compared to water. Since more massive disks are likely to form planetesimals more efficiently than their lower mass counterparts they will have a higher C/O ratio inside of the iceline.

In the far-infrared, Herschel has detected a number of molecules including cold water in the outer disks of TW Hya (Hogerheijde et al. 2011) and DG Tau (Podio et al. 2013). In both cases the water emission comes from a region of the disk where grain temperatures are low enough for this molecule to be expected to be frozen onto dust grains. Its presence in the gas can be explained by photodesorption driven by the stellar UV field (Willacy and Langer 2000; Dominik et al. 2005). Hogerheijde et al. (2011) found that the predicted flux from photodesorbed water to be higher than observed, and suggested that the answer to this discrepancy could be grain coagulation. As the grains grow they decouple from the gas motions and sink towards the midplane, taking their ices with them. This removes water ice from the molecular layer, leaving less to be photodesorbed, and thus reducing the gas phase abundance.

Emission lines from some molecules in the disk can be optically thin in the (sub)-millimeter and so in theory these wavelengths can probe the whole of the disk. In practice, however, resolution constraints restrict these observations to $\gtrsim$ 30–100 AU from the protostar, although ALMA allows observations to within a few AU of the closest protostars e.g. ALMA Partnership et al. (2015). Observations by both interferometers and single dish facilities have detected the emission from molecules in the warm molecular region and their distributions can be mapped using interferometers. Relatively simple molecules have been observed including CO (and its isotopologues), H_2CO, CS, C_2H, HCN, CN, HNC, DCN, HC_3N, c-C_3H_2, N_2H^+, HCO^+, DCO^+, H_2D^+ (for a recent review see Dutrey et al. 2014). The fractional abundances relative to H_2 are generally lower than observed in molecular clouds because of the combined effect of depletion in the cold dense midplane and photodissociation in the surface layers. Both processes reduce the gas phase molecular abundances.

Table 1 Molecules observed in protostellar disks and the processes they trace

Process	Molecules
Temperature	CO, ^{13}CO, $C^{17}O$, $C^{18}O$, H_2CO, H_2O, OH, H_2
Density	CO, H_2CO, HCO^+, HC_3N
Ionization	HCO^+, N_2H^+, CH^+
Velocity	CO, CS, HC_3N
Photodissociation	CN, HCN, HNC, OH, H_2O, C_2H
Photodesorption	H_2O, OH
Grain surface chemistry	H_2CO, complex organics, CO_2, C_2H_2
High temperature chemistry	Complex organics
Deuteration	HD, DCO^+, DCN, H_2D^+

In addition to the composition of the disk, the detection of molecules can also provide information about the physical conditions under which they exist and the chemical processes that are active (Table 1). Observations of molecular deuteration are especially important since they trace the thermal history of the disk (see also Sect. 3.1). Deuterated molecules, with their lower column densities, also provide a tracer of the gas in the midplane, e.g., ALMA observations of DCO^+ have been used to trace the location of the CO iceline in a Herbig AeBe disk (Mathews et al. 2013) and H_2D^+ is a potential tracer of the ionization level in the midplane (Ceccarelli et al. 2004).

2.3 Protostellar Disk Modeling

Chemical models of protostellar disks vary in their complexity, but they all predict the three-layer structure beyond the iceline as shown in Fig. 2. The chemical simulations require as their basis, a background disk model to provide the density, temperature and UV field as a function of location in the disk. These can be provided by hydrodynamical models of a $1 + 1D$ steady-state α-disk model in vertical hydrostatic equilibrium such as those of D'Alessio et al. (2005);[1] (see also Fig. 1). These physical disk models are computationally intensive and are generally run independently of the chemical simulations. Current chemical models are based on detailed chemical kinetics and can contain hundreds of species (in both the gas and the ices) and thousands of reactions. These models include the ionizing effects of UV photons, cosmic rays, X-rays and short-lived radionuclides such as ^{26}Al. Some models have also focused on isotopic chemistry such as deuterium (e.g., Aikawa and Herbst 1999; Woods and Willacy 2007; Willacy and Woods 2009; Kavelaars et al. 2011; Furuya et al. 2013; Albertsson et al. 2014) or carbon and oxygen isotopes (Woods and Willacy 2009, see also Sect. 3.2). The simplest models assume that the conditions in the disk did not change with time and ignore dynamical motions, although a few also consider advection e.g., Willacy et al. (1998); Aikawa et al. (1999); Woods and Willacy (2007); Woods and Willacy (2009); Willacy and Woods (2009). More recent models have included processes such as grain growth and settling, e.g., Aikawa and Nomura (2006); Fogel et al. (2011); Vasyunin et al. (2011); Akimkin et al. (2013), dynamical motions, (e.g. Ilgner et al. 2004; Willacy et al. 2006; Semenov et al. 2006; Semenov and Wiebe 2011; Ilgner and Nelson 2006; Turner et al. 2006; Aikawa 2007; Hersant et al. 2009; Heinzeller et al. 2011) and

[1] http://www.cfa.harvard.edu/youngstars/dalessio/.

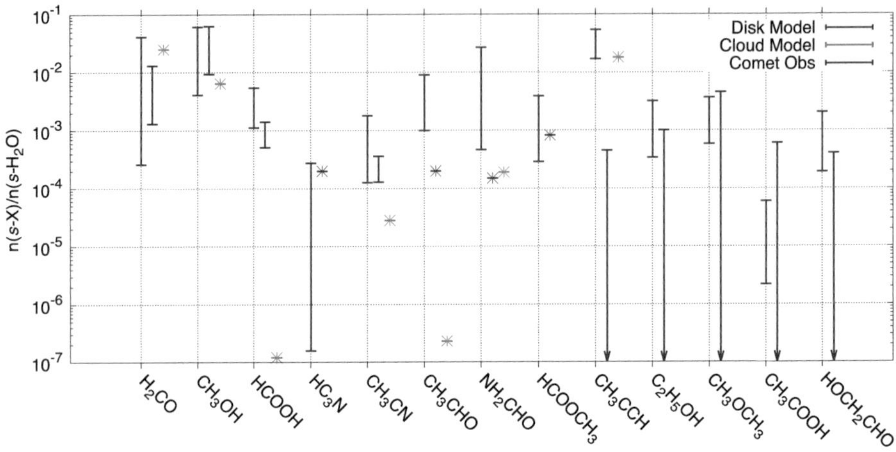

Fig. 3 Complex organics can be formed by chemistry in the ice mantles of dust grains in a protostellar disk. This figure shows the predicted range of abundances of some of these molecules relative to water ice (*red lines*) compared to those derived from cometary comae (*blue lines*; Bockelée-Morvan et al. 2004; Crovisier 2006; Crovisier et al. 2004). The *green asterisks* are the results from the initial molecular cloud model used as the input for the disk model. From Walsh et al. (2014)

hydrodynamical evolution of the disk structure itself (e.g. Dodson-Robinson et al. 2009; Yu et al. 2014). Here we highlight a few recent results.

2.3.1 Formation of Complex Organics in Protostellar Disks

Of particular interest to the chemistry of comets is the work of Walsh et al. (2014) who modeled the possible formation of complex organics through ice chemistry in protostellar disks. Figure 3 shows the range of abundances they find in ices located outside of 20 AU in a solar-type disk. The large molecules form by sequential addition of atoms to existing mantle molecules, e.g., ethylene glycol can form by adding hydrogen, oxygen and carbon atoms to CO:

$$\mathrm{CO} \xrightarrow{\mathrm{H}} \mathrm{HCO} \xrightarrow{\mathrm{C}} \mathrm{HC_2O} \xrightarrow{\mathrm{O}} \mathrm{OCCHO}$$

$$\mathrm{OCCHO} \xrightarrow{\mathrm{H}} \mathrm{CHOCHO} \xrightarrow{\mathrm{2H}} \mathrm{HOCH_2CHO} \xrightarrow{\mathrm{2H}} (\mathrm{HOCH_2})_2$$

(Charnley and Rodgers 2008, see Fig. 6). The models of Walsh et al. suggest that chemistry in disks will greatly increase the abundance of these large molecules relative to what can be created in the parent cloud. The predicted abundances relative to H_2O are consistent in many cases with observations of comets (Bockelée-Morvan et al. 2004; Crovisier 2006; Crovisier et al. 2004). If these molecules can be desorbed they will also be present in the disk gas, and should be detectable by ALMA providing a test of the model.

2.3.2 The Effects of Dynamics on the Chemistry

There are strong links between the disk chemistry and its dynamics. The temperature, density and radiation field in a disk control the chemistry, but the chemistry (and grain evolution) in turn controls the dynamics of the disk through the ionization structure. Models have found

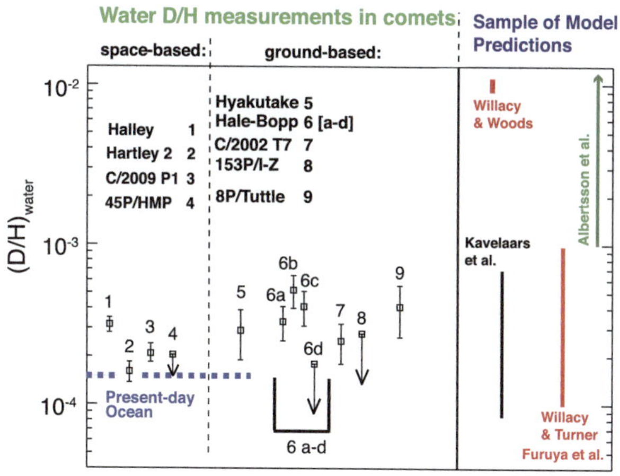

Fig. 4 Current disk models can only account for the observed D/H ratios in comets if they include mixing. The non-mixing models of Willacy and Woods (2009) and T. Millar (private communication) predict D/H ratios that are consistently higher than the observations. Albertsson et al. (2011) treat the evolution of the molecular cloud to the hot core stage only and predict $(D/H)_{water} > 10^{-3}$. Turbulent mixing is included in the disk models of Kavelaars et al. (2011) (radial mixing) and Willacy and Turner (2015) (vertical mixing at 5 AU) and these produce D/H_{water} in agreement with the observations. Furuya et al. (2013) predict similar D/H ratios to Willacy and Turner (2015). References: (1) Eberhardt et al. (1995), (2) Hartogh et al. (2011), (3) Bockelée-Morvan et al. (2012), (4) Lis et al. (2013), (5) Bockelée-Morvan et al. (1998) (6a-d) Meier et al. (1998) and Crovisier et al. (2004), (7) Hutsemékers et al. (2008), (8) Biver et al. (2006), (9) Villanueva et al. (2009). Adapted from Bonev et al. (2014)

that better agreement with observations of the molecular layer in the outer disk can be obtained if vertical mixing is included, e.g., Willacy et al. (2006); Semenov et al. (2006), and more sophisticated models have now considered 2-D mixing (e.g., Heinzeller et al. 2011; Semenov and Wiebe 2011; Albertsson et al. 2014). Mixing smooths out chemical abundance gradients but also brings together species that might not otherwise exist in the same location, e.g., in the inner disk, vertical mixing combines OH and H_2 in a region where the temperature is high enough for the neutral-neutral reaction, $OH + H_2 = H_2O + H$ to occur (Heinzeller et al. 2011). As a result the abundance of H_2O increases.

In addition to affecting molecular abundances, mixing also changes isotopic ratios. One problem with the non-mixing models is the high deuteration levels predicted in the comet formation zone, e.g., Willacy and Woods (2009). These reflect the high levels set in the parent molecular cloud, whereas those detected in comets tend to be somewhat lower. As D/H ratios track the thermal history of the material this suggests that at least some of the cometary ices formed at warmer temperatures than are found in the interstellar medium (ISM), e.g., Meier and Owen (1999). Figure 4 compares the observed molecular deuteration in cometary water with the predictions of several models. The non-mixing models uniformly over-predict HDO/H_2O, but the addition of mixing (either radial or vertical) brings the ratio into closer agreement with the observations (Bonev et al. 2014).

With vertical mixing this reduction in the deuteration of water is achieved by the destruction and reformation of ice mantles. When dust grains are mixed vertically they can reach the surface layers where high temperatures and/or UV fields desorb their ices. The water molecules are dissociated and the resulting oxygen, hydrogen and deuterium atoms travel back down towards the midplane into more shielded regions where they reform molecules. The molecular deuteration depends on the formation temperature. Since the disk is warmer

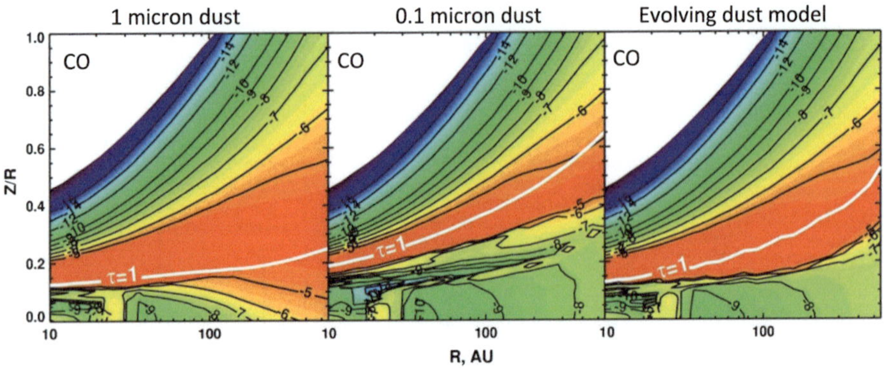

Fig. 5 The CO abundance distribution in a protostellar disk, showing the effects of grain growth. The highest abundances are shown in *red*. Grain growth reduces the opacity in the surface layers, allowing UV to penetrate further and pushing the molecular layer closer to the midplane. With 1 micron dust (*left*) the molecular layer is much thicker, and reaches closer to the midplane than in the model with 0.1 micron dust (*center*). On the *right* are the results from an evolving dust model where the grains are allowed to both grow and sediment. In this case the molecular layer is thicker than in the small grain model, and it moves closer to the midplane. From Vasyunin et al. (2011). ©AAS. Reproduced with permission

than the 10 K parent molecular cloud core, the reformed water ice has a lower D/H ratio. Hence vertical mixing can reduce the initial interstellar D/H ratio in water ice $\sim 10^{-2}$) to the level detected in comets ($\sim 10^{-4}$) (Fig. 4 and see also Furuya et al. 2013). Willacy and Turner (2015) and Furuya et al. (2013) also predict radial variations in the deuteration of water. A radial gradient in HDO/H_2O is also found in the (radial) mixing models of Kavelaars et al. (2011).

2.3.3 Grain Evolution

Grain-grain collisions in protostellar disks lead to efficient grain growth and ultimately result in the formation of planetesimals. As the grains grow they will decouple from the gas motions and sink towards the midplane. This removes grains from the upper layers of the disk, reducing the opacity and allowing UV to penetrate deeper, which increases photodissociation and heating of the newly irradiated layers, and moves the molecular layer closer to the midplane. An additional effect of grain growth is a reduction in the freezeout rate, which depends on the surface area of the grains. Figure 5 shows the effect on the chemistry of CO in the models of Vasyunin et al. (2011) (see also Aikawa and Nomura 2006; Walsh et al. 2014). Vasyunin et al. found that CO, CN and SO, as well as the ratio of C_2H_2/HCN, should be good tracers of the grain growth.

2.4 Gas-Dust Chemistry and Icelines in Protoplanetary Disks

Icelines in the early nebula are defined as radii at which specific molecules undergo a shift in abundance from the gas phase to the condensed phase. There are three of specific interest related to H_2O, CO_2 and CO and these occur at radii where the temperature is approximately 150 K, 47 K and 20 K respectively (Öberg et al. 2011). Recent observations by ALMA have allowed icelines to be identified in protoplanetary disks. For example, in HD 163926 (Qi et al. 2011; de Gregorio-Monsalvo et al. 2013;

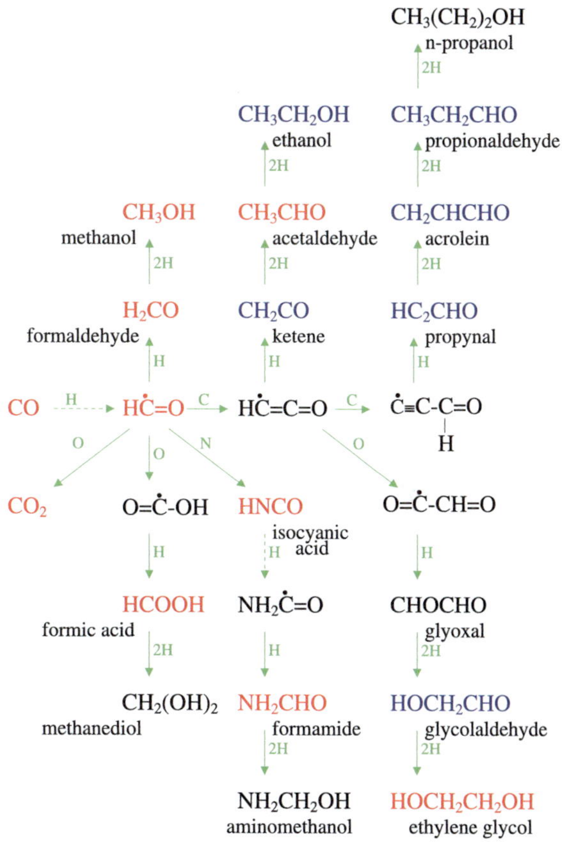

Fig. 6 Grain surface reaction scheme forming organics via atom addition reactions to CO on cold dust. Molecules in *red* are detected both in comets and in interstellar/protostellar sources, whereas those in *blue* are found only in the latter. Species in *black* are not yet detected in space. The addition of deuterium atoms rather than hydrogen atoms to grain species can result in high levels of molecular deuteration. Adapted from Charnley and Rodgers (2008)

Mathews et al. 2013). The icelines associated with H_2O, CO_2, and CO are regions of dramatic chemical change and these could be reflected in the molecular composition of the gas and ice that is transported and mixed, both vertically and radially, in the evolving disk.

The chemical signature of such an environment may be imprinted in cometary and meteoritic composition if the Sun formed in a stellar cluster containing at least one massive star (Adams 2010; Mumma and Charnley 2011). Nearby OB stars strongly affect the positions of the CO_2 and CO icelines and the chemical structure of the disk (Walsh et al. 2013). The CO iceline occurs at tens of AU from the central protostar and is most accessible to observations at millimeter and submillimeter wavelengths. The physical conditions here most closely resemble those in the densest molecular cloud cores and so the related gas-dust chemistry could also be similar (Henning and Semenov 2013). This chemistry can lead to formation of many complex organic molecules on cold dust grains, as well as potentially large isotopic fractionation in both gas phase and grain-surface reactions.

For example, outside the CO iceline, CO-rich ices on dust grains can undergo (tunneling) addition reactions with H (and D) atoms accreted from the gas (Tielens 1983; Charnley 1997). Atoms of O, C and N will also accrete and these processes can lead to rich organic chemistry (Fig. 6; Charnley 1997; Charnley and Rodgers 2008; Herbst and van Dishoeck 2009). Many of the proposed hydrogenation sequences have now been demonstrated in surface chemistry experiments (Watanabe and Kouchi 2008; Theulé et al. 2013) and such an ice chemistry explains the presence of the distinctive and inter-related suite of

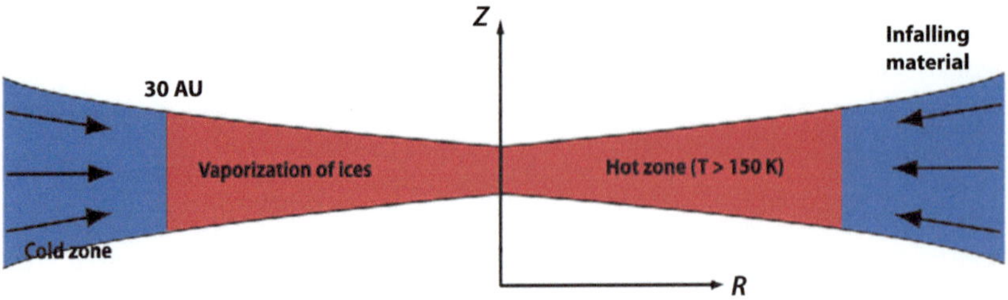

Fig. 7 Two reservoirs of ices in the PSN. A first reservoir (cold zone), located at distances higher than ~ 30 AU, is constituted from amorphous ices coming from ISM. The second reservoir (hot zone), located within the $\sim$30 AU, is made from volatiles initially in the form of amorphous ices that were transported from ISM towards the inner and hot part of the disk. When reaching regions with temperatures higher than ~ 150 K, these ices vaporized. During the cooling of the disk, volatiles located in the inner 30 AU condensed again but in crystalline forms, including both pure crystalline ices and clathrates (see text)

organic molecules detected around massive and low-mass protostars (in hot cores and hot corinos, respectively), where the ice mantles have been evaporated into the hot gas (Herbst and van Dishoeck 2009). Figure 6 indicates that many of the relevant organic molecules could also be present in comets and that, if correct, ketene, ethanol, propynal, acrolein, propionaldehyde and propanol should be searched for in suitably bright apparitions. Several of the organic compounds formed in the scheme of Fig. 6 may be detectable in protoplanetary disks with ALMA. Vertical transport of ice-covered dust to the less well-shielded layers of the disk, as well as (inward) radial transport, can lead to these organic compounds being released by evaporation and/or photodesorption. This is probably the origin of the H_2CO detected in HD 163296, which lies outside of the CO iceline at about 160 AU (Qi et al. 2013b).

2.5 Two Distinct Reservoirs of Water Ice in the Protosolar Nebula

Formation scenarios of the protosolar nebula (PSN) invoke two main reservoirs of ices present in the disk midplane and that took part in the production of icy planetesimals (see Fig. 7). The first reservoir, located in the inner region of the PSN, contains ices (dominated by H_2O, CO, CO_2, CH_4, N_2 and NH_3) originating from the ISM, which, due to their near solar vicinity, were initially vaporized. The ice vaporization distance never exceeds ~ 30 AU from the Sun, depending on the total source luminosity and characteristics of the collapsing cloud (Chick and Cassen 1997). With time, the decrease of temperature and pressure conditions allowed the water in this reservoir to condense at ~ 150 K (at typical nebula pressure conditions) in the form of microscopic crystalline ice (Kouchi et al. 1994), leaving negligible water in the vapor phase to condense at lower temperatures where amorphous ice would be expected. It is postulated that a substantial fraction of the volatile species were trapped as clathrates during this condensation phase as long as free water ice was available within 30 AU in the outer solar nebula (Mousis et al. 2000) and there was enough time to overcome the kinetics of clathration. On the other hand, the remaining volatiles that were not enclathrated (due to the lack of available water ice or a low kinetics of clathration) probably formed pure condensates at lower temperatures in this part of the nebula (Mousis et al. 2012a, 2012b). The other reservoir, located at larger heliocentric distances, is composed of ices originating from the ISM that did not vaporize when entering into the disk. In this reservoir, water ice was essentially in the amorphous form and the other volatiles remained trapped in the amorphous matrix (Owen et al. 1999; Notesco and Bar-Nun 2005).

Interestingly, it has been shown recently that amorphous ice could form from photodesorption and freeze-out of water molecules near the surface layers of the PSN (Ciesla 2014). In these conditions, the transport of the icy grains throughout the outer solar nebula would lead crystalline ice to be lost and reformed as amorphous ice. In turn, amorphous ice generated on the disk's surface could be transformed to crystalline as it migrates through the protoplanetary disk (Ciesla 2014). This cycling scenario needs to be further investigated if one wants to assess its role in the formation or destruction of the two original icy reservoirs. Also, the different transport mechanisms must be simultaneously taken into account at the different stages of the disk's evolution to depict the abundance of water ice in the two reservoirs as a function of the heliocentric distance. For example, as described in Sect. 2.3.2, it has been shown that vertical mixing can decrease the abundance of water inside the $\sim$ 15 AU of the disk. In contrast, it has also been proposed that the outward vapor diffusion induced by the presence of the iceline with its subsequent condensation in a narrow location (with a width around 0.5 AU) would deplete the region inside the iceline of vapor down to subsolar values, and increase substantially the ice abundance at the iceline location by factors larger than 10 times the protosolar value (Ali-Dib et al. 2014). In addition, recent works depicting the chemical evolution of both solids and gases from the pre-stellar core to the protostar and circumstellar disk phases (Visser et al. 2009b; Visser et al. 2011; Harsono et al. 2013) should be taken into account to update the representation of the two water ice reservoirs present in the PSN. In these cases, midplane ices may have experienced multiple desorption events during the evolution of the disk. Nevertheless, if one follows this classical picture, icy solids that formed at heliocentric distances less than $\sim$30 AU mainly agglomerated from a mixture of clathrates and pure condensates, whose ratio depends on the amount of available crystalline water and its clathration efficiency (Mousis et al. 2009b). In contrast, solids produced at higher heliocentric distances (i.e., in the cold outer part of the solar nebula) were formed from primordial amorphous ice originating from the ISM. Therefore, depending where the outer solar system bodies formed, clathrates may have been agglomerated in comets (Marboeuf et al. 2010a, 2010b, 2012), and in the building blocks of the giant planets (Gautier et al. 2001; Alibert et al. 2005a, 2005b; Mousis et al. 2012a, 2009b) and in their surrounding satellite systems (Lunine and Stevenson 1985; Mousis 2004; Mousis and Gautier 2004; Mousis et al. 2009a). Regardless the possible presence of clathrates, there is today indirect evidence that comets formed from crystalline water ice because their deuterium-to-hydrogen ratios measured in H_2O are substantially lower ($\sim$ 6–12 times the protosolar value) than the most deuterium-rich primitive meteorite ($\sim$ 35 times the protosolar value), whose level of deuteration is expected to be close to the one found in the ice infalling from the presolar cloud (Kavelaars et al. 2011; Hartogh et al. 2011; Ceccarelli et al. 2014). This difference can only been explained by the isotopic exchange between the deuterium-rich water and molecular hydrogen in the gas phase (Kavelaars et al. 2011; Hartogh et al. 2011; Ceccarelli et al. 2014). Hence, in order to enable this reaction, amorphous ice entering the disk from ISM must have been vaporized prior recondensing again in crystalline form, otherwise the isotopic exchange did not occur.

3 Fractionation in Protostellar Disks

3.1 The Deuterium Fractionation: The Link Between Comets and the Earliest Phases of the Solar System and Interstellar Chemistry

The origin of the volatile ices found in comets is a central issue for understanding the formation and early evolution of the solar system. There are three possibilities. First, one view is that *all* the ices are comprised of pristine interstellar molecules (e.g. Irvine et al. 1996). There is indeed a strong similarity with the many known interstellar molecules believed to form on interstellar grains, although there are differences in the relative abundances of the volatile ices (Mumma and Charnley 2011). Until recently this idea was widely considered to be untenable (see Sect. 3.1.2 and Cleeves et al. 2014a). Second, at the other extreme, it has been advocated that any interstellar molecules experienced total obliteration and 'chemical reset' in the disk (Pontoppidan and Blevins 2014). In this case all the volatile molecules observed in comets must have been formed by nebular chemistry.

Comets and meteorites contain both high-temperature processed materials, such as crystalline silicates, and matter that apparently formed at very low temperatures, as indicated by the isotopic enrichments measured in D and ^{15}N (Mumma and Charnley 2011). These characteristics are most naturally explained by the large-scale turbulent transport of dust and gas from different nebular chemical environments into the comet-formation zones (Ciesla and Charnley 2006). The recognition that non-equilibrium models of nebular chemistry are essential, particularly in the comet-forming regions, has motivated many detailed studies of protoplanetary disk chemistry, as recently reviewed by Henning and Semenov (2013) (see also Sect. 2.3.2). Turbulence leads to an outward radial diffusion of material from the hot inner nebula and an inwards advection of material from the cooler outer nebula. It also leads to vertical mixing and thus to midplane material experiencing the enhanced UV and X-ray fluxes of the upper disk layers (Glassgold et al. 1997). Hence, in this third partially-mixed scenario some specific volatile molecules and isotopic ratios could be remnant interstellar matter, or derive from warm and cold regions of the nebula (as reviewed by Pontoppidan et al. 2014). Deciding between such molecular origins has been problematic.

3.1.1 The Molecular Deuteration Process

Molecular deuteration provides a potential clue to the history of cometary volatiles. Although deuterium in the Universe is a tiny fraction with respect to the hydrogen (D/H $=[2.535 \pm 0.05] \times 10^{-5}$: Pettini and Cooke 2012), its abundance in interstellar molecules has an enormous diagnostic power, because it may be much larger that the elemental D/H ratio. This generally occurs in cold ($\leq$ 30–50 K) environments and in trace species, namely molecules with abundances smaller than $\sim 10^{-5}$.

The first key point to keep in mind is that, in cold gas, the major deuterium reservoir is HD and only a small fraction of deuterium is locked in other species. An exception to this rule is represented by the water, whose abundance can reach $\sim 10^{-4}$ and, consequently, water might trap a substantial fraction of deuterium.

The second key point is that the deuterium atoms are "extracted" from the HD molecules because the cosmic rays that hit H and H_2 create H_3^+ ions, which in turn react with HD and form H_2D^+ ions.

$$H_3^+ + HD \rightleftharpoons H_2D^+ + H_2 + 220\ K \tag{1}$$

(see also Fig. 6). The reverse reaction has a small energy barrier, which cannot be overcome at low ($\leq$ 20–30 K) temperatures, so that the H_2D^+/H_3^+ abundance ratio becomes larger than the D/H elemental ratio. The kinetics of Reaction (1) at low temperatures is also strongly influenced by the spin state of the reacting hydrogen molecules, i.e., whether they are predominantly in the *ortho* or *para* states. The higher internal energy of *ortho*-H_2 means that the reverse process can be driven efficiently if the H_2 *ortho*/*para* ratio is large. Thus, *ortho*-H_2 acts as a "poison" for interstellar deuteration which requires that most of the H_2 be converted to *para*-H_2 (see Pagani et al. 2011). Reaction (1) is the first step towards the

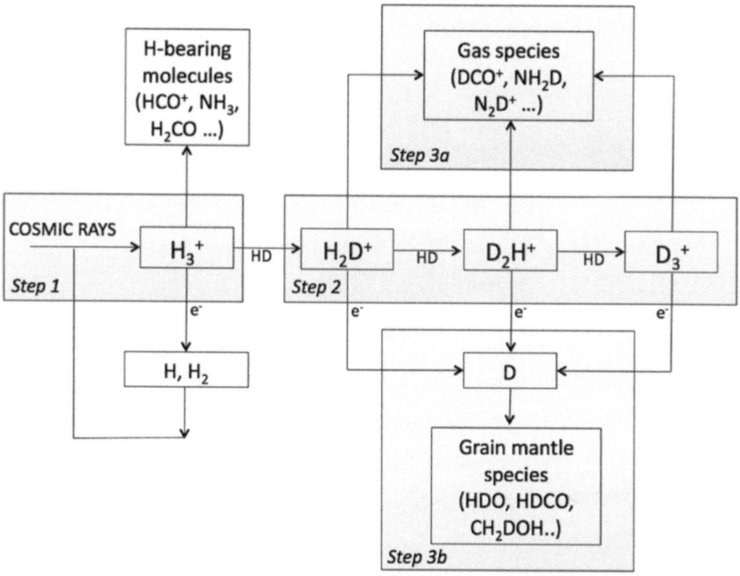

Fig. 8 Schema of the molecular deuteration process in cold gas (Ceccarelli et al. 2014). Molecular deuterium occurs through three basic steps: 1) formation of H_3^+ ions from the interaction of cosmic rays with H and H_2; 2) formation of H_2D^+ (HD_2^+ and D_3^+) from the reaction of H_3^+ (H_2D^+ and HD_2^+) with HD; 3) formation of other D-bearing molecules from reactions with H_2D^+ (HD_2^+ and D_3^+) in the gas phase (Step 3a) and on the grain mantles (Step 3b)

deuteration of the trace species, as H_2D^+ reacts with all neutral molecules and passes, once every three times, the D atom to the product species. This is schematically shown in Fig. 8. A similar process can also occur through the CH_3^+ ions at slightly higher temperatures.

In practice, therefore, a molecular deuteration enhanced with respect to the D/H elemental abundance is a thermometer of the gas temperature.

The third and final key point to keep in mind is that neutral molecules can freeze-out into the grain mantles and, in this way, they can be transmitted through the different stages of the star and planet formation. A sketch of this overall process is shown in Fig. 9. Therefore, ice species formed during the very first phases might be conserved and "reappear" in the gas at much later stages: for example when comets and planets are formed. Conversely, the molecular deuteration in comets and planets can help us to understand in what epoch the relevant species was formed, at least in principle. In this sense, molecular deuteration can be thought of as an Ariadne's thread which links the various phases of solar system evolution. For a more detailed review see Ceccarelli et al. (2014).

3.1.2 Deuteration in the Solar Nebula

Deuteration in disks is driven mainly via the reaction of H_2D^+ ions (Reaction (1)). At low temperatures the reverse process is inhibited and this leads to enhanced deuteration in gaseous molecules through ion-molecule isotope exchange reactions. Elevated atomic D/H ratios also pertain and lead to significant deuteration in grain-surface reactions, such as those depicted in Fig. 10 and in the surface reduction of accreted O atoms to form HDO. Condensation of CO leads to enhanced abundances of gaseous H_2D^+ and to significant deuterium fractionation beyond the CO iceline, as traced by DCO^+ (Mathews et al. 2013). Enhanced

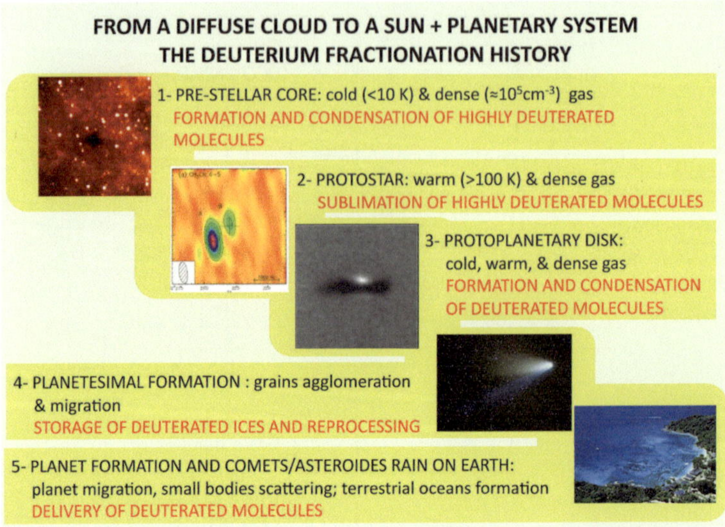

Fig. 9 Schematic summary of the different phases of the formation of a solar-type star and planets, highlighting the formation and fate of deuterated molecules (adapted from Ceccarelli et al. 2014). Likely, the same process gave birth to the solar system

D/H ratios in simple molecules, such as water, ammonia and methane, can be produced both in ion-molecule reactions and grain-surface additions of H and D to accreted atoms (Ceccarelli et al. 2014). At the CO iceline the grain-surface chemistry depicted in Fig. 10 can lead to very high D/H ratios and multiply deuterated molecules (Charnley et al. 1997; Ceccarelli et al. 2014). Observations show that simple deuterated species like DCN and by DCO^+ are present in disks (Qi et al. 2008) and we would expect that more complex fractionated organics, e.g., HDCO, D_2CO, CH_3OD and CH_2DOH, to also be present. Between the CO and CO_2 icelines (see Sect. 2.4), large deuterium enrichments can persist in organic molecules since ion-molecule deuteration involving C_2HD^+ and CH_2D^+ can persist up to higher temperatures (~ 70 K and ~ 50 K) than H_2D^+ (~ 30 K) (Millar et al. 1989). It is possible that the D/H ratios of several organic molecules can become re-fractionated even in the warm inner nebula, prior to the condensation of acetylene and methane, both of which will occur outside the CO_2 iceline. This means that deuteration can still occur in cold gas that is mixed radially inwards, and be initiated in warmer gas coming from the inner nebula (Willacy 2007; Albertsson et al. 2014).

However, the assumption of interstellar rates of cosmic-ray driven ionization (ζ) in the cold regions of the outer midplane has recently been called into question. Cleeves et al. (2013) showed that the effects of either deflection by stellar winds or magnetic field topology (through the magnetic mirror effect) could significantly attenuate the penetration of cosmic rays into magnetized protoplanetary disks, and substantially reduce ζ by several orders of magnitude at the midplane. In this picture the only sources of ionization close to the midplane are scattered X-rays and the decay of short-lived radionuclides. In fact, Dolginov and Stepinski (1994) had previously pointed out that inhomogenous random magnetic fields could also significantly inhibit the transport of cosmic rays throughout the disk and lead to very low values of ζ.

Cleeves et al. (2014a) developed this scenario further by pointing out that effectively extinguishing cosmic ray ionization from the nebular midplane would greatly reduce the

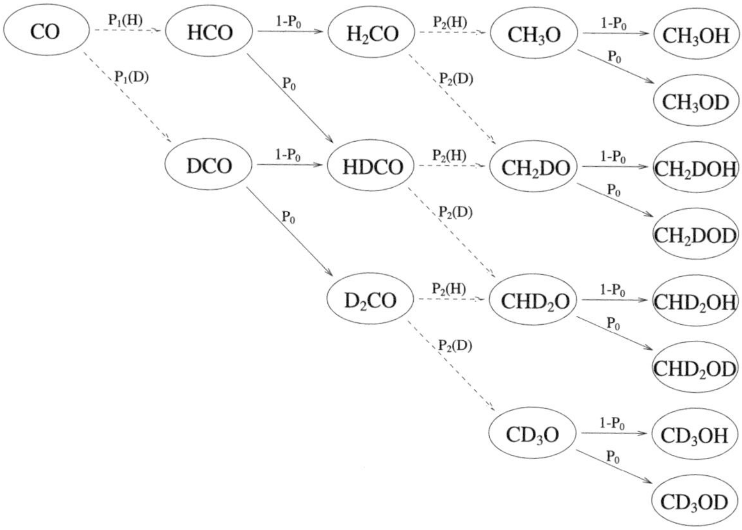

Fig. 10 Grain surface deuteration pathways associated with methanol formation from hydrogenation of CO ice. The probabilities for each reaction pathway, P_i, assume a statistical steady state and are related to the atomic D/H ratio of the accreting gas and the energy barrier for addition of an H or D atom (from Charnley et al. 1997)

efficacy of Reaction (1) and consequently the generation of enhanced molecular D/H ratios by nebular ion-molecule chemistry. They demonstrated that if the enhanced interstellar HDO/H_2O ratios initially available were lost, then nebular chemistry could not regenerate those currently measured in the solar system, concluding that as much as 50 % of the Earth's oceans and perhaps *all* cometary water is of interstellar origin. Low values of ζ will also reduce the atomic D/H ratio in the disk midplane and so reduce the D/H ratios and the level of multiple deuteration in the molecules formed in the scheme of Fig. 10. This may account for the upper limits on HDCO, CH_2DOH and CH_3DOD in comet Hale-Bopp (Crovisier et al. 2004) relative to the large observed interstellar D/H ratios measured in these molecules, which could have been lost or modified upon incorporation into the nebula. Although these molecules could be partially reformed in the cold outer nebula, this process, and any related deuteration, may be less efficient than in molecular clouds. Cometary CH_3OH/H_2O ratios are typically $\sim$ 1–4 %, whereas $\sim$ 5–15 % are typical for the ISM, and can be as high as 40 % (Mumma and Charnley 2011). Cleeves et al. (2014b) also modeled the general protoplanetary disk chemistry for a variety of assumed ζ values and identified molecular ions that may eventually allow the ionization structure to be determined. If these spatial distributions are observable with ALMA they could provide valuable information on the actual role of cosmic rays in nebular chemistry.

3.1.3 The Message from Molecular Deuteration

Figure 11 shows the molecular deuteration in the different objects of the solar system and in objects which are believed to eventually form a solar-like system. It gives us the following major messages:

- All reported solar system objects, including Earth, possess a molecular deuteration larger, by more than a factor 10, than the D/H elemental ratio of the environment where the Sun

Fig. 11 D/H ratio in molecules observed in solar system objects and in objects which are believed to eventually form a solar-like system. The *filled symbols* refer to water measures whereas *open symbols* refer to organic matter. The different objects are marked (note: CC means Carbonaceous Chondrites, IOM Insoluble Organic Matter, and SOM Soluble Organic Matter). References to the various measures can be found in Ceccarelli et al. (2014)

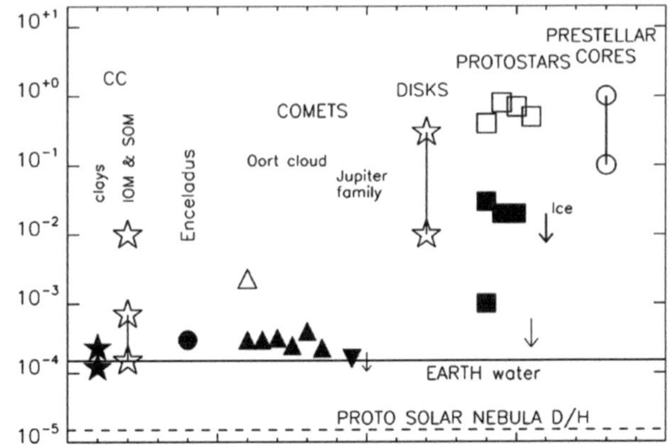

was born, and that is represented by the line marked with "Proto Solar Nebula". This implies that the process of molecular deuteration was at work and it was at least partially inherited from the earlier phases of the solar system formation era.

- There is a general trend, with the objects that formed earlier being also the ones with the largest molecular deuteration, in agreement with the scenario that the process started in very cold conditions. This implies that there was only a partial transmission between the subsequent steps of Fig. 9. Besides, earlier objects in the plot also correspond to material whose distance from the central heating source is larger. Therefore, the trend also tells us that there was not a general and substantial remixing of the different regions (outer and inner, where molecular deuteration is different because of the different temperatures). Exactly how much material was passed through each step or remixed remains, however, unanswered.
- The organic matter is systematically more enriched in deuterium atoms than the water. This is true in protostellar objects, comets and meteorites. However, while this result is robust for the protostellar objects and meteorites, where several measurements of the deuteration have been made, it is less certain for the comets, where so far the only organic species for which molecular deuteration has been measured is HCN. The reason for this dichotomy in the molecular deuteration is completely unclear. In protostellar objects, it is believed that this reflects the different phases of formation of water and organics, with water being formed very early when the gas is not very cold, while the organics were synthesized later from frozen CO. It is possible that a similar process is at the origin of the different deuteration in meteorites and comets. Alternatively, they may conserve a "core" of the protostellar molecular deuteration.
- Molecular deuteration in a given molecule has a strong temperature dependence at low temperatures. This is true for both gas phase and icy grain surface formation. In ice mantles the higher binding energy of deuteration atoms at < 20 K leads to significant enhancements of deuterated species that form via hydrogenation reactions (see also Fig. 10). These low temperatures correspond to what modelers predict for the cold midplane in the giant planet formation zone of the protoplanetary disk. Measurements of the molecular deuteration in comets provide a potential tracer of the thermal conditions under which they formed. For further discussion see Bockelée-Morvan et al. (2015, this issue).
- Molecular deuteration in a given molecule has a strong temperature dependence at low temperatures. This is true for both gas phase and icy grain surface formation. In ice mantles the higher binding energy of deuteration atoms at < 20 K leads to significant enhance-

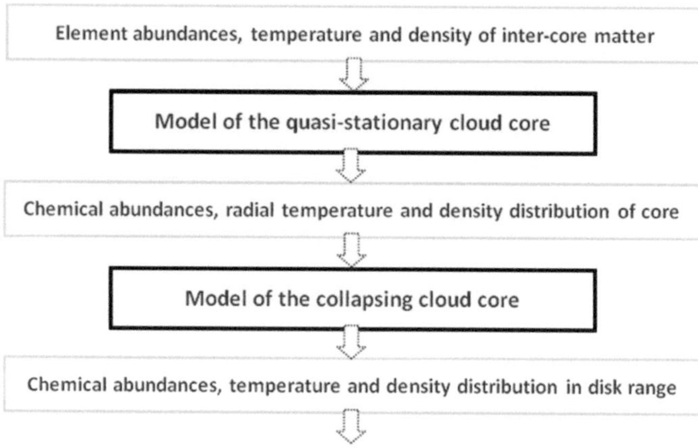

Fig. 12 Flowchart illustrating the evolution of the two precursor stages (the quasi-stationary cloud core and the collapsing core) to the solar nebula. Evolution in these stages will affect the composition of material incorporated into the solar nebula and hence the composition of planetary bodies that form within it

ments of deuterated species that form via hydrogenation reactions (see also Fig. 10). Measurements of the molecular deuteration in comets provide a potential tracer of the thermal conditions under which they formed. For further discussion see Bockelée-Morvan et al. (2015, this issue).

3.1.4 *A Model of the Deuteration of Water in the Early Stages of the Evolution of the Solar Nebula*

Molecular deuteration potentially provides a means of tracing the history of solar system material and its links to the parent molecular cloud. In this section we present the results of a specific model of the evolution of the deuteration of water during the collapse of the parent molecular cloud, since the composition of the protosolar nebula depends in part on the chemical history of this core (Fig. 12). The evolution of the core has been modeled recently by Aikawa et al. (2012); Wakelam et al. (2012); Taquet et al. (2014) and Tornow et al. (2014a, 2014b).

In these models the cloud core is initially described as a Bonner-Ebert sphere (Aikawa et al. 2008) which is embedded in a constant density gas-shell with a visual extinction of up to 4 magnitudes (Wakelam et al. 2012), and irradiated by an external UV field. It is allowed to evolve for between 1 and 6 Myrs (Tornow et al. 2014a). Both the gas phase[2] and grain[3] chemistry is followed. Since the core is only partially shielded, photo-chemistry has a clear effect on the radial profile of the relative water abundance in the outer core region (Fig. 13). At the end of the quasi-stationary core stage the temperature of the gas and dust decreases to 8–9 K in the center for boundary values between 13–15 K. Consequently, water ice builds up on the cold grains, reaching to a fairly constant relative abundance of $\sim 10^{-4}$ within 10^5–10^6 years. In contrast, Fig. 13 shows a clear radial gradient for the gas phase water abundance. Relatively low ratios (about 0.01–10 ppm) of gas to dust phase abundance were obtained for the outer core range with up to 25 ppm in the inner region. Caselli et al. (2012)

[2]Nahoon_public_apr11.tar.gz at http://kida.obs.u-bordeaux1.fr/models/.

[3]disk_chemistry_OSU08ggs_UV.zip at http://www.mpia.de/homes/semenov.

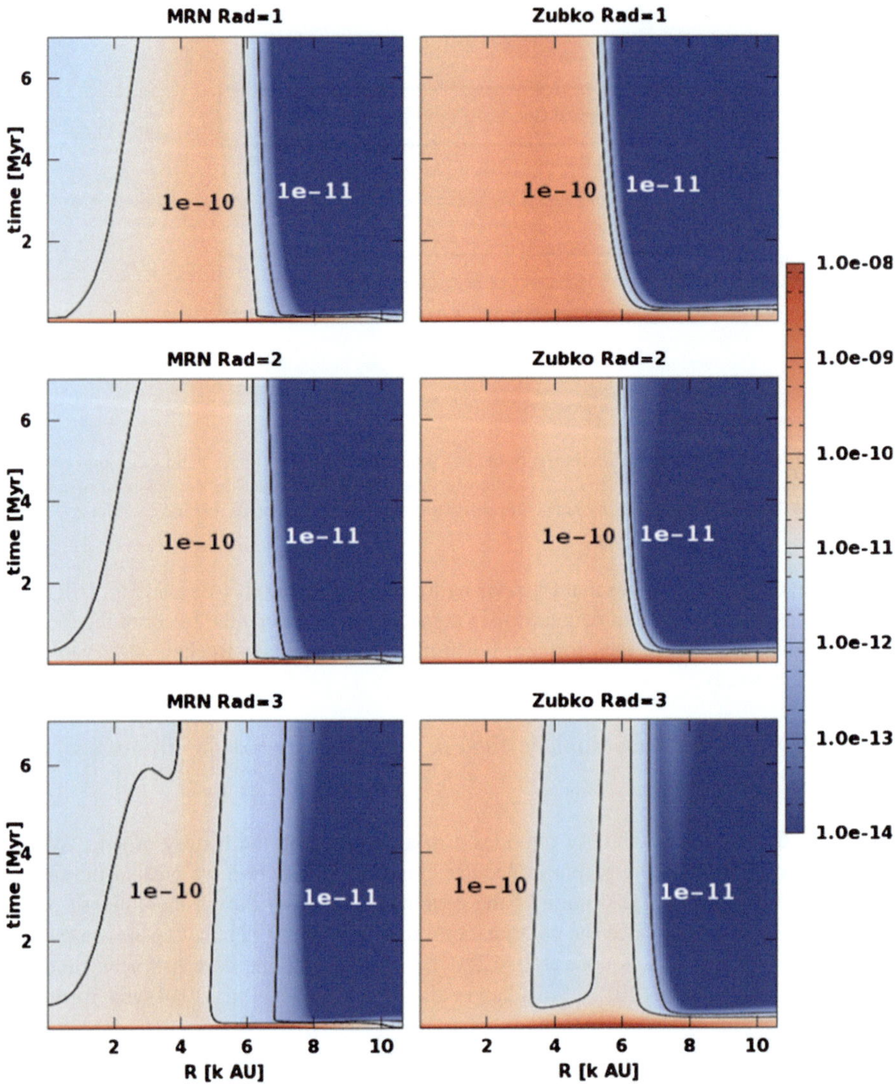

Fig. 13 Relative gas phase abundance of water versus time and radius (1 kAU = 10^3 AU). The intensity of the standard interstellar radiation field (IRSF) is scaled by Rad = 1, 2, 3. Two grain distributions are used. MRN refers to Mathis et al. (1977) and Zubko refers to Zubko et al. (2004). The mean grain size is 5.5×10^{-5} cm and 3.5×10^{-5} cm for MRN and Zubko respectively. The cosmic ray ionization rate is taken to be 10^{-17} s^{-1}. The related gas and grain chemistry is modeled using a combined code (Tornow et al. 2014a)

determined higher ratios due to a higher efficiency of water photodesorption; a consequence of the larger grain size and smaller photodesorption yield used in the Tornow et al. models. If the deuterium chemistry is included according to the method of Albertsson et al. (2013) the temporal and radial distribution of the gas phase HDO/H_2O ratio can be derived. It reaches a maximum of about 0.02 in the medium shielded core region, and a minimum of 10^{-4} in the low shielded outer boundary region (Fig. 14). In contrast, the dust phase ratio of HDO/H_2O

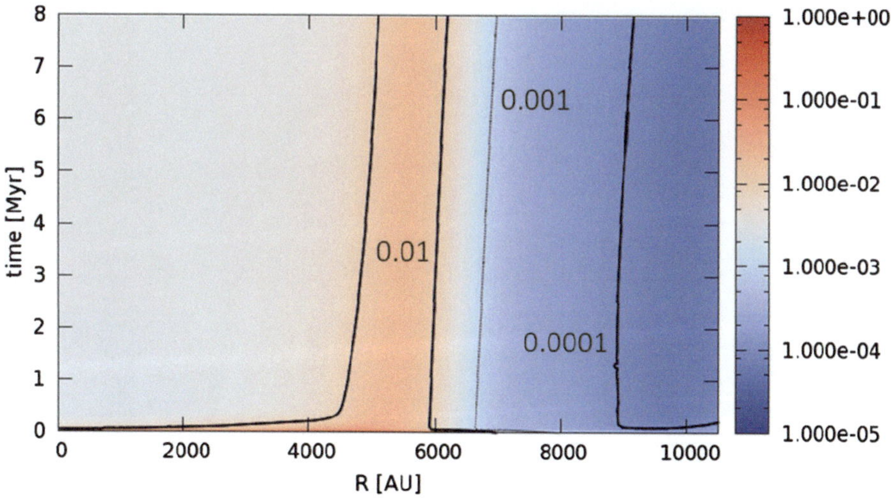

Fig. 14 Temporal and radial HDO/H_2O distribution in the gas phase (Tornow et al. 2014b) for the standard ISRF and MRN grain size (see Fig. 13)

is clearly larger and only varies between 0.01 and 0.1. Its maximum is located in the well shielded inner core region at very early times ($\sim$ (2–3) $\times$ 10^5 years).

The collapse of the cloud core forms a protostar surrounded by an optically thick disk and an outer envelope (Saigo et al. 2008; Schönke and Tscharnuter 2011). During this stage the chemical evolution can be followed along a Lagrangian path from the outer core boundary to the center of the core (Tornow et al. 2014b). Figure 15 shows the HDO/H_2O ratio along each member of the Lagrangian path set. Water formed in the quasi-stationary core stage is transported into the region of terrestrial planet formation. Signatures of large-scale, supersonic water inflows have been seen with Herschel (Mottram et al. 2013) who determined that these flows occur on the core scale ($\leq 10^4$ AU), e.g., IRAS 15398 and L1527 or at least on the envelope scale ($\geq 3 \times 10^4$ AU), e.g., NGC1333-IRAS4A. Based on their data Mottram et al. suggest an outside-in collapse, in agreement with the model shown in Fig. 15.

The gas phase deuteration of water varies widely, depending on location and evolutionary time. At 1 AU it ranges from 0.001 for 0.1–0.2 Myrs, to 0.1 for 0.3 Myrs and to 0.03 for times later than 0.4 Myrs. A minimum HDO/H_2O ratio near 0.001 extends from the radial region around 0.1 AU at early collapse times to nearly 9 AU at later times. The predicted HDO/H_2O ratios of Wakelam et al. (2012) and Tornow et al. (2014b) are slightly lower than those of Aikawa et al. (2012) but larger than the ratios observed by Persson et al. (2014). An over-estimation of the water deuteration may be caused by a low collapse temperature which would suggest perfect cooling. Including the ortho-to-para ratio of H_2 can also lead to lower HDO/H_2O ratios (Taquet et al. 2013, 2014). In addition, water formed during the quasi-stationary core stage must be destroyed by the UV radiation of the interstellar radiation field, nearby stellar sources or X-rays (producing secondary UV photons in a similar way as cosmic rays) and UV photons from the protostar (France et al. 2014). As a result, photodissociation of deuterium-rich water molecules ($HDO + h\nu \longrightarrow D + OH$, or $H + OD$), together with 1-D radial or vertical mixing enable a subsequent reformation of water by hot neutral reactions in the inner disk plane (Furuya et al. 2013).

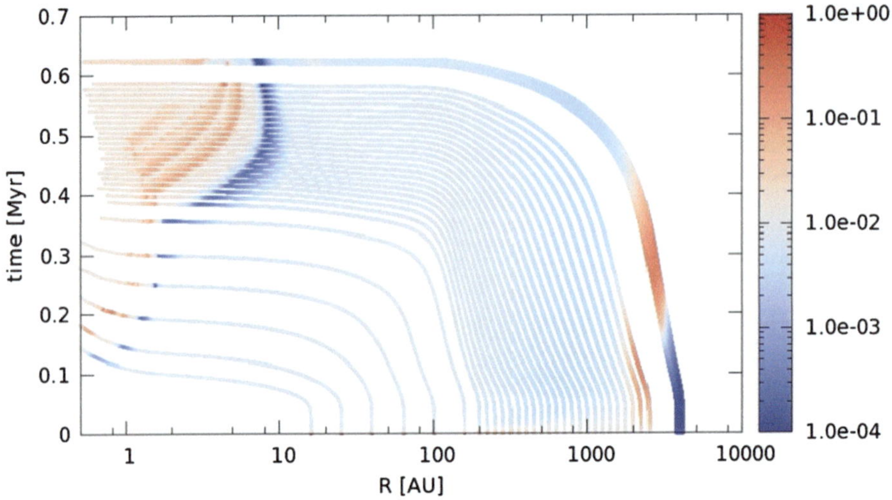

Fig. 15 HDO/H_2O ratio in the gas phase on Lagrangian paths. At 1 AU the HDO/H_2O ratio varies between 10^{-3} and 0.1, while in the dust phase it is much smoother, and only varies between 0.03 and 0.05 (Tornow et al. 2014a, 2014b)

3.2 Fractionation of Carbon and Nitrogen in the Solar Nebula

Isotopic enrichment of ^{15}N in cometary organics (the nitriles CN and HCN) and ammonia is commonly observed (Jehin et al. 2009; Bockelée-Morvan et al. 2015, this issue). Chemical models have shown that fractionation in low-temperature ion-molecule reactions can produce large ^{15}N enrichments in interstellar clouds (Rodgers and Charnley 2008; Wirström et al. 2012). However, a recent study by Roueff et al. (2015) has demonstrated that several key reactions may in fact possess energy barriers, leading to doubts as the efficacy of ion-molecule processes in N fractionation. Of particular interest for comets is that ^{15}N fractionation in the outer nebula should be similarly impaired by low values of the cosmic ray ionization rate, ζ. Thus, as found by Cleeves et al. (2014a) for water deuteration, the low $^{14}N/^{15}N$ ratios commonly measured in both Oort Cloud and Jupiter Family comets is strongly suggestive of an interstellar origin.

The $^{12}C/^{13}C$ ratios measured thus far in simple cometary molecules (CO, HCN, CN) are solar (Jehin et al. 2009) and Bockelee-Morvan (these proceedings). In regions of the nebula where water has condensed, the gaseous C/O elemental ratio can increase to order unity (Öberg et al. 2011). Because water, OH and O_2 recycle carbon nuclei back into CO (Langer et al. 1984), ion-molecule fractionation in regions with $C/O \sim 1$ will act instead to channel carbon nuclei into simple organic molecules. Thus, as the sole carbon fractionation process involves the $^{13}C^+$ exchange reaction with ^{12}CO (Langer et al. 1984), CO condensation may in fact be the reason for the almost uniformity of cometary $^{12}C/^{13}C$ ratios measured to date. On the other hand, the condensed CO is predicted to produce distinctive $^{12}C/^{13}C$ fractionation patterns in the organic molecules formed in the grain-surface reaction scheme of Fig. 6 (Charnley et al. 2004). These involve molecules whose ^{13}C isotopologues have not yet been detected in comets (e.g., H_2CO, CH_3OH) but for which the predicted fractionation has been confirmed by observations of protostellar envelopes (Wirström et al. 2011). In this case, once again, ion-molecule pathways to ^{13}C fractionation in the nebula are predicated on the presence of significant flux of cosmic rays.

If ion-molecule reactions indeed play a greatly reduced role in the cooler parts of the disk then the major sources of nebular fractionation will involve UV photochemistry. Isotope selective photodissociation of CO has been proposed as an explanation of the oxygen isotope ratios measured in meteorites (Lyons and Young 2005). Isotopic enrichment of ^{13}C, ^{15}N and D though isotope-selective photodissociation of isotopologues of CO, N_2 and HD has also been considered (Woods and Willacy 2009; Heays et al. 2014; Cleeves et al. 2014a). However, these processes will only occur in thin upper layers of the disk and so will be negligible at the midplane. Thus, to be viable, these processes require efficient downward vertical mixing in the disk. On the other hand, ISP occurring in the over layers of the presolar core, where the interstellar UV field is attenuated, could in principle produce isotopic anomalies in the oxygen and carbon isotopes (Yurimoto and Kuramoto 2004; Lee et al. 2008; Charnley and Rodgers 2009). This environment could also ultimately produce significant fractionation in nitrogen-bearing species but this has not yet been demonstrated quantitatively.

3.3 Solar Nebula Chemistry and the Origin of Oxygen Isotopic Anomalies in the Solar System

Oxygen is a very common element that is a major constituent of many minerals and rocks. It exists in three stable isotopes: ^{16}O, ^{17}O and ^{18}O. The fractionation between these isotopes is not consistent across all solar system samples, as might be expected if the oxygen isotopes were homogeneously distributed across the solar nebula, and if only mass-dependent fractionation processes were active. Figure 16 is a three-isotope plot of the oxygen isotopic compositions of primitive solar system materials such as chondrites, chondrules, shows that their relative isotopic compositions are different compared to those of terrestrial samples (McKeegan et al. 2011; Yurimoto et al. 2007; Clayton 1993) The terrestrial line is characteristic of mass-dependent fractionation, where ^{18}O is twice as fractionated as ^{17}O relative to ^{16}O, whereas the steeper, slope-1 line, of the primitive materials suggests it is the result of mass-independent fractionation.

There are three possible mechanisms that could achieve this mass-independent effect (Yurimoto et al. 2007). Firstly the isotope anomalies could represent an inherited heterogeneity in solar system materials, where the solids are rich in ^{16}O and the gas in the rarer isotopes. This could occur either from nucleosynthesis in other stars (Clayton et al. 1973) or from nuclear reactions triggered by energetic particles produced by the Sun or by galactic cosmic rays (Lee 1978). These nuclear processes could produce fluctuations in the ^{16}O abundance. The lack of observed isotopic anomalies in other elements suggests this mechanism is unlikely (Clayton 1993). Additionally, ^{16}O presolar grains are rare in meteorites (Nagashima et al. 2004; Nittler 2003).

A second suggestion is chemical fractionation within the solar nebula. Differences in the reaction rates of symmetric molecules, e.g. $^{16}O^{16}O$, compared to asymmetric ones, e.g. $^{16}O^{17}O$, can lead to the observed isotopic distributions, since the most abundant isotope is almost entirely contained in the symmetric molecules, while the rarer isotopes are in the asymmetric ones (Heidenreich and Thiemens 1983). However, the exact reactions by which this effect might be achieved are uncertain (Yurimoto et al. 2007).

The final possibility is isotope-selective photodissociation of CO, either in the parent cloud (Yurimoto and Kuramoto 2004) or at the surface of the protosolar disk (Young 2007; Lyons and Young 2005) could also result in enhancements of the rare isotopes of oxygen. CO dissociates when irradiated by far-ultraviolet photons with wavelengths between

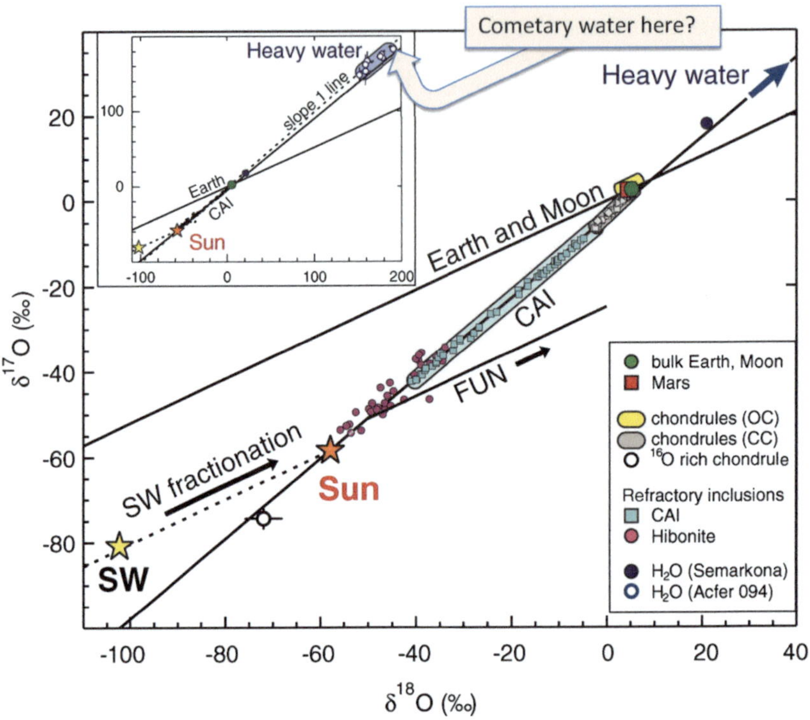

Fig. 16 A three-isotope plot for oxygen in the solar system. The Earth and Moon lie along a fractionation line with slope 0.52. This line is consistent with fractionation of a single homogenous source, by processes that depend on the mass of the isotopes, so that ^{18}O is nearly twice as fractionated as ^{17}O. Data from meteorites and the Genesis mission falls along a slope-1 line indicating they contain material that has undergone some kind of mass-independent fractionation. The Genesis results suggest that much of the rocky material in the inner solar system was enriched in ^{17}O and ^{18}O relative to ^{16}O before the accretion of the first planetesimals. Hence rocky planets are not made of the average nebular material that formed the Sun. From McKeegan et al. (2011) (reprinted with permission)

91.2 nm and 110 nm. The dissociation process occurs in two steps, with CO first entering a bound excited state before dissociating. As a result the CO absorption spectrum consists of a large number of narrow lines at specific wavelengths determined by the vibrational and rotational levels involved. The change in mass between $C^{16}O$ and its isotopologues $C^{17}O$ and $C^{18}O$ slightly shifts the lines in the absorption spectra, and consequently the spectra for the different isotopologues do not overlap. Hence when CO is dissociated, either in the protostellar disk or in the molecular cloud core, the different isotopologues do not shield each other (see, for example, Visser et al. 2009a). $C^{16}O$ will become optically thick closest to the surface, in a region where $C^{17}O$ and $C^{18}O$ can still dissociate. When coupled with turbulent mixing that can transport the ^{17}O and ^{18} towards the colder, shielded midplane of the disk this can result in the formation of water molecules that are enhanced in the rarer isotopes and which could later be incorporated into CAIs and other meteoritic material.

The in situ measurements that Rosetta will make will directly probe the isotopic composition of the volatile component of a comet. This will test the viability of the potential causes of the oxygen isotopic anomalies and provide new clues as to the origin of the cometary material and the conditions under which it formed.

4 The Origin of the Nitrogen Deficiency in Comets

Nitrogen is one of the most abundant elements in the PSN (Asplund et al. 2009). Its abundance was measured in the giant planets atmospheres in which its stable thermodynamic form is NH_3. The nitrogen abundance is found moderately to substantially enriched, compared to its solar value, in Saturn and Jupiter (Mousis et al. 2015), but is loosely constrained in Uranus and Neptune (de Pater et al. 1989; de Pater and Richmond 1989; Ali-Dib et al. 2014). A puzzling revelation was the apparent severe depletion of nitrogen in comets. Mass spectrometry measurements of 1P/Halley's dust composition allowed to find that the nitrogen-to-oxygen ratio is depleted by a factor ~ 3 with respect to the solar value (Jessberger and Kissel 1991). Subsequent analysis showed that the ice part of the comet is depleted in nitrogen by a factor up to 75 (Wyckoff et al. 1991). This apparent depletion appears controversial because the N_2 molecule, thought to be the main nitrogen-bearing volatile in the whole PSN (Lewis and Prinn 1980; Mousis et al. 2002), has a molecular mass close to that of the very abundant CO molecule, implying that the mass spectrometer aboard the Giotto spacecraft did not have enough resolution to separate the two molecules (Eberhardt et al. 1987).

More recently, Cochran et al. (2000) used the 2DCoude spectrograph on the 2.7-m Harlan J. Smith telescope of McDonald Observatory to observe the N_2^+/CO_2^+ in comets 22P/1995 S1 (deVico) and C/1995 O1 (Hale-Bopp), and found values around $\sim 1 \times 10^{-4}$. These values hint at a severe depletion in N_2 since the N_2^+/CO^+ ratio is usually equal to N_2/CO. This is surprising specially that N_2 and CO have very close condensation temperatures (Fray and Schmitt 2009). Moreover, efforts were made to find N_2 using the Far Ultraviolet Spectroscopic Explorer, but with no success (Bockelée-Morvan et al. 2004). All these observations, and many more, motivated theorists to try to explain this depletion.

In Iro et al. (2003), the authors assumed that volatiles are accreted by comets as clathrates (Lunine and Stevenson 1985), and explain the nitrogen depletion by the low efficiency of N_2 trapping in water ices. One of this work's conclusions is that N_2 can be trapped only if water ice is present in abundance higher than 2.8 $\times$ solar value. This is due to the relative ease by which CO can be trapped in comparison with N_2, so this last can be clathrated only if there is enough water left after all the present CO was captured. Since N_2 is trapped in crystalline water ice at temperatures lower than CO (Mousis et al. 2010), their work implied that comets might have formed at temperatures higher than 45 K, allowing the clathration of CO but not N_2. This temperature lower limit is one of shortcomings of this model since in many protoplanetary disks models, temperature is thought to decrease below this value (Hueso and Guillot 2005). The recent observation of the CO iceline in TW Hya also supports these models (Qi et al. 2013a). Another caveat is the reliability of the Iro et al. (2003)'s determination of the N_2/CO ratio in clathrates since the potential parameters describing the guest–clathrate interaction in their statistical model correspond to guest–guest interactions (case of pure solutions) and not to the guest–water interactions usually used in such models (Mousis et al. 2012b). Also, the model is unable to explain the thick N_2 ice cover on Triton and Pluto, both bodies thought to be formed in the comets region (Lellouch et al. 2011).

Motivated by the shortcomings of Iro et al. (2003), Mousis et al. (2012b) introduced another statistical model based on an appropriate set of potential parameters and that seems to address these issues. In this model, volatiles are also accreted as clathrate-hydrates, except N_2 which, due to its poor trapping propensity, is assumed to condense at ~ 20 K in the PSN and subsequently trapped in pure crystalline form in comets. Later radiogenic heating by the decay of short-lived nuclides will heat up these bodies to temperatures enough to liberate the trapped N_2, but not CO. In Triton and Pluto, the gravity is sufficiently important to prevent the gaseous N_2 from escaping, and let it condense on the surface and form the observed cover. In the case of comets however, the gravity is negligible and the gaseous N_2 will escape, leaving comets depleted in nitrogen. A shortcoming of this model is that no radiogenic particles were found by stardust (Flynn 2008).

A recent model, applied to the description of Uranus and Neptune formations, also provides an alternative solution. Ali-Dib et al. (2014) proposed that the diffusive redistribution of vapor will remove the gaseous N_2 from the region below its iceline, possibly explaining the probable nitrogen depletion in Uranus and & Neptune if they formed in this region. One can invoke a similar argument for the nitrogen depletion in comets. In this case, the CO-rich comets will need to form between the CO and N_2 icelines, in order to accrete CO in solid form (since its vapor will also be depleted below its iceline) but not N_2. The relatively small distance between the two icelines ($\sim$ 4–5 AU) is however a caveat for this model since it imposes the formation of all comets in this restricted area of the outer PSN.

Finally, all these models took the N_2 depletion at face value. As discussed above, the N_2 molecule's mass is very close to CO, and the two molecules can be distinguished only using an instrument with a very high spectral resolution, such as ROSINA on board the Rosetta spacecraft (Gulkis and Alexander 2008). If this (or another) instrument is able to detect important amounts of N_2 in a comet, this might resolve at least part of the problem. Since important chemical differences are known to exist between Oort Cloud comets and Jupiter Family comets (Mumma and Charnley 2011), positive detections or definitive exclusion of N_2 in both types of populations are needed before drawing final conclusions.

5 Spin Temperature as a Cosmogonic Indicator in Comets

The volatile composition, D/H ratio in molecular volatiles and the spin temperature, as given by the ortho-to-para ratio in water or other molecules, are considered to be possible cosmogonic indicators for comets and as such may be interpreted as providing information about the formation history of the molecules in question. In this section, we discuss spin temperature as a possible cosmogonic indicator for comets.

Molecules that contain identical hydrogen nuclei (like H_2O and CH_4) display nuclear spin isomers. Water, for example, has two spin species referred to as ortho-H_2O if the nuclear spins of the hydrogen atoms are parallel and para-H_2 if they are anti-parallel. The lowest energy level of para-H_2O lies 23.8 cm^{-1} ($\sim$ 34 K) below the lowest ortho level, so when water is formed at temperatures below about 50 K, the formation of para states is preferred and the ortho-para ratio (OPR) is less than the statistical equilibrium value of 3. Likewise for methane, statistical equilibrium corresponds to abundance ratios A:E:F of 5:2:9 and statistical equilibrium is reached for temperatures above about 60 K.

OPRs have been routinely measured in water in comets since the apparition of comet 1P/Halley (Mumma et al. 1987). This is currently done by fitting synthetic fluorescent emission models independently for each spin state as shown in Figs. 17 and 18. Corresponding spin temperatures are determined by placing the OPR on a theoretical curve such as that for water shown in Fig. 19. OPR is also commonly inferred for ammonia by measuring the OPR for NH_2, thought to be formed primarily via photodissociation of ammonia (Kawakita et al. 2004; Shinnaka et al. 2011). More recently, spin temperature has been measured in cometary methane for comets C/2007 N3 Lulin (Gibb et al. 2012), C/2007 W1 (Villanueva et al. 2011), C/2004 Q2 (Bonev et al. 2009; Kawakita and Kobayashi 2009), and C/2001 Q4 (Kawakita et al. 2005).

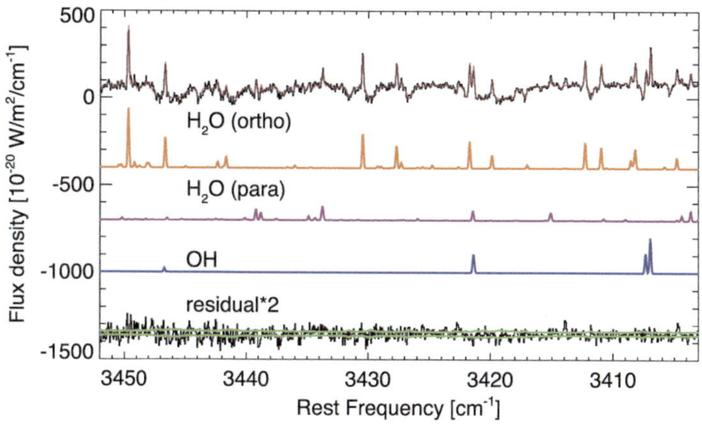

Fig. 17 Spectrum of comet C/2007 N3 (Lulin) on 31 Jan 2009 with the sum of the telluric and best-fit synthetic fluorescent emission model (*red*) overplotted. The best fit water ortho (*orange*) and para (*purple*) models are shown. Also shown is OH prompt emission (*olive*). To illustrate the quality of the fit, the residual (scaled by a factor of two) with the best-fit models subtracted is shown. The 1-sigma error envelope is overplotted in *green*. The best fit is for statistical equilibrium ($T_{\mathrm{spin}} > 60$ K)

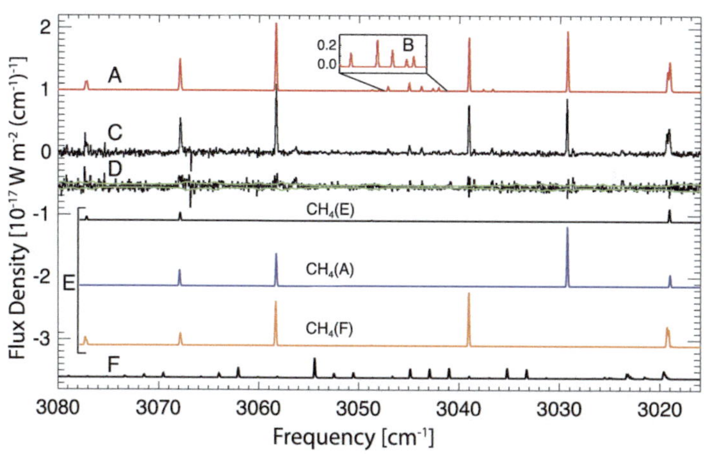

Fig. 18 Residual spectrum of comet C/2007 N3 (Lulin) on 31 Jan 2009 (**C**) with the sum of the best-fit synthetic fluorescent emission model (*red*) overplotted (**A**) and offset vertically for clarity. The OH prompt emission is shown in the *inset* (**B**). To illustrate the quality of the fit, the residual (scaled by a factor of two) with the best-fit models subtracted is shown in (**D**) with the 1-sigma error envelope overplotted in *green*. Beneath the residuals, individual fluorescent emission models are shown for the three spin components of CH_4 (**E**) and CH_3D (**F**) expanded vertically for clarity. The best fit is for statistical equilibrium ($T_{\mathrm{spin}} > 60$ K). Adapted from Gibb et al. (2012)

The meaning of spin temperatures has been the subject of much debate. For an isolated molecule of water or methane in the gas phase, for example, nuclear spin conversion is strongly forbidden. For this reason, spin temperature has been considered a cosmogonic indicator in comets that can provide information on the formation temperature (Crovisier 1994; Mumma et al. 1987). Observational evidence supports the suggestion that OPRs do not change once volatiles are released into the gas phase. For example, recent results map-

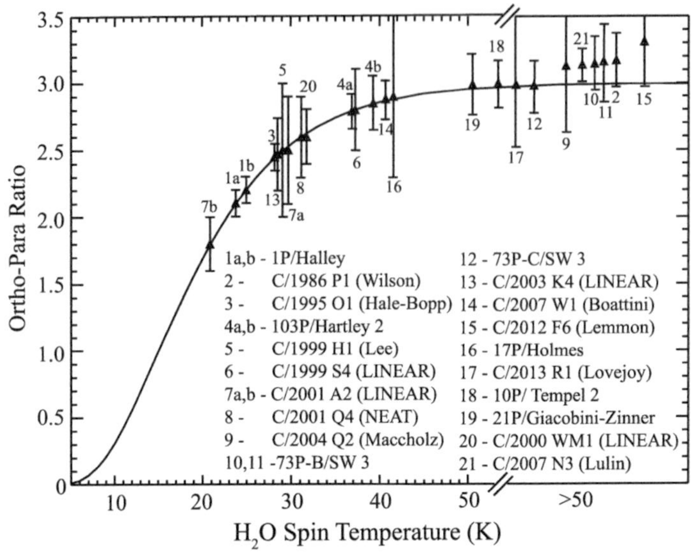

Fig. 19 OPR vs. Spin Temperature in water with measurements for individual comets. Individual OPR measurements are from Mumma et al. (1988: 1a, 1b, 2); Crovisier et al. (1997: 3, 1999: 4a, b); Dello Russo et al. (2005: 5, 6, 7a, 7b, 2007: 11, 12, 2008: 16); Kawakita et al. (2006: 8); Bonev et al. (2009: 9, 2008: 10); Woodward et al. (2007: 13); Villanueva et al. (2011: 14); Paganini et al. (2014: 15, 2015: 17, 2012: 18); DiSanti et al. (2013: 19); Radeva et al. (2010: 20); Gibb et al. (2012: 21)

ping the spin and rotational temperatures in cometary comae are consistent with a scenario in which the OPR is unchanged by coma processes (Bonev et al. 2008). Also, the OPR for NH_2 does not correlate with the heliocentric distance (and hence rotational temperature) of the comet, nor does it correlate significantly with abundances of other volatiles (Shinnaka et al. 2011).

However, it has also been suggested that the spin temperature of water depends on the evaporation history of the ice (Buntkowsky et al. 2008). Hama et al. (2011) found experimentally that the OPR of water desorbed from ice after being vapor deposited at 8 K or formed via irradiation of a CH_4/O_2 mixture at 8 K was consistent with statistical equilibrium, suggesting that OPR may equilibrate during the thermal desorption process or that energetically produced water does not undergo nuclear spin conversion (to a lower OPR) in water ice. Both formation processes for water in the Hama et al. (2011) experiment were energetic, and the authors note that the spin temperature of water molecules produced by non-energetic processes must also be studied before definitive conclusions can be drawn, especially since cometary volatiles may have formed by low-energy processes. Nonetheless, such studies raise the possibility that cometary OPRs may not, in fact, be cosmogonic.

Can the measurements of OPR in comets shine any light on this question? We show the measurements of cometary OPR in water to date in Fig. 19. Water was chosen due to the greater number of OPR measurements than for other molecules in comets. Early measurements suggested a clustering near $T_{spin} \sim 30$ K, leading many authors to conclude a formation temperature of ~ 30 K for a large population of comets. Similar conclusions have been drawn from studies based on CH_4 (Kawakita et al. 2006) and NH_2 (Shinnaka et al. 2011), particularly given the lack of correlation between spin temperature and other properties of the observed comets (Shinnaka et al. 2011). However, it must be noted that among the 20 comets measured to date, nearly all of them are within 1- or 2-sigma of the statistical equilibrium value, excepting the measurements reported for 1P/Halley, one of the two measurements for C/2001 A2, and C/1995 O1 (Hale-Bopp). Likewise, half of all measurements are at or above the statistical equilibrium value. It is also notable that no measurements exist below a spin temperature of 20 K, and this observation holds for methane and ammonia spin temperatures as well.

Possible interpretations of spin temperatures measured in comets are therefore (1) that it is a cosmogonic indicator that gives the formation temperature of the molecule in question and therefore perhaps the formation conditions of the comet, (2) that it is a measurement that, while unchanged by coma processes, is set by processes occurring in the solid ice during or prior to sublimation. Current data are not sufficient to draw firm conclusions at this point. Even if further observations result in most comets consistent with statistical equilibrium, we may still be sampling cosmogonic material since it is known that comets contain material processed at high temperatures, such as crystalline silicates, which are interpreted to originate from turbulent radial mixing in the solar nebula (Bockelée-Morvan et al. 2002). Mixing of volatile material that formed at high temperatures in the inner disk to the comet forming region would have resulted in a higher OPR than for uncontaminated material that formed in the cold midplane. Clearly a great deal more work needs to be done to understand the significance of molecular spin states in cometary material.

6 On the Origin of Neutral Sodium in Comet Tails

The presence of sodium D line emission has been confirmed in a large number of comets close to perihelion since it was first reported more than a century ago (Bredichin 1882). Observations of comet C/1995O1 Hale-Bopp during the spring of 1997 led to the discovery of a new tail connected with the sodium D line emission. This neutral sodium gas tail is entirely different from the previously known ion and dust tails, and its associated source is unclear. Up to now, only physical reasons have been advanced to rationalize its origin. It has been proposed that this third type of tail is shaped by radiation pressure due to resonance fluorescence of sodium atoms (Cremonese et al. 1997). Further possibilities were considered and then rejected, as photo-sputtering and/or ion sputtering of nonvolatile dust grains (Ip and Jorda 1998). In the same way, collisions between the cometary dust and very small grains were also questioned (Ip and Jorda 1998). The scenario presented in this review is completely different since entirely based upon chemical grounds. It is shown that the Na^+ ions, washed out of the refractory material during the hydration phase, loose their positive charge to evolve progressively into neutral species with their migration towards the surface of cometary ices.

Ellinger et al. (2015) follow the chemical path of sodium, starting from the incorporation into the ice until the final transformation into a neutral atom when released from the sublimating crystalline (see Sect. 2.5) cometary ice. The modeling of bulk, surface structures and reactivity of ices has been essentially developed for environmental applications (Casassa and Pisani 2002; Casassa et al. 2005; Calatayud et al. 2003) Levering on these results, high-level numerical simulations based on first principle periodic density functional theory (DFT) were used to describe the solid structure of the ice in the form of apolar hexagonal ice (Bussolin et al. 1998) composed of bi-layers of water molecules.

A code, specifically designed for the study of periodic systems, namely, the Vienna *ab initio* simulation package (VASP) (Kresse and Hafner 1994a, 1994b) was used to carry out all the calculations. The generalized gradient approximation (GGA) was employed in the form of the Perdew and Wang exchange-correlation functional (Perdew et al. 1992) coupled

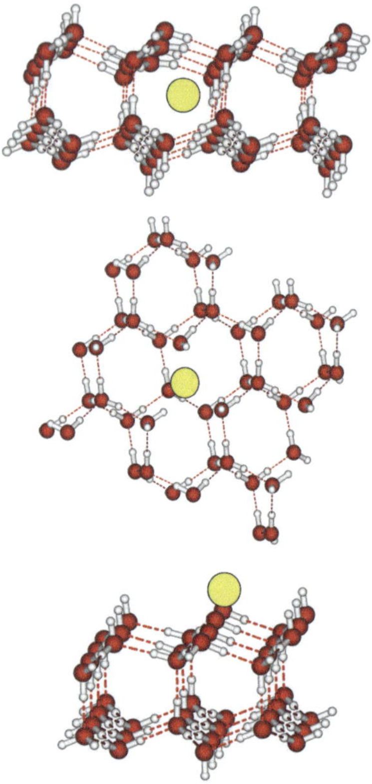

Fig. 20 Typical structures of Na-H_2O_{ice} arrangements. *Top*: Na in the cavity inside hexagonal ice. *Middle*: Na replacing H_2O in the crystal. *Bottom*: Na adsorbed on the surface of the ice. *Color code*: *yellow* (Na); *red* (O); *white* (H)

to the Grimme correction (Grimme et al. 2010) to take care of the long-range van der Waals interactions. The evolution of the ionic character of the sodium atom as it reaches the ice surface was obtained by a topological analysis (Silvi and Savin 1994) performed by means of the TOPMOD package (Llusar et al. 1999) and its recent extension to periodic systems (Kozlowski and Pilmé 2011).

A first calculation on isolated Na(H_2O) showed that the interaction of Na with one H_2O is similar to that between two H_2O molecules in the H_2O dimer (≈ 0.3 eV), which gives a hint in view of a possible replacement of a molecule of water by a sodium atom in organized structures. In liquid water, microsolvatation of sodium atoms has been studied on small clusters of Na$(H_2O)n$ ($n = 1$–6) to mimic the water environment. Both the structures in which Na is positioned on the surface and the encapsulated structure in which Na is surrounded by a solvation shell are minima on the potential surface. They are very close in energy (Hashimoto and Morokuma 1994).

In the ice, the periodic structure of the crystal imposes strong geometrical constraints. The most stable structure, not found in the liquid, is the one where Na takes the place of a water molecule in the lattice (middle structure in Fig. 20) when the ice is formed ($\Delta E \approx$ 1 eV). Contrary to microsolvatation, the stability of Na encapsulated in the bulk of the crystal ($\Delta E \approx 0.5$ eV), shown in the top structure in Fig. 20) differs from that of the adsorbed Na according to the position on the surface. For example, the adsorption energies vary from ($\Delta E \approx 0.6$ to ≈ 0.5 and ≈ 0.1 eV) when the adsorption takes place over the center of a surface hexagon, over a surface oxygen (bottom structure in Fig. 20) and over an OH dangling bond, respectively.

The most relevant point to this study is that the charge q_{Na} on the sodium atom depends strongly on the position with respect to the ice surface, as does the adsorption energy. When constrained to stay in the cavity inside the hexagonal lattice, Na remains largely ionized ($q_{\mathrm{Na}} \approx +0.9$) which is close to the value of $\approx +0.8$ when Na takes the place of an H_2O molecule in the bulk. When Na is adsorbed on top of an oxygen of the surface layer, $q_{\mathrm{Na}} \approx +0.2$, a value similar to that found for the isolated $Na(H_2O)$ complex. Then, as the ice is progressively eroded by the radiation field, Na eventually reaches the surface where it can desorb as a neutral atom when the upper layer of the ice vaporizes.

7 Summary

The observed compositions of comets are the end product of a chemical history that began in the cold parent molecular cloud core. The change in physical conditions experienced by the core material as the core collapsed and the protosun and protosolar disk formed and evolved led to chemical changes that are reflected in the cometary volatiles that we observe today. Each stage of evolution left its imprint in the mixing ratios of molecules observed in comets and disentangling their effects requires interaction between comet and disk researchers. The combination of the two fields of study provide a powerful tool in understanding the drivers that determine comet composition and conversely what the variations in comet composition can tell us about the relative formation conditions of these comets.

Protostellar disks provide complementary information to the comet observations. Their chemistry is complex, covering ice chemistry in the cold, shielded midplane, warm gas phase chemistry and hot, highly ionized chemistry. Comets form from the ices in the midplane, yet dynamical processes ensure that this region does not evolve in isolation but rather is the end product of material that has been processed in a variety of different environments and then brought together in the cometary ices. Establishing the origin of cometary ices therefore requires not only studying the midplane but also an understanding the chemical and physical evolution of the disk as a whole. By studying disks we can observe the chemistry and physics that shaped the solar system as they shape other protoplanetary systems.

Molecular deuteration is a link that might tie together the disparate objects of the solar system and their formation history and give us the picture of the whole solar system formation history. However, at present this link has still weak connections that need to be reinforced. The advent of the powerful new facilities, as for example ALMA and IRAM-NOEMA, will certainly allow this field to progress rapidly in the next few years.

Rosetta will provide detailed ground-truths for the organic and isotopic makeup of cometary volatiles. The new insights obtained into the chemical processes which likely produced these molecules—interstellar or nebular—will allow us to constrain models of the formation and evolution of the early solar system. Its instruments will characterize the composition of the comet 67P/Churyumov-Gerasimenko and distinguish between the molecules native to the comet nucleus (parent species) and those generated by chemistry or by fragmentation in the comae (daughter species). The observations it will make will provide new insights into the formation conditions of the comet. They will also elucidate the degree to which interstellar material can survive its journey into the solar nebula and from there into comets.

Several measurements will characterize the its thermal history, e.g., measurements of molecular spin temperatures (Gulkis and Alexander 2008) and deuteration and of the abundance of argon (Hassig et al. 2015). The relative abundances of elements can tell us something about the dominant molecular form of each element at the time of formation—if an element is depleted relative to others then the major carrier of that element must have been too volatile to be condensed at the time of formation, or must have desorbed in the intervening time. In particular Rosetta measurements may help to elucidate the mystery of the lack of nitrogen in comets (Altwegg et al. 2015).

A key measurement will be the ratio of oxygen isotopes in water. Meteoritic water has a very different ratio of the oxygen isotopes compared to oxygen isotopes in the Sun. Rosetta provides our best opportunity to directly detect the isotope ratio from the early solar system, and in so doing will distinguish between the various theoretical models of how the oxygen isotope fractionation occurred.

In summary, Rosetta's observations will provide new insights into the chemical processes that resulted in the formation of cometary volatiles and consequently into the conditions that were present in the solar nebula at their time of formation.

Acknowledgements The work of K. Willacy and C. Alexander was carried out at the Jet Propulsion Laboratory, California Institute of Technology, under a contract with the National Aeronautics and Space Administration. S.B. Charnley and S.N. Milam acknowledge the support of the Goddard Center for Astrobiology. Support for K. Willacy, S.B. Charnley and S.N. Milam was partially provided by the NASA Origins of Solar Systems Program. E. Gibb was supported by the NASA Exobiology and Evolutionary Biology (grant NNX11AG44G) and NSF Planetary Astronomy (grant AST-1211362) programs. O. Mousis and C. Ceccarelli acknowledge support from CNES. Support for O. Mousis was provided by the A*MIDEX project (no. ANR-11-IDEX-0001-02) funded by the "Investissements d'Avenir" French government program managed by the French National Research Agency (ANR). M. Ali-Dib was supported by a grant from the city of Beançon. E.S. Wirström was supported by the Swedish National Space Board.

References

F.C. Adams, The birth environment of the solar system. Annu. Rev. Astron. Astrophys. **48**, 47 (2010)

M.F. A'Hearn, L.M. Feaga, H.U. Keller, H. Kawakita, D.L. Hampton, J. Kissel, K.P. Klaasen, L.A. McFadden, K.J. Meech, P.H. Schultz, J.M. Sunshine, P.C. Thomas, J. Veverka, D.K. Yeomans, S. Besse, D. Bodewits, T.L. Farnham, O. Groussin, M.S. Kelley, C.M. Lisse, F. Merlin, S. Protopapa, D.D. Wellnitz, Cometary volatiles and the origin of comets. Astrophys. J. **758**, 29 (2012)

Y. Aikawa, Cold CO gas in protoplanetary disks. Astrophys. J. **656**, L93 (2007)

Y. Aikawa, E. Herbst, Deuterium fractionation in protoplanetary disks. Astrophys. J. **526**, 314 (1999)

Y. Aikawa, T. Nomura, Physical and chemical structure of protoplanetary disks with grain growth. Astrophys. J. **642**, 1152 (2006)

Y. Aikawa, T. Umebayashi, T. Nakano, S.M. Miyama, Evolution of molecular abundances in protoplanetary disks with accretion flows. Astrophys. J. **519**, 705 (1999)

Y. Aikawa, V. Wakelam, R.T. Garrod, E. Herbst, Molecular evolution and star formation: From prestellar cores to protostellar cores. Astrophys. J. **674**, 984 (2008)

Y. Aikawa, V. Wakelam, F. Hersant, R.T. Garrod, E. Herbst, From prestellar to protostellar cores II. Time dependence and deuterium fractionation. Astrophys. J. **760**, 40 (2012)

V. Akimkin, S. Zhukovska, D. Wiebe, D.A. Semenov, Ya. Pavlyuchenkov, A. Vasyunin, T. Birnstiel, Th. Henning, Protoplanetary disk structure with grain evolution: The ANDES model. Astrophys. J. **766**, 8 (2013)

T. Albertsson, D.A. Semenov, Th. Henning, Modeling deuterium fractionation in cold and warm molecular environments with large chemical networks. Astrophys. J. Suppl. Ser. **207**, 27 (2011)

T. Albertsson, D.A. Semenov, A.I. Vasyunin, Th. Henning, E. Herbst, New extended deuterium fractionation model: Assessment at dense ISM conditions and sensitivity analysis. Astrophys. J. Suppl. Ser. **207**, 27 (2013)

T. Albertsson, D.A. Semenov, Th. Henning, Chemodynamical deuterium fractionation in the early solar nebula: The origins of water on Earth and in asteroids and comets. Astrophys. J. **784**, 39 (2014)

M. Ali-Dib, O. Mousis, J.-M. Petit, J.I. Lunine, The measured compositions of Uranus and Neptune from their formation on the CO iceline. Astrophys. J. **793**, 9 (2014)

Y. Alibert, O. Mousis, W. Benz, On the volatile enrichments and composition of Jupiter. Astrophys. J. **622**, L145 (2005a)

Y. Alibert, O. Mousis, C. Mordasini, W. Benz, New Jupiter and Saturn formation models meet observations. Astrophys. J. **626**, L57 (2005b)

ALMA Partnership et al., First results from high angular resolution ALMA observations toward the HL Tau region. Astrophys. J. Lett. (2015, in press). arXiv:1503.02649

K. Altwegg, H. Balsiger, A. Bar-Nun, J.J. Berthelier, A. Bieler, P. Bochsler, C. Briois, U. Calmonte, M. Combi, J. De Keyser, P. Eberhardt, B. Fiethe, S. Fuselier, S. Gasc, T.I. Gombosi, K.C. Hansen, M. Hässig, E. Kopp, A. Korth, A. Jäckel, L. LeRoy, U. Mall, B. Marty, O. Mousis, E. Neefs, T. Owen, H. Rème, M. Rubin, T. Sémon, C. Tzou, H. Waite, P. Wurz, Comet 67P/Churyumov-Gerasimenko, a Jupiter family comet with a high D/H ratio. Science **347**, 387 (2015)

M. Asplund, N. Grevesse, A.J. Sauval, P. Scott, The chemical composition of the Sun. Annu. Rev. Astron. Astrophys. **47**, 481 (2009)

S.A. Balbus, J.F. Hawley, A powerful local shear instability in weakly magnetized disks. I—Linear analysis. II—Nonlinear evolution. Astrophys. J. **376**, 214 (1991)

N. Biver, D. Bockelée-Morvan, J. Crovisier, D.C. Lis, R. Moreno, P. Colom, F. Henry, F. Herpin, G. Paubert, M. Womack, Radio wavelength molecular observations of comets C/1999 T1 (McNaught-Hartley), C/2001 A2 (LINEAR), C/2000 WM_1 (LINEAR) and 15P/Ikeya-Zhang. Astron. Astrophys. **449**, 1255 (2006)

D. Bockelée-Morvan, D. Gautier, D.C. Lis, K. Young, J. Keene, T. Phillips, T. Owen, J. Crovisier, P.F. Goldsmith, E.A. Bergin, D. Depois, A. Wootten, Deuterated water in comet C/1996 B2 (Hyakutake) and its implications for the origin of comets. Icarus **133**, 147 (1998)

D. Bockelée-Morvan, D. Gautier, F. Hersant, J.-M. Huré, F. Robert, Turbulent radial mixing in the solar nebula as the source of crystalline silicates in comets. Astron. Astrophys. **384**, 1107 (2002)

D. Bockelée-Morvan, J. Crovisier, M.J. Mumma, H.A. Weaver, The composition of cometary volatiles, in *Comets II* (2004), p. 391

D. Bockelée-Morvan, N. Biver, B. Swinyard, M. de Val-Borro, J. Crovisier, P. Hartogh, D.C. Lis, R. Moreno, S. Szutowicz, E. Lellouch, M. Emprechtinger, G.A. Blake, R. Courtin, C. Jarchow, M. Kidger, M. Küppers, M. Rengel, G.R. Davis, T. Fulton, D. Naylor, S. Sidher, H. Walker, Herschel measurements of the D/H and $^{16}O/^{18}O$ ratios in water in the Oort-cloud comet C/2009 P1 (Garradd). Astron. Astrophys. **544**, 15 (2012)

D. Bockelée-Morvan, U. Calmonte, S. Charnley, J. Duprat, C. Engrand, A. Gicquel, M. Hässig, E. Jehin, H. Kawakita, B. Marty, S. Milam, A. Morse, P. Rousselot, S. Sheridan, E. Wirström, Cometary isotopic measurements. Space Sci. Rev. (2015, this issue). doi:10.1007/s11214-015-0156-9

B.P. Bonev, M.J. Mumma, H. Kawakita, H. Kobayashi, G.L. Villanueva, IRCS/Subaru observations of water in the inner coma of comet 73P-B/Schwassman-Wachmann 3: Spatially resolved rotational temperatures and ortho-para ratios. Icarus **196**, 241 (2008)

B.P. Bonev, M.J. Mumma, E.L. Gibb, M.A. DiSanti, G.L. Villanueva, K. Magee-Sauer, R.S. Ellis, Comet C/2004 Q2 (Machholz): Parent volariles, a search for deuterated methane, and constraint on the CH_4 spin temperature. Astrophys. J. **699**, 1563 (2009)

B.P. Bonev, E. Gibb, G. Villanueva, M. DiSanti, L. Paganini, M. Mumma, J. Keane, K. Meech, K. Wright, E. Sudholt, M. Besic, OH* prompt emission and near-infrared searches for HDO in the moderately bright comets C/2007 N3 Lulin and 103P/Hartley 2, in *Asteroids, Comets and Meteors* (2014)

T. Bredichin, Les vapeurs du sodium dans la comète de Wells. Astron. Nachr. **102**, 207 (1882)

G. Buntkowsky, H.-H. Limbach, B. Walaszek, A. Adamczyk, Y. Xu, H. Breitzke, A. Schweitzer, T. Gutmann, M. Wachtler, J. Frydel, Th. Emmler, N. Amadeu, D. Tietze, B. Chaudret, Mechanisms of dipolar ortho/para H_2O conversion in ice. Z. Phys. Chem. **222**, 1049 (2008)

G. Bussolin, S. Casassa, P. Pisani, P. Ugliengo, Ab initio study of HCl and HF interaction with crystalline ice. I. Physical adsorption. J. Chem. Phys. **108**, 9516 (1998)

M. Calatayud, D. Courmier, C. Minot, Ionization of HCl and HF in ice: a periodic DFT study. Chem. Phys. Lett. **369**, 287 (2003)

J.S. Carr, J.R. Najita, Organic molecules and water in the planet formation region of young circumstellar disks. Science **319**, 1504 (2008)

S. Casassa, C. Pisani, Interaction of HOCl with a chlorinated ice surface to produce molecular clorine: An ab-initio study. J. Chem. Phys. **116**, 9856 (2002)

S. Casassa, M. Calatayud, K. Doll, K.C. Minot, C. Pisani, Proton ordered cubic and hexagonal periodic models of ordinary ice. Chem. Phys. Lett. **409**, 110 (2005)

P. Caselli, E. Keto, E.A. Bergin, M. Tafalla, Y. Aikawa, T. Douglas, L. Pagani, U.A. Yildiz, F.F.S. van der Tak, C.M. Walmsley, C. Codella, B. Nisini, L.E. Kristensen, E.F. van Dishoeck, First detection of water vapor in a pre-stellar core. Astrophys. J. Lett. **L37**, 5 (2012)

C. Ceccarelli, C. Dominik, B. Lefloch, P. Caselli, E. Caux, Detection of H_2D^+: Measuring the midplane degree of ionization in the disks of DM Tauri and TW Hydrae. Astrophys. J. **607**, L51 (2004)

C. Ceccarelli, P. Caselli, D. Bockelee-Morvan, O. Mousis, S. Pizzarello, F. Robert, D. Semenov, Deuterium fractionation: the Ariadne's thread from the pre-collapse phase to meteorites and comets today, in *Protostars and Planets VI*, ed. by H. Beuther, R.S. Klessen, C.P. Dullemond, Th. Henning (University of Arizona Press, Tucson, 2014). arXiv:1403.7143

S.B. Charnley, On the nature of interstellar organic chemistry, in *Astronomical and biochemical origins and the search for life in the Universe*, vol. 89, ed. by C. Batalli Cosmovici, S. Bowyer, D. Werthimer (Editrice Compositori, Bologna, 1997)

S.B. Charnley, S.D. Rodgers, Interstellar reservoirs of cometary matter. Space Sci. Rev. **138**, 59 (2008)

S.B. Charnley, S.D. Rodgers, Clouds, clumps, cores and comets—a cosmic chemical connection? Adv. Geosci. **15**, 211 (2009)

S.B. Charnley, A.G.G.M. Tielens, S.D. Rodgers, Deuterated methanol in the Orion Compact Ridge. Astrophys. J. **482**, L203 (1997)

S.B. Charnley, P. Ehrenfreund, T.J. Millar, A.C.A. Boogert, A.J. Markwick, H.M. Butner, R. Ruiterkamp, S.D. Rodgers, Observational tests for grain chemistry: Posterior isotopic labeling. Mon. Not. R. Astron. Soc. **347**, 157 (2004)

K.M. Chick, P. Cassen, Thermal processing of interstellar dust grains in the primitive solar environment. Astrophys. J. **477**, 398 (1997)

F.J. Ciesla, The phases of water ice in the solar nebula. Astrophys. J. **784**, L1 (2014)

F.J. Ciesla, S.B. Charnley, The physics and chemistry of nebular evolution, in *Meteorites and the Early Solar System II*, ed. by D.S. Lauretta, H.Y. McSween Jr. (University of Arizona Press, Tucson, 2006), p. 209

R.N. Clayton, Oxygen isotopes in meteorites. Annu. Rev. Earth Planet. Sci. **21**, 115 (1993)

R.N. Clayton, L. Grossman, T.K. Mayeda, A component of primitive nuclear composition in carbonaceous meteorites. Science **182**, 485 (1973)

L.I. Cleeves, F.C. Adams, E.A. Bergin, Exclusion of cosmic rays in protoplanetary disks: stellar and magnetic effects. Astrophys. J. **772**, 5 (2013)

L.I. Cleeves, E.A. Bergin, C.M.O. 'D Alexander, F. Du, D. Graninger, K.I. Öberg, T.J. Harries, The ancient heritage of water ice in the solar system. Science **345**, 1590 (2014a)

L.I. Cleeves, E.A. Bergin, F.C. Adams, Exclusion of cosmic rays in protoplanetary disks. II. Chemical gradients and observational signatures. Astrophys. J. **794**, 123 (2014b)

A.L. Cochran, W.D. Cochran, E.S. Barker, N_2^+ and CO^+ in Comets 122P/1995 S1 (deVico) and C/1995 O1 (Hale-Bopp). Icarus **146**, 583 (2000)

G. Cremonese, H. Boehnhardt, J. Crovisier, H. Rauer, A. Fitzsimmons, M. Fulle, J. Licandro, D. Pollacco, G.P. Tozzi, R.M. West, Neutral sodium from comet Hale-Bopp: A third type of tail. Astrophys. J. **490**, L199 (1997)

J. Crovisier, The water molecule in comets—fluorescence mechanisms and thermodynamics of the inner coma. Astron. Astrophys. **130**, 361 (1994)

J. Crovisier, The molecular composition of comets and its interrelation with other small bodies of the Solar System, in *IAU Symp 229 "Asteroids, Comets, Meteors"*, ed. by L. Daniela, M. Sylvio Ferraz, F.J. Angel (2006), p. 113

J. Crovisier, K. Leech, D. Bockelée-Morvan, T.Y. Brooke, M.S. Hanner, B. Altieri, H.U. Keller, E. Lellouch, The spectrum of comet Hale-Bopp (C/1995 O1) observed with the Infrared Space Observatory at 2.9 AU from the Sun. Science **275**, 1904 (1997)

J. Crovisier, Th. Encrenaz, E. Lellouch, E. Bockelée-Morvan, B. Altieri, K. Leech, A. Salama, M.J. Griffin, Th. de Graauw, E.F. van Dishoeck, R. Knacke, T.Y. Brooke, ISO spectroscopic observations of short-period comets, in *The Universe as Seen by ISO*, ed. by P. Cox, M.F. Kessler. ESA SP-427 (ESA, Noordwijk, 1999), p. 161

J. Crovisier, D. Bockelée-Morvan, P. Colom, N. Biver, D. Despois, D.C. Lis, The composition of ices in comet C/1995 O1 (Hale-Bopp) from radio spectroscopy: Further results and upper limits on undetected species. Astron. Astrophys. **418**, 1141 (2004)

P. D'Alessio, B. Merin, N. Calvet, L. Hartmann, B. Montesinos, WWW database of models of accretion disks irradiated by the central star. Rev. Mex. Astron. Astrofís. **41**, 61 (2005)

I. de Gregorio-Monsalvo et al., Unveiling the gas-and dust disk structure in HD 16329 using ALMA observations. Astron. Astrophys. **557**, 133 (2013)

I. de Pater, M. Richmond, Neptune's microwave spectrum from 1 mm to 20 cm. Icarus **80**, 1 (1989)

I. de Pater, P.N. Romani, S.K. Atreya, Uranus deep atmosphere revealed. Icarus **82**, 288 (1989)

N. Dello Russo, B.P. Bonev, M.A. DiSanti, M.J. Mumma, E.L. Gibb, K. Magee-Sauer, R.J. Barber, J. Tennyson, Water production rates, rotational temperatures, and spin temperatures in comets C/1999 H1 (Lee), C/1999 S4, and C/2001 A2. Astrophys. J. **621**, 537 (2005)

N. Dello Russo, R.J. Vervack, H.A. Weaver, N. Biver, D. Bockelée-Morvan, J. Crovisier, C.M. Lisse, Compositional homogeneity in the fragmented comet 73P/Schwassmann-Wachmann 3. Nature **448**, 172 (2007)

N. Dello Russo, R.J. Vervack, H.A. Weaver, M.M. Montgomery, R. Deshpande, Y.R. Fernández, E.L. Martin, The volatile composition of comet 17P/Holmes after its extraordinary outburst. Astrophys. J. **680**, 793 (2008)

M.A. DiSanti, B.P. Bonev, G.L. Villanueva, M.J. Mumma, Highly depleted ethane and mildly depleted methanol in comet 21P/Giacobini-Zinner: Application of a new empirical ν_2-band model for CH_3OH near 50 K. Astrophys. J. **763**, 1 (2013)

S.E. Dodson-Robinson, K. Willacy, P. Bodenheimer, N.J. Turner, C.A. Beichman, Ice lines, planetesimal composition and solid surface density in the solar nebula. Icarus **200**, 672 (2009)

A.Z. Dolginov, T.F. Stepinski, Are cosmic rays effective for ionization of protoplanetary disks? Astrophys. J. **427**, 377 (1994)

C. Dominik, C. Ceccarelli, D. Hollenbach, M. Kaufman, Gas-phase water in the surface layers of protoplanetary disks. Astrophys. J. **635**, 85 (2005)

C.P. Dullemond, C. Dominik, The effect of dust settling on the appearance of protoplanetary disks. Astron. Astrophys. **421**, 1075 (2004)

A. Dutrey, D. Semenov, E. Chapillon, U. Gorti, S. Guilloteau, F. Hersant, M. Hogerheijde, M. Hughes, G. Meeus, H. Nomura, V. Piétu, C. Qi, V. Wakelam, Physical and chemical structure of planet-forming disks probed by millimeter observations and modeling, in *Protostars and Planets VI*, ed. by H. Beuther, R. Klessen, C. Dullemond, Th. Henning (University of Arizona Press, Tuscon, 2014), p. 317

P. Eberhardt, D. Krankowsky, W. Schutte, U. Dolder, P. Limmerzahl, J.J. Berthelier, J. Woweries, U. Stubbemann, R.R. Hodges, J.H. Hoffman, J.M. Illiano, The CO and N_2 abundance in comet P/Halley. Astron. Astrophys. **187**, 481 (1987)

P. Eberhardt, M. Reber, D. Krankowsky, R.R. Hodges, The D/H and $^{18}O/^{16}O$ ratios in water from comet P/Halley. Astron. Astrophys. **302**, 301 (1995)

Y. Ellinger, F. Pauzat, O. Mousis, A. Guilbert-Lepoutre, F. Leblanc, M. Ali-Dib, M. Doronin, E. Zicler, A. Doressoundiram, Neutral sodium in cometary tails as a remnant of early aqueous alteration. Astrophys. J. Lett. **801**, L30 (2015)

G.J. Flynn, Physical, chemical, and mineralogical properties of comet 81P/Wild 2 particles collected by Stardust. Earth Moon Planets **102**, 447 (2008)

J.K.J. Fogel, T.J. Bethell, E.A. Bergin, N. Calvet, D. Semenov, Chemistry of a protoplanetary disk with grain settling and Lyα radiation. Astrophys. J. **726**, 29 (2011)

K. France, E. Schindhelm, E.A. Bergin, E. Roueff, H. Abgrall, High-resolution ultraviolet radiation fields of classical T Tauri stars. Astrophys. J. **127**, 20 (2014)

N. Fray, B. Schmitt, Sublimation of ices of astrophysical interest: A bibliographic review. Planet. Space Sci. **57**, 2053 (2009)

K.Y. Furuya, Y. Aikawa, H. Nomura, F. Hersant, V. Wakelam, Water in protoplanetary disks: Deuteration and turbulent mixing. Astrophys. J. **779**, 11 (2013)

D. Gautier, F. Hersant, O. Mousis, J.I. Lunine, Enrichments in volatiles in Jupiter: A new interpretation of the Galileo measurements. Astrophys. J. **550**, L227 (2001)

E.L. Gibb, B.P. Bonev, G. Villanueva, M.A. DiSanti, M.J. Mumma, E. Sudholt, Y. Radeva, Chemical composition of comet C/2007 N3 (Lulin): Another "atypical" comet. Astrophys. J. **750**, 102 (2012)

A.E. Glassgold, J. Najita, J. Igea, X-ray ionization of proto-planetary disks. Astrophys. J. **480**, 344 (1997)

S. Grimme, J. Antony, S. Ehrlich, H. Krieg, A consistent and accurate ab initio parameterization of density functional dispersion correction (DFT-D) for the 94 elements H-Pu. J. Chem. Phys. **132**, 154104 (2010)

S. Gulkis, C. Alexander, Composition measurements of a comet from the Rosetta Orbiter spacecraft. Space Sci. Rev. **138**, 259 (2008)

T. Hama, N. Watanabe, A. Kouchi, M. Yokoyama, Spin temperature of water molecules desorbed from the surfaces of amorphous solid water, vapor-deposited and produced from photolysis of a CH_4/O_2 solid mixture. Astrophys. J. Lett. **738**, 15 (2011)

D. Harsono, R. Visser, S. Bruderer, E.F. van Dishoeck, L.E. Kristensen, Evolution of CO lines in time-dependent models of protostellar disk formation. Astron. Astrophys. **555**, 45 (2013)

P. Hartogh, D.C. Lis, D. Bockelée-Morvan, M. de Val-Borro, N. Biver, M. Küppers, M. Emprechtinger, E.A. Bergin, J. Crovisier, M. Rengel, R. Moreno, S. Szutowicz, G.A. Blake, Ocean-like water in the Jupiter-family comet 103P/Hartley 2. Nature **478**, 218 (2011)

K. Hashimoto, K. Morokuma, Ab-initio molecular-orbital study of $Na(H_2O)N$ ($N = 1$–6) clusters and their ions—comparison of electronic-structure of the surface and their interior complexes. J. Am. Chem. Soc. **116**, 11436 (1994)

M. Hassig, K. Altwegg, J. Berthelier, U. Almonte, T. Gombosi, J. De Jeyser, B. Fiethe, S. Fuselier, L. LeRoy, M. Rubin, The capabilities of ROSINA/DFMS to measure argon isotopes at comet 67P/Churyumov-Gerasimenko. Planet. Space Sci. **105**, 175 (2015)

J.F. Hawley, S.A. Balbus, A powerful local shear instability in weakly magnetized disks. Astrophys. J. **376**, 223 (1991)

J.F. Hawley, C.F. Gammie, S.A. Balbus, Local three-dimensional simulations of an accretion disk hydromagnetic dynamo. Astrophys. J. **464**, 690 (1996)

A.N. Heays, R. Visser, R. Gredel, W. Ubachs, B.R. Lewis, S.T. Givson, E.F. van Dishoeck, Isotope selective photodissociation of N_2 by the interstellar radiation field and cosmic rays. Astron. Astrophys. **562**, 61 (2014)

J.E. Heidenreich III, M.H. Thiemens, A non-mass-independent isotope effect in the production of ozone from molecular oxygen. J. Chem. Phys. **78**, 892 (1983)

D. Heinzeller, H. Nomura, C. Walsh, T.J. Millar, Chemical evolution of protoplanetary disks—the effects of viscous accretion, turbulent mixing and disk winds. Astrophys. J. **731**, 115 (2011)

Th. Henning, D. Semenov, Chemistry in protoplanetary disks. Chem. Rev. **113**, 9016 (2013)

E. Herbst, E.F. van Dishoeck, Complex organic interstellar molecules. Annu. Rev. Astron. Astrophys. **47**, 427 (2009)

F. Hersant, V. Wakelam, A. Dutrey, S. Guilloteau, E. Herbst, Cold CO in circumstellar disks. On the effects of photodesorption and vertical mixing. Astron. Astrophys. **493**, L49 (2009)

M.R. Hogerheijde, E.A. Bergin, C. Brinch, L.I. Cleeves, J.K. Fogel, G.A. Blacke, C. Dominik, D.C. Lis, G. Melnick, D. e Neufeld, O. Panić, J.C. Pearson, L. Kristensen, U.A. Yildiz, E.A. Dishoeck, Detection of the water reservoir in a forming planetary system. Science **354**, 338 (2011)

R. Hueso, T. Guillot, Evolution of protoplanetary disks: Constraints from DM Tauri and GM Aurigae. Astron. Astrophys. **442**, 703 (2005)

D. Hutsemékers, J. Manfroid, E. Jehin, J.M. Zucconi, C. Arpigny, The $^{16}OH/^{18}OH$ and OD/OH isotope ratios in comet C/2002 T7 (LINEAR). Astron. Astrophys. **490**, 31 (2008)

M. Ilgner, R.P. Nelson, On the ionization fraction in protoplanetary disks. II. The effect of turbulence mixing on gas-phase chemistry. Astron. Astrophys. **445**, 223 (2006)

M. Ilgner, R.P. Nelson, Turbulent transport and its effect on the dead zone in protoplanetary discs. Astron. Astrophys. **483**, 815 (2008)

M. Ilgner, Th. Henning, A.J. Markwick, T.J. Millar, Transport processes and chemical evolution in steady accretion disk flows. Astron. Astrophys. **415**, 643 (2004)

S. Inutsuka, T. Sano, Self-sustained ionization and vanishing dead zones in protoplanetary disks. Astrophys. J. **628**, L155 (2005)

W.-H. Ip, L. Jorda, Can the sodium tail of comet Hale-Bopp have a dust-impact origin? Astrophys. J. **496**, L47 (1998)

N. Iro, D. Gautier, F. Hersant, D. Bockelée-Morvan, J.I. Lunine, An interpretation of the nitrogen deficiency in comets. Icarus **161**, 511 (2003)

W.M. Irvine, D. Bockelée-Morvan, D.C. Lis, H.E. Matthews, N. Biver, J. Crovisier, J.K. Davies, W.R.F. Dent, D. Gautier, P.D. Godfrey, J. Keene, A.J. Lovell, T.C. Owen, T.G. Phillips, H. Tauer, F.P. Schloerb, M. Senay, K. Young, Spectroscopic evidence for interstellar ices in comet Hyakutake. Nature **383**, 418 (1996)

E. Jehin, J. Manfroid, D. Hutsemékers, C. Arpigny, J.-M. Zucconi, Isotopic ratios in comets: Status and perspectives. Earth Moon Planets **105**, 167 (2009)

E.K. Jessberger, J. Kissel, Chemical properties of cometary dust and a note on carbon isotopes, in *IAU Colloq. 116: Comets in the Post-Halley Era*, vol. 167 (1991), p. 1075

J.J. Kavelaars, O. Mousis, J.-M. Petit, H.A. Weaver, On the formation location of Uranus and Neptune as constrained by dynamical and chemical models of comets. Astrophys. J. Lett. **734**, L30 (2011)

H. Kawakita, H. Kobayashi, Formation conditions of icy materials in comet C/2004 Q2 (Machholz). II. Diagnostics using nuclear spin temperatures and deuterium-to-hydrogen ratios in cometary molecules. Astrophys. J. **693**, 388 (2009)

H. Kawakita, J.-i. Watanabe, R. Furusho, T. Fuse, M.T. Capria, M.C. De Sanctis, G. Cremonese, Spin temperatures of ammonia and water molecules in comets. Astrophys. J. **601**, 1152 (2004)

H. Kawakita, J.-I. Watanabe, R. Furusho, T. Fuse, D.C. Boice, Nuclear spin temperature and deuterium-to-hydrogen ratio of methane in comet C/2001 Q4 (NEAT). Astrophys. J. Lett. **623**, 49 (2005)

H. Kawakita, N. Dello Russo, R. Furusho, T. Fuse, J.-i. Watanabe, D.C. Boice, K. Sadakane, N. Arimoto, M. Ohkubo, T. Ohnishi, Ortho-to-para ratios of water and ammonia in comet C/2001 Q4 (NEAT): Comparison of nuclear spin temperatures of water and methane. Astrophys. J. **643**, 1337 (2006)

A. Kouchi, T. Yamamoto, T. Kozasa, T. Kuroda, J.M. Greenberg, Conditions for condensation and preservation of amorphous ice and crystallinity of astrophysical ices. Astron. Astrophys. **290**, 1009 (1994)

D. Kozlowski, J. Pilmé, J. Comput. Chem. **32**, 3207 (2011)

G. Kresse, J. Hafner, Ab-initio molecular-dynamics simulation of the liquid-metal amorphous-semiconductor transition in germanium. Phys. Rev. B **49**, 14251 (1994a)

G. Kresse, J. Hafner, Norm-conserving and ultrasoft pseudopotentials for first-row and transition elements. J. Phys. Condens. Matter **6**, 8245 (1994b)

W.D. Langer, T.E. Graedel, M.A. Frerking, P.B. Armentrout, Carbon and oxygen isotope fractionation in dense interstellar clouds. Astrophys. J. **27**, 581 (1984)
T. Lee, A local proton irradiation model for isotopic anomalies in the solar system. Astrophys. J. **224**, 217 (1978)
J.-E. Lee, E.A. Bergin, J.R. Lyons, Oxygen isotope anomalies of the Sun and the original environment of the solar system. Meteorit. Planet. Sci. **43**, 1351 (2008)
E. Lellouch, C. de Bergh, B. Sicardy, H.U. Käufl, A. Smette, High resolution spectroscopy of Pluto's atmosphere: Detection of the 2.3 μm CH_4 bands and evidence for carbon monoxide. Astron. Astrophys. **530**, L4 (2011)
J.S. Lewis, R.G. Prinn, Kinetic inhibition of CO and N_2 reduction in the solar nebula. Astrophys. J. **238**, 357 (1980)
D.C. Lis, N. Biver, D. Bockelée-Morvan, P. Hartogh, E.A. Bergin, G.A. Blake, J. Crovisier, M. de Val-Borro, E. Jehin, M. Küppers, J. Manfroid, R. Moreno, M. Rengel, S. Szutowicz, A Herschel study of D/H in water in the Jupiter-family comet 45P/Honda-Mrkos-Pajdusáková and prospects for D/H measurements with CCAT. Astrophys. J. **774**, L3 (2013)
R. Llusar, A. Beltran, J. Andrès, B. Silvi, Topological analysis of electron density in depleted homopolar chemical bonds. J. Comput. Chem. **20**, 1517 (1999)
J.I. Lunine, D.J. Stevenson, Thermodynamics of clathrate hydrate at low and high pressures with application to the outer solar system. Astrophys. J. Suppl. Ser. **58**, 493 (1985)
J.R. Lyons, E.D. Young, CO self-shielding as the origin of oxygen isotope anomalies in the early solar nebula. Nature **435**, 193 (2005)
U. Marboeuf, O. Mousis, J.-M. Petit, B. Schmitt, Clathrate hydrates formation in short-period comets. Astrophys. J. **708**, 812 (2010a)
U. Marboeuf, O. Mousis, J.-M. Petit, B. Schmitt, A.L. Cochran, H.A. Weaver, On the stability of clathrate hydrates in comets 67P/Churyumov-Gerasimenko and 46P/Wirtanen. Astron. Astrophys. **525**, 144 (2010b)
U. Marboeuf, B. Schmitt, J.-M. Petit, O. Mousis, N. Fray, A cometary nucleus model taking into account all phase changes of water ice: Amorphous, crystalline, and clathrate. Astron. Astrophys. **542**, 82 (2012)
G.S. Mathews, P.D. Klassen, A. Juhász, D. Harsono, E. Chapillon, E.F. van Dishoeck, D. Espada, I. de Gregorio-Monsalvo, A. Hales, M.R. Hogerheijde, J.C. Mottram, M.G. Rawlings, S. Takahasi, L. Testi, ALMA imaging of the CO snowline of the HD 163296 disk with DCO^+. Astron. Astrophys. **557**, 132 (2013)
J.S. Mathis, W. Rumpl, K.H. Nordsieck, The size distribution of interstellar grains. Astrophys. J. **217**, 425 (1977)
K.D. McKeegan, A.P.A. Kallio, V.S. Heber, G. Jarzebinski, P.H. Mao, C.D. Coath, T. Kunihiro, R.C. Wiens, J.E. Nordholt, R.W. Moses, D.B. Reisenfeld, A.J.G. Jurewicz, D.S. Burnett, The oxygen isotopic composition of the Sun inferred from captured solar wind. Science **332**, 1528 (2011)
R. Meier, T.C. Owen, Cometary deuterium. Space Sci. Rev. **1**, 99 (1999)
R. Meier, T.C. Owen, D.C. Jewitt, H.E. Matthews, M. Senay, M.N. Biver, D. Bockelée-Morvan, J. Crovisier, D. Gautier, Deuterium in comet C/1995 O1 (Hale-Bopp): Detection of DCN. Science **279**, 1707 (1998)
T.J. Millar, A. Bennett, E. Herbst, Deuterium fractionation in dense interstellar clouds. Astrophys. J. **340**, 906 (1989)
J.C. Mottram, E.F. van Dishoeck, M. Schmalzl, L.E. Kristensen, R. Visser, M.R. Hogerheijde, S. Bruderer, Waterfalls around protostars. Infall motions toward Class 0/1 envelopes as probed by water. Astron. Astrophys. **558**, 126 (2013)
O. Mousis, Modeling the thermodynamical conditions in the Uranian subnebula—implications for regular satellite composition. Astron. Astrophys. **413**, 373 (2004)
O. Mousis, D. Gautier, Constraints on the presence of volatiles in Ganymede and Callisto from an evolutionary turbulent model of the Jovian subnebula. Planet. Space Sci. **52**, 361 (2004)
O. Mousis, D. Gautier, D. Bockelée-Morvan, F. Robert, B. Dubrulle, A. Drouart, Constraints on the formation of comets from D/H ratios measured in H_2O and HCN. Icarus **148**, 513 (2000)
O. Mousis, D. Gautier, D. Bockelée-Morvan, An evolutionary turbulent model of Saturn's subnebula: Implications for the origin of the atmosphere of Titan. Icarus **156**, 162 (2002)
O. Mousis, J.I. Lunine, C. Thomas, M. Pasek, U. Varboeuf, Y. Alibert, V. Ballenegger, D. Cordier, Y. Ellinger, F. Pauzat, S. Picaud, Clathration of volatiles in the Solar Nebula and implications for the origin of Titan's atmosphere. Astrophys. J. **691**, 1780 (2009a)
O. Mousis, U. Marboeuf, J.L. Lunine, Y. Alibert, L.N. Fletcher, G.S. Orton, F. Pauzat, Y. Ellinger, Determination of the minimum masses of heavy elements in the envelopes of Jupiter and Saturn. Astrophys. J. **696**, 1348 (2009b)
O. Mousis, J.I. Lunine, S. Picaud, D. Cordier, Volatile inventories in clathrate hydrates formed in the primordial nebula. Faraday Discuss. **147**, 509 (2010)

O. Mousis, J.I. Lunine, N. Madhusudhan, T.V. Johnson, Nebular water depletion as the cause of Jupiter's low oxygen abundance. Astrophys. J. **751**, L7 (2012a)

O. Mousis, A. Guilbert-Lepoutre, J.I. Lunine, A.L. Cochran, J.H. Waite, J.M. Petit, P. Rousselot, The dual origin of the nitrogen deficiency in comets: Selective volatile trapping in the nebula and postaccretion radiogenic heating. Astrophys. J. **757**, 146 (2012b)

O. Mousis et al., Scientific rationale of Saturn's in situ exploration. Planet. Space Sci. **104**, 29 (2015)

M.J. Mumma, S.B. Charnley, The chemical composition of comets–emerging taxonomies and natal heritage. Annu. Rev. Astron. Astrophys. **49**, 471 (2011)

M.J. Mumma, H.A. Weaver, H.P. Larson, The ortho-para ratio of water vapor in comet P/Halley. Astron. Astrophys. **187**, 419 (1987)

M.J. Mumma, W.E. Blass, H.A. Weaver, H.P. Larson, in *The Formation and Evolution of Planetary Systems*, Baltimore, ed. by H.A. Weaver, F. Paresce, L. Danly (Cambridge University Press, Cambridge, 1988), p. 157

K. Nagashima, A.N. Krot, H. Yurimoto, Stardust silicates from primitive meteorites. Nature **428**, 921 (2004)

J.R. Najita, J.S. Carr, K.M. Pontoppidan, C. Salyk, E.F. van Dishoeck, G.A. Blake, The HCN-water ratio in the planet formation region of disks. Astrophys. J. **766**, 134 (2013)

L.R. Nittler, Presolar stardust in meteorites: recent advances and scientific frontiers. Earth Planet. Sci. Lett. **209**, 259 (2003)

G. Notesco, A. Bar-Nun, A ~ 25 K temperature of formation for the submicron ice grains which formed comets. Icarus **175**, 546 (2005)

K.I. Öberg, R. Murray-Clay, E.A. Bergin, The effects of snowlines on C/O in planetary atmospheres. Astrophys. J. **743**, L16 (2011)

T. Owen, P. Mahaffy, H.B. Niemann, S. Atreya, T. Donahue, A. Bar-Nun, I. de Pater, A low-temperature origin for the planetesimals that formed Jupiter. Nature **402**, 269 (1999)

L. Pagani, E. Roueff, P. Lesaffre, Ortho-H_2 and the age of interstellar dark clouds. Astrophys. J. **739**, L35 (2011)

L. Paganini, M.A. DiSanti, M.J. Mumma, G.L. Villanueva, B.P. Bonev, J.V. Keane, E.L. Gibb, H. Boehnhard, K.J. Meech, The unexpectedly bright comet C/2012 F6 (Lemmon) unveiled at near-infrared wavelengths. Astron. J. **147**, 15 (2014)

J.P. Perdew, J.I. Chevary, S.H. Vosko, K.A. Jackson, M.R. Pederson, D.J. Singh, C. Fiolhais, Atoms, molecules, solids and surfaces—applications of the generalized gradient approximation for exchange and correlation. Phys. Rev. B **46**, 6671 (1992)

M.V. Persson, J.K. Jorgensen, E.F. van Dishoeck, D. Harsono, The deuterium fractionation of water on solar system scales in deeply-embedded low-mass protostars. Astron. Astrophys. **563**, 74 (2014)

M. Pettini, R. Cooke, A new precise measurement of the primordial abundance of deuterium. Mon. Not. R. Astron. Soc. **425**, 2477 (2012)

L. Podio, I. Kamp, C. Codella, S. Cabrit, B. Nisini, C. Dougados, G. Sandell, J.P. Williams, L. Testi, W.-F. Thi, P. Woitke, R. Meirjink, M. Spaans, G. Aresu, F. Ménard, C. Pinte, Water vapor in the protoplanetary disk of DG Tau. Astrophys. J. Lett. **766**, L5 (2013)

K.M. Pontoppidan, S.M. Blevins, The chemistry of planet forming regions is not interstellar. Faraday Discuss. **168**, 49 (2014)

K.M. Pontoppidan, C.P. Dullemond, E.F. van Dishoeck, G.A. Blake, A.C.A. Boogert, N.J. Evans II, J.E. Kessler-Silacii, F. Lahuis, Ices in the edge-on disk CRBR 2422.8-3423: Spitzer spectroscopy and Monte Carlo radiative transfer modeling. Astrophys. J. **622**, 463 (2005)

K.M. Pontoppidan, C. Salyk, G.A. Blake, H.A. Käufl, Spectrally resolved pure rotational lines of water in protoplanetary disks. Astrophys. J. **722**, L173 (2010)

K.M. Pontoppidan, C. Salyk, E.A. Bergin, S. Brittain, B. Marty, O. Mousis, K.I. Öberg, Volatiles in protoplanetary disks, in *Protostars and Planets VI*, ed. by H. Beuther, R. Klessen, C. Dullemond, Th. Henning (University of Arizona Press, Tucson, 2014), p. 363

C. Qi, D.J. Wilner, Y. Aikawa, G.A. Blake, M.R. Hogerheijde, Resolving the chemistry in the disk of TW Hydrae. I. Deuterated species. Astrophys. J. **681**, 1396 (2008)

C. Qi et al., Resolving the CO snow line in the disk around HD 163296. Astrophys. J. **740**, 84 (2011)

C. Qi, K.I. Öberg, D.J. Wilner, P. D'Alessio, E.A. Bergin, S.M. Andrews, G.A. Blake, M.R. Hogerheijde, E.F. van Dishoeck, Imaging of the CO snow line in a Solar Nebula analog. Science **341**, 630 (2013a)

C. Qi, K.I. Öberg, D.J. Wilner, H_2CO and N_2H^+ in protoplanetary disks: Evidence for a CO-ice regulated chemistry. Astrophys. J. **765**, 34 (2013b)

Y.L. Radeva, M.J. Mumma, B.P. Bonev, M.A. DiSanti, G.L. Villanuea, K. Magee-Sauer, E.L. Gibb, H.A. Weaver, The organic composition of comet C/2000 WM1 (LINEAR) revealed through infrared spectroscopy. Icarus **206**, 764 (2010)

S.D. Rodgers, S.B. Charnley, Nitrogen isotopic fractionation of interstellar nitriles. Astrophys. J. **689**, 1448 (2008)

E. Roueff, J.C. Loison, K.M. Hickson, Isotopic fractionation of carbon, deuterium and nitrogen: a full chemical study. Astron. Astrophys. **576**, 99 (2015)
K. Saigo, K. Tomisaka, T. Matsumoto, Evolution of first cores and formation of stellar cores in rotating molecular cloud cores. Astrophys. J. **674**, 997 (2008)
C. Salyk, K.M. Pontoppidan, G.A. Blake, F. Lahuis, E.F. van Dishoeck, N.J. Evans II, H_2O and OH gas in the terrestrial planet-forming zones of protoplanetary disks. Astrophys. J. **676**, L49 (2008)
J. Schönke, W. Tscharnuter, Protostellar collapse of rotating cloud cores. Covering the complete first accretion period of the stellar core. Astron. Astrophys. **526**, 139 (2011)
D. Semenov, D. Wiebe, Chemical evolution of turbulent protoplanetary disks and the solar nebula. Astrophys. J. Suppl. Ser. **196**, 25 (2011)
D. Semenov, D. Wiebe, Th. Henning, Gas-phase CO in protoplanetary disks: A challenge for turbulent mixing. Astrophys. J. **647**, L57 (2006)
Y. Shinnaka, H. Kawakita, H. Kobayashi, E. Jehin, J. Manfroid, D. Hutsemékers, C. Arpigny, Ortho-to-para abundance ratio (OPR) of ammonia in 15 comets: OPRs of ammonia versus $^{14}N/^{15}N$ ratios in CN. Astrophys. J. **729**, 81 (2011)
B. Silvi, A. Savin, Classification of chemical bonds based on topological analysis of electron localization functions. Nature **371**, 683 (1994)
V. Taquet, P.S. Peters, C. Kahane, C. Ceccarelli, A. López-Sepulcre, C. Toubin, D. Duflot, L. Wiesenfeld, Water ice deuteration: A tracer of the chemical history of protostars. Astron. Astrophys. **550**, 127 (2013)
V. Taquet, S.B. Charnley, O. Sipila, Multilayer formation and evaporation of deuterated ices in prestellar and protostellar cores. Astrophys. J. **791**, 1 (2014)
P. Theulé, F. Duvernay, G. Danger, F. Borget, J.B. Bossa, B. Vinogradoff, F. Mispelaer, T. Chiavassa, Thermal reactions in interstellar ice: A step towards molecular complexity in the interstellar medium. Adv. Space Res. **52**, 1567 (2013)
A.G.G.M. Tielens, Surface chemistry of deuterated molecules. Astron. Astrophys. **119**, 177 (1983)
C. Tornow, P. Gast, I. Pelivan, S. Kupper, E. Kührt, U. Motschmann, Water formation in early solar nebula: I. Quasi-stationary cloud core. Planet. Space Sci. **98**, 216 (2014a)
C. Tornow, P. Gast, U. Motschmann, S. Kupper, E. Kührt, I. Pelivan, Water formation in early solar nebula: II. Collapsing cloud core. Planet. Space Sci. **98**, 233 (2014b)
N.J. Turner, K. Willacy, G. Bryden, H.W. Yorke, Turbulent mixing in the outer solar nebula. Astrophys. J. **639**, 1218 (2006)
N.J. Turner, T. Sano, N. Dziourkevitch, Turbulent mixing and the dead zone in protostellar disks. Astrophys. J. **659**, 729 (2007)
N.J. Turner, S. Fromang, C. Gammie, H. Klahr, G. Lesur, M. Wardle, X. Bai, Transport and accretion in planet-forming disks, in *Protostars and Planets VI*, ed. by H. Beuther, R. Klessen, C. Dullemond, Th. Henning (University of Arizona Press, Tucson, 2014), p. 411
A.I. Vasyunin, D.A. Wiebe, T. Birnstiel, S. Zhukovska, Th. Henning, C.P. Dullemond, Impact of grain evolution on the chemical structure of protoplanetary disks. Astrophys. J. **727**, 76 (2011)
G.L. Villanueva, M.J. Mumma, B.P. Bonev, M.A. DiSanti, E.L. Gibb, H. Böhnhardt, M. Lippi, A sensitive search for deuterated water in Comet 8P/Tuttle. Astrophys. J. **690**, 5 (2009)
G.J. Villanueva, M.J. Mumma, M.A. DiSanti, B.P. Bonev, E.L. Gibb, K. Magee-Sauer, G.A. Blake, C. Salyk, The molecular composition of comet C/2007 W1 (Boattini): Evidence of a peculiar outgassing and a rich chemistry. Icarus **216**, 227 (2011)
R. Visser, E.F. van Dishoeck, J.H. Black, The photodissociation and chemistry of CO isotopologues: Applications to interstellar clouds and circumstellar disks. Astron. Astrophys. **503**, 323 (2009a)
R. Visser, E.F. van Dishoeck, S.D. Doty, C.P. Dullemond, The chemical history of molecules in circumstellar disks. I. Ices. Astron. Astrophys. **495**, 881–897 (2009b)
R. Visser, S.D. Doty, E.F. van Dishoeck, The chemical history of molecules in circumstellar disks. II. Gas-phase species. Astron. Astrophys. **534**, 132 (2011)
V. Wakelam, C. Vastel, Y. Aikawa, A. Coutens, S. Bottinelli, E. Caux, Chemical modeling of water deuteration in IRAS 16293-2422. Mon. Not. R. Astron. Soc. **445**, 2854 (2012)
C. Walsh, T.J. Millar, H. Nomura, Molecular line emission from a protoplanetary disk irradiated externally by a nearby massive star. Astrophys. J. **766**, L23 (2013)
C. Walsh, T.J. Millar, H. Nomura, E. Herbst, S. Widicus Weaver, Y. Aikawa, J.C. Laas, A.I. Vasyunin, Complex organic molecules in protoplanetary disks. Astron. Astrophys. **563**, 33 (2014)
N. Watanabe, A. Kouchi, Ice surface reactions: A key to chemical evolution in space. Prog. Surf. Sci. **83**, 439 (2008)
K. Willacy, The chemistry of multiply deuterated molecules in protoplanetary disks. I. The outer disk. Astrophys. J. **660**, 44 (2007)
K. Willacy, W.D. Langer, The importance of photoprocessing in protostellar disks. Astrophys. J. **544**, 903 (2000)

K. Willacy, N.J. Turner, Turbulent mixing and the abundance of water ice at the snow-line of a protostellar disk (2015, in preparation)
K. Willacy, P.M. Woods, Deuterium chemistry in protoplanetary disks. II. The inner 30 AU. Astrophys. J. Lett. **703**, 479 (2009)
K. Willacy, H.H. Klahr, T.J. Millar, Th. Henning, Gas and grain chemistry in a protoplanetary disk. Astron. Astrophys. **338**, 995 (1998)
K. Willacy, W.D. Langer, M. Allen, G. Bryden, Turbulence-driven diffusion in protoplanetary disks: Chemical effects in the outer regions. Astrophys. J. **644**, 1202 (2006)
E.S. Wirström, W.D. Geppert, A. Hjalmarson, C.M. Persson, J.H. Black, P. Bergman, T.J. Millar, M. Hamberg, E. Vigren, Observational tests of interstellar methanol formation. Astron. Astrophys. **533**, 24 (2011)
E.S. Wirström, S.B. Charnley, M.A. Cordiner, S.N. Milam, Isotopic anomalies in primitive solar system matter: Spin-state-dependent fractionation of nitrogen and deuterium in interstellar clouds. Astrophys. J. **757**, L11 (2012)
P.M. Woods, K. Willacy, Benzene formation in the inner regions of protostellar disks. Astrophys. J. **655**, L49 (2007)
P.M. Woods, K. Willacy, Carbon isotope fractionation in protoplanetary disks. Astrophys. J. **693**, 1360 (2009)
C.E. Woodward, M.S. Kelley, D. Bockelée-Morvan, R.D. Gehrz, Water in comet C/2003 K4 (LINEAR) with Spitzer. Astrophys. J. **671**, 1065 (2007)
S. Wyckoff, S.C. Tegler, L. Engel, Nitrogen abundance in comet Halley. Astrophys. J. **367**, 641 (1991)
E.D. Young, Time-dependent oxygen isotopic effects of CO self-shielding across the solar protoplanetary disk. Earth Planet. Sci. Lett. **262**, 468 (2007)
M. Yu, K. Willacy, S. Dodson-Robinson, N. Turner, N.J. Evans II, Peering into the giant planet forming region of an MRI-active disk with rare CO isotopologues. Astrophys. J. (2014, submitted)
H. Yurimoto, K. Kuramoto, Molecular cloud origin for the oxygen isotope heterogeneity in the solar system. Science **305**, 1763 (2004)
H. Yurimoto, K. Kuramoto, A.N. Krot, E.R.D. Scott, J.N. Cuzzi, M.H. Thiemens, J.R. Lyons, Origin and evolution of oxygen isotopic compositions of the solar system, in *Protostars and Planets V* (University of Arizona Press, Tucson, 2007)
V. Zubko, E. Dwek, R.G. Arendt, Interstellar dust models consistent with extinction, emission, and abundance constraints. Astrophys. J. Suppl. Ser. **152**, 211 (2004)

Space Sci Rev (2015) 197:191–269
DOI 10.1007/s11214-015-0223-2

Origin and Evolution of the Cometary Reservoirs

Luke Dones[1] · Ramon Brasser[2] · Nathan Kaib[3] · Hans Rickman[4,5]

Received: 13 February 2015 / Accepted: 5 October 2015 / Published online: 24 November 2015

Abstract Comets have three known reservoirs: the roughly spherical Oort Cloud (for long-period comets), the flattened Kuiper Belt (for ecliptic comets), and, surprisingly, the asteroid belt (for main-belt comets). Comets in the Oort Cloud were thought to have formed in the region of the giant planets and then placed in quasi-stable orbits at distances of thousands or tens of thousands of AU through the gravitational effects of the planets and the Galaxy. The planets were long assumed to have formed in place. However, the giant planets may have undergone two episodes of migration. The first would have taken place in the first few million years of the Solar System, during or shortly after the formation of the giant planets, when gas was still present in the protoplanetary disk around the Sun. The Grand Tack (Walsh et al. in Nature 475:206–209, 2011) models how this stage of migration could explain the low mass of Mars and deplete, then repopulate the asteroid belt, with outer-belt asteroids originating between, and outside of, the orbits of the giant planets. The second stage of migration would have occurred later (possibly hundreds of millions of years later) due to interactions with a remnant disk of planetesimals, i.e., a massive ancestor of the Kuiper Belt. Safronov (Evolution of the Protoplanetary Cloud and Formation of the Earth and the Planets, 1969) and Fernández and Ip (Icarus 58:109–120, 1984) proposed that the giant planets would have migrated as they interacted with leftover planetesimals; Jupiter would have moved slightly inward, while Saturn and (especially) Uranus and Neptune would have moved outward from the Sun. Malhotra (Nature 365:819–821, 1993) showed that Pluto's orbit in the 3:2 resonance with Neptune was a natural outcome if Neptune captured Pluto into resonance while

We thank the NASA Cassini Data Analysis Program for support of some of the work described here.

✉ L. Dones
luke@boulder.swri.edu

1 Southwest Research Institute, 1050 Walnut St., Suite 300, Boulder, CO 80302-5142, USA

2 Earth-Life Science Institute, Tokyo Institute of Technology, 2-12-1-IE-1 Ookayama, Meguro-ku, Tokyo, 152-8550, Japan

3 H.L. Dodge Department of Physics and Astronomy, University of Oklahoma, 440 W. Brooks St., Norman, OK 73019, USA

4 Department of Physics and Astronomy, Uppsala University, 75120 Uppsala, Sweden

5 PAS Space Research Center, Bartyckya 18A, 00716 Warszawa, Poland

it migrated outward. Building on this work, Tsiganis et al. (Nature 435:459–461, 2005) proposed the Nice model, in which the giant planets formed closer together than they are now, and underwent a dynamical instability that led to a flood of comets and asteroids throughout the Solar System (Gomes et al. in Nature 435:466–469, 2005b). In this scenario, it is somewhat a matter of luck whether an icy planetesimal ends up in the Kuiper Belt or Oort Cloud (Brasser and Morbidelli in Icarus 225:40–49, 2013), as a Trojan asteroid (Morbidelli et al. in Nature 435:462–465, 2005; Nesvorný and Vokrouhlický in Astron. J. 137:5003–5011, 2009; Nesvorný et al. in Astrophys. J. 768:45, 2013), or as a distant "irregular" satellite of a giant planet (Nesvorný et al. in Astron. J. 133:1962–1976, 2007). Comets could even have been captured into the asteroid belt (Levison et al. in Nature 460:364–366, 2009). The remarkable finding of two "inner Oort Cloud" bodies, Sedna and 2012 VP_{113}, with perihelion distances of 76 and 81 AU, respectively (Brown et al. in Astrophys. J. 617:645–649, 2004; Trujillo and Sheppard in Nature 507:471–474, 2014), along with the discovery of other likely inner Oort Cloud bodies (Chen et al. in Astrophys. J. Lett. 775:8, 2013; Brasser and Schwamb in Mon. Not. R. Astron. Soc. 446:3788–3796, 2015), suggests that the Sun formed in a denser environment, i.e., in a star cluster (Brasser et al. in Icarus 184:59–82, 2006, 191:413–433, 2007, 217:1–19, 2012b; Kaib and Quinn in Icarus 197:221–238, 2008). The Sun may have orbited closer or further from the center of the Galaxy than it does now, with implications for the structure of the Oort Cloud (Kaib et al. in Icarus 215:491–507, 2011).

We focus on the formation of cometary nuclei; the orbital properties of the cometary reservoirs; physical properties of comets; planetary migration; the formation of the Oort Cloud in various environments; the formation and evolution of the Kuiper Belt and Scattered Disk; and the populations and size distributions of the cometary reservoirs. We close with a brief discussion of cometary analogs around other stars and a summary.

Keywords Comets · Long-period comets · Jupiter-family comets · Main-belt comets · Oort Cloud · Kuiper Belt · Comets: individual: 67P/Churyumov-Gerasimenko

1 From the Beginning

1.1 A Brief Pre-history of the Solar System

We begin with a few words about the formation of the Sun and the disk of gas and dust (the solar nebula) that surrounded it. We base much of our discussion on the review chapters in *Protostars and Planets VI* (Beuther et al. 2014). Pfalzner et al. (2015) provide an excellent overview of the formation of the Solar System, including constraints from meteorites and the evolution of protoplanetary disks. Andrews (2015) gives a very thorough review of the literature on the evolution of solids ("dust") in protoplanetary disks. The book by Armitage (2010) covers the astrophysics of planet formation; the author periodically releases an abridged, but updated, version of this material (Armitage 2014). Armitage (2015) has also published a more technical discussion of physical processes in protoplanetary disks from his 2015 Saas-Fee lectures (http://isdc.unige.ch/sf2015/program). Older reviews are given in a free collection of articles from a 2007 Nobel Symposium (Piskunov et al. 2008; http://iopscience.iop.org/1402-4896/2008/T130).

Molecular clouds are the densest, coldest part of the interstellar medium, and consist primarily of H_2 and He. A typical Giant Molecular Cloud (GMC) in the solar neighborhood, such as the Orion Molecular Cloud Complex, has a mass of order 10^5 solar masses, a density of 100 molecules/cm^3, and a temperature of 15 K (Stahler and Palla 2004). GMCs contain much denser cores, which appear to be the typical sites of star formation, with sizes of 0.01–0.05 pc (2000–10,000 AU) and densities exceeding 10^4 molecules/cm^3. The cores are clustered, and most lie within filaments (Myers 2009; Molinari et al. 2014).

Protostellar disks are seen around many young stars, including the 3–10 Myr old pre-main sequence star TW Hydrae. This star is surrounded by a massive gas disk that extends beyond 100 AU, along with mm-sized "dust" out to ~60 AU and μm-sized dust further out (Andrews et al. 2012). Qi et al. (2013) infer the presence of a CO "snow line" at 30 AU in the TW Hydrae disk. Formation of such large disks is not a simple consequence of a core's collapse, as magnetic fields can remove angular momentum (Li et al. 2014). In models by Tomida et al. (2015), the initial disk has a radius of 5 AU and grows as the accretion of gas continues.

Most stars form in clusters within GMCs, with some 90 % being born within "embedded" clusters, i.e., clusters still surrounded by gas and dust, which typically survive for 10 Myr (Lada and Lada 2003; Allen et al. 2007; Li and Adams 2015). The Trapezium in the Orion Nebula is a nearby embedded cluster. Only about 10 % of embedded clusters survive to become open clusters, gravitationally bound stellar aggregates that usually survive 100 Myr to 1 Gyr (Li and Adams 2015). The Hyades and Pleiades in Taurus are nearby open clusters. Clusters disperse due to various forms of "feedback," e.g., protostellar outflows, stellar winds, and supernova explosions (Krumholz et al. 2014). Protostellar disks disperse in ~3 Myr for dust and 2 Myr for gas (Cloutier et al. 2014), with loss from the disk first dominated by accretion onto the star, and later by photoevaporation by ultraviolet radiation from the star, which clears the disk from the inside out (Alexander 2014; Alexander et al. 2014). Because of rapid disk dispersal, planetesimals must form, and the giant planets must accrete their gas, in only a few million years. After this time, many stars are surrounded by "debris disks" of small particles created by collisions between planetesimals (as in the asteroid belt) or released by comets (Greaves et al. 2004, 2005; Su et al. 2007; Moro-Martín et al. 2007; Hillenbrand et al. 2008; Wyatt 2008; Lawler et al. 2009; Matthews et al. 2014; Moro-Martín et al. 2015). The star Fomalhaut, which is estimated to be 400 Myr old, is surrounded by a complex debris disk which extends beyond 150 AU from the star (Kalas et al. 2013).

1.2 Formation of Small Bodies and Planets

Cometary nuclei, or, informally, "comets" are small bodies composed of volatile "ices" (H_2O, CO, CO_2, CH_3OH, H_2CO, NH_3, etc.) and refractory grains including rock-forming minerals such as olivine and pyroxene as well as organics (long-chain aliphatic hydrocarbons, aromatic nitriles, etc.). The mass of refractory materials ("dust") is comparable to the mass in volatiles, and many comets (e.g., 1P/Halley, 9P/Tempel 1) are better described as icy dirtballs than dirty snowballs (Keller 1989).

Comets are the most numerous macroscopic primitive bodies in our Solar System. As examples of primitive bodies, the 2011 planetary decadal survey (Space Studies Board 2012) lists "asteroids, comets, Kuiper belt objects (KBOs), the moons of Mars, and samples—meteorites and interplanetary dust particles (IDPs)—derived from them."[1] To this list we

[1] We generally use the term "Kuiper Belt Objects" in a broad sense to mean small bodies with semi-major axes outside of Neptune's orbit, excepting inner Oort Cloud bodies like Sedna. "KBOs" thus include classical and resonant KBOs and Scattered Disk Objects. In most cases we follow the terminology of Gladman et al. (2008), except that they refer to bodies beyond Neptune as Trans-Neptunian Objects (TNOs). See Bierhaus and Dones (2015) for a summary of different types of KBOs and Sect. 6.1 for a comparison of dynamical classification schemes.

can add the Trojan asteroids of Jupiter and Neptune and the irregular satellites of the giant planets.

Broadly speaking, comets and asteroids are thought to be the building blocks of the planets. The most popular model for the formation of all the planets is core accretion (Raymond et al. 2009), with the giant planets accreting gas once their cores grow massive enough (Lissauer et al. 2009; Boley 2009). However, steady growth by collisions from tiny grains to planetesimals to the cores of the giant planets is difficult, if not impossible, because particles are likely to stick to each other only at very low speeds. At higher speeds, particles will bounce or fragment (Güttler et al. 2010; Johansen et al. 2014, 2015b; Testi et al. 2014). In addition, gas in the solar nebula orbits a bit more slowly than solids. This produces a drag force that can cause rapid inward drift of particles, possibly driving them into the Sun (Weidenschilling 1977; Nakagawa et al. 1986). The drift rate peaks for meter-sized bodies in a minimum-mass solar nebula. This "radial drift barrier" may not be a problem for realistic protoplanetary disks, however (Laibe et al. 2012).

Safronov (1969) and Goldreich and Ward (1973) proposed that planetesimals could form in the solar nebula via gravitational instability (GI). GI requires that the density of solids exceed a critical density; this condition requires that solids must be confined to a very thin layer. It was questioned whether such a thin disk could form within a turbulent nebula (Weidenschilling and Cuzzi 1993). However, both "passive" and "active" mechanisms can concentrate particles within the turbulent gas in certain regions of the nebula, allowing planetesimals to form via GI (Chiang and Youdin 2010; Pfalzner et al. 2015) or other means. Passive mechanisms include concentration of particles in large vortices that spontaneously form in the nebula (Barge and Sommeria 1995; Barranco and Marcus 2005; Raettig et al. 2015) and pile up of mm-sized particles ("chondrules") between km-scale eddies (Cuzzi et al. 2001, 2008). Active mechanisms include the streaming instability (Youdin and Goodman 2005), in which the relative motion between gas and solids leads to the formation of filaments in which planetesimals can form. For both passive and active concentration mechanisms, planetesimals with sizes of 10 or 100 km up to 1000 km, increasing with distance from the Sun, are likely to form (Hogan and Cuzzi 2007; Johansen et al. 2007, 2012; Cuzzi et al. 2010; Chambers 2010; Johansen and Klahr 2011). In fact, asteroids, KBOs, and the Trojans of Jupiter and Neptune have "knees" in their size distributions near diameters of 100 km that, according to coagulation modeling, are likely to be primordial (see Fig. 1). Thus asteroids and Kuiper Belt Objects may have been "born big" (Morbidelli et al. 2009a; Johansen et al. 2012) [see Sect. 6.3].

The need to form the cores of the giant planets within the lifetime of the protoplanetary disk, thought to be a few Myr (Cloutier et al. 2014), presents a severe challenge for theories of planet formation (Levison et al. 2010). Several recent papers address this issue by invoking accretion of "pebbles," very small solid bodies that interact strongly with gas in the solar nebula (Zsom et al. 2010; Güttler et al. 2010), by the planetesimals formed by the mechanisms discussed above (Lambrechts and Johansen 2012, 2014; Chambers 2014; Kretke and Levison 2014).[2] Formation of a giant planet system like our own is not a general result of

[2]The term "pebble accretion" is used in studies of planet formation that assume planetesimals have already formed (e.g., Morbidelli et al. 2015; Bitsch et al. 2015). If "pebbles" are present in the solar nebula, the planetesimals may be able to grow rapidly into planetary "embryos," the cores of the giant planets, or the terrestrial planets. The planetesimals, in turn, might consist of pebbles that clumped together through the streaming instability or other mechanisms that concentrate particles (Wahlberg Jansson and Johansen 2014). Within low-mass planetesimals, pebbles will collide at low speeds, so fragmentation will not occur and the planetesimal will remain a low-density "pebble pile" (Wahlberg Jansson and Johansen 2014; Hopkins 2014). Jacobson and Walsh (2015) define a pebble as a particle with Stokes number between 0.01 and 1. (These

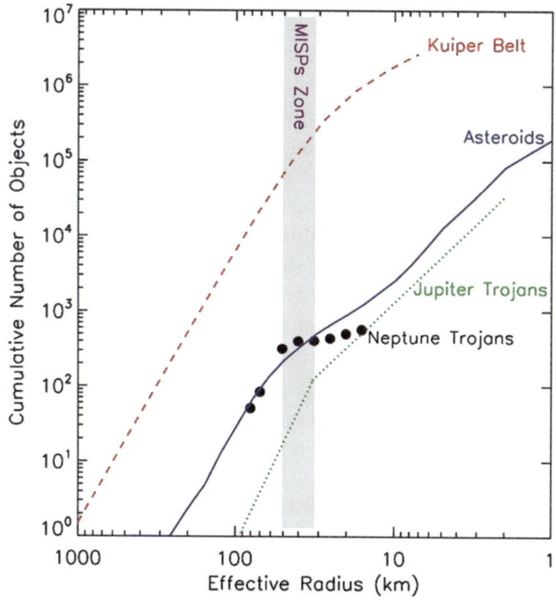

Fig. 1 Cumulative size distributions of several small-body populations in the Solar System from Sheppard and Trujillo (2010). The size of the bodies decreases to the right. From most to least numerous, the figure shows Kuiper Belt Objects (*dashed red line*), main-belt asteroids (*solid blue line*), Trojans of Neptune (*filled circles*), and Trojans of Jupiter (*dotted green line*). All four small populations have similar steep slopes for bodies with radii $\gtrsim 50$ km, and all have a change in their size distribution (sometimes called a knee, elbow, rollover, or flattening) at slightly smaller sizes. Sheppard and Trujillo (2010) refer to the knee around 35–50 km in radius (70–100 km in diameter) as the zone of Missing Intermediate-Sized Planetesimals (MISPs) and argue that its consistent location is related to the formation of these bodies, rather than collisional evolution. Morbidelli et al. (2009a) have reached the same conclusion from modeling of the size distribution of main-belt asteroids. New observations and modeling of the size distribution of the Kuiper Belt since 2010 have strengthened the conclusion that its size distribution flattens (see Fig. 17)

simulations to date. For example, Kretke and Levison (2014) found "our simulations tend to form hundreds of Mars- and Earth-mass objects between 4 and 10 AU," a situation known as oligarchic growth (Kokubo and Ida 1998). However, Levison et al. (2015) have shown that if pebbles form during an extended period within the lifetime of the solar nebula, a few of the embryonic planets can grow into gas giants. Nonetheless, the details of how the gas giants Jupiter and Saturn, the ice giants Uranus and Neptune, and ice-rich planetesimals, i.e., cometary nuclei, formed remain uncertain.

Stokes numbers correspond to particle radii $r \approx 0.8$ cm–80 cm in the asteroid belt and 0.07–7 cm at 30 AU (Carrera et al. 2015), so the astrophysical use of the word "pebble" differs from its geological meaning.) The Stokes number of a particle in a flow, St, is a measure of the particle's "stopping time" τ_s due to its interaction with the gas, relative to the particle's orbital period. Specifically, $St = \Omega \tau_s$, where Ω is the particle's orbital frequency and $\Omega = 2\pi/T$, where T is the particle's orbital period. When $St \ll 1$, the particle follows the flow, while for $St \gg 1$, the particle does not "feel" the gas. Carrera et al. (2015) find that the streaming instability can operate for $0.003 \leq St \leq 0.3$, with the range depending on the solid-to-gas ratio in the disk. The biggest particles inferred in protoplanetary disks or seen to coagulate in experiments are only mm–cm sized. Carrera et al. (2015) therefore focus on formation of planetesimals from chondrules (typically $r \approx 0.3$ mm), since half or more of the mass in many types of chondritic meteorites is in the form of chondrules. Johansen et al. (2015a) consider the growth of asteroids, Kuiper Belt Objects, and planetary embryos by accretion of chondrules.

The comet nuclei we observe today are thought to have formed as planetesimals of ice and rock, sometimes called "cometesimals," beyond the snow line in the solar nebula. Using a one-dimensional model of particle coagulation and setting (i.e., "a series of levels at a single heliocentric distance", Weidenschilling 2004), Weidenschilling (1997) found that cometesimals should be mechanically weak, with substantial macroporosity and microporosity. Weidenschilling (1997) also found that internal structure would have a variety of scales, "but with some tendency for preferential sizes in the range $\sim$10–100 m".

In the past decade, the idea of a fundamental scale for planetesimals of order 100 km has become popular, as discussed above. In that view, asteroids and cometary nuclei smaller than 100 km would be collisional fragments (Morbidelli et al. 2009a). However, Weidenschilling (2011) maintains that the size-frequency distribution (SFD) of asteroids, which is much better determined than the SFD of cometary nuclei, may still be consistent with the traditional view of accretion from sub-km planetesimals.

Johansen et al. (2015a) find that, starting with pebbles $\approx$25 cm in radius, the largest planetesimal that would form in the asteroid belt by the streaming instability would have a diameter of about 200 km, assuming a minimum-mass solar nebula. The size distribution of the planetesimals formed in these simulations is a power law up to a sharp cutoff, with no "bump". However, if "planetesimals are born within an ocean of small [$\approx$ mm-sized] chondrules that do not directly participate in planetesimal formation," the planetesimals can accrete the chondrules. The size distribution of the resulting "asteroids" more closely resembles the observed main-belt size distribution, which has bumps at diameters of $\approx$120 and 400 km. Johansen et al. (2015a) note that, while their initial conditions are based on simulations of the streaming instability, "planetesimals with characteristic sizes of 100 km" that formed by another mechanism would grow by accreting chondrules in a similar way.

Blum et al. (2014, 2015) argue that the very low tensile strength of comets indicates that they via formed by the streaming instability (Youdin and Goodman 2005; Johansen et al. 2007), and Groussin et al. (2015) reach similar conclusions for 67P/Churyumov-Gerasimenko, the target of the Rosetta mission, from OSIRIS observations. Wahlberg Jansson and Johansen (2014) develop a simple model for the collapse of gravitationally bound "pebble clouds" of various masses. They predict that clouds that collapse to planetesimals with diameters $\lesssim$ 40 km (such as 67P) remain "pebble piles" that consist of primordial pebbles. The cometary nucleus should resemble the porous "primordial rubble pile" discussed by Weissman (1986). Morbidelli and Rickman (2015) find that 67P is likely to be a collisional fragment; whether the Rosetta observations are consistent with this conclusion is not yet clear.

The OSIRIS camera on Rosetta imaged features called "goosebumps" or "dinosaur eggs" within the interior of an active pit in the Seth region of the nucleus of 67P. These boulders "exhibit a characteristic scale of about 3 m" (Sierks et al. 2015). Weissman and A'Hearn (2015) inferred that goosebumps "are the basic building blocks of the cometary nucleus," and suggested that ice condensation onto silicate particles (Ros and Johansen 2013) might explain why these building blocks were larger than the mm-to-cm-sized grains expected to be able to stick together when they collide in the solar nebula (Blum et al. 2014, 2015). Kretke and Levison (2015), by contrast, argue that the particles in the coma of "hyperactive" comet 103P/Hartley 2, whose radii, assuming the particles are bright and icy, range up to 30 cm (Kelley et al. 2013, 2015b), are primordial pebbles. Comet 103P has a terrestrial D/H value (Hartogh et al. 2011; see Sect. 3.4) and is rich in CO_2 and poor in CO (A'Hearn et al. 2011; Weaver et al. 2011). Kretke and Levison (2015) argue that these observations suggest a warmer formation location for 103P than for most comets. Closer to the Sun, bigger pebbles can be concentrated by instabilities. If the smaller particles seem in the comae of many comets are also pebbles, these comets presumably formed further from the Sun.

1.3 Planetary Migration

The gravitational interaction between a massive body (i.e., a planet or satellite) and a disk of gas and/or solid bodies transfers angular momentum between the body and the disk. Goldreich and Tremaine (1980) pointed out that both the orbit of Jupiter and the structure of the solar nebula might evolve rapidly due to such interactions. With the discovery of the first extrasolar planet orbiting a Sun-like star, 51 Pegasi b, only 0.05 AU from its primary (Mayor and Queloz 1995), interest in planet migration through interactions with a gas disk soared (Lin et al. 1996; Ward 1997). The investigation of planetary migration due to interactions with gas (Papaloizou and Terquem 2006; Lubow and Ida 2011; Baruteau et al. 2014) and planetesimals (Murray et al. 1998; Levison et al. 2007, 2011; Kirsh et al. 2009; Ormel et al. 2012; Minton and Levison 2014; Bonsor et al. 2014) remains a thriving area of research.

In his classic book, Safronov (1969) remarked that the ejection of planetesimals from the Solar System by the giant planets, as occurs during the formation of the Oort Cloud, should cause a recoil that "entails grave consequences related to the redistribution of angular momentum," i.e., substantial orbital migration of the planets. Fernández and Ip (1984) modeled the recoil using a Monte Carlo Öpik–Arnold code. They found that Jupiter would move in toward the Sun a little, Saturn would move away from the Sun a little, and Uranus and Neptune would move outward a lot. However, their calculations relied on many uncertain assumptions. The key step forward was taken by Malhotra (1993). It had long been known that Pluto completes two orbits for every three of Neptune. In this 3:2 resonance, the orbits are phased so that when Pluto and Neptune reach conjunction, Pluto is near its aphelion distance of 49 AU, whereas Neptune orbits at 30 AU. Thus close approaches between Pluto and Neptune cannot take place (Cohen and Hubbard 1965; Williams and Benson 1971). However, no one knew how Pluto had ended up in such an orbit. Using both perturbation theory and orbital integrations, Malhotra (1993) showed that Pluto could have been captured into the 3:2 resonance with Neptune and acquired its orbital eccentricity of 0.25 if it had formed in a circular orbit around the Sun and Neptune had migrated outward by at least 5 AU. In other words, the planetary migration studied by Fernández and Ip (1984) explained Pluto's unusual orbit!

As it happens, the first Kuiper Belt Object had been discovered the year before Malhotra's paper appeared. Between acceptance and publication of Malhotra (1993), KBOs that also occupied the 3:2 resonance were discovered. Two decades later, we know of KBOs that occupy the 3:2, 5:2, 2:1, 3:1, 5:1, 4:3, 5:3, 7:3, 5:4, and 7:4 resonances with Neptune (Gladman et al. 2012; Pike et al. 2015), as well as Neptune Trojans in the 1:1 resonance (Sheppard and Trujillo 2006). The relative populations of the resonances and the detailed orbital distributions within them provide sensitive probes of the details of Neptune's migration (Hahn and Malhotra 2005; Parker 2015). The idea that the giant planets all migrated due to interactions with planetesimals is a new paradigm in our understanding of the histories of small bodies—comets and asteroids—in our Solar System (Hahn and Malhotra 1999; Minton and Malhotra 2009; DeMeo and Carry 2014). We will discuss the influential Nice model, a detailed scenario for planetary migration and its effect on the small body populations, in Sect. 4.3.

This review will discuss the orbital distributions of the cometary reservoirs and the processes that shaped them. Our understanding of comets a decade ago is described well by Festou et al. (2004) and Fernández (2005); we primarily focus on developments since then. In Sect. 2 we describe the observed orbital properties of the nearly isotropic comets (long-period and Halley-type comets), the ecliptic comets (Jupiter-family, Encke-type, and Centaur comets), and the recently discovered main-belt comets. Section 3 examines the physical properties of comets. Section 4 explores the migration of the giant planets in the context of the Grand Tack and the Nice model, and the possible relation of the Nice model to the Late Heavy Bombardment of the Moon. Section 5 discusses the orbital evolution that stored icy planetesimals from the planetary region in the Oort Cloud. Section 6 does the same for the Kuiper Belt/Scattered Disk. Section 7 estimates the populations of the Kuiper Belt/Scattered Disk and the Oort Cloud. In Sect. 8, we briefly describe observations that argue for the presence of comets around other stars. Section 9 summarizes our main results. We do not discuss Rosetta results on ecliptic comet 67P/Churyumov-Gerasimenko in detail, as they have only begun to appear during the preparation of this article (e.g., Taylor et al. 2015; Bibring et al. 2015; Vincent et al. 2015, and a special issue of *Astronomy and Astrophysics* (http://www.aanda.org/articles/aa/abs/2015/11/contents/contents.html)).

We refer the reader to the other chapters in this volume for discussions that bear on the International Cometary Workshop's goal of elucidating the role of comets in tracing the formation and evolution of the Solar System. Bockelée-Morvan et al. (2015), Gudipati et al. (2015), Mandt et al. (2015), and Willacy et al. (2015) discuss cometary isotope measurements and models in more detail, among many other topics. Guilbert-Lepoutre et al. (2015) explore physical evolution of comets; Cochran et al. (2015) treat the composition of gas and dust in comets; and Lamy et al. (2015) seek to constrain the subsurface properties of cometary nuclei with radar observations.

2 Cometary Reservoirs: Orbital Properties

Comets have been seen for as long as people have looked at the sky. Bailey et al. (1990) and Yeomans (1991) explore ideas of comets through Comet Halley's most recent perihelion passage in 1986. Chinese, Japanese, and Korean historical records of comets and meteor showers have been compiled by Xu et al. (2000) and Pankenier et al. (2008). The *Handbook of Archaeoastronomy and Ethnoastronomy* (Ruggles 2015) discusses comets passim. For example, Medupe (2015) discusses San paintings on rock that appear to be comets (Ouzman 2010), while Ziółkowski (2015) describes Andean observations of C/1533 M1, "the comet of Ataw Wallpa," in July 1533, the month when the last Inka emperor was killed.

Gary Kronk has completed four volumes cataloguing comets appearing from ancient times through 1959 (Kronk 1999, 2003, 2007, 2009). A fifth volume by Kronk and Maik Meyer cataloguing comets from 1960 through 1982 has appeared (Kronk and Meyer 2010), and a sixth volume covering 1983 through 1999 is planned. The first observations listed by Kronk (1999) refer to a comet in 674 BC in Babylonian tablets and comets in 632 BC and 612 BC in Chinese records (Ho 1962). Every return of Comet Halley back to 240 BC has been observed (Stephenson and Yau 1985; Kronk 1999). Observations of comets before 2300 BC have appeared in catalogues, but Kronk (1999) states that some of the dates listed are uncertain by more than a century.

According to Aristotle (384–322 BC), the Pythagoreans (named for Pythagoras, ca. 570–495 BC), believed that there was only one comet and that it was a planet that appeared infrequently near the horizon, like Mercury (Yeomans 1991). Aristotle himself regarded comets as "meteorological," i.e., "sublunar," phenomena, and this remained the dominant view through the Middle Ages (Fernández 2005).

Johannes Kepler's study of Tycho Brahe's data on Mars led him to conclude that the orbit of Mars around the Sun was an ellipse. Kepler published this result in his *Astronomia Nova* in 1609. A decade later he stated that all the planets follow elliptical orbits around the Sun (this is now known as Kepler's first law). According to Applebaum (1996), "by 1666 ellipses and non-uniform motion were well-known to most astronomers". In 1607 Thomas Harriot and William Lower observed a comet (Halley's comet, in fact). They were among the first readers of *Astronomia Nova*, and in 1610 Lower wrote a letter to Harriot in which he suggested that comets too follow elliptical orbits (see Thony Christie's discussion at https://thonyc.wordpress.com/2014/07/15/comets-and-heliocentricity-a-rough-guide/). The pace of the "quest for elliptical orbits" (Ruffner 2010) quickened after 1650. Johannes Hevelius observed many comets between 1652 and 1677, and the cover of his *Cometographia* (1668) shows him indicating that comets follow curved orbits (Chapin 1952; Habashi 2014). Pierre Petit, Adrien Auzout, Giovanni Borelli, Robert Hooke, John Flamsteed, and others debated the nature of comets during this time (Westfall 1983; Heidarzadeh 2008; Eicher 2013).[3]

Isaac Newton was the first to develop a way to determine the orbit of a comet, assumed to be a parabola, from three observations (Yeomans 1991). Newton fit a parabolic orbit for the Great Comet of 1680. More than a century later, Encke computed an orbit for the same comet (now known as C/1680 V1), including planetary perturbations. Encke calculated a semi-major axis in the range 300–600 AU (Hind 1852), so Newton's assumption of a parabolic orbit was a good approximation.

Another bright comet appeared in 1682. Edmond Halley computed a parabolic, then an elliptical, orbit for this comet, and suggested that it might be the same comet seen in 1531 and 1607. The identification was not trivial, because the time between successive perihelion passages for this comet typically changes by a year or two due to perturbations by the giant planets (Yeomans et al. 1986). Halley predicted that the comet would reappear in late 1758 or early 1759 (Kronk 1999). In fact, the comet was first recovered by a German farmer and amateur astronomer, Palitzsch, on Christmas 1758, although it was not recognized to be the periodic comet now known as 1P/Halley for another month (Yeomans 1991).

In 1819, Encke showed that a comet discovered in 1818 could be linked with comets seen in 1805, 1795, and 1786 (Kronk 2003). This, the second periodic comet, is now known as 2P/Encke. Halley and Encke have very different orbits. Halley has had an orbital period of 74–79 years during the two millennia it has been observed, and reaches aphelion near 35 AU from the Sun, beyond the orbit of Neptune. Encke's orbital period is only 3.3 years, and its aphelion distance is 4.1 AU, so its entire orbit is interior to that of Jupiter. Halley follows a retrograde orbit around the Sun with an inclination of 162° to the ecliptic, while Encke's orbit is prograde with an inclination of 12°.

As more comets were discovered, some could be fit with periodic orbits, but most could only be fit by parabolic orbits, meaning that the comet passed well outside the planetary region or that observations were very limited. Lardner (1853) summarized the catalogue of Hind (1852) as follows: 207 comets in all, of which 40 followed elliptic (i.e., periodic) orbits, 160 followed parabolic orbits, and 7 followed hyperbolic orbits.[4] Of the 40 periodic comets, 13—including 2P/Encke, the now-crumbled 3D/Biela (see Sect. 3.3), and the happily rediscovered D/1819 W1 = 289P/Blanpain—had aphelion distances within the orbit of Saturn and small inclinations. Six comets, including 1P/Halley and 23P/Brorsen-Metcalf, had semi-major axes near Uranus's orbit and much larger inclinations, although Halley was the only retrograde comet of the group. Twenty-one comets, including C/1680 V1, had semi-major axes beyond the planets, with 11 being retrograde and 10 being prograde.

[3] We think an anonymous referee for calling our attention to the fascinating history of ideas about comets in the 17th century.

[4] The "hyperbolic" comets were almost certainly on highly eccentric elliptical ("near-parabolic") orbits. See Sect. 2.1 for discussion.

Lardner's three categories of periodic comets roughly correspond to what we now call Jupiter-family comets (JFCs), Halley-type comets (HTCs), and returning long-period comets (LPCs) (Levison 1996). The comets on parabolic and hyperbolic orbits, in turn, are akin to "new" long-period comets (Oort 1950; Levison 1996). The traditional dividing line between JFCs and HTCs is an orbital period (P) of 20 years, with JFCs having $P < 20$ years (semi-major axes $a < 7.4$ AU), and HTCs having 20 years $\leq P \leq$ 200 years (7.4 AU $\leq a \leq$ 34.2 AU). Levison (1996) (also see Carusi et al. 1987) developed a new classification in which comets with Tisserand parameters (T) with respect to Jupiter less than 2 were called "nearly isotropic comets" (NICs), and those with $T > 2$ were called "ecliptic comets".[5] Nearly isotropic comets include HTCs, which Levison (1996) took to have semi-major axes $a < 40$ AU, and LPCs ($a > 40$ AU). In this taxonomy, Encke-type comets, JFCs and Centaurs (or Chiron-types, as Levison called them) are subclasses of ecliptic comets. In most cases, the status of a comet as a Jupiter-family or Halley-type comet is the same based on the old dividing line at a period of 20 years or the new division at $T = 2$, but a few comets with $P < 20$ years (e.g., 96P/Machholz, $a = 3.03$ AU, $T = 1.942$, perihelion distance $q = 0.12$ AU, inclination $i = 58°$) that were formerly considered JFCs are now classified as HTCs (Levison 1996). Gladman et al. (2008), Tancredi (2014), and Jewitt et al. (2015) provide revised versions of Levison's taxonomy.

The term "short-period comet" is still used occasionally, but usage is inconsistent, with some authors using the term to mean comets with $P < 200$ years, while for others, short-period comets are those with $P < 20$ years and "intermediate-period comets" are those with 20 years $\leq P \leq$ 200 years. (The latter definition is equivalent to the orbital period-based classification into JFCs and HTCs above.) In any case, use of a dynamical taxonomy based on the Tisserand parameter is preferable, since comets more often stay in the same class than in a taxonomy based on orbital period (Levison 1996). Also, the upper limit of 200 years for periodic comets is no longer very meaningful, as comet 153P/Ikeya-Zhang has an orbital period in excess of 300 years, its last two perihelion passages having occurred in 1661, when it was probably observed by Hevelius, and 2002 (Hasegawa and Nakano 2003).

As of August 23, 2015, there are 327 numbered periodic comets, the most recent being 327P/Van Ness (also known as P/2002 Q1 and P/2015 P2). In addition, there are 326 unnumbered periodic comets, the latest being P/2015 P4 PanSTARRS (see Seiichi Yoshida's web page at http://www.aerith.net/comet/catalog/index-periodic.html). These totals include, in order of decreasing numbers, Jupiter-family, Halley-type, and other (Centaur, Encke-type, and main-belt) comets. Because comets with shorter periods (i.e., Encke-types and JFCs) have more opportunities for discovery, these numbers do not reflect the real proportions of different types of comets. The Oort Cloud is thought to be the main source of long-period and Halley-type comets (Sect. 2.1), while the Kuiper Belt/Scattered Disk is believed to be the primary source of ecliptic comets (Sect. 2.2). The main-belt comets are likely to have orbited near their current locations for billions of years, though some could have been implanted in the asteroid belt from more distant regions of the Solar System (Levison et al. 2009; Walsh et al. 2011). In Fig. 2, we show the quasi-stable reservoirs of small bodies in the Solar System, with the main-belt comets located within the asteroid belt.

Most comets in the historical record are long-period comets, i.e., they have only been seen at one perihelion passage. The most recent *Catalogue of Cometary Orbits* (Marsden

[5]For a comet or asteroid with semi-major axis a, eccentricity e, and inclination i with respect to the orbital plane of a planet with semi-major axis a_P, the Tisserand parameter $T = a_P/a + 2\sqrt{a(1-e^2)/a_P}\cos i$. We consider the Tisserand parameter of comets with respect to Jupiter, for which $a_P \sim 5.2$ AU. The Tisserand parameter is an approximation to the Jacobi integral, which is a constant of motion for the circular restricted three-body problem (Carusi et al. 1995).

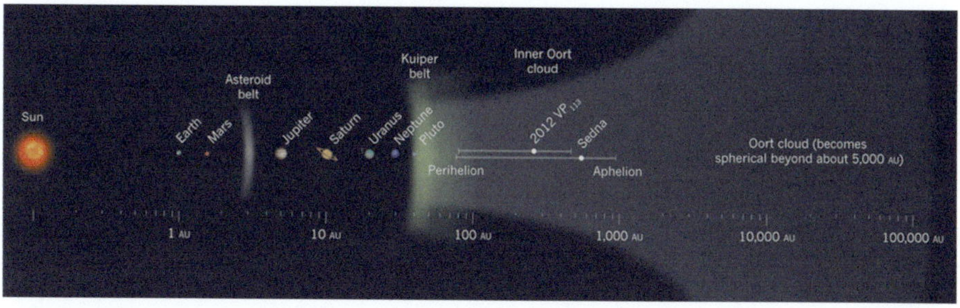

Fig. 2 Diagram showing the distances from the Sun of quasi-stable small-body populations we discuss, from Schwamb (2014). Most main-belt asteroids (and all known main-belt comets) have semi-major axes, a, between 2 and 3.5 AU, although there are also asteroids (not shown) librating around the 3:2 resonance with Jupiter (Hildas, $a \approx 3.7$–4.2 AU) and the 1:1 resonance with Jupiter (Trojans, $a \approx 5.0$–5.4 AU). Continuing outward from the Sun, we have Trojans of Neptune ($a = 30$ AU); Kuiper Belt Objects ($a \approx 35$–48 AU), including "Plutinos" like Pluto in the 3:2 resonance ($a \approx 39$ AU); the inner Oort Cloud (generally taken to be $a < 20{,}000$ AU) and the outer Oort Cloud ($20{,}000$ AU $< a \lesssim 100{,}000$ AU). If the Oort Cloud formed when the Sun was a field star as it is now (i.e., not part of a cluster of stars), models predict an inner edge of the Oort Cloud about 3000 AU from the Sun. The Scattered Disk, an outward extension of the Kuiper Belt, is not shown (see Sect. 6.1 for a discussion of dynamical classes in the Kuiper Belt/Scattered Disk). The discovery of Sedna and 2012 VP_{113} with semi-major axes of hundreds of AU and perihelion distances of $\approx$80 AU suggests that the Sun formed in a cluster (see Sect. 5.2)

and Williams 2008) lists 3708 apparitions of 2844 different comets through June 2008, including 200 numbered periodic comets. For most single-apparition comets, only parabolic orbits could be determined. The "Small-Body Search" at JPL's Solar System Dynamics web site yields 3366 comets as of October 29, 2015. Since the launch of the Solar and Heliospheric Observatory (SOHO) in December 1995, most comet discoveries have been of sungrazers found by SOHO (see Maik Meyer's web page at http://www.comethunter.de/). On September 13, 2015, the SOHO mission discovered its 3000th comet.[6] Knight et al. (2010) estimate the diameters of SOHO comets to range from 10 to 70 meters, with the lower cutoff probably an artifact of incompleteness. SOHO comets are thus much smaller than typical comets observed from the Earth (see Sect. 7). Dones et al. (2004) plot the very different perihelion and inclination distributions of long-period comets discovered historically (through 1995) and in the ~7 years after that.

Bob Johnston's web site lists 5454 known comets as of July 4, 2015, including "318 numbered, 3559 with provisional designations, 1577 without official designations" (http://www.johnstonsarchive.net/astro/sslist.html). Johnston's list is the most complete to our knowledge.[7]

[6] See https://www.nasa.gov/feature/goddard/esa-nasa-solar-observatory-discovers-its-3000th-comet and http://sungrazer.nrl.navy.mil/.

[7] Johnston explains his total as follows (W.R. Johnston, personal communication, 2015): "The comet numbers are my attempt to count all known comets, whether or not they have received IAUC/MPC designations. The categories include:

- Numbered comets—those periodic comets with permanent numbers, excluding Chiron and Echeclus which are counted among Centaurs.
- Provisional designations—those with official designations other than permanent numbers, old or new style, both long- and short-period, and including 15 letter-designated fragments of numbered comets.
- Without designations—known or suspected comets without IAUC designations, includes: SOHO/STEREO comets; distinct ancient/historical comets identified by Gary Kronk in the *Cometography* volumes but not

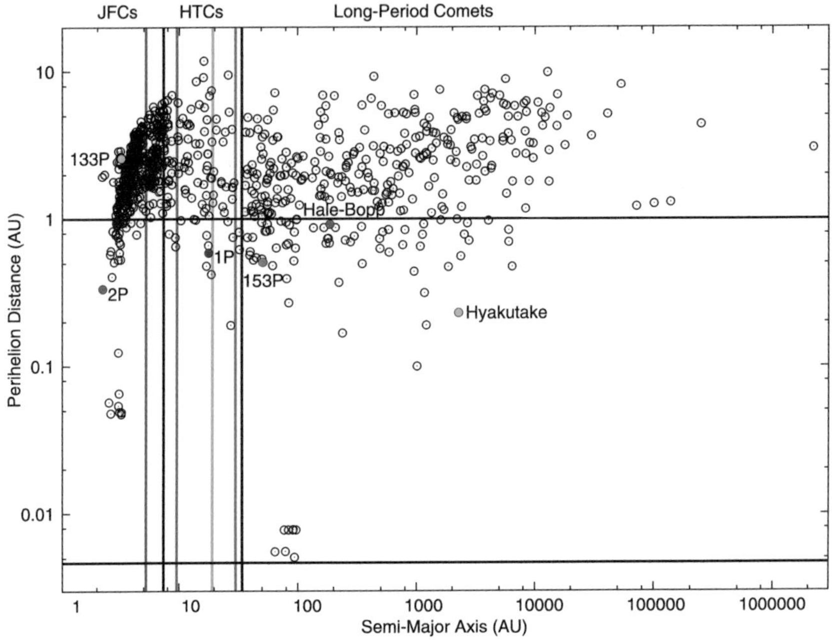

Fig. 3 Osculating perihelion distance (q) vs. semi-major axis (a) for a sample of 1188 comets from the JPL Solar System Dynamics web site. The figure shows vertical lines at the semi-major axes of Jupiter, Saturn, Uranus, and Neptune ($\approx$5.2, 9.5, 19.2, and 30.1 AU, respectively) and at the traditional divides between Jupiter-family comets and Halley-type comets ($20^{2/3} \approx 7.4$ AU) and between Halley-type comets and long-period comets ($200^{2/3} \approx 34.2$ AU). The figure also has horizontal lines at 1 AU and at the solar radius of $r_\odot = 4.65 \times 10^{-3}$ AU. Comets with perihelion distances q less than Earth's aphelion distance of 1.017 AU are Earth-crossing, while those with $q < r_\odot$ will collide with the Sun. The comets with $a \sim 100$ AU and $q < 0.01$ AU are Kreutz sun-grazers, while the "sun-skirting" comets with $a \sim 3$ AU and $q \sim 0.04$ AU are probably ultimately derived from 96P/Machholz (Sekanina and Chodas 2005)

In Figs. 3–5, we show scatter plots of osculating perihelion distance (q) vs. semi-major axis (a); inclination to the ecliptic (i) vs. a; and Tisserand parameter with respect to Jupiter (T) vs. q, respectively, for a selection of 1188 comets from the JPL Small-Body Database (http://ssd.jpl.nasa.gov/sbdb_query.cgi). In each figure we indicate the location of six comets of interest. In order of increasing semi-major axis, these are 2P/Encke; 133P/Elst-Pizarro; 1P/Halley; 153P/Ikeya-Zhang; C/1995 O1 Hale-Bopp; and C/1996 B2 Hyakutake.

The clump of comets in Fig. 3 with semi-major axes between 3 and 7 AU consists of Jupiter-family comets; the main-belt comet 133P/Elst-Pizarro is on the edge of the band. The other five comets highlighted in the plot are Earth-crossing—2P/Encke, 1P/Halley, 153P/Ikeya-Zhang (the first periodic comet with an orbital period in excess of 200 years), and the two great comets of the 1990s, Hale-Bopp and Hyakutake. Figure 3 also shows a

identified with numbered/provisional comets; and "X" comets. Of those without designations, 456 are ones listed by Kronk from pre-modern observations, most of the remainder are SOHO comets.

Given that a number of SOHO comets have made identified returns, I do not delete sungrazers that are suspected but not proven to have disintegrated on perihelion passage (though … many, possibly most, have). Similarly, some comets given "D" designations have been rediscovered so I don't automatically exclude these. I do exclude those known to have collided with Jupiter (25 Shoemaker-Levy 9 fragments) or the Sun (the latter based on a calculated perihelion below the Sun's surface)."

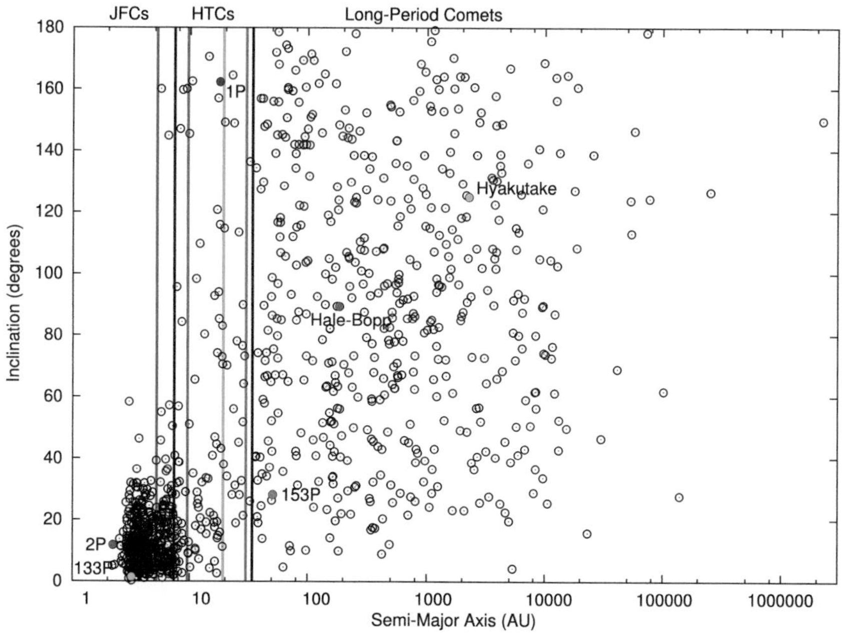

Fig. 4 Osculating orbital inclination (i) vs. a for the sample of comets shown in Fig. 3. The same vertical lines as in Fig. 3 are indicated

number of Halley-type and long-period comets with $q > 5$ AU. Historically, long-period comets were generally only discovered if their perihelion distance was well inside Jupiter's orbit, but C/2003 A2 Gleason was active near its perihelion at 11.4 AU. Figure 4 shows that the Jupiter-family comets also clump at small semi-major axes and inclinations. Halley-type comets have a wide range of inclinations, both prograde and retrograde, and long-period comets have a roughly isotropic distribution of inclinations. Dynamical modeling indicates that the JFCs must derive from a low-inclination source, thought to be the Kuiper Belt/Scattered Disk, while the HTCs and LPCs derive from the Oort Cloud (see Sect. 5).

Figure 5 shows the Tisserand parameters, T, of the comets vs. their perihelion distances. The Tisserand parameter is useful for classification because a comet's value of T is usually somewhat more stable than is the comet's perihelion distance. The Jupiter-family comets cluster at T slightly less than 3, indicating that they encounter Jupiter at low velocities and hence are strongly focused by Jupiter's gravity, causing frequent orbital changes (see, e.g., Maquet 2015 and Fig. 8 for comet 67P). (Encounter velocities are approximately $\sqrt{3-T}$ times Jupiter's orbital speed for comets with $T < 3$.) Halley-type and long-period comets encounter Jupiter at much higher speeds.

2.1 Nearly Isotropic Comets: Long-Period and Halley-Type Comets

To investigate the nature of the Oort Cloud, the reservoir for long-period and Halley-type comets, we need to determine the original orbits of a large number of comets before they entered the planetary region and to quantify selection effects, i.e., biases, in their discovery (Everhart 1967a; Kresák 1982; Wiegert and Tremaine 1999; Horner and Evans 2002; Francis

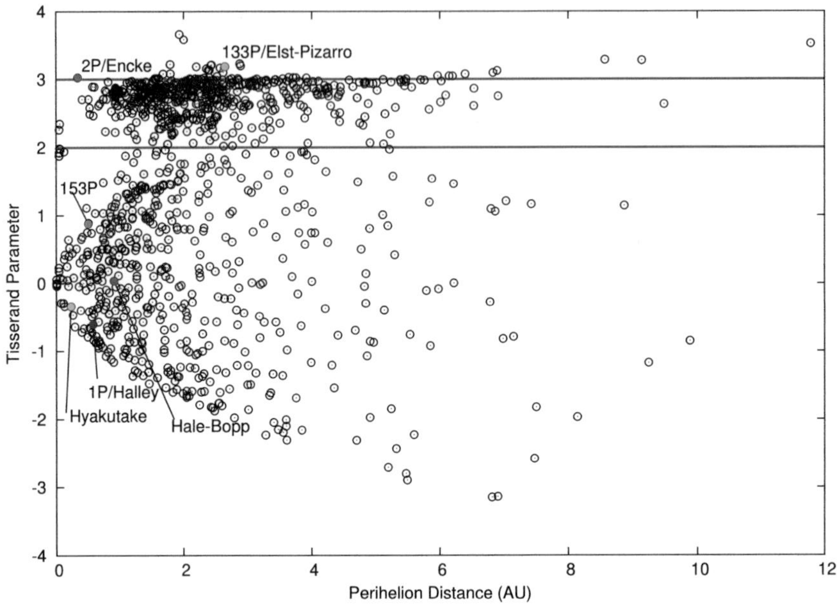

Fig. 5 Tisserand parameter with respect to Jupiter (T) vs. q for the sample of comets shown in Figs. 3 and 4. Horizontal lines are shown at $T = 2$ and $T = 3$. Jupiter-family comets are often taken to have $2 < T < 3$, while long-period and Halley-type comets have $T < 2$

2005). Unfortunately, this is a difficult task for two reasons. First, for LPCs with original semi-major axes $a_{\rm orig} \gtrsim 20{,}000$ AU and perihelion distances $q \lesssim 10$ AU, the change in the comet's semi-major axis due to planetary perturbations is typically comparable to its original value (Everhart 1968; Fernández and Brunini 2000; Fouchard et al. 2013). This means that the comet is likely to be perturbed either into a much smaller, more tightly bound orbit or into a much bigger, possibly unbound (hyperbolic) orbit. By contrast, comets with $a_{\rm orig} \lesssim 20{,}000$ AU and $q \lesssim 10$ AU are less likely to become observable (which usually requires that the comet's perihelion be reduced to $q \lesssim 3$ AU, where water ice begins to sublime) because of the "Jupiter barrier" (or Jupiter-Saturn barrier). This "barrier" was long assumed to work as follows. Galactic tides (i.e., the difference in gravitational attraction on a comet and the Sun due to the mass of the Galaxy, primarily nearby matter in the galactic disk; see Fig. 6) cause a very long-period oscillation in q for long-period comets. Tides are a bigger effect on comets with larger a; the magnitude of the change in q per orbit of the comet, Δq, is proportional to $a^{7/2}$, with the exact value depending on the orientation (inclination and argument of perihelion) of the orbit. Typical values of Δq are 9 AU for $a = 30{,}000$ AU and 2 AU for $a = 20{,}000$ AU (Dybczyński and Królikowska 2011). Consider two comets whose perihelion distances are decreasing due to galactic tides. Both comets start with $q = 12$ AU and have identical orbital orientations, but one comet has $a = 30{,}000$ AU and the other has $a = 20{,}000$ AU. The first comet will have a perihelion distance $q - \Delta q = 3$ AU on its next passage through the planetary region (if the planets did not eject it) and a reasonable chance of being discovered. The second comet will have $q - \Delta q = 10$ AU and virtually no chance of being discovered. In order to eventually reach $q = 3$ AU, the second comet must pass through the planetary region several times; each time it has a ~50 % chance of being ejected from the Solar System. Thus comets with smaller a require a longer (and less likely)

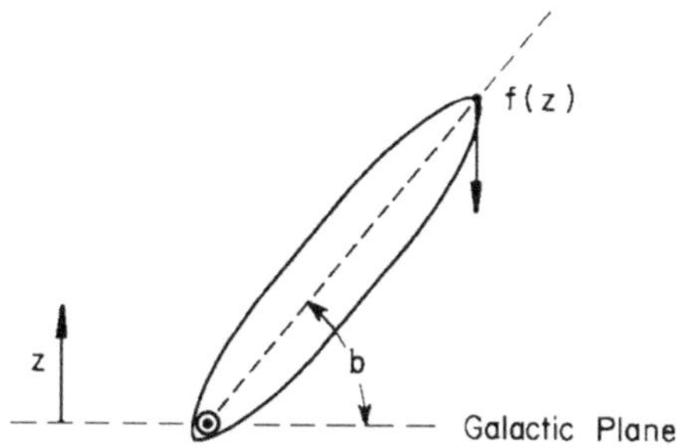

Fig. 6 Illustration (not to scale) of the orbit of a comet on a highly eccentric orbit inclined to the galactic plane by an angle b, from Torbett (1986). The force $f(z)$ is the "vertical" force due to nearby matter in the galactic disk, where z is the distance from the galactic plane. The difference between the force at the comet's position and at the Sun's position is the vertical tide. The Sun is shown at the midplane of the disk for clarity. The Sun's actual distance from the midplane is smaller than the local scale height of the disk (Joshi 2007). At distances this close to the midplane, the vertical tide is proportional to the difference in z values of the comet and the Sun to good accuracy

run of "good luck" in order to become observable (Hills 1981; Levison et al. 2001; Kaib and Quinn 2009; Dybczyński and Królikowska 2011).

Hills (1981) suggested that the Jupiter barrier produced such a strong selection effect that the apparent "inner edge" of the Oort Cloud at ~20,000 AU was an "observational artifact". Comets in an inner cloud with smaller semi-major axes (3000–20,000 AU), which might be much more numerous than comets in the classical outer cloud, would only become visible when a passing star traversed the inner cloud, causing a comet "shower". Hills (1981) calculated that while there would be a steady flux of comets with $a = 20{,}000$ AU, comet showers would bring in observable comets with semi-major axes of 10,000 and 5000 AU only 9 % and 0.8 % of the time, respectively (also see Heisler et al. 1987 and Heisler 1990).

The relative populations of the inner and outer clouds are not known observationally. Hills (1981) estimated that the mass of the inner cloud might be 200 times greater than that of the outer cloud, but models in which the Sun formed in its current galactic environment (e.g., Dones et al. 2004) suggest that there should be comparable numbers of comets in the two regions.[8] In any case, detailed studies indicate that the Jupiter-Saturn barrier is more porous than the heuristic argument above would suggest (Kaib and Quinn 2009; Dybczyński and Królikowska 2011, 2015; Fouchard et al. 2014). This is one reason why the estimate by Hills (1981) of the population of the inner cloud relative to the outer is likely too high.

A second factor that complicates the determination of original orbits of long-period comets is that orbital fits that neglect nongravitational forces systematically overestimate the original semi-major axes. This effect is biggest for comets with very large semi-major axes and small perihelion distances (Marsden et al. 1973, 1978; Królikowska 2001, 2006). Thus the number of comets with accurate original orbits is surprisingly small. For example, Wiegert and Tremaine (1999) based their study of the evolution of long-period comets on the 289 such comets observed well enough (Quality Class I or II, Marsden et al. 1978) that original elements could be computed in the 1993 *Catalogue of Cometary Orbits* (Marsden and Williams 1993). (426 original and future orbits are given in Marsden and Williams 2008.) In light of these issues, a careful effort to derive original orbits

[8]The term "inner Oort Cloud" is now used for bodies like Sedna with perihelion distances outside the orbits of the planets and semi-major axes of hundreds of AU (as opposed to thousands of AU for the inner cloud discussed by Hills 1981). The existence of "Sednas" is not expected in models in which the Sun formed as an isolated star. See Sects. 5 and 6 for further discussion.

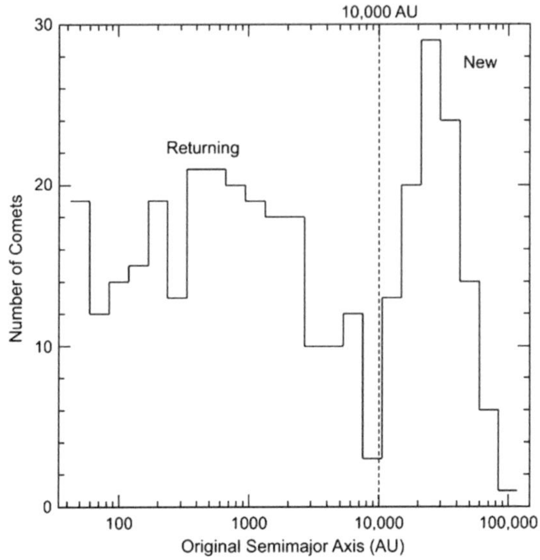

Fig. 7 Distribution of original semi-major axes, a_{orig}, of long-period comets from the *Catalogue of Cometary Orbits* (Marsden and Williams 2003). Comets are often classified as dynamically "new" or "returning," depending on whether a_{orig} is greater than or less than 10,000 AU, respectively. However, this classification is crude. To determine whether a particular comet is "new", it must be integrated backward one orbit, under the influence of the Sun and planets, galactic tides, and possibly nongravitational forces and nearby stars. According to the orbital integrations of Dybczyński and Królikowska (2015), most comets with $a_{\text{orig}} < 25{,}000$ AU are returning. In their sample, only comets with $a_{\text{orig}} > 50{,}000$ AU are certain to be new. Figure from Dones et al. (2004)

for long-period comets by Malgorzata Królikowska and her colleagues (Królikowska et al. 2014) [http://ssdp.cbk.waw.pl/LPCs/near_parabolic_comets_main.html] is of great importance. Thus far, they have determined original and future orbits for 119 comets (each calculated 250 AU from the Sun[9]) with a dynamical model including the eight planets, Pluto, and relativistic effects. For 45 of the comets, nongravitational force parameters were derived. For each of the 119 comets, 5001 "virtual comets" (i.e., "clones") with orbits within the uncertainties of the observations were integrated (Królikowska 2014). In addition, Dybczyński and Królikowska (2015) have studied the effect of the galactic potential and 90 nearby stars on the past and future orbits of a set of near-parabolic comets.

A histogram of the original orbital energies of long-period comets shows a spike of bodies with very large semi-major axes, a_{orig}, and a long tail of comets with smaller values of a_{orig} (see Fig. 7). The "Oort spike" is often taken to consist of comets with $a_{\text{orig}} > 10^4$ AU (or, equivalently, $1/a_{\text{orig}} < 100 \times 10^{-6}$ AU^{-1}). Comets in the Oort spike are often assumed to be dynamically "new," i.e., entering the planetary region for the first time since they were placed in the Oort Cloud (e.g., Oort 1950; Marsden et al. 1978). Since the estimated population of the cloud is proportional to the rate at which new comets become visible (Sect. 7.2), it is important to determine how many comets are really new. The work of Królikowska and Dybczyński (2010), Królikowska et al. (2014), and Dybczyński and Królikowska (2015) has

[9]Todorovic-Juchniewicz (1981) found that the orbit of a long-period comet can be approximated well by a barycentric orbit neglecting planetary perturbations beyond distances of 150 to 200 AU from the Sun. The Polish group therefore tabulates orbital elements before and after a comet traverses the planetary region at a distance of 250 AU.

shown that many comets with $a_{orig} > 10^4$ AU are *not* new. In their studies, most comets with a_{orig} between 10,000 and 25,000 AU ($1/a_{orig}$ between 40×10^{-6} and 100×10^{-6} AU^{-1}) are returning, so they cannot be considered pristine. According to their work, only comets with $a_{orig} > 50{,}000$ AU ($1/a_{orig} < 20 \times 10^{-6}$ AU^{-1}) are certain to be new.

No interstellar comet, i.e., a comet that was not bound to the Solar System before it entered the planetary region, has yet been discovered. Dybczyński and Królikowska (2015) state that C/2007 W1 (Boattini) is "the first serious candidate" for an interstellar comet, with an original orbital energy $1/a_{orig} = (-42.75 \pm 2.34) \times 10^{-6}$ AU^{-1}. This corresponds to a velocity at infinity $v_\infty = 0.2$ km/s. A true interstellar comet would be likely to have v_∞ of order 18 km/s, the Sun's velocity with respect to the Local Standard of Rest (McGlynn and Chapman 1989). Comets with small v_∞ will be gravitationally focused by the Sun, and thus there will be some bias favoring discovery of slower-moving comets (Moro-Martín et al. 2009). However, few stars (and hypothetical extrasolar comets) approach the Solar System with velocities $v_\infty \lesssim 10$ km/s (García-Sánchez et al. 2001; Bailer-Jones 2015). Slightly "hyperbolic" original orbits for comets are more likely an artifact of imperfect modeling of non-gravitational forces (Królikowska 2001) or omission of the effects of unknown stellar perturbers (Dybczyński and Królikowska 2015). Unfortunately, detection of interstellar comets is unlikely in the near future, even with the Large Synoptic Survey Telescope (Moro-Martín et al. 2009). However, the lack of hydrogen in the atmospheres of warm helium-atmosphere (DB) white dwarfs can place interesting upper limits on the abundance of interstellar comets that might have collided with the white dwarfs (Jura 2011).

2.2 Ecliptic Comets: Centaurs, Jupiter-Family Comets, and Encke-Type Comets

Though long-period and short-period comets manifestly have different orbital distributions, a number of studies in the 1970s investigated whether long-period comets might, in rare cases, evolve onto short-period orbits. For instance, Everhart (1972) showed that long-period (near-parabolic) comets could be "captured" onto short-period orbits if they lost energy through repeated encounters with Jupiter, particularly if they began with small inclinations and perihelion distances near Jupiter's orbit. However, Joss (1973) found that capture of long-period comets into short-period orbits was too inefficient to result in the observed number of short-period comets. (See the book by Bailey et al. 1990 for discussion of historical concepts of comet dynamics, including ideas about capture.) The difficulty of supplying short-period comets from the Oort Cloud inspired Fernández (1980) to investigate whether "a comet belt beyond Neptune" could instead be the source.

The definition of the subclasses of ecliptic comets has to do with dynamical coupling to Jupiter. Only the Jupiter-family comets (JFCs) can have close encounters with Jupiter, experiencing the resulting major orbital transformations, while Centaurs have perihelia beyond Jupiter's orbit, and Encke-types have aphelia far inside Jupiter's perihelion distance. In the orbits characteristic of each subclass one can find both comets and inactive objects referred to as asteroids or minor planets. For the Centaurs, this likely has to do with the perihelion distance, which is often too large to allow the usual cometary activity, and with relatively minor differences in surface composition.[10] However, in the bulk Centaurs are

[10]The activity of an individual Centaur may bear no clear relationship to its distance (r) from the Sun. A coma around Chiron was first detected in 1989 when Chiron was inbound at $r \sim 12$ AU. Activity decreased through perihelion at 8.5 AU in 1996 until 1999, and then increased in 2000 and 2001 when Chiron was 10–11 AU from the Sun (Duffard et al. 2002). Chiron appears to have been *more* active when it was near aphelion at 19 AU in the early 1970s (Bus et al. 2001). Many comets, including Centaurs such as Chiron and

thought to be as ice-rich as comets independent of whether they are able to show activity or not. The JFCs and Encke-types only comprise active objects, although the level of activity varies a lot (see Sect. 3), but their inactive counterparts (often called asteroids in cometary orbits, Kresák 1980; Hahn and Rickman 1985; Hartmann et al. 1987; Jewitt 1996; Weissman et al. 2002; Fernández et al. 2005; Kim et al. 2014; see Yan Fernández's list at http://www.physics.ucf.edu/~yfernandez/lowtj.html) may mostly be of cometary nature too.

In terms of orbital elements, the JFCs are mainly characterized by low inclinations (about 30° or less) and aphelion distances close to Jupiter's mean distance from the Sun (see Figs. 3 and 4). The latter is the historic reason for their name. In addition, an imprint of Jupiter's influence is seen in their distribution of arguments of perihelia (Quinn et al. 1990). These cluster near 0° and 180°, meaning that one node is usually close to Jupiter's orbit. This property makes them vulnerable to orbital changes through close encounters with the planet, and backward integrations of newly discovered JFCs frequently indicate such close encounters in the recent past, which often changed the orbits considerably (Carusi et al. 1985; Yeomans et al. 2005; Królikowska and Szutowicz 2006).

Tancredi (1995, 1998) finds that Jupiter-family comets typically have chaotic orbits due to close encounters with planets (especially Jupiter). The median Lyapunov time for the JFCs he investigated is $\tau = 60$ years. In other words, the uncertainty in the orbit of a typical JFC grows with time t at a rate proportional to $\exp(t/\tau)$. Even if the orbit of a JFC is known to seven significant digits, the uncertainty will be of order unity 1000 years later if no more observations are made. The divergence of possible orbits due to chaos is shown for comet 67P/Churyumov-Gerasimenko in Fig. 8. Dynamically, 67P is a typical JFC, with a wide range of possible orbits only hundreds of years ago (Królikowska 2003; Groussin et al. 2007; Maquet 2015; Guzzo and Lega 2015). The range of possible orbits becomes visible in the bottom panel of Fig. 8 prior to a distant encounter with Jupiter in 1923 (0.92 AU, or $\approx$2.7 Hill radii of Jupiter).

The namesake of the Encke-types, comet 2P/Encke, is famous for its small aphelion distance—only 4.1 AU. This makes the orbit inert to short-term changes due to Jupiter. Placing comets on Encke-like orbits appears to require the action of nongravitational forces (Steel and Asher 1996; Levison et al. 2006a). However, in the observational record there are indeed several objects (both comets and asteroids) occupying the range of aphelion distances from 2P/Encke to the usual JFCs. The Centaurs, when counting all of them independent of activity, are interesting in that their inclination distribution is mostly consistent with being JFC precursors in line with common opinion, but they also host some objects whose inclinations are much larger. This may mean a bimodal origin and dynamical history (see Sect. 5.5).

2.3 Main-Belt Comets/Active Asteroids

The third, and most recently discovered, reservoir of comets is the asteroid belt (Bertini 2011; Jewitt et al. 2015). The story begins with an unexceptional outer-belt asteroid, 1979 OW_7, that was first seen in 1979. Sixteen years later, comet P/1996 N2 was discovered by Elst and Pizarro. It was quickly realized that the newly discovered comet was likely to be the same object as 1979 OW_7, observed three revolutions of the Sun later. This body is now known both as asteroid (7968) and as comet 133P/Elst-Pizarro. Its semi-major axis,

29P/Schwassmann-Wachmann 1, undergo outbursts in which they brighten by several magnitudes or more (Gronkowski and Wesołowski 2015). Chiron probably was in outburst near its last aphelion passage (Prialnik et al. 2004).

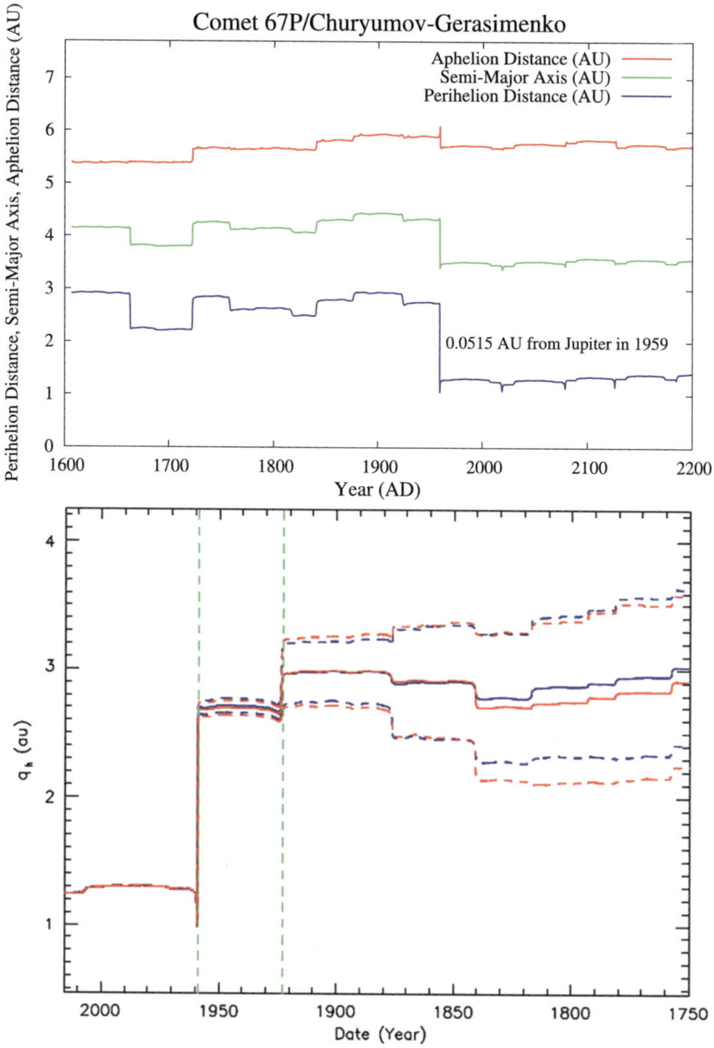

Fig. 8 The semi-major axis (*green*), perihelion distance (*blue*), and aphelion distance (*red*) of comet 67P from JPL/Horizons (http://ssd.jpl.nasa.gov/horizons.cgi) from 1599 to 2201 (top panel) and the perihelion distance of 67P (*red* and *blue curves*) from Maquet (2015) from 2015 back to 1740 (bottom panel). Both panels show that 67P's perihelion distance was reduced from 2.7 AU to 1.3 AU due to a close encounter to within 0.05 AU of Jupiter in 1959. The top panel shows 67P's nominal orbit from a numerical integration including a model for nongravitational forces. The bottom panel shows *vertical dashed green lines* showing the time of the close encounter with Jupiter in 1959 and a more distant encounter (0.92 AU) in 1923. The *blue lines* show the mean ± 1 standard deviation values of the perihelion distance for an ensemble of 1000 orbits, including nongravitational forces, calculated by Maquet (2015). The *red lines* show the perihelion distance for another 1000 orbits without nongravitational forces. The orbit of 67P is chaotic, which manifests itself as a large uncertainty in the comet's perihelion distance (and other orbital elements) prior to 1923

perihelion distance, and orbital inclination are 3.16 AU, 2.65 AU, and 1.4°, respectively (see Figs. 3–5), placing it within the family associated with Themis, a large (200-km-diameter) C-type asteroid (Toth 2000; Nesvorný et al. 2015). When the dust tail of Elst-Pizarro was first seen in 1996, the comet was slightly past perihelion. It was first proposed that the tail could have resulted from an impact with a tiny asteroid. Further analysis indicated that dust had been released for a period of weeks or months before discovery, inconsistent with a single impact, and that 133P rotated rapidly with a period of 3.5 hours (Hsieh et al. 2004). Furthermore, Elst-Pizarro displayed a tail again in 2002 and 2007, near the times of its next two perihelion passages (Hsieh et al. 2004, 2010). This left two possibilities—that 133P was a barely active former JFC that had recently evolved into a orbit in the asteroid belt through nongravitational forces, or that Elst-Pizarro had resided in the outer belt for billions of years but still managed to release small amounts of dust. In the latter case, mass loss might either be due to sublimating near-surface ice (as for a typical comet) or through some mechanism related to 133P's rapid rotation that did not require the presence of ice.

More main-belt comets (MBCs) have been discovered since 2005. Several groups have examined vast numbers of asteroids for cometary activity (Hsieh 2009; Gilbert and Wiegert 2009, 2010; Waszczak et al. 2013; Hsieh et al. 2015). These surveys suggest that there are, at most, some 100 main-belt cometary nuclei with diameters $d > 1$ km. This upper limit is roughly a factor of 10 smaller than estimates of the population of active JFCs of this size (Di Sisto et al. 2009; Brasser and Wang 2015). Yan Fernández lists ten main-belt comets (although he notes that the mechanism causing them to lose mass is not known), as well three "[l]ikely impacted or spun-up asteroids," P/2010 A2 LINEAR, (596) Scheila, and P/2012 F5 Gibbs (http://www.physics.ucf.edu/~yfernandez/cometlist.html#mbc). In HST images, P/2010 A2 shows a point-like nucleus and an X-shaped structure within its dust trail (Jewitt et al. 2010). Jewitt et al. (2010) interpret the observations in terms of either an impact into a previously unknown small asteroid ($d \approx 0.12$ km), or as the result of "rotational bursting" due to spin-up via the YORP effect (Vokrouhlický et al. 2015). Similarly, Jewitt et al. (2011) find that a dust outburst from the 113-km asteroid (596) Scheila was likely due to an impact. Jewitt (2012) introduced the term "active asteroids" for all MBCs, on the grounds that calling them "comets" can be misinterpreted "to mean that the active objects are necessarily icy. In most cases, we don't know if they are icy. In some cases we know that they are not" (http://www2.ess.ucla.edu/~jewitt/mbc.html). Jewitt et al. (2015) states that the best candidates for ice sublimation among active asteroids are (1) Ceres (see below), 133P/Elst-Pizarro, 238P/Read (Hsieh et al. 2011), and 313P/Gibbs (Jewitt et al. 2015).

The first main-belt body on which ice and organics were detected was (24) Themis, the parent body of the collisional family that contains the first two known MBCs, 133P/Elst-Pizarro and 176P/LINEAR (Rivkin and Emery 2010; Campins et al. 2010). The first definitive detection of *outgassing* from a main-belt body was from Herschel observations of (1) Ceres, the largest asteroid. Küppers et al. (2014) found that Ceres emits 6 kg/s of water vapor, which requires only 0.6 km^2 of exposed ice on its surface. These findings suggest that some asteroids, particularly in the outer belt, have retained subsurface age for the age of the Solar System and become weakly active (Prialnik and Rosenberg 2009)—typically only near perihelion, when the predicted sublimation rate is much higher. In some of these cases, impacts by small asteroids may have triggered the onset of activity (Hsieh et al. 2015). Other MBCs probably contain no ice at all, with their dust trails produced by collisions between asteroids (e.g., for the inner-belt object P/2010 A2 (LINEAR)), spin-up (as for P/2013 R3 Catalina-PANSTARRS) (Jewitt et al. 2014; Hirabayashi et al. 2014), or other mechanisms. As an extreme example, Jewitt and Li (2010) describe asteroid (3200) Phaethon, the parent of the Geminid meteoroids, as a "rock comet" which releases dust when it reaches temperatures of $\sim$1000 K near perihelion ($q = 0.14$ AU). See Jewitt (2013) and Jewitt et al. (2015) for further discussion.

3 Physical Properties of Comets

Comet nuclei are notoriously difficult to observe, and this has hampered ground-based exploration for a long time. The concept of a solid nucleus was championed in the classic paper by Whipple (1950) but was long regarded with skepticism by some scientists. After all, no nucleus had in fact been observed, and the evidence was indirect. The non-gravitational effects in the orbital motions of periodic comets offered the most convincing case for solid nuclei with a sublimating ice component, giving rise to asymmetric outgassing and an associated jet force (Whipple 1950).

The lack of resolved images of comet nuclei did not prevent theorists from proposing models for the structure of these objects and the workings of the non-gravitational force. Thus, for instance, it was realized that the dust observed in the comae and tails of comets might get concentrated at the very surfaces of comet nuclei, since the pressure of the sublimating gas might not be able to lift the larger grains into space against the tiny gravity of the nuclei. This led to the concept of *dust mantles* (e.g., Brin and Mendis 1979). On a different theme, from a comparison of the vapor pressure curves of cosmochemically important substances with the typical activity of comets as a function of heliocentric distance (Hübner 1965; Delsemme and Miller 1971), it was concluded that H_2O should be the dominant volatile constituent of comet nuclei. Hence, early models of the non-gravitational force (e.g., Marsden et al. 1973) were based on a theoretically derived dependence of the sublimation rate of H_2O ice on heliocentric distance.

Other developments concerned the structure that comet nuclei were expected to have if the nuclei were identified with icy planetesimals originating from beyond the snow line in the solar nebula (Donn and Rahe 1982). However, all such models needed the test of observations that had yet to be made. Those had to await the 1986 apparition of comet 1P/Halley and the associated observing campaigns, including a fleet of spacecraft that flew by the comet near Halley's perihelion passage. This was the first revolution in comet science after the historic year 1950, when both the Oort Cloud (Oort 1950) and the solid nucleus (Whipple 1950) were introduced. Since then there has been further rapid development due to both ground-based and space-based observations, and we shall now focus on a few of the most important topics.

3.1 Density and Porosity

Applying a thermal model for the surface layers of a rotating comet nucleus (Rickman and Froeschlé 1983) to describe the pattern of H_2O outflow and calculate the momentum transfer efficiency of outgassing that regulates the jet force, Rickman (1986, 1989) estimated the mass of the nucleus of 1P/Halley from the observed perihelion delay using Gauss's equations. The volume of the nucleus, as established from the Giotto imaging (Keller et al. 1986), together with this mass, indicated a surprisingly low bulk density. Uncertainties over the asymmetry of H_2O production about the perihelion passage implied a wide error bar on the result, stated as 0.28–0.65 g/cm^3, but this whole range is considerably less than 1 g/cm^3 (although see Peale 1989). Similar results were obtained using the same method by Sagdeev et al. (1987). Recent developments of the method have led to density estimates of 0.18–0.30 g/cm^3 for 19P/Borrelly (Davidsson and Gutiérrez 2004), <0.5 g/cm^3 for 67P/Churyumov-Gerasimenko (Davidsson and Gutiérrez 2005), <0.6–0.8 g/cm^3 for 81P/Wild 2 (Davidsson and Gutiérrez 2006) and 0.45 ± 0.25 g/cm^3 for 9P/Tempel 1 (Davidsson et al. 2007). Independent support for a low density, albeit with a large uncertainty, came from the analysis of the ejecta plume from the nucleus of 9P/Tempel 1 created by the Deep Impact mission (Richardson et al. 2007). Yet another method, numerical modeling of the clumping of the fragments of Shoemaker-Levy 9 after its tidal disruption, yields a pre-disruption density of 0.3–0.4 g/cm^3 (Movshovitz et al. 2012), smaller than the initial determination of $\approx$0.6 g/cm^3 by Asphaug and Benz (1996).

The first direct determination of a comet bulk density comes from the Rosetta exploration of 67P/Churyumov-Gerasimenko (the mass from the Radio Science Investigation, and the volume from OSIRIS imaging). The initial result 0.470 ± 0.045 g/cm^3 (Sierks et al. 2015) has been adjusted to 0.515 ± 0.020 g/cm^3 after the previously unseen southern hemisphere was imaged (Jorda et al. 2015). The bulk porosity implied by this density is of obvious interest. The dust/gas ratio in comet comae had been estimated from observed gas and dust production rates, but the latter suffered from an incorrect assumption about the size distribution of the grains. At radii less than a few microns, a steep slope (i.e., few large grains relative to smaller grains) had been correctly deduced from visual and near-IR observations (Hanner 1984). However, extrapolating this to much larger sizes would prove to be inaccurate in view of the abundance of large grains identified by the Halley flybys (McDonnell et al. 1986; Simpson et al. 1987). Thus, the typically estimated dust/gas ratios in comets increased from $\sim$0.1 to $\sim$1, in agreement with the "interstellar" model of cometary material (Greenberg 1987).

The compact density of a 1:1 mix of ice and cometary refractories—the latter including a significant fraction of organic materials (Kissel et al. 1986)—would be about 1.5 g/cm^3. Hence, the Halley results indicated a bulk porosity of $\sim$70 % for the nucleus. Similar estimates for short-period comets in general were obtained by Rickman et al. (1987) using observed non-gravitational effects and light curves. A bulk porosity of this order is consistent with accretion speeds up to about 20 m/s for objects of cometary size (Davidsson et al. 2015), and this is in turn consistent with either pebble swarm collapse (Johansen et al. 2015a) or hierarchical accretion (Weidenschilling 2008) as the mode of formation (Rickman et al. 2015). Jutzi and Asphaug (2015) show that impacts between similar-sized, porous bodies at speeds up to $\approx 2 v_{\mathrm{vesc}}$, where the mutual escape velocity v_{vesc} of 67P's lobes is 0.7 m/s, can result in bilobed shapes as seen for comet Churyumov-Gerasimenko. If the two lobes came together very early on, it is not clear whether they could survive later impacts.

The bulk porosity of a comet does not necessarily constrain the local porosity of the surface layer of a comet nucleus, and radar observations have indicated a much lower porosity for those layers in some comets (Harmon et al. 2004; Lamy et al. 2015). Bulk porosities also do not reveal the scale of the porosity. Kofman et al. (2015) estimate the porosity of the head region of 67P to be 75–85 %, similar to the 70–80 % estimated for the nucleus as a whole (Sierks et al. 2015). Active pits on 67P imply that the nucleus is heterogeneous within the top few hundred meters below the surface (Vincent et al. 2015). The large microporosity predicted by the interstellar model (Greenberg and Hage 1990) would account for the observed comet density, but a significant macroporosity as predicted by the primordial rubble-pile model (Weissman 1986) may also be present. The nature of the cometary porosity is important for the transport of heat and vapor inside the nuclei, and thus for issues regarding how chemically pristine comets are and for interpretation of observed volatile abundances in comets (Prialnik et al. 2004; Bockelée-Morvan et al. 2004; Mandt et al. 2015).

3.2 Size and Activity

Determining the sizes of comet nuclei is a matter of relevant instrumentation, which historically was not available. Thus, in old times, the concept of typically km-sized objects was based on expectation from observed gas production rates and non-gravitational effects.

However, the first direct results on individual nucleus sizes that did not involve assumptions about the outgassing led to an important change in the picture. The close-up imaging of the 1P/Halley nucleus in 1986, combined with ground-based photometry, revealed that this object was larger and darker than had been thought. This, in turn, led to another interesting insight—the water production rate was much less than would have been the case if the whole surface had been able to sublimate freely into space.

The fact that comets in general produce much less gas per unit surface area of their nuclei than free sublimation of ice would yield has been verified by subsequent observations. This largely concerns short-period comets for logistical reasons (ease of advance planning; availability of ephemerides) and means that the Jupiter-family comets have been well explored. The emerging picture is one of very dark objects (typical geometric albedo $p_V \sim 4\ \%$), while the level of outgassing varies quite a lot between individual comets (Lamy et al. 2004). Among the least active nuclei in the Jupiter family are those of 28P/Neujmin 1 and 49P/Arend-Rigaux, with outgassing rates of no more than 0.1 % relative to free sublimation near their perihelia, while at the other extreme some comets—notably 21P/Giacobini-Zinner and 46P/Wirtanen—produce water at about 100 % or more with respect to free sublimation from the whole nucleus. Note that a value larger than 100 % indicates either that the simple free sublimation model is inadequate to describe the outgassing, or that water is produced in large quantities from icy grains or boulders in the region surrounding the nucleus. A case in point is the small comet 103P/Hartley 2 (Meech et al. 2011).

One may ask if these properties are intrinsic to the respective nuclei or if they vary with time for each particular comet. There seems to be some truth in both possibilities. On the one hand, there is clear evidence that some Jupiter-family comets have undergone periods of dormancy (Kresák 1987), when their activity was much lower than otherwise so that they passed unobserved, and others have shown jumps in gas production rate between subsequent perihelia without any significant perturbation of the orbit in between. Moreover, Rickman et al. (1991) found evidence from the statistics of non-gravitational effects in Jupiter-family comets that JFCs become particularly active after major decreases of the perihelion distance, whereupon the activity subsides on a time scale of about ten orbital revolutions. On the other hand, there also appears to be a trend for the largest nuclei to be the most inactive and for hyperactivity to occur among the smallest nuclei (Tancredi et al. 2006).

Another fundamental question about comet activity is whether it is determined by the relative size of active spots or of the extent to which the outgassing is quenched due to the depth of the source layer. Here, again, the answer seems ambiguous. On the one hand, the Deep Impact exploration of the 9P/Tempel 1 nucleus revealed that the amount of spectrally identifiable H_2O ice at the surface was far too small to account for the observed H_2O outgassing from this comet (Sunshine et al. 2006). Hence, active spots in the sense of freely sublimating ice cannot be important for its activity. However, the concept of active regions in the sense of the outgassing being confined to certain latitudes or otherwise restricted terrains is definitely of great relevance for explaining the asymmetric gas production curves or light curves of comets. By this we understand that the outgassing rate peaks either before or after perihelion passage, although the global insolation of course peaks at perihelion. This may mean that the substance driving the activity (usually, H_2O ice) is closest to the surface in some specific region on the nucleus.

Thus, the lack of activity may be explained by the burial of the active substance under an inert mantle. Several theories for the growth of such a quenching and insulating layer (usually called a dust mantle) have been proposed (Brin and Mendis 1979; Rickman et al. 1990; Kuehrt and Keller 1994). However, the presence of hyperactivity needs a different explanation. The most obvious idea is that there is a distributed source of water molecules in the coma so that only part of the observed H_2O production originates from the nucleus. This was proposed to explain the high H_2O production rate of comet 46P/Wirtanen as a consequence of the production of icy grains due to fragmentation of the nucleus surface layers near perihelion passage (Rickman and Jorda 1998).

The close approach of the dynamically new long-period comet C/2013 A1 Siding Spring to Mars in October 2014 allowed unique observations, including the first resolved images (138 m/pixel) of the nucleus of a comet arriving directly from the Oort Cloud (http://www.uahirise.org/releases/siding-spring/). The Mars Reconnaissance Orbiter HiRISE team is determining the best way to remove residual coma from their images in order to place limits on the size of the comet's nucleus (Alfred McEwen and Alan Delamere, personal communication, 2015). Oort and Schmidt (1951) found that new comets are typically dustier and have a shallower trend of brightness with distance from the Sun (i.e., a smaller value of the photometric index ν, Eq. (4)) than "old" (returning) comets. Although their idea that new comets undergo "the quick loss of a frosting of super-volatile materials" (Whipple 1978) is questionable (Wiegert and Tremaine 1999), the relationship between the brightnesses and size of long-period comet nuclei remains unknown. The HiRISE results could thus provide important clues for issues such as the mass of the Oort Cloud.

3.3 Population Properties

Fragmentation is another process that plays a role in the physical evolution of comet nuclei. In 1846 one of the first known short-period comets, 3D/Biela, was observed to split into several major fragments. The comet reappeared in 1852 but was never seen again. Its debris is still observed annually as the Andromedid meteors. Thus, a major splitting event may mark the end of a comet's lifetime. However, there is also the case of comets 42P/Neujmin 3 and 53P/Van Biesbroeck, which have a common origin in the splitting of a precursor comet in 1846 (Carusi et al. 1984), but have both appeared many times since, including in 2015. This shows that, as far as comet lifetimes and end states are concerned, major splitting is a two-edged sword.

In any case, minor splits have been observed frequently enough (Chen and Jewitt 1994) to qualify as a potentially very important mass loss mechanism—thus limiting the lifetimes of comets. A recent study of comet evolution by Di Sisto et al. (2009) included both H_2O sublimation and splitting events as erosional mechanisms, finding that the latter are likely to dominate and effectively set the number of orbits that Jupiter-family comets may survive to values much less than sublimation alone would imply. Such results warrant close attention because of their repercussions on the capture rate from the Scattered Disk into the Jupiter family that will keep the latter in a steady state (Brasser and Wang 2015).

Obviously, the size distribution of the Jupiter-family comet nuclei is of great importance, because it is affected by the evolutionary processes that have shaped those nuclei. It also reflects the original size distribution of the nuclei before they were captured into the inner Solar System and thus carries information about the shape of the size distribution of the unobserved, smaller members of the Scattered Disk.

Many papers have dealt with the determination of a power-law fit to the cumulative size distribution $N(D)$ of the diameters, D, of the nuclei of Jupiter-family comets. However, there is no clear consensus from these investigations. A power-law index near 2.0 (i.e., $N(D) \propto D^{-2.0}$) was favored by Snodgrass et al. (2011), Weiler et al. (2011) and Fernández et al. (2013), while Solontoi et al. (2012) found a value of 2.4 and Tancredi et al. (2006) obtained 2.7. The reported error bars amounted to 0.2 to 0.3. A recent analysis by Belton (2015) found 2.2 by a straightforward collection of data and 2.8 by applying a correction

(Meech et al. 2004) for discovery bias. Craters on young surfaces on the regular satellites of Jupiter and Saturn (e.g., Europa and parts of Enceladus) indicate an index of about 1 for sub-km impactors (Zahnle et al. 2003; Dones et al. 2009; Bierhaus et al. 2012). Exactly how these results relate to the size distribution of the source population in the Scattered Disk remains to be convincingly shown. It also remains to be determined whether the latter distribution is compatible with an origin in a collisional cascade (Farinella and Davis 1996) or with a primordial mechanism of planetesimal formation (Weidenschilling 2008; Wahlberg Jansson and Johansen 2014).

3.4 Isotopic Ratios

Isotope ratios in comets were first used to demonstrate that comets are true members of the Solar System rather than being captured from interstellar space. In this regard, it was an important finding from the Giotto Neutral Mass Spectrometer (Eberhardt et al. 1995) that the oxygen ratio $^{16}O/^{18}O$ in water from comet 1P/Halley is identical to the terrestrial value (close to 500). Likewise, measurements of $^{12}C/^{13}C$ in cometary CN proved consistently to be close to the Solar System ratio of 89 (Wyckoff et al. 2000).

More recently, attention has focused on the D/H ratio and how it compares with the Vienna Standard Mean Ocean Water (VSMOW) value 1.56×10^{-4} representing the Earth's ocean, since this has an obvious bearing on the cometary contribution to the terrestrial H_2O inventory. The number of comets for which this ratio has been determined is still small, and as a result, opinions have shifted rapidly. A major issue has been whether Oort Cloud comets and ecliptic comets cluster around different values. In particular, comets C/1995 O1 (Hale-Bopp) and C/1996 B2 (Hyakutake), as well as 1P/Halley, which probably also originated in the Oort Cloud, showed internally consistent values of $D/H \approx 3 \times 10^{-4}$, about twice the VSMOW value. Oort Cloud comet C/2009 P1 (Garradd) has a smaller D/H value, $(2.06 \pm 0.22) \times 10^{-4}$, but it is still about 2σ above the D/H of SMOW (Bockelée-Morvan et al. 2012). Evidence from JFCs had to wait until the Herschel observations of 103P/Hartley 2 (Hartogh et al. 2011), which unexpectedly showed a terrestrial D/H value, $(1.61 \pm 0.24) \times 10^{-4}$. Lis et al. (2013) used Herschel to place a 3σ upper limit of $D/H < 2.0 \times 10^{-4}$ for Jupiter-family comet 45P/Honda-Mrkos-Pajdušáková. However, Rosetta measurements in comet 67P/Churyumov-Gerasimenko (Altwegg et al. 2015) yielded $D/H = (5.3 \pm 0.7) \times 10^{-4}$, about three times VSMOW.

Thus, it now appears that no systematic difference can be found between the nearly-isotropic and ecliptic comets, but that there are significant variations overall and within each group. Before the importance of the migration of the giant planets was appreciated, the nearly-isotropic and ecliptic comets were thought to have formed in largely distinct regions. Broadly speaking, the nearly-isotropic comets would have formed in the region of the giant planets, while the ecliptic comets would have formed in the Kuiper Belt. In this picture, the formation of ecliptic comets further from the Sun might have revealed itself as a higher D/H value (Hartogh et al. 2011). However, the overlap of D/H values for nearly-isotropic and ecliptic comets suggests both types of comet formed within the same broad region, as predicted by Brasser and Morbidelli (2013). The proportions of comets formed at different distances from the Sun are not yet clear.

4 Planetary Migration

Recognition that planetary migration was not a smooth process has profound implications for where we find small bodies today and where they have been in the past. Broadband photometry, spectroscopy, isotope ratios, and ortho-para ratios (spin temperatures) are among the ways that observers have attempted to constrain where cometary nuclei formed (see, e.g., Mumma and Charnley 2011). The materials incorporated into comets undergo chemical processing and isotopic fractionation in both the Sun's parent molecular cloud and in the solar nebula (Visser et al. 2009, 2011; van Dishoeck et al. 2014; Willacy et al. 2015; Mandt et al. 2015). The Stardust mission to Jupiter-family comet 81P/Wild 2 revealed the presence of crystalline silicate grains that require high temperatures to form, implying tiny refractory particles were transported from the inner Solar System to beyond the snow line, where they were incorporated into cometary nuclei (Brownlee et al. 2006; McKeegan et al. 2006; Nakamura et al. 2008; Brownlee 2014). Models of planetary migration show that many planetesimals (e.g., cometary nuclei, Trojans, Kuiper Belt Objects) too moved both inward and outward from the Sun as they interacted gravitationally with the giant planets.

Specifically, the giant planets may have undergone two episodes of migration. The first would have taken place during or shortly after giant planet formation, when gas was still present in the protoplanetary disk. The second would have occurred later (possibly hundreds of millions of years later) due to interactions with a remnant disk of planetesimals, i.e., a massive ancestor of the Kuiper Belt. In Sects. 4.1 and 4.3–4.6 we discuss the Grand Tack and the Nice model, which try to determine the orbital evolution of the planets and minor planets (asteroids and comets) during these two phases of Solar System history. The implications of one or both of these episodes of migration for the nature of comets are discussed in the chapters by Bockelée-Morvan et al. (2015), Gudipati et al. (2015), Guilbert-Lepoutre et al. (2015), and Mandt et al. (2015).

4.1 The Grand Tack

The existence of "hot Jupiters" orbiting other stars strongly suggests that jovian-mass planets can migrate significant distances via gravitational interactions with the gas disk from which the planets formed. (See Seager 2010; Howard 2013; Lissauer et al. 2014; Winn and Fabrycky 2015; and Laughlin and Lissauer 2015 for general reviews of extrasolar planets, and Baruteau et al. 2014 and Davies et al. 2014 for reviews of disk-planet interactions in planetary systems.) In the case of massive planets, this likely occurs through Type II migration, in which the gravity of the planet in question has opened up a gap in the disk (Papaloizou and Lin 1984). Although Type II migration proceeds at a much slower pace than Type I (pre-gap-opening) migration, simulations show that migration is consistently inward in the case of an isolated planet (e.g., Ward 1986; Lin and Papaloizou 1986). However, Masset and Snellgrove (2001) found that with two planets simultaneously migrating in a gas disk, they can get locked into resonance and the direction of migration can be reversed. Morbidelli and Crida (2007) confirmed this result and showed that such behavior could be expected for Jupiter and Saturn early in our Solar System's history. Essentially, Jupiter begins migrating inward until Saturn catches up with Jupiter and the two get locked into a 3:2 mean-motion resonance (MMR). In this configuration, the inner disk torques the resonant planetary configuration much more strongly than the torquing from the outer disk, and the planets begin migrating outward in resonance.

Initially, it was thought that Jupiter and Saturn could not have migrated outward very far in this resonant configuration, since this implies that Jupiter once orbited substantially closer to the asteroid belt and terrestrial planet region (Morbidelli and Crida 2007). However, the possibility of significant migration was explored in more depth in Walsh et al. (2011). This scenario, called the Grand Tack model, proposes that Jupiter initially begins migrating inward from just beyond the snow line near 4 AU (see Fig. 9). This migration essentially continues unabated until Jupiter reaches 1.5 AU, at which point Saturn becomes trapped in

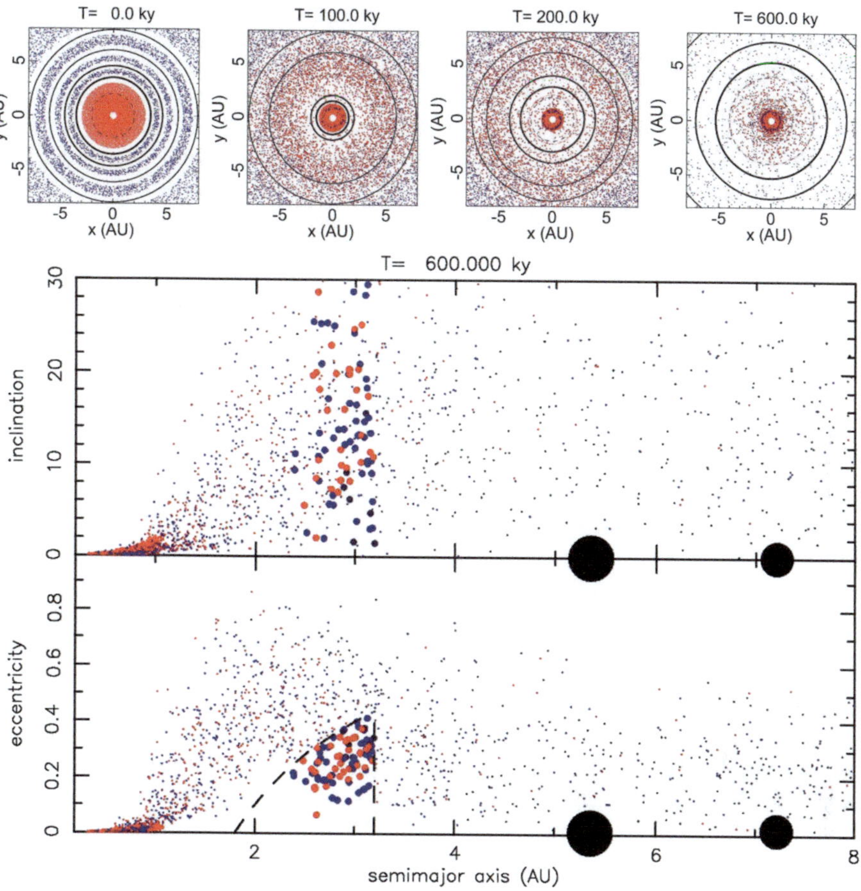

Fig. 9 Top: Snapshots of the reference simulation of the Grand Tack by Walsh et al. (2011), from http://www.boulder.swri.edu/~kwalsh/GrandTack.html. The four panels show the planetary orbits and asteroids at the starting time and 10^5, 2×10^5, and 6×10^5 years into the simulation. The *red dots* are planetesimals ("asteroids") that begin less than 3 AU from the Sun, and are proxies for S-type asteroids. The *blue dots* begin between the orbits of the giant planets (3.5–8 AU from the Sun) and outside them (8–13 AU), and are proxies for C-type asteroids. At the beginning of the simulation, Jupiter is assumed to have formed at 3.5 AU with its present-day mass. Saturn's 30 $M_\oplus$ core is initially at ~ 4.5 AU and grows to 60 $M_\oplus$ in 10^5 years as Jupiter migrates inward. Saturn's core is assumed to remain at 4.5 AU during this phase. Similarly, the cores of Uranus and Neptune begin about 6 and 8 AU from the Sun, respectively, and grow from initial masses of 5 $M_\oplus$ without migrating. Once Saturn reaches 60 $M_\oplus$ its inward migration begins, and is much faster than that of the fully-grown Jupiter. When Saturn catches Jupiter, Saturn is trapped in the 3:2 resonance; this is assumed to occur when Jupiter is at 1.5 AU. The direction of migration then reverses, and the two planets migrate outward together. Jupiter and Saturn capture Uranus and Neptune into resonance and these planets are then pushed outwards as well. Saturn, Uranus and Neptune reach their full masses at the end of the migration when Jupiter reaches 5.4 AU. The final orbital configuration of the Grand Tack simulations with the giant planets in resonance with each other and Saturn, Uranus, and Neptune interior to their current locations provides the initial conditions for the later events in recent versions of the Nice model (Sects. 4.4–4.6). Bottom: The orbital inclinations (top) and eccentricities (bottom) of the planetesimals remaining following the Grand Tack, from Walsh et al. (2012). The *black dots* indicate Jupiter and Saturn. The *dotted line* in the eccentricity plot roughly outlines the current asteroid belt. Planetesimals within the asteroid belt boundaries are enlarged in both the inclination and eccentricity plots. The *red dots*, representing S-type asteroids, dominate the inner belt, while the *blue dots*, representing C-type asteroids, dominate the outer belt

the 3:2 MMR with Jupiter and the planets reverse migration direction. This outward migration takes Jupiter out to $\sim$5 AU, at which point the migration is slowed by the dissipation of the gaseous component of the protoplanetary disk.

The work of Walsh et al. (2011) showed that a significant inward, then outward migration by Jupiter and Saturn provides an intriguing explanation for some curious features of the inner Solar System. First, terrestrial planet formation simulations consistently form planets exterior to Earth that are much more massive than the real Mars (Chambers 2001; Raymond et al. 2009), which has a mass $\approx 0.1\ M_\oplus$. Hansen (2009) demonstrated that this problem could be solved if the terrestrial planets formed from a narrow disk of material between 0.7 and 1.0 AU rather than a wide disk extending out to a few AU. However, these initial conditions were physically unmotivated in Hansen (2009).[11] In the Grand Tack model, Jupiter's inward migration truncates the outer edge of the disk of rock planetesimals, effectively generating the initial conditions suggested in Hansen (2009), and the terrestrial planets that eventually form near 1.5 AU in the Grand Tack model often have masses and eccentricities close to the values for Mars.

In addition, the Grand Tack model provides a novel explanation for a number of characteristics of the asteroid belt. First, the migration of Jupiter inward and then outward through the asteroid belt severely depletes its mass, and this explains why so little mass is found there today ($\lesssim 10^{-3}\ M_\oplus$). Secondly, the increased water content observed for bodies in the outer asteroid belt compared to the inner belt has traditionally been interpreted as marking the location of the snow line. For instance, Hayashi (1981) assumed ice would sublime at 170 K. In an optically thin solar nebula, this would place the snow line at 2.7 AU, in the middle of the main belt. However, the Grand Tack presents an alternative explanation. During Jupiter's initial inward migration, its gravity decimates most of the primordial asteroid belt. However, a tiny fraction of this rocky primordial material is scattered to larger semi-major axes. Once Jupiter begins migrating back out, it reimplants a small fraction of this material into the orbital space occupied by the observed inner asteroid belt. Also during the outward migration phase, the giant planets encounter small bodies in the primordial Kuiper Belt, as well as planetesimals originally orbiting between the giant planets. A small fraction of these are dynamically passed between the giant planets before being implanted in the region of the outer asteroid belt. This neatly explains why the observed inner asteroid belt is dominated by S-type asteroids (rocky material in the simulations), while the outer asteroid belt is observed to be dominated by C-type asteroids (distant icy planetesimals in the simulations). One consequence of the Grand Tack model is that C-type asteroids should be much more similar in composition to comets than to bodies in the inner asteroid belt.

The Grand Tack is newer than the Nice model (which we discuss in Sect. 4.3), and hence is not as fully developed. The "tack point," i.e., the innermost distance to which Jupiter migrates, is a free parameter in the model, and is taken to be 1.5 AU to explain the small mass of Mars. Raymond and Morbidelli (2014) address several criticisms of the Grand Tack. The first concerns the migration dynamics of the tack. The planets' degree of migration and rate of gas accretion are both proportional to the local gas disk surface density, so it is not clear how Saturn could ever catch up to Jupiter if the two planets formed and migrated in the same region of the Solar System. However, if Jupiter formed first, changing conditions of

[11] Motivated by a model for the formation of super-Earths, Izidoro et al. (2014) proposed that the surface mass density of solids could have declined very rapidly with distance in the region of the terrestrial planets and the asteroid belt. This model explains the small mass of Mars, but predicts asteroid eccentricities and, particularly, inclinations that are smaller than the observed values (Izidoro et al. 2015). Levison et al. (2015b) have proposed yet another model, a variant of their "viscous stirred pebble accretion" idea (Levison et al. 2015), to explain the small mass of Mars and the asteroid belt.

the gas disk could allow for a faster migration rate of Saturn (Bitsch et al. 2014). In addition, for Jupiter and Saturn to reverse migration direction (i.e., tack) their mass ratio needs to be between 2 and 4. (Jupiter's mass is 3.34 times that of Saturn.) This ratio must be maintained to place the planets near their current orbits, yet they presumably continue to grow as they migrate outward. Our understanding of giant planet growth remains limited, and it is not clear if this is a significant issue for the Grand Tack. Finally, Alexander et al. (2012) and Davis et al. (2014) point out significant isotopic differences between C-type asteroids and satellites of Saturn as well as Oort Cloud comets, even though the Grand Tack suggests that these populations should form in the same inter-giant planet regions.

However, the formation of Saturn's satellites in an circumplanetary disk may impart unique isotopic gradients on the satellites. Furthermore, Brasser and Morbidelli (2013) argue that most Oort Cloud comets originate in the trans-Neptunian region rather than among the giant planets. Thus, the significance of these isotopic discrepancies for the viability of the Grand Tack is debatable. Ongoing modeling of the formation of the terrestrial planets (Rubie et al. 2015; Kaib and Cowan 2015; Ogihara et al. 2015) and geochemical studies such as determination of the timing of aqueous alteration of chondritic asteroid parent bodies (Doyle et al. 2015) will help to determine the viability of the Grand Tack.

4.2 The Late Heavy Bombardment

When the Apollo and Luna missions returned samples of the Moon between 1969 and 1976, researchers noted the rarity of rocks older than 4 billion years (Gyr). The Moon is thought to have formed some 4.5 Gyr ago (e.g., Bottke et al. 2015). Tera et al. (1974) interpreted a peak in isotopic recrystallization ages of lunar rocks some 3.9 Gyr ago as a "terminal lunar cataclysm," a brief [compared with the 500 Myr or more since the Moon formed, Carlson et al. 2014] cratering episode that might have destroyed older rocks. Tera et al. (1974) noted "This cataclysm is associated with the Imbrium impact and very possibly the formation of Crisium and Orientale and possibly several other major basins in a narrow time interval ($\sim 2 \times 10^8$ yr or less)."

Most dynamical mechanisms that would cause a heavy bombardment on the Moon (e.g., some change in the planets' orbits that would destabilize the orbits of asteroids and/or comets) would produce a bombardment of the Earth and other terrestrial planets, and perhaps the entire Solar System (Wetherill 1975). The Nice model (Sect. 4.3), which was conceived three decades after the Moon landings, provided the first really plausible mechanism for producing a widespread late bombardment event. Planets, satellites, and small bodies would have undergone heavy cratering and, in some cases, global effects such as differentiation (Barr and Canup 2010), loss of ice through impact vaporization (Nimmo and Korycansky 2012), and catastrophic disruption (Movshovitz et al. 2015). In the Nice model, comets would strike the Earth and Moon at a median time of 3.7 Myr (90 % within $\approx$30 Myr) after the Jupiter-Saturn resonance crossing, while asteroids would do so at a median time of 10–50 Myr (90 % by $\approx$50–150 Myr) after the resonance crossing (Gomes et al. 2005b).

Unfortunately, the interpretation of the lunar record remains ambiguous (Baldwin 1974; Hartmann 1975, 2003; Ryder 1990; Haskin 1998; Haskin et al. 1998; Cohen et al. 2000; Hartmann et al. 2000; Levison et al. 2001; Stöffler and Ryder 2001; Neukum et al. 2001; Ivanov et al. 2002; Kring and Cohen 2002; Koeberl 2003; Strom et al. 2005, 2015; Bottke et al. 2007, 2012; Chapman et al. 2007; Ćuk et al. 2010, 2011; Malhotra and Strom 2011; Ćuk 2012; Marchi et al. 2012, 2013; Norman and Nemchin 2014). The Apollo and Luna missions only visited near-equatorial regions of the near side of the Moon. It was realized later that all the landing sites might be dominated by far-flung ejecta from the impacts that created the Imbrium, Serenitatis, and perhaps Nectaris basins (Spudis 1996; Spudis et al. 2011). The clustering of sample ages may not be very meaningful if it only reflects two or three events.

The reality of a lunar cataclysm depends heavily on when the Nectaris basin formed. In the chronology of Wilhelms et al. (1987), Imbrium formed 3.85 Gyr ago and Nectaris formed 3.92 Gyr ago, implying a "spike" of a dozen or so impact basin-forming events in a 70-Myr period. However, others suggest that Nectaris may be some 4.2 Gyr old (Fischer-Gödde and Becker 2011). Apollo 16 landed some 600 km northwest of the center of Nectaris and 1600 km southeast of Imbrium. Unfortunately, the provenance of the impact melts sampled by the Apollo 16 astronauts is ambiguous, and Cohen et al. (2015) estimate that "Imbrium and Serenitatis ejecta make up the majority of the basin impact melt at the Apollo 16 surface, but a significant amount of Nectaris melt should also be present." Recent studies suggest that basin formation on the Moon extended back to at least 4.1 or 4.2 Gyr ago, and the idea of a single cataclysm $\approx$3.9 Gyr ago has lost favor (Morbidelli et al. 2012; Marchi et al. 2014; Norman and Nemchin 2014; Hartmann 2014). For instance, Morbidelli et al. (2012) argue for a "sawtooth," i.e., a "moderate uptick" of impacts 4.1–4.2 Gyr ago, with the 2500 km South Pole-Aitken basin, the largest and oldest definite impact structure on the Moon, forming at least 4.3 Gyr ago. Hopkins and Mojzsis (2015), by contrast, find that the Moon suffered "... '[p]icket fence'-like multiple cataclysms," with an event recorded in zircons 4.24 Gyr ago possibly recording the South Pole-Aitken impact.

In short, the Nice model provides an intriguing mechanism to bring about a cataclysm, but it is not clear whether one or more spikes in impact flux occurred on the Moon and elsewhere in the Solar System. Sampling within the South Pole-Aitken basin will help to determine the basin's age (Hurwitz and Kring 2014). This constraint will be of great value in unraveling early lunar history, and potentially in determining when the giant planets migrated to their present-day configuration.

Regardless of the exact chronology of lunar bombardment, the idea that planetary systems can become dynamically unstable long after their planets form is now well-established (Levison et al. 1998; Laskar and Gastineau 2009; Batygin et al. 2015). Lisse et al. (2012) argue that the vast quantities of warm dust observed around the $\sim$1.4-Gyr old star η Corvi were produced by an "dynamical stirring of their parent Kuiper Belt," i.e., an event akin to that proposed in the Nice model. We now turn to the Nice model and its descendants.

4.3 The Nice Model

The smooth migration model of Malhotra (1993, 1995) [Sect. 1.3] in explaining the eccentricity of Pluto turned out to have two difficulties. The first is that the migration of the giant planets induced by the scattering of planetesimals leaves the planets with eccentricities that are lower than their current values; this is especially so for Jupiter and Saturn. Second, there was no inherent mechanism that caused the giant planets to begin migrating. These issues needed to be addressed in order to construct a consistent history of the evolution of the Solar System.

Thommes et al. (1999) proposed a scenario in which Uranus and Neptune grew as cores between Jupiter and Saturn, were scattered due to planet-planet gravitational interactions to near their current locations, and had their random velocities damped by interactions with a remnant planetesimal disk. This model avoided the problem of forming the ice giants at their current locations, with long orbital time scales and low (estimated) solid surface density at the time of formation (e.g., Levison and Stewart 2001). The work of Thommes et al. (1999) provided the first strong numerical evidence that the outer Solar System could

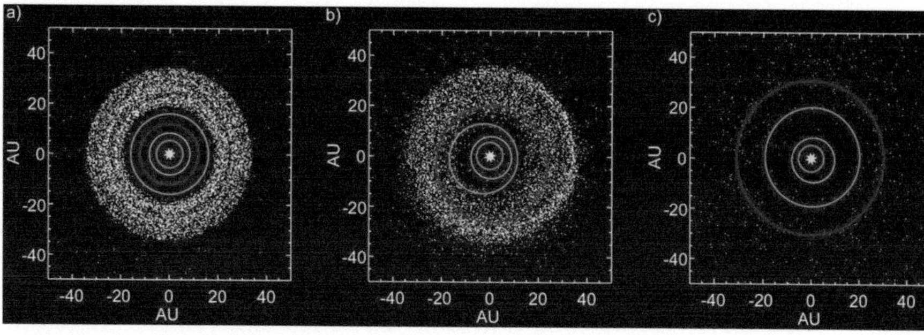

Fig. 10 Left panel: The orbits of Jupiter, Saturn, Uranus, Neptune, and the planetesimal disk (i.e., the massive ancestor of the Kuiper Belt) before Jupiter and Saturn cross their 2:1 resonance. in the Nice model of Tsiganis et al. (2005). Middle: After the resonance crossing, the orbits of Uranus (*light blue*) and Neptune (*dark blue*) become eccentric. Neptune crosses much of the "Kuiper Belt," and planetesimals are scattered throughout the Solar System. Right: The planets have now settled down into a stable configuration, and most planetesimals have been ejected from the Solar System by the planets. Also see the animation at http://www.skyandtelescope.com/sky-and-telescope-magazine/beyond-the-printed-page/chaos-in-the-early-solar-system/. Figure from https://commons.wikimedia.org/wiki/File:Lhborbits.png by Dr. Mark Booth (https://en.wikipedia.org/wiki/User:AstroMark). Licensed under the Creative Commons Attribution-Share Alike 3.0 Unported license

have undergone considerable rearrangement from the time that the giant planets achieved their current masses, and in a manner much more violent than suggested by Fernández and Ip (1984) and Malhotra (1993, 1995).

These developments led a group of researchers in the city of Nice, France to construct a new model for the dynamical evolution of the Solar System. The essence of the model is that the giant planets formed in a considerably more compact configuration, in which Saturn started just inside the 2:1 mean-motion resonance with Jupiter, and the ice giants Uranus and Neptune were located at 11–13 AU and 13.5–17 AU. The giant planets were surrounded by a massive planetesimal disk situated roughly between 15 and 30 AU with a total mass of about 35 $M_\oplus$ (Tsiganis et al. 2005). Gravitational interactions of planetesimals with Neptune caused the planets to migrate on a very long time scale, with the outer three planets gaining angular momentum while Jupiter lost angular momentum. The repartitioning of energy and angular momentum caused Jupiter and Saturn to cross their 2:1 mean-motion resonance. This crossing excited the eccentricities of the gas giants, which in turn forced the orbits of the ice giants to become eccentric and inclined. The system became violently unstable, with the gas giants scattering Uranus and Neptune to near their present locations (in some cases, after switching their order), and planetesimals flying all through the Solar System (Tsiganis et al. 2005). An illustration of this scenario, now called the "Nice model," is shown in Fig. 10. The Nice model has many nice properties, such as often placing the planets on their current orbits with the correct eccentricities, and (possibly) explaining the Late Heavy Bombardment (Gomes et al. 2005b; Sect. 4.2) and the broad inclination distribution of Jupiter's Trojan population (Morbidelli et al. 2005), while typically preserving the regular satellite populations during the planetary encounters (Tsiganis et al. 2005; Deienno et al. 2011).[12] At the same time that Trojans are captured in the Nice model, Hildas and other

[12]Deienno et al. (2011) find that the uranian satellites out to Oberon, the outermost regular moon, would likely survive encounters with another ice giant and massive planetesimals during the Nice model, but hypothetical moons only a bit further from Uranus would have been destabilized. Gomes et al. (2012) find that if Neptune had a close encounter with Jupiter, Triton would have a 65 % chance of remaining bound to Neptune. If

outer-belt asteroids can be captured (Levison et al. 2009). In their model, D- and P-type asteroids originated beyond 15 AU, while C-types are indigenous to the main belt (or were implanted very early on in the Grand Tack model, Walsh et al. 2011). Migration models to date do not predict the (rare) D-type asteroids found in the inner main belt (DeMeo et al. 2014; DeMeo and Carry 2014).

Even though the original Nice model enjoyed many successes, it shared a problem with the models of Malhotra (1993, 1995): there was no clear explanation for the initial conditions. In addition, the effect of the migration of the giant planets on the terrestrial planets was not discussed. This led to several follow-up models with refined initial conditions and evolution of the giant planets, which we discuss in the next few subsections.

4.4 The Nice 2 Model

Ideally, if one wanted to investigate a scenario like the Nice model in which the planetary orbits become unstable, one would start with the solar nebula, with its gas and dust. One would calculate how and where the planets formed, where they migrated while the nebula was present, and where any small bodies left over would reside. Of course, we can't yet perform such an ambitious feat, so the original Nice model makes some rather arbitrary assumptions. For example, it assumes that the giant planets were fully formed, but that small bodies just a little further from the Sun didn't grow into planets. Furthermore, the location of the inner edge of the planetesimal disk beyond the planets (the ancestor of today's Kuiper Belt) was chosen so that the system would go unstable after many hundreds of millions of years, to agree with the 600 Myr delay between the formation of the planets ($\sim$4.5 Gyr ago) and the purported lunar cataclysm ($\sim$3.9 Gyr ago). The inventors of the Nice model vowed to improve the model by starting with more realistic initial conditions.

The "Nice 2" model (Levison et al. 2011) started from simulations of how the giant planets would have moved around due to the presence of gas in the solar nebula (Morbidelli et al. 2007). The simulations of Morbidelli et al. (2007) indicated that the planets probably would have been packed even more tightly than in the original Nice model. Typically, the planets were in a "multi-resonant" configuration; for example, in the case that Levison et al. (2011) focused on, Jupiter is in a 3:2 resonance with Saturn, Saturn is in a 3:2 resonance with "Ice I" (i.e., Uranus or Neptune—the inner ice giant sometimes evolves to be the outer ice giant), and Ice I is in a 4:3 resonance with Ice II. (For brevity, we will use the notation "3:2 3:2 4:3" for a configuration like this.) The four giant planets begin with semi-major axes between 5.4 and 11.7 AU, as compared with 5.2–30.1 AU now. This tight configuration can be stable for a long time because of the resonances. As in the original Nice model, a disk of small bodies ("planetesimals") is assumed to be present outside the orbits of the planets, but the Nice 2 model assumes a bigger "gap" between the planets and the disk, and the disk is assumed to consist of Pluto-mass bodies. In the original Nice model, the disk was represented by massless "test particles". Levison et al. (2011) found that even if the "Plutos" didn't leak out of the disk, their gravitational interactions with each other would feed back into the planetary system, pushing it into an even tighter configuration and eventually causing an instability. A long delay between the formation of the planets and the onset of the instability was less dependent on fine-tuned initial conditions than in the

Uranus had a close encounter with Jupiter, all of the regular satellites would remain bound to Uranus about 60 % of the time. The orbital properties of the giant planet satellite systems can be used to constrain the nature of the planetary encounters that took place in the Nice model and its progeny such as the "jumping Jupiter" model (Deienno et al. 2014; Nesvorný et al. 2014; Cloutier et al. 2015).

original Nice model. In their simulations, Levison et al. (2011) find that "the giant planets become unstable after a time ranging between 300 Myr and 1 Gyr in ~25 % of the cases, for a wide range of disk masses and locations of the disk inner edge."[13] However, the Nice 2 model still did not address a fundamental problem: the orbital stability of the terrestrial planets. The stability of the terrestrial planets during a late episode of giant planet migration depends on the degree to which migration excites the terrestrial planets' eccentricities and inclinations. We now turn to that topic.

4.5 The Jumping Jupiter Model

The evolution of the eccentricities (e_i) and longitudes of perihelia (ϖ_i) of the planets in the Solar System can approximately be described by the Laplace–Lagrange solution of the secular equations of motion (e.g., Murray and Dermott 1999):

$$\begin{aligned} e_i \cos \varpi_i &= \sum_{k=1}^{8} M_{i,k} \cos \alpha_k \\ e_i \sin \varpi_i &= \sum_{k=1}^{8} M_{i,k} \sin \alpha_k, \end{aligned} \tag{1}$$

where the index i in Eq. (1) refers to the planet under consideration. Here the $M_{i,k}$ are coefficients that depend on the planets' masses and semi-major axes, and $\alpha_k = g_k t + \beta_k$. The g_k are called proper frequencies of the secular perihelion-eccentricity motion of the planets. The frequencies g_1 to g_4 are proper to the terrestrial planets and the frequencies g_5 to g_8 characterize the giant planets from Jupiter to Neptune. The frequencies g_k can be classified into two groups: a fast group with values exceeding 17″/yr, and a slow group with values below 8″/yr.

When Jupiter was closer to Saturn, the value of the g_5 frequency, which is proper to Jupiter, was higher than it is today, because the strength of the mutual perturbations depends only on the masses and the *ratio* of the semi-major axes. Figure 11 plots g_5 as a function of the period ratio between Saturn and Jupiter. The values of g_1, g_2, g_3, and g_4 are shown as horizontal lines. As Jupiter and Saturn drift apart, Jupiter's proper precession frequency decreases and crosses the proper frequencies associated with the terrestrial planets (Brasser et al. 2009; Agnor and Lin 2012). This is called a secular resonance crossing. Such a crossing can significantly modify the amplitudes of the proper modes of the terrestrial planets, i.e., their eccentricities.

The effect of these resonances was reported by Brasser et al. (2009), who studied the response of the terrestrial planets to late giant planet migration, and by Agnor and Lin (2012). Both Brasser et al. (2009) and Agnor and Lin (2012) demonstrated that the crossing of the resonances $g_5 = g_2$ and $g_5 = g_1$ caused the greatest eccentricity increase for Mercury, Venus and Earth because the resonances were crossed slowly. The crossings occur when the Saturn-to-Jupiter period ratio is 2.15 and 2.25, respectively (see Fig. 11). The typical migration timescale of Jupiter and Saturn due to planetesimal scattering is $\tau = 5$ Myr (Tiscareno and Malhotra 2003; Morbidelli et al. 2010). Such a slow migration causes the eccentricities of the terrestrial planets to far exceed their current values. Agnor and Lin (2012) were aware of the difficulty of keeping the terrestrial system stable as the gas giants migrated. For this

[13] Reyes-Ruiz et al. (2015) find shorter instability timescales (<70 Myr) using a different code (Mercury 6.5) to investigate this problem.

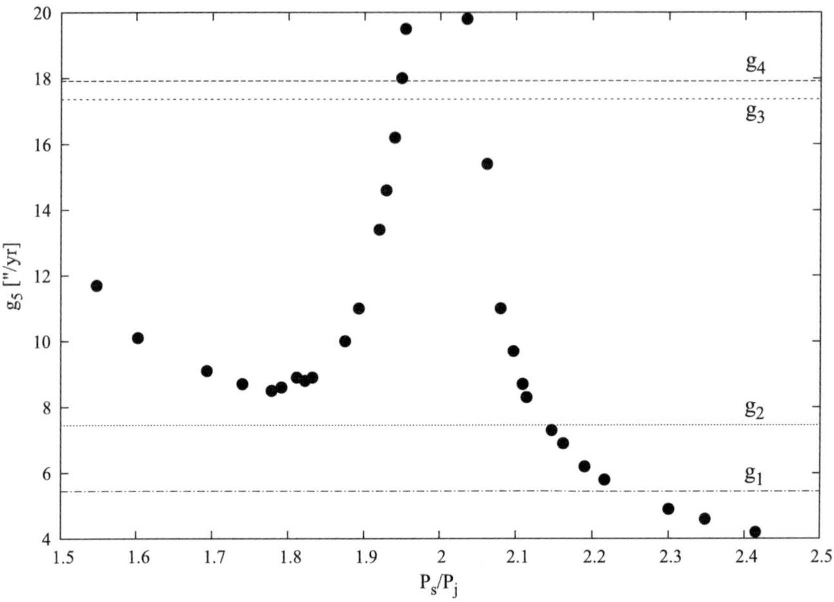

Fig. 11 Evolution of proper frequency g_5 as a function of the ratio of the orbital periods of Saturn and Jupiter. Figure from Brasser et al. (2009)

reason they advocated an early episode of giant planet migration, occurring before the terrestrial planets had fully formed. This epoch of giant planet migration would have occurred after the gas disk had dissipated, and thus it would have followed immediately after the Grand Tack had taken place (Walsh and Morbidelli 2011).

Brasser et al. (2009) suggested that the migration time could be reduced if Jupiter scattered one of the ice giants outwards. Energy and angular momentum conservation cause Jupiter to move inwards as it scatters the ice giant outwards, thereby increasing the period ratio with Saturn. This model, in which the typical time scale of Jupiter's migration through scattering in this way is less than 0.1 Myr, was dubbed the "jumping-Jupiter" scenario. It has been supported through work on the asteroid belt by Minton and Malhotra (2009), Morbidelli et al. (2010) and Walsh and Morbidelli (2011), who demonstrated that the migration of Jupiter had to be fast, whether this occurred late or right after the gas disk had dissipated. The encounters between Jupiter and an ice giant would be consistent with the capture of the irregular satellites. Indeed, Nesvorný et al. (2007) demonstrated that the orbital structure of the irregular satellites of Saturn, Uranus and Neptune is consistent with small body capture during planet-planet encounters. However, in their study they left open whether the same mechanism could have been responsible for Jupiter's irregular satellites, since they had no encounters between an ice giant and Jupiter in the simulations, which were based upon the original Nice model (Sect. 4.3). Now that it appears that such an encounter is required, it follows that most likely Jupiter captured its irregular satellite population in the same manner. Capture of jovian irregulars is natural in jumping-Jupiter scenarios (Nesvorný et al. 2014).

Further refinements and support for the jumping-Jupiter mechanism were presented in Agnor and Lin (2012) and Brasser et al. (2013). Agnor and Lin (2012) studied the origin and evolution of the Angular Momentum Deficit (AMD), a global measure of eccentricity and inclination (Laskar 1997), of the terrestrial planets. Agnor and Lin (2012) concluded that late migration of Jupiter and Saturn had to occur with an e-folding time scale $\tau \ll 1$ Myr, otherwise the AMD of the terrestrial planets would be incompatible with its current value. From numerical simulations Agnor and Lin (2012) find that the excitation of the eccentricities of the terrestrial planets scales as $\tau^{1/2}$ for large τ, but when $\tau \lesssim 0.1$ Myr the excitation imposed on the terrestrial planets is independent of τ because the excitation is impulsive rather than adiabatic. In other words, for values of τ shorter than 0.1 Myr the AMD increase of the terrestrial planets is independent of τ and is equal to a constant value. This led Brasser et al. (2013) to quantify the AMD increase in the terrestrial planets during a late episode of giant planet migration. They placed the terrestrial planets in various jumping-Jupiter giant planet migration simulations and concluded that the AMD increase in the terrestrial planets is typically comparable to the current value, provided that the orbits of the terrestrial planets initially had much lower eccentricities and inclinations than they do today.

In summary, an episode of late giant planet migration can only be made consistent with the orbits of the terrestrial planets and the structure of the asteroid belt if the migration was of the jumping Jupiter variety, i.e., if Jupiter scattered an ice giant and its recoil made it migrate inwards on a time scale shorter than 0.1 Myr. Unfortunately, the jumping-Jupiter scenario suffers from a problem that is difficult to overcome: the probability of successfully reproducing the current Solar System is less than 5 %. This posed difficulties in advocating the jumping-Jupiter model. In the next subsection we explore the consequences of this low probability.

4.6 Models with More than Four Giant Planets

The low probability of success of reproducing the current orbital properties of the giant planets from the jumping-Jupiter scenario demanded further investigation. The first one to do so was Nesvorný (2011), who suggested that the outer Solar System could have contained five giant planets—two gas giants, three ice giants—one of which was ejected during the instability. These giant planets would have formed in a quintuple resonant chain. Nesvorný (2011) ran a small set of simulations of four-planet and five-planet configurations. Batygin et al. (2012) carried out an independent investigation of systems with five giant planets (see Fig. 12).

Nesvorný measured the success of his simulations with four criteria. First, the final outcome needs to have four giant planets (two gas giants, two ice giants). Second, the semi-major axes of the four surviving planets need to be within 20 % of their current values, with their eccentricities below 0.11 and inclinations below 2°. Third, Jupiter's eccentricity eigenmode e_{55} has to be larger than 0.022 (half the current value; see Table 2 of Morbidelli et al. 2009c). Finally, the Saturn-Jupiter period ratio needs to have evolved from less than 2.1 to more than 2.3 in less than 1 Myr. Nesvorný (2011) concluded that the five-planet cases generally fared better than the four-planet cases, with extended configurations better than compact ones.

Nesvorný and Morbidelli (2012) carried out giant planet simulations with a wide variety of initial conditions, always starting from a multiple resonant configuration. They ran seven sets of four-planet initial conditions, seven five-planet ones and four cases with six planets (two gas giants, four ice giants). They used the same criteria of success as Nesvorný (2011) and drew several important conclusions.

First, Nesvorný and Morbidelli (2012) were unable to simultaneously match all their criteria starting with four planets with a probability greater than 1 %. For this reason, they rejected the four-planet model. Second, the five-planet models generally fare much better, but the outcomes vary. Generally, compact five-planet configurations—such as those with orbital periods in the ratios 3:2 3:2 4:3 4:3—have a lower success rate than looser configurations such as 3:2 3:2 2:1 2:1. The bottleneck appears to be matching the amplitude of

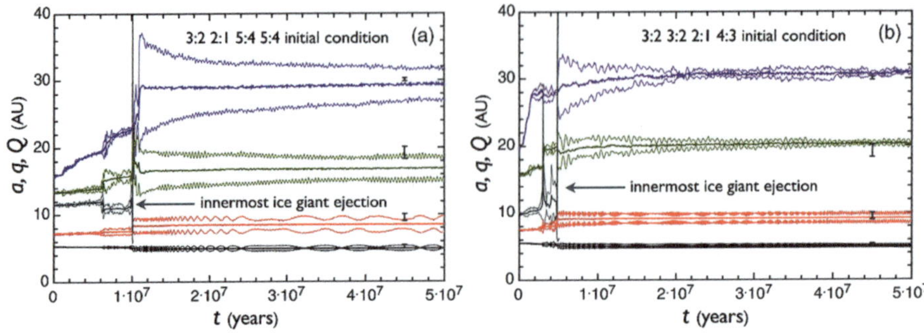

Fig. 12 Evolution of semi-major axis, perihelion distance, and aphelion distance for two systems of five planets (Jupiter, Saturn, and three ice giants) with a massive planetesimal disk outside their orbits, from Batygin et al. (2012). In both panels, Jupiter and Saturn begin in a 3:2 resonance. In the left panel, Saturn and the inner ice giant are in a 2:1 MMR, while both pairs of ice giants are in 5:4 MMRs. In the right panel, Saturn and the inner ice giant are in a 3:2 MMR, the inner pair of ice giants is in a 2:1 MMR, and the outer pair of ice giants is in a 4:3 MMR. In each panel, the innermost ice giant is ejected, leaving behind four planets with orbits similar to those of the outer planets in our Solar System. For the two cases shown here, the planetary system goes unstable in $\lesssim$10 Myr. Morbidelli and Crida (2007), Nesvorný (2011), and Nesvorný and Morbidelli (2012) have investigated systems of five and six giant planets that are stable in isolation (without an external disk) for hundreds of Myr to >1 Gyr. Whether the instability can be delayed for many hundreds of Myr, as in the original Nice model, for five- or six-planet systems with an external disk is not yet clear

e_{55}, which is more difficult in compact configurations than in loose ones. It is not clear why this is the case, since Nesvorný and Morbidelli (2012) typically used the same disk mass of 35 $M_\oplus$ for many sets of simulations, but they suspect that the system is too compact and the instability becomes too violent to match all constraints. Increasing the disk mass does not help because massive disks damp the eccentricities of the gas giants too much and drive Jupiter and Saturn too far apart. Changing the mass of the fifth planet does not help either. However, chains that are separated too far do not work because then Neptune ends up beyond 30 AU. Adding a sixth planet does not improve the statistics very much either, nor does starting Jupiter and Saturn in the 2:1 resonance, which may be incompatible with hydrodynamical simulations (Pierens and Nelson 2008).

Ultimately, Nesvorný and Morbidelli (2012) find that the best chances of matching all criteria come from the 3:2 3:2 3:2 3:2 and 3:2 3:2 2:1 3:2 chains. However, a recent study has cast some doubt on this finding. Brasser and Lee (2015) have performed a very large number of simulations of three resonant configurations: the resonant Nice chain with four planets in 3:2 4:3 4:3, a compact five-planet case in 3:2 3:2 4:3 4:3, and a loose five-planet case 3:2 3:2 2:1 3:2. The aim of their study was to determine whether it was possible to simultaneously reproduce the current obliquities of the giant planets and their current orbital spacing during the instability.[14] It is thought that Saturn's obliquity of 27° is a result of capture in a secular spin-orbit resonance with Neptune's nodal eigenfrequency s_8 (Ward and Hamilton 2004; Hamilton and Ward 2004). Brasser and Lee (2015) tried to determine under which conditions it was possible to tilt Saturn to its current obliquity of 27° and keep Jupiter's obliquity below 5° (Jupiter's obliquity is 3°). They employed similar but more relaxed criteria for success than Nesvorný and Morbidelli (2012), but found similar outcomes. However, they conclude that the loose case 3:2 3:2 2:1 3:2 is incompatible with the obliquities of the gas giants,

[14]Vokrouhlický and Nesvorný (2015) have independently used the planetary instability models of Nesvorný and Morbidelli (2012) to investigate the excitation of the obliquities of Jupiter and Saturn.

even though, like Nesvorný and Morbidelli (2012), they find that it best matches the current orbital properties of the outer Solar System. Instead, they find that the compact chain 3:2 3:2 4:3 4:3 best matches both the orbital properties and the obliquities of the gas giants, though with a very low probability (∼0.3 %). In addition, they also determine that the four-planet resonant Nice case works almost as well, contradicting the result of Nesvorný and Morbidelli (2012). However, note that Brasser and Lee (2015) performed many more simulations and also employed more relaxed criteria than Nesvorný and Morbidelli (2012).

Kaib and Chambers (2015) perform orbital integrations of five and six giant planet systems from Nesvorný and Morbidelli (2012) with a planetesimal disk. After the planets' orbits become unstable (usually within 100 Myr), they add the terrestrial planets to the simulations. They find a probability of 1 % or less that the orbital architectures of both the terrestrial and giant planets can be reproduced, and suggest that the giant-planet instability in Nice-like models may not have caused the Late Heavy Bombardment (LHB). Indeed, the authors of the Nice model published their work on the possible link between the Nice model and the LHB as a separate paper precisely because the timing of the instability was (and remains) uncertain (Gomes et al. 2005b).

In summary, a model with five giant planets in an initially compact configuration (Nesvorný and Morbidelli 2012), with one ice giant then being ejected, is most successful at matching the current orbits and obliquities of the outer Solar System. However, it is not yet clear whether the four-planet resonant Nice model (Morbidelli et al. 2007) can be ruled out.

5 Formation and Evolution of the Oort Cloud

The Oort Cloud may have formed in two stages (Brasser et al. 2008). First a tightly bound cloud of bodies like Sedna would have taken shape when the Sun was still embedded in its birth cluster (Sects. 5.2 and 5.3; Brasser and Wang 2015). After the Sun left the cluster, a more loosely bound cloud, the classical Oort Cloud, would form, its population peaking about 1 billion years later and declining very gradually thereafter (Sect. 5.1; Dones et al. 2004; Lewis et al. 2013).

5.1 Oort Cloud Formation in the Current Galactic Environment

As we describe in Sect. 1.2, cometary nuclei are thought to be small solid bodies ("planetesimals") that formed beyond the snow line—slightly interior to, between, and slightly outside of the orbits of the giant planets.[15] Most small bodies in these regions that did not originally cross the orbit of a planet would have been excited onto planet-crossing orbits due to distant gravitational perturbations.

The long-period comets we observe are understood to be the small lucky fraction of planetesimals that became planet-crossing, but managed to find a quasi-stable niche in the Oort Cloud. (From now on, we will refer to these planetesimals as "comets".) First, a planet pumps up the orbital energy (i.e., semi-major axis, a) of a comet, while the comet's perihelion distance q remains in the planetary region. (Once the comet's orbit becomes highly

[15] Shannon et al. (2015b) simulate the formation of the Oort Cloud with the planets in their present-day orbits and the Sun in its current galactic environment. They estimate that some 4 % of the bodies in the Oort Cloud should be "asteroids," i.e., bodies that formed within 2.5 AU of the Sun.

eccentric, q is nearly constant.) A comet's orbit can continue to grow until the comet became unbound from the Solar System, and thereafter wanders interstellar space. This was probably the fate of most comets, especially those in the vicinity of Jupiter and Saturn.

However, the same factor that limits the Sun's reach—the gravitational effect of stars and other matter in the solar neighborhood—provides a possible stabilizing mechanism. Once the comet's orbit becomes large enough, passing stars affect it (Öpik 1932; Oort 1950). In relative terms, stars change cometary perihelion distances (q) much more than they change the overall size of the orbit (a). If passing stars can lift a comet's perihelion from the planetary region before the planets can eject it from the Solar System, the comet will reach an orbit in the Oort Cloud. The characteristic size of the cloud is set by the condition that the timescale for changes in a is comparable to the timescale for changes in q due to passing stars. Assuming the density of matter in the present-day galactic environment of the Sun, this condition yields a cloud of comets with semi-major axes on the order of 10,000 to 100,000 AU (Tremaine 1993; Brasser and Duncan 2008). The trajectories of the stars are randomly oriented in space, so stellar perturbations eventually cause the comets to attain a nearly isotropic velocity distribution, with a median inclination i to the ecliptic of 90° and a median eccentricity $e = 1/\sqrt{2} \approx 0.7$.[16] Figures 13 and 14 show simulations of the time evolution of the Oort Cloud that assume the Sun has always been an isolated ("field") star. Later, passing stars reduce the perihelion distances of a small fraction of these comets so that they re-enter the planetary region and potentially become observable.

This outline of comet dynamics is similar to that described by Oort (1950). Developments since then include (1) recognition of the role of the smooth long-term effect of the total amount of nearby galactic matter, i.e., the "galactic tide," as a perturber at least as important as passing stars; (2) the likely formation of the Sun in an embedded cluster; (3) migration of the giant planets; and (4) possible radial migration of the Solar System within the Galaxy. We describe these developments in the remainder of Sect. 5. See Dones et al. (2004), Chaps. 4–6 of Fernández (2005), Rickman (2014), and Higuchi and Kokubo (2015) for further discussion of Oort Cloud dynamics and Morbidelli (2005) for a review of both Oort Cloud and Kuiper Belt dynamics.

5.2 Oort Cloud Formation and Evolution in a Cluster

The Sun likely formed as a member of a star cluster. Lada and Lada (2003) review embedded clusters, i.e., star clusters embedded within Giant Molecular Clouds (GMCs) and largely obscured from view in visual light. They estimate, based on the general star formation rate and the abundance of embedded clusters in our part of the Galaxy, that most nearby stars are born in such clusters. Embedded clusters usually dissolve within 10 Myr (Lada 2010; Dukes and Krumholz 2012) [see Sect. 1.1]. However, some embedded clusters evolve into gravitationally bound open clusters, which can survive much longer (Li and Adams 2015). Lamers et al. (2005) find that the lifetime against disruption $t_{\rm dis}$ of a star cluster increases with the mass of the cluster M as $t_{\rm dis} \propto M^{0.62}$. In the solar neighborhood, this relation yields $t_{\rm dis} = 310 \pm 120$ Myr for a 1000 $M_\odot$ cluster and 1300 ± 500 Myr for a 10,000 $M_\odot$ cluster. The number of stars in different embedded clusters is such that the likelihood for any star to be born in a cluster with ~100, 1000, or 10,000 members is about the same (Lada and Lada 2003). Therefore, for the Sun's birth cluster, one may consider all these possibilities.

[16] For an isotropic distribution, $\cos i$ is uniformly distributed between -1 and 1, so the median value of $\cos i = 0$, corresponding to $i = 90°$, and e^2 is uniformly distributed between 0 and 1, so the median value of e^2 is $\frac{1}{2}$ and the median value of $e = 1/\sqrt{2}$ (Jeans 1919).

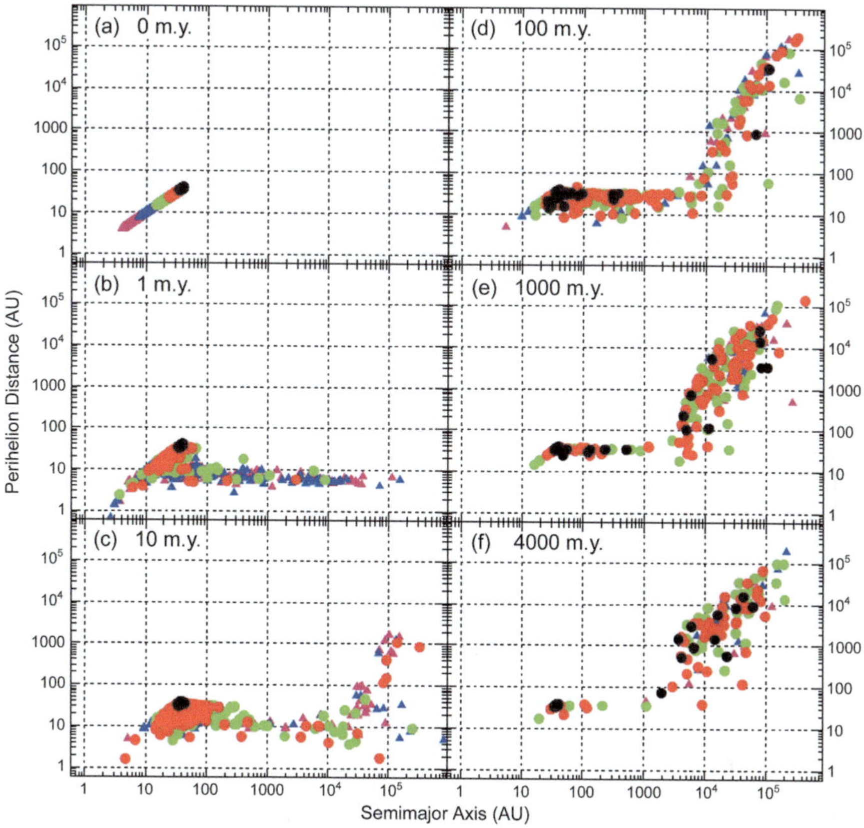

Fig. 13 Scatter plot of osculating barycentric pericenter distance q vs. osculating barycentric semi-major axis a at six times in a simulation of Oort Cloud formation, from Dones et al. (2004). The Sun is assumed to be a single star in its current galactic environment throughout the simulation. The points are color-coded to reflect the region in which the simulated comets formed. Jupiter region comets (initial semi-major axes between 4 and 8 AU) are *magenta triangles*; Saturn region comets (8–15 AU) are *blue triangles*; Uranus region comets (15–24 AU) are *green circles*; Neptune region comets (24–35 AU) are *red circles*; Kuiper Belt comets (35–40 AU) are *black circles*. (**a**) Initial conditions for the simulation. (**b**) 1 Myr into the simulation. (**c**) 10 Myr into the simulation. (**d**) 100 Myr. (**e**) 1000 Myr. (**f**) Final results for the simulation, at 4000 Myr, i.e., at roughly the present time. The population of the Oort Cloud peaks about 800 Myr into the simulation. Note that in (**f**), there is a nearly empty gap for semi-major axes between $\approx$200 and 1000 AU. This region is not expected to be populated if the Sun was always a single star, but can be if the Sun were once a member of a cluster (see Sect. 5.2)

Oort Cloud formation in a cluster has been considered for two decades (Gaidos 1995; Eggers et al. 1997). Fernández (1997) made the interesting suggestion that a dense stellar environment might assist cloud formation by allowing comets scattered by the gas giants to be stored inside the cloud due to the high frequency of close stellar encounters and the tidal action of surrounding gas. These comets would populate the cloud in a broad energy range, primarily in an inner Oort Cloud with typical semi-major axes of a few thousand AU. Fernández (1997) thought an important advantage would be the increased storage efficiency, leading to less severe requirements on the number of icy planetesimals in the region of giant planet formation. He suggested that the dense environment would naturally occur if the Sun formed in a molecular cloud or open cluster.

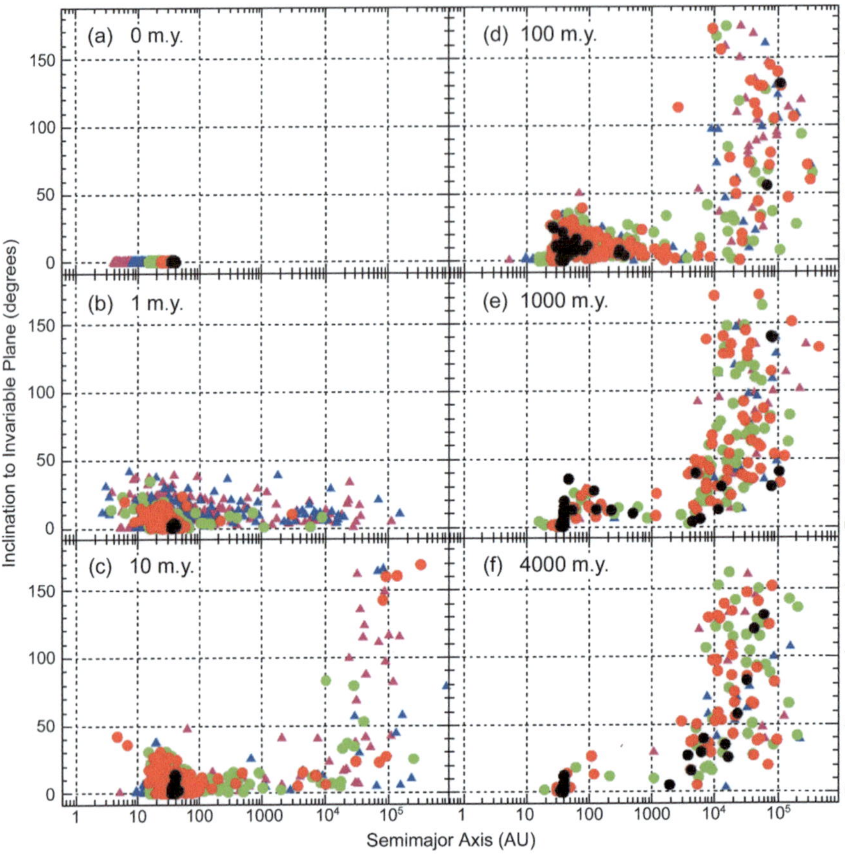

Fig. 14 Scatter plot of osculating inclination to the invariable plane vs. semi-major axis at six times in the simulation shown in Fig. 13. (**a**) Initial conditions for the simulation. (**b**) 1 Myr. (**c**) 10 Myr. (**d**) 100 Myr. (**e**) 1000 Myr. (**f**) Final results for the simulation, at 4000 Myr. At the end of the calculation, the inclination distribution is nearly isotropic, even in most of the inner Oort Cloud. Figure from Dones et al. (2004)

Around this time the importance of embedded clusters became better known. After the discovery of Sedna in 2003 (Brown et al. 2004), the time was ripe to consider those clusters as a suitable birthplace for the Sun in order to allow the storage of scattered objects into orbits with semi-major axes less than 1000 AU. Morbidelli and Levison (2004) found a stellar passage at a velocity of 1 km/s, decoupling Sedna from a Scattered Disk orbit, to be the preferred explanation for its strange orbit, noting that this velocity is typical of stellar encounters within clusters, but they were not specific about the kind of cluster.

Brasser et al. (2006) modeled the formation of an Oort Cloud including a population of Sednas in the framework of an embedded cluster with the Sun as a member. The results showed the formation of an inner Oort Cloud including a population of Sednas, and while Brasser et al. (2007) were less optimistic after the inclusion of gas drag from the solar nebula, the problem concerned the scattering of planetesimals by the planets and did not involve the cluster stars. It thus seems that, apart from the problem of gas drag opposing the formation of a Scattered Disk, a primordial Oort Cloud satisfying the Sedna constraint may have been formed if the Sun was born in an embedded cluster.

Fernández and Brunini (2000) followed up on the suggestion of Fernández (1997) by simulating the formation of the Oort Cloud as it would occur inside a stellar cluster. Their main result was that the stellar density of the cluster is important in determining the structure of the cloud, with denser clusters causing more concentrated clouds (i.e., cometary semi-major axes are typically smaller in denser clusters). Kaib and Quinn (2008) took an important further step in extending the simulations of cluster origin for the age of the Solar System, taking into account that the Sun would only have spent a small amount of time inside the cluster. They used the same kind of birth cluster, namely, an open cluster, but considered some amount of residual gas. Due to the stochastic nature of the closest stellar passages within such a cluster, they found a significant amount of scatter in the resulting Oort Cloud concentrations.

Brasser et al. (2008) also looked into the further evolution of the Oort Cloud after the Sun left the birth cluster—in this case, the embedded cluster of their earlier studies. The inner Oort Cloud formed in the cluster was found to be largely inert, with just a small leakage of comets into the outer halo during this subsequent phase. Together with the efficiency of storage into the inner cloud, this implied an uncomfortably large mass in the initial, comet-forming region and, thus, a problem explaining Oort Cloud formation. The continued emplacement of comets into the cloud from the Scattered Disk, noted by Kaib and Quinn (2008), was seen as a possible remedy along with unmodeled, dramatic perturbations during rare encounters with Giant Molecular Clouds.

The first models of Oort Cloud formation in clusters assumed that the solar birth cluster had a modest stellar density and was short-lived. For example, Fernández and Brunini (2000) focused on clusters with 10 and 25 stars/pc^3, with a few calculations for a "superdense" cluster with 100 stars/pc^3, and assumed the clusters dissolved completely in 100 Myr. In fact, the mean stellar density in a cluster is of order 100 stars/pc^3, with only a weak dependence on the number of stars in the cluster (Adams 2010).

Evidence for short-lived radionuclides such as ^{26}Al (half-life 0.72 Myr) and ^{60}Fe (2.6 Myr) in the oldest solid material of the Solar System has implications for the solar birth cluster. Aluminum-26 could have been injected by either a nearby supernova explosion or from winds from a Wolf-Rayet star (Davis and McKeegan 2014). Iron-60 has often been taken as evidence for a nearby supernova just before the formation of the Solar System, but a downward revision in the initial abundance of ^{60}Fe is consistent with inheritance from the "galactic background" (Tang and Dauphas 2012). In other words, though the ^{60}Fe is still thought to have originated in supernovae, there need not have been a nearby explosion. Likewise, Young (2014) argues that the abundances of ^{26}Al, ^{36}Cl, ^{41}Ca, and ^{60}Fe are the legacy of supernovae and Wolf-Rayet stars in the parent molecular cloud, rather than from "unusual proximal stellar sources". Gounelle and Meynet (2012) discuss a model in which the Sun formed as one of $\approx$600 stars in a dense shell 5–10 pc from a massive star whose wind provides ^{26}Al. Iron-60, in turn, originates in one or more supernova explosions from an earlier generation of stars.

Adams (2010) gives a comprehensive discussion of several aspects on the population of the solar birth cluster. He primarily focuses on supernovae as the source of short-lived radionuclides. Adams (2010) estimates the preferred mass of the exploding star to be 25 $M_\odot$. Because of the rarity of such massive stars, this implies the cluster had $N \gtrsim 1000$ stars. The small eccentricities and inclinations of the planets provide another constraint (Adams and Laughlin 2001; Malmberg et al. 2007). If a star had passed within $\approx$400 AU of the Sun, Neptune's eccentricity would have been excited to $\approx$0.1, and a passage $\approx$200 AU from the Sun could have ejected Neptune from the Solar System. The non-occurrence of these disasters implies $N \lesssim 10{,}000$ stars. Happily, a slightly more distant encounter with a passing star may explain the orbits of Sedna and other inner Oort Cloud bodies. This constraint implies $1000 \lesssim N \lesssim 10{,}000$. Adams (2010) concludes "our Solar System is likely to have formed within a moderately large cluster with $N = 10^3$–10^4... The necessary cluster systems are relatively common, with the Trapezium Cluster in Orion being the closest analog."

The open cluster M67 is estimated to be 4 Gyr old. In the N-body model of Hurley et al. (2005), the cluster began with a mass of $\approx$19,000 $M_\odot$ (36,000 stars) and now has a mass of 2000 $M_\odot$. Önehag et al. (2011) identified a star in M67 that is very similar to the Sun, and stated that "[a] common origin of both stars as members of M67 is conceivable, albeit not likely, considering their different Galactic orbits." Pichardo et al. (2012) inferred that the high relative velocity between the Sun and M67 made a common origin extremely improbable (although see the discussion in Pfalzner et al. (2015)). Nonetheless, the Sun could have spent $\approx$100 Myr in a cluster like M67 without disturbing the orbits of the planets (Adams 2010). Spending its entire lifetime in such an environment would likely result in severe truncation of the Oort Cloud.

5.3 Capture of Comets from Other Stars in the Birth Cluster

Simulations of Oort Cloud formation have suffered from a difficult problem: they are unable to reproduce the inferred ratio between the population of the Scattered Disk (Luu et al. 1997; Duncan and Levison 1997) and the Oort Cloud. The Scattered Disk is widely believed to be the source of the Jupiter-family comets [JFCs] (Duncan and Levison 1997). The JFCs are a set of comets whose Tisserand parameters with respect to Jupiter satisfy $2 < T < 3$ (Levison 1996). Care has to be taken with this definition because this criterion would erroneously classify some comets as JFCs when they have $T < 3$ due to (i) a high inclination but an orbit entirely outside that of Jupiter or (ii) very low-inclination objects that have their perihelia close to Jupiter but very large semi-major axes (Gladman et al. 2008). Thus we follow Gladman et al. (2008) and require that a JFC must also have $q < 7.35$ AU so that the object is most likely dynamically controlled by Jupiter.

From the current population of JFCs, Duncan and Levison (1997) and Levison et al. (2008) estimate there are 6×10^8 bodies in the Scattered Disk with diameter $d > 2$ km or absolute total magnitude $H_T < 9$. Their estimate of the Scattered Disk population was approximately a factor of 150 to 750 lower than recent estimates of the Oort Cloud population from LPCs (Francis 2005) with total absolute magnitude $H_T < 11$. Compensating for the difference in H_T used in the derivations for the Oort Cloud ($H_T < 11$) and Scattered Disk ($H_T < 9$) populations is difficult because JFCs and LPCs appear to have strongly different distributions of H_T in the range from 9 to 11 (Fernández et al. 1999; Fernández and Sosa 2012; Francis 2005). On the other hand, numerical simulations tend to yield an Oort Cloud to Scattered Disk population ratio (at fixed object size) of 5 (Kaib and Quinn 2008) to 20 (Dones et al. 2004; Brasser et al. 2010). Thus, simulations consistently underestimate the observed Oort Cloud to Scattered Disk population ratio by at least an order of magnitude!

Two solutions have been proposed to remedy this problem: forming part of the Oort Cloud while the Sun was still in its birth cluster (Fernández and Brunini 2000; Brasser et al. 2006, 2012b; Kaib and Quinn 2008), or forming the Oort Cloud by the capture of comets from other stars (Levison et al. 2010, based on the idea of Zheng et al. 1990). The Zheng et al. (1990) study suffered from simplifying assumptions and a low capture efficiency and thus Levison et al. (2010) revisited the idea.

Levison et al. (2010) numerically integrated the orbits of comets and stars embedded in a spherically symmetric distribution of gas and stars, with a volume density that scaled as r^{-1}, where r is the distance from the center of the cluster, up to a cutoff radius R_c which ranged from 0.3 to 1 pc, while the number of stars was varied between 30 to 300. The gas of the embedded cluster had a fixed density for 3 Myr, after which it decayed away over 10,000 years. The star formation efficiency was typically assumed to be 20 % and each star was surrounded by a disk of comets on elongated orbits with perihelia near 30 AU and semi-major axes from 1000 to 5000 AU. These initial conditions were chosen to mimic scattering of comets by unseen planets. The integration time was typically 10 Myr to 50 Myr, and the simulations were stopped when the median distance between stars reached 2 pc.

Levison et al. (2010) determine the fraction of comets that ended up bound around each star with semi-major axis $a < 50{,}000$ AU. In their baseline simulations, Levison et al. (2010) assumed the number of comets initially orbiting each star was independent of stellar mass. The comets were captured either by direct exchange when two stars had a close encounter, or by random capture in which some free-floating comets in the cluster happened to have velocity vectors nearly parallel to one of the escaping stars and thus became bound to that star. This mechanism is similar to a mechanism for the formation of wide binaries in the Galaxy (Kouwenhoven et al. 2010). This capture process is aided by the expansion of the cluster itself. The typical capture efficiency, scaled by the original protoplanetary disk (f_c), was typically a few percent, with a maximum value of 26 %. Taking the Solar System as a guideline, Levison et al. (2010) suggest that the Sun could have captured $2 \times 10^{13} f_c$ comets, and with $f_c \sim 3$ %, this is enough to capture the whole Oort Cloud from free-floating comets in the cluster. Scaling for the number of comets in the Saturn-Neptune zone of the Solar System, Levison et al. (2010) suggest that over 90 % of Oort Cloud comets may be of extrasolar origin. Levison et al. (2010) used a second line of reasoning that suggested $\frac{1}{3}$ to $\frac{2}{3}$ of the Oort Cloud was extrasolar. In any case, these estimates are uncertain because the mass of the protoplanetary disk, the typical masses and spacing of planets that scatter comets around other stars, and the average mass of a comet are all poorly known.

5.4 Formation of the Oort Cloud at Different Distances from the Galactic Center

One of the underlying assumptions in models of Oort Cloud formation had long been that the local galactic environment remained essentially unchanged until the present epoch (e.g., Duncan et al. 1987), at least after the Sun left its birth cluster, However, recent developments in galactic dynamics suggest that this assumption may not be correct. Sellwood and Binney (2002) found that stars orbiting near the corotation resonance of spiral arms (which are transient with varying pattern speed over galactic lifetimes) can exchange angular momentum with the arms, transporting the stars to other regions of the galactic disk. During this transport many stars maintained low peculiar velocities similar to the Sun's. In simulations modeling the formation and evolution of Milky Way-like galaxies, Roškar et al. (2008) observed this radial stellar migration as well. They suggested that over half of the stars in the present-day solar neighborhood may have been born in substantially different areas of the galactic disk (exterior or interior to the annulus in the disk where the Sun currently orbits).[17] In addition, they found the most common migration scenario is for solar neighborhood stars to have been born closer to the Galactic Center. This calls into question the idea that the Sun's galactic environment (orbit) has not changed significantly during the last 4.5 Gyr.[18]

[17] Genzel et al. (2010) argue that the distance to the Galactic Center is $R_0 = 8.15$ kpc. Bash (1986) estimates that the Sun's current orbit takes it from 0.995 R_0 to 1.145 R_0, i.e., from 8.11 kpc to 9.33 kpc if we adopt the value of R_0 from Genzel et al. (2010).

[18] There is no consensus on the direction or amount of the Sun's migration. For instance, Martínez-Barbosa et al. (2015) find that the Sun has probably undergone little migration, but if it has done so, it was prob-

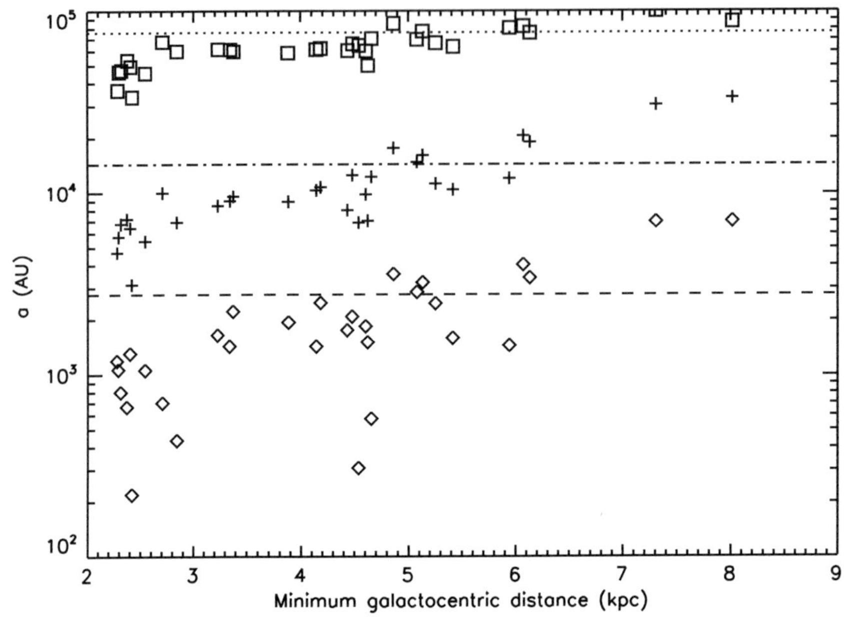

Fig. 15 Plot of $a_{\rm min}$ (*diamonds*), $a_{\rm max}$ (*squares*), and median semi-major axis (*crosses*) vs. the minimum galactocentric distance for each of the solar analogs in Roškar et al. (2008). $a_{\rm min}$ is the semi-major axis interior to the orbits of 95 % of the bodies in the Oort Cloud, and $a_{\rm max}$ is the semi-major axis exterior to 95 % of all Oort Cloud bodies. In addition, we show $a_{\rm min}$ (*dashed line*), $a_{\rm max}$ (*dotted line*) and the median a (*dash-dot-dash line*) for a control simulation that assumes a fixed galactic environment similar to the present solar neighborhood. Figure from Kaib et al. (2011)

Such evolution could have a substantial effect on the structure and evolution of the Oort Cloud. Using an analytical model for the local galactic tide and stellar perturbations, Brasser et al. (2010) modeled the formation of the Oort Cloud at a variety of galactocentric distances between 2 and 20 kpc. They found that the Oort Cloud trapping efficiency is always near ∼4 % and is virtually independent of the Sun's galactocentric distance. In addition, they found that the inner edge of the Oort Cloud (in terms of semi-major axis) can vary between ∼2000 and ∼30,000 AU. This is strongly dependent on the galactocentric distance, since the local density of galactic matter (and therefore strength of tidal and stellar perturbations) declines exponentially with increasing galactocentric distance.

However, Brasser et al. (2010) assumed a static galactic environment, whereas Roškar et al. (2008) predicted that it should vary over the lifetime of the Sun. Kaib et al. (2011) modeled Oort Cloud formation around a migrating Sun by actually approximating the galactic environmental variation experienced by stars found on solar-like galactic orbits at the end of the simulations of Roškar et al. (2008). Of this star sample, it was found that over 60 % formed interior to 7 kpc, and over 87 % spent part of their lives within 6 kpc of the Galactic Center. This migration was found to have significant consequences for the Oort Cloud's structure. First, the median semi-major axis varied by about an order of magnitude depending on the minimum galactocentric distance attained by the Sun (see Fig. 15). In addition,

ably born further out, near 11 kpc, while Halle et al. (2015) find that stars in the solar neighborhood "may have experienced very limited churning from the inner disc." Hayden et al. (2015) find that "about 30 % of the stars in our Galaxy have traveled a long way from the orbits in which they were born" (http://www.sdss3.org/press/20150730.farfromhome.php).

the trapping efficiency fluctuated by about a factor of 4, depending on *when* the Sun attained its minimum galactocentric distance. If the Sun spent its early life far from the Galactic Center, a very loosely bound Oort Cloud was constructed, which would then be mostly stripped away when the Sun migrated into a denser galactic environment. In contrast, a more tightly bound Oort Cloud was formed if the Sun spent its early life nearer the Galactic Center, and this cloud was largely unaffected by the weaker galactic perturbations incurred if the Sun migrated outward later. Finally, Kaib et al. (2011) found that some solar analogs could have comet clouds whose inner edges resided near or inside Sedna's orbit. However, the efficiency of populating this region is low with respect to the modern Scattered Disk. Every Sedna-size body on a Sedna-like orbit today would require between 5 and 100 Sedna-size bodies in the modern Scattered Disk. At present, only two such objects are known in the Scattered Disk [Eris and (225088) 2007 OR_{10} "Snow White"] (Brown et al. 2005; Schwamb et al. 2009). Meanwhile, it is likely that many more large objects remain to be discovered on Sedna-like orbits.

5.5 Origin of High-Inclination Centaurs—Oort Cloud or Kuiper Belt?

The Centaurs are a class of small objects, such as Chariklo and Chiron, that follow unstable orbits in the region of the giant planets. Typically, these objects have semi-major axes of tens of AU and perihelion distances (q) among the giant planets. From dynamical studies it is believed that the Centaurs originate from the Kuiper Belt (KB)/Scattered Disk (SD) and form the bridge between these KBOs and the Jupiter-family comets (JFCs). This conclusion was motivated by dynamical studies demonstrating that KBOs regularly evolve onto Centaur orbits (Tiscareno and Malhotra 2003; Emel'yanenko et al. 2005; Di Sisto and Brunini 2007) and even all the way down to JFCs (Levison and Duncan 1997; Duncan and Levison 1997). All three studies of Centaurs found that the inclinations (i) of the Centaurs tend to remain low: Tiscareno and Malhotra (2003) report a characteristic time-weighted mean Centaur inclination of 16°, while Emel'yanenko et al. (2005) and Di Sisto and Brunini (2007) find comparable values. None of these studies recorded any Centaurs with $i > 60°$. Is this upper value in agreement with an origin in the KBO region? Brown (2001) computes the debiased inclination distribution for KBOs and suggests the functional form $\sin(i)\exp(-i^2/2\sigma^2)$, where $\sigma = 13° \pm 5°$. If the Centaurs originate from the KBO region, they should roughly retain the same inclination distribution. We can then compute the probability of finding a Centaur with $i > 60°$, i.e., $p(i > 60°)$, because it is just the complementary cumulative inclination distribution of the Centaur population. Assuming $\sigma = 13°$ results in a probability $p(i > 60°) \sim 2 \times 10^{-5}$. From this low value it is not surprising that the dynamical studies did not record any Centaurs with $i > 60°$. It appears that for a Centaur to obtain an even higher inclination, an external agent is needed.

Given the very low probability for a Centaur originating from the KBO region to reach an inclination above 60°, it is surprising that more than 10 Centaurs with inclinations $i > 60°$ are known, with six on retrograde orbits. (The JPL Small-Body Database lists 328 Centaurs in all as of October 31, 2015.) The only known reservoir of small bodies that has many objects on such high-inclination orbits is the Oort Cloud (Dones et al. 2004), and thus it is probable that these high-inclination Centaurs originate from the Oort Cloud and mimic its inclination distribution. If, instead, these high-inclination and retrograde Centaurs had originated from the KBO region, we should have detected tens of thousands of low-inclination Centaurs for every retrograde one. This is not the case and thus the presence of the high-inclination Centaurs indicates there is an alternative source.

Several studies (Levison 1996; Wiegert and Tremaine 1999; Levison et al. 2006b) have demonstrated that Jupiter and Saturn can extract objects from the Oort Cloud and decrease these bodies' semi-major axes (a) to several tens of AU. This process works as follows: an Oort Cloud object that passes through perihelion among the giant planets may experience an increase or reduction in its semi-major axis, depending on the closest approach distance to the planets and the planets' phasing at perihelion. This process repeats itself at roughly constant perihelion distance with random phasing. Eventually, a few objects will have their semi-major axes reduced sufficiently to become Centaurs. Thus, it is viable that the high-inclination and retrograde Centaurs with $q < 15$ AU were either extracted and pulled in solely by Jupiter and Saturn, or they could have been decoupled by Uranus and Neptune and subsequently passed to Saturn and Jupiter, which then pulled them all the way down to small semi-major axes. However, there are three objects that must have undergone a different evolution because they have $q > 15$ AU and $i > 70°$. These bodies are beyond the gravitational control of Saturn and thus fall into the domain of Uranus and Neptune. One of these, 2008 KV42 "Drac," is retrograde. Its dynamics was studied by Gladman et al. (2009), who suggested two scenarios for its origin: (i) either it followed a rare pathway from the Kuiper Belt or Oort Cloud to become a Centaur (also see Tiscareno and Malhotra 2003; Emel'yanenko et al. 2005; and Di Sisto and Brunini 2007), or (ii) it points to a new reservoir of bodies with semi-major axes of hundreds of AU and large inclinations. However, each scenario has problems. In particular, the Kuiper Belt mechanism in the first scenario is very inefficient, and there is no observational evidence for the reservoir in the second (Schwamb et al. 2010).

These difficulties led Brasser et al. (2012a) to carry out numerical simulations of the evolution of 2.5 million Oort Cloud comets under the gravitational influence of the Sun, giant planets, galactic tide, and passing stars over the age of the Solar System. They showed that the only viable source of the high-inclination Centaurs, particularly those with $q > 15$ AU, is the Oort Cloud. These high-i, high-q Centaurs had to be decoupled from the Oort Cloud by the gravitational perturbations of Uranus and Neptune alone, because an encounter with Saturn or Jupiter would have reduced their perihelia below 15 AU.

Brasser et al. (2012a) demonstrated that the Oort Cloud will only dominate over KBOs as a source for Centaurs above a certain inclination, which they determined to be 70°. Their model for the inclination and perihelion distribution of these objects is depicted in Fig. 16. The three objects found so far fall in the region with the highest probability density.

By making use of the observational data of Schwamb et al. (2010), Brasser et al. (2012a) estimated there are between 1 and 200 Centaurs with absolute magnitude $H < 8$, inclination $i > 70°$ and perihelion $q > 15$ AU. This can be refined a bit further by considering the following. Morbidelli et al. (2009b) estimate that there were originally some 10^8 objects in the trans-Neptunian disk with $H < 8$, assuming all these objects had albedos of 4.5 %. Taking a 3 % efficiency for Oort Cloud formation in the current galactic environment (Dones et al. 2004; Kaib and Quinn 2008; Brasser and Morbidelli 2013) implies there are $\approx 3 \times 10^6$ Oort Cloud objects with $H < 8$. Brasser et al. (2012a) found a steady-state fraction $\sim 10^{-5}$ of Oort Cloud bodies on high-inclination, high-perihelion orbits, so we expect some 30 Centaurs with inclination $i > 70°$ and perihelion $q > 15$ AU of the same size. Further discoveries of these objects are needed to be able to constrain the number of Oort Cloud objects with $H < 8$.

The Centaur population with inclinations below 70° is dominated by a Scattered Disk origin (Brasser et al. 2012a). Petit et al. (2011) give an "order of magnitude estimate" of the number of Scattered Disk Objects with absolute magnitude $H_g < 9.16$ (diameter > 100 km for 5 % albedo) as 5000 (+5000, −3000). According to Di Sisto and Brunini (2007), the steady-state ratio of the number of Centaurs to the number of bodies in the Scattered Disk is 0.037, so the expected number of Centaurs in the same magnitude range is 200 (+200, −100). The JPL Small-Body Database (http://ssd.jpl.nasa.gov/sbdb_query.cgi) lists

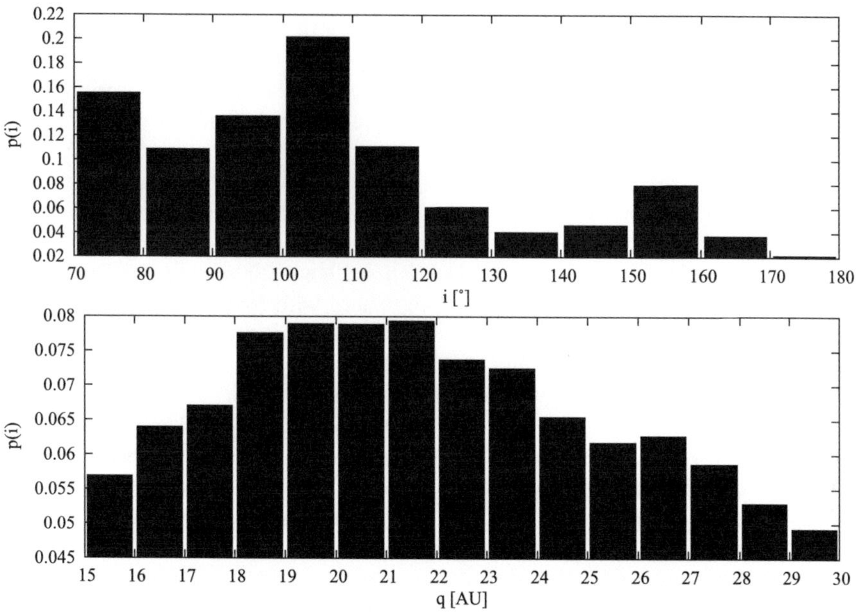

Fig. 16 Top panel: The steady-state inclination distribution of the high-inclination, high-perihelion distance Centaurs. Bottom panel: The steady-state perihelion distribution of the high-inclination, high-perihelion distance Centaurs. Figure from Brasser et al. (2012a)

40 Centaurs with $H_g \lesssim 9.16$. Given that discovery of Centaurs is quite incomplete, this is reasonable agreement, albeit with large uncertainties.

6 Formation and Evolution of the Kuiper Belt

6.1 Dynamical Classes and Resonant Structure

The trans-Neptunian region shows a rich dynamical structure. The classification of objects in this region is a difficult problem (Gladman et al. 2008; Lykawka 2012). We give an overview of several of the existing schemes.[19]

It is generally agreed that the trans-Neptunian region consists of least three overlapping reservoirs. The main constituent is the Kuiper Belt [as defined by the first discoveries of KBOs in the early 1990s][20] (Kuiper 1951), which extends in semi-major axis from just beyond the orbit of Neptune at 30 AU and ends near 50 AU (Allen et al. 2001), possibly at the 2:1 resonance with Neptune at 48 AU (Levison et al. 2008). Objects in the Kuiper Belt are generally dynamically stable over the age of the Solar System (Duncan et al. 1995) and do not, in most cases, cross the orbit of Neptune. The second reservoir is the Scattered

[19] An animation by Alex Parker showing the rough relative sizes and true orbital motion of all trans-Neptunian objects with semi-major axes greater than Neptune's as of 2014 can be seen at https://vimeo.com/96874127. See http://www.alexharrisonparker.com/datavisualization/ for other visualizations by Parker.

[20] See http://www.icq.eps.harvard.edu/kb.html for a discussion of how Kuiper never predicted the present-day existence of the belt that bears his name, while others, such as Leonard, Whipple, and Cameron, came closer to doing so.

Disk (SD) [or "scattering objects" (Gladman et al. 2008)], traditionally thought to contain objects with semi-major axes $a > 50$ AU and perihelia near Neptune (Duncan and Levison 1997). The third is the Oort Cloud, long assumed to start thousands of AU from the Sun (e.g., Duncan et al. 1987), but now believed to begin at semi-major axes of only hundreds of AU (Brown et al. 2004).

The Kuiper Belt can be further subdivided into the resonant population, which has objects in a mean-motion resonance with Neptune—such as the plutinos in the 3:2 and the twotinos in the 2:1 (Gladman et al. 2012)—and the classical Kuiper Belt Objects, which sit between the 3:2 and 2:1 resonances with Neptune (Delsanti and Jewitt 2006; Petit et al. 2011). There is also a small stable niche between the 4:5 resonance with Neptune at 34.6 AU and the 3:2 (Gladman et al. 2008). The classical belt has a dynamically "cold" component with inclinations generally below 4° (Brown 2001; Petit et al. 2011) and a "hot" component, with inclinations going above 40° (Brown 2001; Kavelaars et al. 2008).

These classifications did not encompass all possibilities. The SD was only defined for objects with $a > 50$ AU, but objects with low eccentricities and semi-major axes just beyond 50 AU were not considered. Elliot et al. (2005) define objects as Scattered Disk Objects (SDOs) when their eccentricities exceed 0.2 and they are non-resonant. Volk and Malhotra (2008) consider an SDO to have a perihelion distance $q > 33$ AU and $a > 50$ AU, while Brasser and Morbidelli (2013) consider all objects with $q > 30$ AU to be SDOs, but only if they have $a < 1000$ AU. Ultimately, the only way to correctly identify an SDO is to determine whether it shows diffusion in semi-major axis on a rapid time scale, e.g., in 10 Myr (Gladman et al. 2008), as any classification based on osculating eccentricity, semi-major axis and perihelion distance is prone to uncertainty.

Beyond semi-major axes of ≈50 AU, there is another class of bodies, the detached objects (Emel'yanenko et al. 2003; Delsanti and Jewitt 2006), also called the high-perihelion Scattered Disk (HPSD) (Gomes et al. 2005a) or extended Scattered Disk (Gladman et al. 2002). These bodies are "detached" in the sense that their perihelion distances are far enough outside of Neptune's orbit ($q \gtrsim 38$ AU) that they interact less strongly with Neptune than do bodies with smaller perihelion distances. (148209) 2000 CR_{105} ($a = 230$ AU, $q = 44$ AU) was the first detached object to be discovered. Gladman et al. (2008) modify the criteria of Elliot et al. (2005) and define detached objects as non-scattering Trans-Neptunian Objects (TNOs) with eccentricities $e > 0.24$ and semi-major axes below 2000 AU. Recently, the detached objects have been further subdivided into extended Scattered Disk objects and inner Oort Cloud objects (Brasser and Schwamb 2015). The latter, which so far consists of only four objects—(90377) Sedna, 2004 VN_{112}, 2010 GB_{174}, and 2012 VP_{113}—are bodies with semi-major axes $a > 250$ AU and perihelia $q > 45$ AU.

In summary, there is no consensus about classification of trans-Neptunian objects based on their semi-major axes, eccentricities, and inclinations. The classification scheme of Gladman et al. (2008), based on numerical simulations, is the correct manner to identify scattering and resonant objects, but it suffers from some arbitrary cuts in eccentricity and a fuzzy edge between the SDOs and the classical objects.

6.2 Dynamical Sculpting of the Kuiper Belt

Even before Jewitt and Luu discovered (15760) 1992 QB_1, the first KBO other than Pluto (Jewitt and Luu 1993), the region near and perhaps outside Neptune's orbit was suspected to have once contained much more mass than it does at present. The planetesimal-driven migration of Neptune by ≈7 AU (Fernández and Ip 1984; Hahn and Malhotra 1999) requires a mass of small bodies comparable to the mass of Neptune (17 $M_\oplus$). Stern (1991) argued from the statistics of presumed impact events in the outer Solar System [the obliquities of Uranus and Neptune and the formation of the Pluto-Charon binary (Canup 2005)] and from the capture of Triton by Neptune that the 20–50 AU region had once contained $\approx$1000 Pluto or Triton-sized bodies, corresponding to 2–4 $M_\oplus$. If the minimum-mass solar nebula is extrapolated to the 30–50 AU region, the mass in solids there would have been $\approx$20–50 $M_\oplus$ (Kenyon and Luu 1998). Simulations of the growth of Kuiper Belt Objects (Stern 1995, 1996; Kenyon and Luu 1998, 1999) show that masses of this order and very low eccentricities are required to grow Pluto-sized bodies *in situ* in 100 Myr–1 Gyr. If Pluto formed somewhat closer to the Sun, as its orbit strongly suggests (Malhotra 1993; see Sect. 1.3), similar requirements still apply for the region where it formed. In short, the early Solar System near Neptune is almost universally assumed to have contained $\gtrsim$10 $M_\oplus$ or more in small bodies, a small part of which survived to become today's Kuiper Belt (although see Shannon et al. 2015a for a rare example of a model in which the cold classical Kuiper Belt never had much more mass than it does now).

Malhotra (1995) showed that if Neptune migrated smoothly outward, most small bodies (KBOs) between 30 and 50 AU would have been captured into mean-motion resonances (MMRs) with Neptune, particularly the 3:2 and 2:1 resonances at $\approx$39.4 and 47.8 AU, respectively. These captured KBOs would be transported outward along with Pluto. During migration, eccentricities and inclinations of KBOs would be excited, although reaching inclinations as large as Pluto's is difficult. As it happens, most known KBOs do not occupy MMRs with Neptune, so Malhotra et al. (2000) proposed that the 30–50 AU region might consist of both "classical" non-resonant KBOs that had formed in place and resonant KBOs that had moved outward.

Gomes et al. (2004) argued Neptune would have undergone runaway migration until it reached the outer edge of the planetesimal disk. Since Neptune orbits at 30 AU, the disk must have ended near this location. (Gomes et al. 2004 favor 35 AU as the location for the edge of the disk.) Gomes (2003) proposed a novel mechanism to explain the hot population of the Kuiper Belt. In his scenario, Neptune scatters planetesimals, exciting their eccentricities and inclinations, and migrates into the disk. Nearly all the planetesimals encounter Neptune repeatedly. However, the planetesimals' eccentricities oscillate due to the effects of mean-motion, secular, and Kozai resonances (Duncan et al. 1995; Gomes et al. 2005a). If Neptune were not migrating, the planetesimals could not decouple from the planet. The migration opens two new avenues. The libration amplitude of a planetesimal's orbit can decrease, allowing the body to reach a stable orbit in a mean-motion resonance. Alternatively, a planetesimal can leave the resonance with a small eccentricity and attain a stable orbit as a hot classical KBO (Nesvorný 2015a). Gomes (2003) estimates the efficiency of the latter mechanism as $\approx$0.2 %, so a 50 $M_\oplus$ disk would yield $\approx$0.1 $M_\oplus$ in hot classicals.

Brown (2001) and Levison and Stern (2001) found that the Kuiper Belt was a superposition of dynamically "hot" and "cold" populations, with the former having a wide range of orbital inclinations and the latter having only small inclinations. Levison and Stern (2001) deduced that the intrinsically brightest, and presumably largest, objects were more likely to have large inclinations. Tegler and Romanishin (2000) reported that "ultrared" KBOs (as Jewitt 2002 calls them) occur only in the cold classicals. (See Barucci et al. 2005 for a broadband color taxonomy of Centaurs and KBOs, Peixinho et al. 2015 for the current status of correlations between optical colors and orbital parameters for KBOs, and Jewitt 2015 for colors of comets and "related bodies".)

Despite these differences between the hot and cold populations, Levison and Morbidelli (2003) proposed that *all* the bodies in the Kuiper Belt formed closer to the Sun than they are now. In particular, Levison and Morbidelli (2003) suggested that the cold KBOs were "pushed outward" by Neptune's 2:1 MMR and dropped out of the resonance when their eccentricities were low. As in the mechanism of Gomes et al. (2004) for producing hot KBOs, the eccentricities of the bodies that become cold KBOs oscillate, but in this case the oscillations are due to a secular resonance embedded within the 2:1 MMR. Levison and Morbidelli (2003) find that $\approx 3\ M_\oplus$ must be librating in the 2:1 resonance for this secular resonance to occur. About 1 % of the bodies would end up as cold classicals, so the mass of this population should be $\approx 0.03\ M_\oplus$, much larger than the mass of $\approx 3 \times 10^{-4}\ M_\oplus$ inferred by Fraser et al. (2014). Other bodies would remain in the 2:1 resonance to this day.

The models of Gomes et al. (2004) and Levison and Morbidelli (2003) were formulated before the Nice model was developed, and assumed the migration of the planets was fairly smooth. Since the planets in the Nice model can "jump" as they encounter each other, Levison et al. (2008) reinvestigated their hypothesis that the entire KBO population initially resided inside 30 AU and was pushed outward by a migrating Neptune. They found a new mechanism that relies on a transient phase when Neptune was scattered by Uranus to have an eccentricity ≈ 0.3. Neptune's eccentricity would then damp in ≈ 1 Myr via dynamical friction with the planetesimals. As Neptune's eccentricity decreased, its resonances would no longer overlap in the Kuiper Belt, and some KBOs would be stranded 40–50 from the Sun. Murray-Clay and Schlichting (2011), Batygin et al. (2011), Dawson and Murray-Clay (2012), and Wolff et al. (2012) also consider a transient eccentric phase for Neptune's orbit.

Nesvorný (2015a,b) propose a model in which Neptune's eccentricity and inclination were never large. Nesvorný (2015a) finds that the inclinations of the hot classical KBOs can be excited to their observed values if Neptune began its outward migration within 25 AU and the migration was slow, with an e-folding timescale $\tau \geq 10$ Myr. Dawson and Murray-Clay (2012) found that capturing the hot classical KBOs typically requires a phase in which Neptune has a moderately eccentric orbit ($e > 0.15$), while a small eccentricity for Neptune is more conducive to survival of the cold classicals. Nesvorný (2015a) found that the mechanism of Gomes (2003) works to capture hot classical KBOs (though probably less efficiently) even if Neptune's eccentricity always stays small, assuming its migration was slow. The companion paper (Nesvorný 2015b) explains the Kuiper Belt "kernel," defined by Petit et al. (2011) as the part of the cold classical population with semi-major axes a between 43.8 and 44.4 AU and eccentricities between 0.03 and 0.08. Nesvorný (2015b) suggests that a concentration of KBOs in this region would have occurred if planetesimals originally located near 40 AU were captured into Neptune's 2:1 resonance and then left the resonance because Neptune's orbit "jumped" by ≈ 0.5 AU. Such a jump might occur due to an encounter between Neptune and an ice giant that was later ejected from the Solar System (Sect. 4.6). In Nesvorný's model, all the cold classical KBOs are presumed to have formed at $a \gtrsim 40$ AU, while most hot classicals formed at $a \lesssim 30$ AU.

6.3 Binaries as a Constraint on the History of the Kuiper Belt

Some 30 % of all cold classical Kuiper Belt Objects are binaries (Stephens and Noll 2006; Noll et al. 2008a), and most cold classicals with absolute magnitudes $H < 6.1$ are binaries (Noll et al. 2014). Binaries in the Kuiper Belt are unique in that the binary components tend to have nearly equal masses, and they are often very widely separated (from 10–2000 times the radius of one of the bodies). The wider separations are a significant fraction of the binary system's Hill radius, R_H; for instance, the components of 2001 QW_{322} are separated by $\sim 0.22\ R_H$ (Parker et al. 2011). Wide binaries are weakly bound systems, and their very existence provides a unique constraint on the dynamical evolution and conditions of the Kuiper Belt over the history of the Solar System.

Petit and Mousis (2004) were the first to use KBO binaries as a probe of the Kuiper Belt's dynamics by determining how easily wide binaries could be disrupted via encounters with other KBOs. They investigated three encounter channels: (i) shattering of one binary component from an energetic collision with a passing KBO; (ii) diffusion of the binary's binding energy via velocity impulses from impacts on the primary and secondary; and (iii) diffusion of the binary's binding energy via velocity impulses from close encounters with passing KBOs. They found that (ii) was much more efficient at unbinding binaries than the other two mechanisms. In addition, the unbinding timescales found for extremely wide KBO binaries such as 1998 WW_{31} and 2001 QW_{322} (Petit et al. 2008) were significantly shorter than the age of the Solar System. They concluded that there must have been ~7 times more extremely wide binaries in the early Solar System.

In the scenario of Levison et al. (2008), KBOs are scattered outward into the cold classical belt during Neptune's eccentric phase. These KBOs likely undergo one or more close encounters with Neptune. Parker and Kavelaars (2010) showed that such encounters are very likely to disrupt the widest KBO binaries. If the existing wide binaries are all primordial, this would require an impossible primordial binary fraction of over 100 % to compensate for the disruption in this scenario. Alternatively, perhaps some of the wide binaries may be a transient population derived from tighter binaries, but it appears unlikely that this is the dominant mechanism for producing them (Nesvorný et al. 2011, Fig. 9). If wide binaries are primordial, it appears that cold classical KBOs must have either formed *in situ* or been implanted via a gentler dynamical pathway. In contrast, encounters with Neptune may naturally explain why the hot component of the Kuiper Belt has a binary fraction that is a factor of ~3 lower than that of the cold classical belt (Noll et al. 2008b).

Petit and Mousis (2004) inferred short binary lifetimes because they assumed a steep power law size distribution with index $q = 4.5$ (see Sect. 7) down to KBOs with diameter 10 km. This conflicts with more recent observations of KBOs, which suggest that the size distribution becomes shallower (i.e., the value of q is smaller) below diameters of 50–100 km (e.g., Fraser et al. 2010). Parker et al. (2011) calculated that 2001 QW_{322} would have a lifetime against "collisional unbinding" of 26 Gyr for $q = 2$ and 2.7 Gyr for $q = 3.5$. (These correspond to survival probabilities of 86 % and 23 %, assuming exponential decay of the starting population for 4 Gyr.) These lifetimes are longer than calculated by Petit and Mousis (2004) because size distributions with smaller values of q have fewer projectiles capable of disrupting binaries. Pan and Sari (2005) suggested that the break in the KBO size distribution is the result of collisional grinding. Nesvorný et al. (2011) repeated the KBO binary lifetime calculations with the size distribution from Fraser et al. (2010). The wider binaries have lower survival rates than the tighter binaries, as do smaller primaries in most cases. For binary separations of 0.001, 0.01, and 0.1 R_H, Nesvorný et al. (2011) found that the primary needs to have a diameter $\gtrsim$60, 100, and 160 km, respectively in order to survive for the age of the Solar System.

More importantly, Nesvorný et al. (2011) showed that while such grinding can produce a rollover in the size distribution, it requires that the early Kuiper Belt possessed a much larger mass in large bodies than it has today. This much larger mass would efficiently disrupt the widest KBO binaries via the same mechanisms as in Petit and Mousis (2004). Since there is no strong gradient of KBO binary fraction with the separation of the binary or the size of the components, they conclude that the existence of ultra-wide KBO binaries rules out a collisional grinding scenario.

7 Population Estimates

7.1 Population and Size Distribution of the Kuiper Belt

Until 1992, the entire known sample of KBOs consisted of one object, Pluto. It was not until 1998 that a single, well-characterized survey detected enough KBOs to allow a size distribution to be estimated with any statistical confidence (Jewitt et al. 1998). Jewitt et al. (1998) fit a single exponential to the luminosity function of the 13 KBOs they discovered:

$$\Sigma(< R) = 10^{\alpha(R-R_0)}, \tag{2}$$

where $\Sigma(< R)$ is the cumulative sky surface density of all KBOs with apparent red magnitude brighter than R. If the albedos of the KBOs, typically assumed to be 4 % in the R band, are independent of their sizes and the bodies detected are all at the same distance, the luminosity function in Eq. (2) can be transformed into a power-law size distribution:

$$\frac{dN}{dD} \propto D^{-q}. \tag{3}$$

In Eq. (3), dN/dD is the differential number distribution of KBOs and D is the diameter of a KBO. The power-law exponent, q, is equal to $5\alpha + 1$. Jewitt et al. (1998) performed a Monte Carlo simulation to account for the range of distances of the bodies they discovered, and found a best fit of $\alpha = 0.58 \pm 0.05$ or $q = 4.0 \pm 0.5$ based on their survey of bodies with R magnitudes from 20 to 23 and previous surveys to $R = 25$.

More recent work (e.g., Gladman et al. 1998) uses a Bayesian analysis to infer the KBO luminosity function. Using a survey with 17 KBO detections, Gladman et al. (2001) found that all surveys to that date could be fit with a slightly steeper luminosity function with $\alpha = 0.69$ ($q \approx 4.4$) for KBOs with $20 \lesssim R \lesssim 27$. Such a steep single power-law cannot continue to arbitrarily small sizes, as this would imply an infinite mass for the KBO population. (This is true for any distribution with $\alpha \geq 0.6$.) These early surveys only detected objects with $D >$ 50–100 km, and as a result the small end of the size distribution remained unconstrained, though a rollover to a shallower slope was expected at some small object size. Evidence for such a rollover came when Bernstein et al. (2004, 2006) used HST to survey a tiny 0.02 deg^2 patch of the invariable plane down to a limiting magnitude of 29.2, far deeper than any ground-based KBO survey. Only three new objects were discovered (with magnitudes between 26 and 28.3), whereas an extrapolation of a size distribution with $\alpha \simeq 0.6$ would predict about 85 new detections. This was clear evidence for a change in the size distribution of the Kuiper Belt at fainter magnitudes. By including the detections of several other well-calibrated ground-based surveys, Bernstein et al. (2004, 2006) argued that the KBO size distribution was well-fit with a double power-law with a break near $D \simeq 100$ km. Other small-body populations in the outer Solar System share a similar "knee" (Fig. 1).

In addition, Bernstein et al. (2004, 2006) found evidence that dynamically hot and cold Kuiper Belt Objects had different size distributions, with the bright end of the cold population being much steeper than the hot [$\alpha \geq 0.97$ vs. $\alpha = 0.66(+0.14, -0.08)$]. However, this result was attained by combining detections from several different surveys, as necessitated by the available data at the time. Fraser and Kavelaars (2009) demonstrate that multiple surveys cannot be combined to reliably compare the size distribution of KBO subpopulations, as calibration uncertainties in the different surveys can manifest themselves as apparent size distribution differences. Fuentes and Holman (2008) find significant differences between the size distributions of hot and cold KBOs using their survey alone, which discovered 82 KBOs. Like Bernstein et al. (2004, 2006), they find that the cold objects have a steeper bright-end power-law [$\alpha = 1.2(+0.3, -0.4)$] than hot KBOs [$\alpha = 0.7(+0.2, -0.1)$]. Similarly, utilizing a single survey with 169 KBO detections, Petit et al. (2011) find $\alpha = 1.2(+0.2, -0.3)$ for cold KBOs and $\alpha = 0.8(+0.3, -0.2)$ for hot KBOs. In summary, cold KBOs appear to have proportionately fewer large bodies than do hot KBOs, but the slope of the size distribution is still uncertain. The biggest KBOs (Pluto, Eris, Makemake, Haumea, etc.) are all members of the hot population (Levison and Stern 2001; Brown 2008; Fraser et al. 2010; Schwamb et al. 2014).

A break in the size distribution is widely expected from models of planetesimal accretion (e.g., Pan and Sari 2005; Kenyon and Bromley 2005; Benavidez and Campo Bagatin 2009). As smaller bodies accrete into the largest bodies of the KBO population, this sets the steep slope of the large end of the size distribution. As accretion progresses, more large bodies are created and the slope becomes flatter (i.e., α becomes smaller). As KBOs' relative velocities are increased via external forcing or self-stirring, collisions eventually become disruptive for small KBOs, and at this point collisional fragmentation controls the size distribution slope at the small end. Thus, such models would predict the break radius of the KBO size distribution marks the transition between these two regimes. Note, however, that Nesvorný et al. (2011) find that a collisional origin of the break is not consistent with the observed properties of binaries among cold classical KBOs. (One reason why the results of Nesvorný et al. 2011 differ from previous work may be their use of more realistic criteria for catastrophic disruption from Leinhardt and Stewart 2009 and Stewart and Leinhardt 2009.) Nesvorný et al. (2011) argue that the break may reflect the process by which KBOs formed. See Sect. 6.3 for further discussion.

In a study of Scattered Disk Objects, Shankman et al. (2013) find that instead of a knee or break in a double power-law size distribution, their detections suggest a sudden decrease ("divot") in the differential size distribution below $D < 100$ km (see Fig. 17). Alexandersen et al. (2014) find a similar divot for Plutinos. The presence of a divot may be consistent with the propagation of a power-law break from $D = 1$ km to $D = 100$ km over the course of planetesimal accretion (Fraser 2009). However, it could also be a signature that planetesimals were born large with $D \sim 100$ km (Morbidelli et al. 2009a), and bodies smaller than this are products of collisional fragmentation, while larger bodies are produced via accretion. The existence of a divot just below $D < 100$ km is also consistent with observations by Sheppard and Trujillo (2010), who find a near-absence of Neptunian Trojans smaller than $D < 100$ km (see Fig. 1). One advantage of a divot is that it can yield an overall size distribution that then picks up in number again at still smaller objects ($1 < D < 10$ km). Volk and Malhotra (2008) show that the best-fit double power-law suggested in Bernstein et al. (2004, 2006) yields a Scattered Disk population of km-sized bodies that is a factor of $\sim$100 too small to produce the population of Jupiter-family comets observed near Earth. However, the size distribution of Shankman et al. (2013), with a divot and a faint-end slope $\alpha_f \sim 0.5$ ($q \sim 3.5$) yields $\sim 2 \times 10^9$ bodies with absolute magnitudes in g-band < 18 ("comet-sized" bodies with diameters $\gtrsim$1.6 km) in the Scattered Disk, more than enough to explain the observed number of JFCs. Similarly, Brasser and Morbidelli (2013) use the known number of Jupiter-family comets with perihelion distances < 2.5 AU to estimate that the Scattered Disk contains $1.7(+3.0, -0.9) \times 10^9$ bodies with diameters $\gtrsim 2.3$ km and Brasser and Wang (2015) find $5.9(+2.2, -5.1) \times 10^9$ for the same quantity.

Occultation experiments are, in principle, the best way to detect comet-sized KBOs. Schlichting et al. (2009, 2012) reported the discovery of two KBOs, each with a diameter $\approx$1 km, in serendipitous occultations of stars in HST Fine Guidance Sensor data. Schlichting et al. (2009, 2012) infer from these observations that enough small KBOs are present to

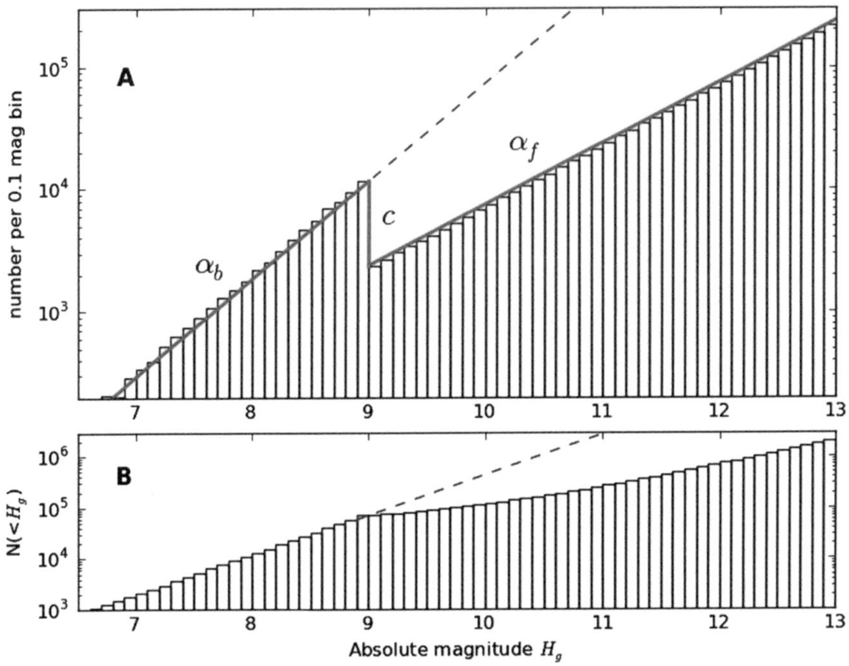

Fig. 17 Histograms of the distribution of absolute magnitudes for the preferred divot solution for Scattering Objects from Shankman et al. (2013). Top panel: The differential distribution (*solid green*), with $\alpha_b = 0.8$ for $H_g < 9$ and $\alpha_f = 0.5$ for $H_g > 9$. The contrast $c \sim 6$ is the ratio of the numbers of bodies per magnitude bin on either side of the divot. Bottom panel: The cumulative distribution. The slope is very flat for magnitudes slightly fainter than the divot, but has steepened by $H_g \approx 13$

supply the Jupiter-family comets. However, all estimates of the number of comet-sized bodies in the Kuiper Belt/Scattered Disk will be highly model-dependent until they are detected in much larger numbers by occultation experiments (see Sect. 9).

7.2 Population of the Oort Cloud

The total population of the Oort Cloud has long been debated. Oort (1950) estimated the cloud contained "about 10^{11} comets of observable size". Determining the population of the cloud is made difficult by two factors. First, we can only infer the number of comets in the cloud through the flux of long-period comets (LPCs; traditionally considered to be comets whose orbital periods exceed 200 years (Sect. 2)) that pass through perihelion close enough to the Sun to become visible to observers from Earth. Long-period comets can be divided into new comets and returning comets. The new comets are a proxy for the total population in the cloud (Wiegert and Tremaine 1999). New comets are traditionally classified as those with semi-major axis $a > 10{,}000$ AU (Wiegert and Tremaine 1999), although that criterion is only a rough guide (Dybczyński and Królikowska 2015) [see Sect. 2.1]. The difficulty arises from the low flux of comets and their inherent faintness, so that determination of the true flux is fraught with large uncertainties, mostly caused by observational incompleteness. The second difficulty arises from the uncertain relation between a comet's absolute magnitude and its size, so that the total population of the Oort Cloud is tied to the smallest size of a comet that one incorporates into the observed sample. We discuss each factor after we briefly explain the agents responsible for making comets visible.

There are two main perturbers of comets in the cloud onto orbits that enter the inner Solar System: passing stars (Weissman 1980; Hills 1981) and the galactic tide (Heisler and Tremaine 1986; Levison et al. 2001). Passing stars usually cause small random deviations in the orbital energy and angular momentum of a comet (Weissman 1980), save for rare exceptions when massive stars pass close to the Sun (Fouchard et al. 2011a). The galactic tide, on the other hand, systematically modifies the orbital angular momenta of the comets without changing the comets' orbital energies (Heisler and Tremaine 1986). In other words, the tide only changes the eccentricity (and thus the perihelion distance).

Oort (1950) discovered that the LPCs have a spike in their energy distribution whose peak corresponds to comets with semi-major axes $a > 10{,}000$ AU.[21] Hills (1981) determined that comets in the spike were overrepresented compared to comets with smaller semi-major axes because of a selection effect unrelated to the fading problem. (See the discussion of the "Jupiter barrier" in Sect. 2.1.) Hills found that the energy perturbations from Jupiter and Saturn were typically larger than the binding energy (proportional to $1/a$) of the comets in the spike, and thus to become visible, a comet had to have its perihelion distance, q, sweep inward across the orbits of Jupiter and Saturn in a single revolution. In other words, for a comet to become visible its perihelion distance had to decrease in a single orbit by $\gtrsim 10$ AU. Comets with more gradual decreases in q would likely be ejected from the Solar System by Jupiter and Saturn before they became visible. The magnitude of the typical perihelion change per orbit (Δq) is much bigger for comets with larger semi-major axes ($\Delta q \propto a^{7/2}$). Hills (1981) calculated that stellar perturbations would be strong enough to torque the perihelion from beyond Saturn to within a few AU of the Sun in a single revolution if the comet's semi-major axis exceeded 20,000 AU. Heisler and Tremaine (1986) obtained nearly the same threshold semi-major axis for the action of the galactic tides.

Heisler and Tremaine (1986) approximated the galactic tide by only using the vertical component (i.e., the differential attraction of the Sun and comet toward the galactic disk), which is an order of magnitude stronger than the radial component. In this approximation the z-component of the comet's orbital angular momentum in the galactic plane is conserved and the comets follow closed trajectories in the $q - \omega$ plane (with q being the perihelion distance and ω being the argument of perihelion). (See, e.g., Breiter and Ratajczak 2005 and Higuchi et al. 2007 for recent discussions.) Including the radial tide due to the differential force from the center of the Galaxy breaks this conservation (Matese and Whitmire 1996; Fouchard et al. 2005, 2006a,b) and the flux of comets to the inner Solar System from the cloud is increased slightly (Levison et al. 2006b). The trajectories that lead comets into the inner Solar System should be quickly depleted, were it not for passing stars to refill them (Rickman et al. 2008; Fouchard et al. 2011a). The synergy between these two perturbing agents ensures there is a roughly steady supply of comets entering the inner Solar System (Fouchard et al. 2011a).

[21] Oort (1950) found that, relative to the number of comets in the spike, fewer "returning" comets than expected were seen. Fitting the energy distribution of long-period comets required Oort to assume that "new or almost new" comets were unusually bright and that comets had a ≈ 1 % chance of being disrupted during each perihelion passage. This "fading" problem persists even now. Wiegert and Tremaine (1999) define fading as "all factors that reduce the intrinsic brightness of the comet near perihelion, and includes splitting into two or more large pieces, disruption into many small pieces, the depletion of volatiles, and the formation of insulating crusts of refractory materials." Wiegert and Tremaine (1999) investigate fading in great detail, and find they can match the observed orbit distribution of long-period comets if the fraction of comets remaining observable after m apparitions $\propto m^{0.6\pm 0.1}$ or if ≈ 95 % of comets survive for only ≈ 6 returns and the rest last indefinitely.

In reality the story is more complex because the Jupiter-Saturn region is not completely opaque (Kaib and Quinn 2009; Fouchard et al. 2011a, 2013), so that comets survive multiple passages through the Jupiter-Saturn region before reaching small perihelion distances and becoming visible (Fouchard et al. 2011b; Królikowska and Dybczyński 2010). In addition there are occasional comet showers, events during which the flux of comets from the Oort Cloud passing through the inner Solar System is greatly enhanced (Hills 1981; Heisler et al. 1987). Usually a comet shower is triggered by a close stellar passage (Rickman et al. 2008; Fouchard et al. 2011a) and lasts a few Myr (Weissman and Hut 1986; Dybczyński 2002). The current Solar System is most likely not experiencing a comet shower, so the flux observed today is the background flux that can be used to estimate the total number of comets in the cloud (Fouchard et al. 2011b).

Even though the perturbations on the cloud are now mostly understood, there remain large uncertainties in its population, in part because of remaining uncertainties in the flux of new comets, but primarily because the conversion between cometary absolute magnitude and the size of the nucleus remains uncertain. There have been many attempts to constrain the total population (Oort 1950; Everhart 1967b; Hills 1981; Weissman 1983, 1996; Wiegert and Tremaine 1999; Francis 2005). The first robust estimate of the flux of LPCs through the inner Solar System was by Everhart (1967b), who estimated a flux of 8000 comets with perihelion distance $q < 4$ AU and absolute magnitude $H_T \leq 10.9$ from 1840 to 1967. (Everhart 1967b states that only 337 long-period comets were discovered during this 127-year period, with most of the 8000 comets judged to be unobservable even under ideal conditions.) Here H_T is given by

$$H_T = V - 5\log \mathcal{D} - 2.5\nu \log r, \tag{4}$$

where in Eq. (4) V is the apparent magnitude of the comet (nucleus with coma), $\mathcal{D}$ is the distance of the comet from Earth in AU, ν is a measure of how the brightness scales with heliocentric distance (the photometric index), and r is the distance of the comet from the Sun in AU. Thus H_T is the magnitude of a comet (nucleus plus coma) if viewed from the Sun, at a distance of 1 AU, and is intended to be an intrinsic property of the comet. Unfortunately, there is a great variability in ν among comets. Its value is 2 for an inactive (asteroidal) body, and >2 for active comets. The first to catalogue this quantity for a large sample of comets was Whipple (1978), who found that for new comets on their inbound leg, $\langle\nu\rangle = 2.44 \pm 0.30$ and on their outbound leg $\langle\nu\rangle = 3.35 \pm 0.27$. For short-period comets ν usually exceeds 3. From a limited sample of LPCs, Sosa and Fernández (2011) find $\langle\nu\rangle \sim 3$. Using $\nu = 4$ yields the commonly used value H_{10}, which is close to the average found by Whipple (1978) for all comets. Some authors have tried to take these variations into account when quoting the flux of LPCs, while others have assumed typical values of ν for different types of comets from Whipple (1978).[22] The most-cited sources estimating the flux using long-term data are Everhart (1967b), Hughes (2001), and Francis (2005). Francis (2005) estimates a flux of 11 long-period comets per year with $q < 4$ AU and $H_T < 10.9$, as opposed to 63 for Everhart (1967b).

[22]The absolute magnitude derived for a comet depends both on the aperture of the telescope used to observe it (since comets are extended sources) and the assumed value of the photometric index ν, which is not always stated. Even though both studies assumed $\nu = 4$, Kresák and Kresáková (1989) derived absolute magnitudes that differed, on average, from those of Vsekhsvyatskij (1958) by 1.9 magnitudes, with a standard deviation of 1.5 magnitudes. For instance, Kresák and Kresáková (1989) derive a mean absolute magnitude for 1P/Halley of 2.8 from its 1835, 1910, and 1986 apparitions, while Hughes (1988), who uses Vsekhsvyatskij's magnitudes, states that "during its last 30 apparitions, [Halley's] H_{10} value has remained sensibly constant at 5.5 ± 0.7."

Everhart (1967b) and Hughes (2001) fit an exponential law to the cumulative absolute magnitude distribution of LPCs [cf. Eq. (2)], but discovered that the slope changes near $H_T \sim 6$–6.5, i.e., there is a break in the distribution. This break could be caused either by observational incompleteness or by a change in the slope of the size distribution of the cometary nuclei. Francis (2005) studied a sample of 51 long-period comets that the LINEAR survey discovered or could have discovered in a 3-year period. His sample is fainter ($3.15 \leq H_T \leq 16.59$, with median 9.65) and more distant ($0.32 \leq q \leq 9.74$ AU, with median 3.2 AU) than those used by Everhart (1967b) and Hughes (2001). Only 12 of the comets had $H_T < 6.5$, so Francis adopted a differential magnitude distribution $dn/dH \propto 10^{\alpha H_T}$ with $\alpha = 0.359$ for $H_T < 6.5$ from Hughes (2001). For fainter comets Francis found $\alpha = 0.00 \pm 0.05$, i.e., the number of comets per unit magnitude is constant. This contrasts with the more common situation in which the number of bodies per unit magnitude increases steadily toward fainter magnitudes (although see the top panel of Fig. 17).

Fernández and Sosa (2012) find that the cumulative H_T distribution $N(< H_T)$ has several segments, with $\alpha = 0.28 \pm 0.10$ for $H_T < 4.0$, $\alpha = 0.56 \pm 0.10$ for $4.0 \leq H_T \leq 5.8$, $\alpha = 0.20 \pm 0.02$ for $5.8 \leq H_T \leq 8.6$, and $\alpha \lesssim 0.2$ for $8.6 \leq H_T \leq 12$. Using the relationship $D = 15.9 \times 10^{-0.13\, H_T}$ (Sosa and Fernández 2011) between a comet's absolute magnitude and the diameter of its nucleus, this corresponds to a cumulative size distribution $N(> D) \propto D^{-\beta}$, where D is the diameter of the nucleus and $\beta \sim 2.15$ for $D > 4.8$ km, $\beta \sim 4.31$ for $2.8\ \text{km} \leq D \leq 4.8$ km, and $\beta \sim 1.54$ for $1.2\ \text{km} \leq D \leq 2.8$ km. The slope must eventually steepen again for the largest comets, as the total amount of mass only converges if $\beta > 3$, assuming the internal density of a comet is independent of the size of the nucleus.

Fernández and Sosa (2012) argue the small value of α for $H_T > 5.8$ implies a real change (flattening) of the size distribution for the smallest cometary nuclei, rather than an artifact due to observational incompleteness. Brasser and Morbidelli (2013) circumvented these uncertainties by calculating the comet flux for LPCs with $H_T < 6.5$, which is thought to be reasonably complete (Hughes 2001). Using the data from Everhart (1967b), Hughes (2001), and Francis (2005), and taking one-third of LPCs to be new,[23] Brasser and Morbidelli (2013) derive a flux of new comets entering the inner Solar System of (0.76 ± 0.33)/year with $q < 2.5$ AU and $H_T < 6.5$.

The flux of new comets through the inner Solar System can be used to obtain the number of comets in the cloud if the fraction per unit time of the Oort Cloud that is visible in the form of new comets is known. The most reliable estimate of this fraction comes from Kaib and Quinn (2009), who report this rate as being $\approx 10^{-11}$/year.[24] By calibrating this probability to the flux of new comets above, Brasser and Morbidelli (2013) conclude that the whole cloud contains $(7.6 \pm 3.3) \times 10^{10}$ comets with $H_T < 6.5$. Likewise, Feng and Bailer-Jones (2015) infer that the Oort Cloud contains some 10^{11} comets.

Using the relationship between H_T and D derived by Sosa and Fernández (2011), $H_T = 6.5$ translates to $D = 2.3$ km. Thus Brasser and Morbidelli (2013) find that the Oort Cloud has a population of $(7.6 \pm 3.3) \times 10^{10}$ comets with diameters $D > 2.3$ km. In recent years the estimated population of the Oort Cloud has decreased substantially. Heisler (1990) inferred there were 5×10^{11} comets with $D > 2.3$ km and $a > 20{,}000$ AU (i.e., in the outer Oort Cloud). Weissman (1996) took the population of the outer Oort Cloud to be 10^{12} and,

[23] Using the traditional definition of a new comet as one with an original semi-major axis >10,000 AU, Wiegert and Tremaine (1999) assume one long-period comet in three is dynamically new, while Fernández and Sosa (2012) find that new comets are 30 ± 10 % of long-period comets with perihelia <1.3 AU.

[24] Kaib and Quinn (2009) estimate that the fraction of Oort Cloud comets reaching orbits with $q < 5$ AU per year is 1.4×10^{-11} and 0.9×10^{-11} for comets from the outer ($a > 20{,}000$ AU) and inner ($a < 20{,}000$ AU) clouds, respectively.

based on Duncan et al. (1987), the population of the inner cloud to be 5 times bigger, for a total Oort Cloud population of 6×10^{12} comets with $D > 2.3$ km. Francis (2005) estimated the outer Oort Cloud contained 2×10^{11} comets with $H_T < 11$ ($d \gtrsim 0.6$ km). The estimate by Brasser and Morbidelli (2013) is smaller still.

8 Cometary Analogs Around Other Stars

Individual comets around stars other than the Sun have, of course, not been imaged to date. However, there is circumstantial evidence for the existence of comets in a number of star systems (Montgomery and Welsh 2012). A notable example is β Pictoris, a ~23 Myr old main-sequence star whose nearly edge-on debris disk was the first to be imaged optically (Smith and Terrile 1984). Circumstellar gas has been detected in the spectrum of the star ever since, including thousands of transient events. These events have been interpreted as being due to transits by "Falling Evaporating Bodies," i.e., comets, passing within 0.5 AU of the parent star. Using eight years of optical spectra of β Pic, Kiefer et al. (2014) detected 493 "cometary clouds". They found that the star has two populations of "exocomets". One ("Population D") is characterized by a larger perihelion distance, aligned orbits (possibly due to the disruption of one or a few parent bodies), and deeper absorption lines. The other population ("S") has a smaller perihelion distance and shallower absorption lines, and may be trapped in a mean-motion resonance with a planet. (β Pictoris has a ~ 7-Jupiter mass planet 9 AU from the star. The possible resonance could be with this planet or with another planet that has yet to be discovered.) The "D" comets' having deeper absorption lines, despite being farther from the star when they transit, suggest they are bigger or "fresher" than the "S" comets (Roberge 2014).

Zuckerman and Song (2012) argue that observations of a CO-rich disk around the 40-Myr old star 49 Ceti is due to "colliding comets" in an extremely massive (400 $M_\oplus$) analog of the Kuiper Belt. Bonsor et al. (2011) find that a "Kuiper Belt-like planetesimal belt" could be the source of elements heavier than helium seen in the atmospheres of one-quarter of DA white dwarfs. Kennedy and Wyatt (2014) discuss 48 nearby stars whose debris disks can be well-modeled by dust with two temperatures, and suggest that many of these systems have two small-body belts, like the asteroid belt and Kuiper Belt in our Solar System.

Boyajian et al. (2015) consider several explanations for aperiodic dips in the flux of a main-sequence F star, KIC 8462852, observed by the Kepler mission. Their preferred scenario involves "the passage of a family of exocomet fragments," which might be a comet shower caused by a known nearby M star.[25]

9 Summary

Key developments in recent years that relate to the evolution of the cometary reservoirs include:

1. Recognition that the giant planets migrated substantially after they formed, possibly in two different episodes described by the Grand Tack and the Nice model.

[25]One dimming event reduced the star's light by 20 %. Wright et al. (2015) state that the light curve is "consistent with a 'swarm' of megastructures," i.e., a Dyson sphere. The SETI Institute is observing KIC 8462852 with the Allen Telescope Array to investigate this possibility (http://www.seti.org/seti-institute/mysterious-star-kic-8462852).

2. The discovery of main-belt comets, as well as ice and organics in C-type asteroids.
3. Better understanding of the size distribution of Kuiper Belt Objects and the Trojans of Jupiter and Neptune, suggesting that there may be a characteristic 100-km scale for planetesimals.
4. A reduced estimate for the population of the Oort Cloud.
5. The discovery of "inner Oort" bodies such as Sedna, which provide evidence that the Sun formed in a large cluster.

We can look forward to further revelations during the Rosetta mission, which will continue through September 2016. The James Webb Space Telescope (JWST), scheduled for launch in 2018, will observe comets at wavelengths from 0.6 to 28.5 μm (Norwood et al. 2014; Milam et al. 2015; Kelley et al. 2015a). Rivkin et al. (2015) list observations aimed at determining the true nature of main-belt comets as an example of a project "taking advantage of multiple aspects of JWST". The Large Synoptic Survey Telescope (LSST) hopes to see first light in 2019. LSST should discover many long-period comets with distant perihelia, thus elucidating the structure of the Oort Cloud. LSST will also measure precise orbits for Kuiper Belt Objects, providing constraints on planetary migration models (Jones and Brown 2015). After its stunning traverse of the Pluto system, New Horizons plans to fly by a small cold classical KBO, 2014 MU_{69}, in January 2019. Eventually, an occultation survey such as TAOS II (http://taos2.asiaa.sinica.edu.tw/) on Earth or Whipple (http://www.lpi.usra.edu/decadal/sbag/topical_wp/CharlesRAlcock.pdf) in space will detect vast numbers of small Kuiper Belt Objects and inner Oort Cloud comets in their native milieus for the first time.

Acknowledgements We thank Bob Johnston for discussions about his database of comets; my tweeps Alessondra Springmann, Sarah Hörst, Chris Granade, Brian Wolven, and especially Andy Kass for help fixing a last-minute BibTeX disaster; Bill Bottke, Robert Jedicke, and Mikael Granvik for data on the size distribution of asteroids; Paul Weissman for sharing his work in advance of publication; Alfred McEwen and Alan Delamere for information on HiRISE observations of comet C/2013 A1; Scott Sheppard, Meg Schwamb, Lucie Maquet, Kevin Walsh, Konstantin Batygin, and Cory Shankman for providing figures; David Nesvorný for discussions about the Kuiper Belt and the Late Heavy Bombardment; the referees for helpful comments; and Nirmala Kumar at Springer and the editors, especially Kathy Mandt, for their patience. The Astrophysics Data System Abstract Service and the arXiv were of great value in searching the literature. We thank Iggy Confidential for motivation during the revisions.

References

F.C. Adams, The birth environment of the Solar System. Annu. Rev. Astron. Astrophys. **48**, 47–85 (2010)

F.C. Adams, G. Laughlin, Constraints on the birth aggregate of the Solar System. Icarus **150**, 151–162 (2001)

C.B. Agnor, D.N.C. Lin, On the migration of Jupiter and Saturn: constraints from linear models of secular resonant coupling with the terrestrial planets. Astrophys. J. **745**, 143 (2012)

M.F. A'Hearn, M.J.S. Belton, W.A. Delamere, L.M. Feaga, D. Hampton, J. Kissel, K.P. Klaasen, L.A. McFadden, K.J. Meech, H.J. Melosh, et al., EPOXI at comet Hartley 2. Science **332**, 1396–1400 (2011)

C.M.O. Alexander, R. Bowden, M.L. Fogel, K.T. Howard, C.D.K. Herd, L.R. Nittler, The provenances of asteroids, and their contributions to the volatile inventories of the terrestrial planets. Science **337**, 721–723 (2012)

R. Alexander, Planet formation in evolving protoplanetary discs, in *Exploring the Formation and Evolution of Planetary Systems*, ed. by M. Booth, B.C. Matthews, J.R. Graham. IAU Symposium, vol. 299 (2014), pp. 179–189

R. Alexander, I. Pascucci, S. Andrews, P. Armitage, L. Cieza, The dispersal of protoplanetary disks, ed. by H. Beuther, R.S. Klessen, C.P. Dullemond, T. Henning (Univ. Arizona Press, 2014), pp. 475–496

M. Alexandersen, B. Gladman, J.J. Kavelaars, J.-M. Petit, S. Gwyn, C. Shankman, A carefully characterised and tracked trans-Neptunian survey, the size-distribution of the Plutinos and the number of Neptunian Trojans. ArXiv e-prints 1411.7953 [astro-ph] (2014)

L. Allen, S.T. Megeath, R. Gutermuth, P.C. Myers, S. Wolk, F.C. Adams, J. Muzerolle, E. Young, J.L. Pipher, The structure and evolution of young stellar clusters, in *Protostars and Planets V*, ed. by B. Reipurth, D. Jewitt, K. Keil (2007), pp. 361–376

R.L. Allen, G.M. Bernstein, R. Malhotra, The edge of the Solar System. Astrophys. J. Lett. **549**, 241–244 (2001)

K. Altwegg, H. Balsiger, A. Bar-Nun, J.J. Berthelier, A. Bieler, P. Bochsler, C. Briois, U. Calmonte, M. Combi, J. De Keyser, et al., 67P/Churyumov-Gerasimenko, a Jupiter family comet with a high D/H ratio. Science **347**, 1261952 (2015)

S.M. Andrews, Observations of solids in protoplanetary disks. ArXiv e-prints 1507.04758 [astro-ph] (2015)

S.M. Andrews, D.J. Wilner, A.M. Hughes, C. Qi, K.A. Rosenfeld, K.I. Öberg, T. Birnstiel, C. Espaillat, L.A. Cieza, J.P. Williams, et al., The TW Hya disk at 870 μm: comparison of CO and dust radial structures. Astrophys. J. **744**, 162 (2012)

W. Applebaum, Keplerian astronomy after Kepler: researches and problems. Hist. Sci. **34**, 451–504 (1996)

P.J. Armitage, Lecture notes on the formation and early evolution of planetary systems (version 4, August 26, 2014). ArXiv e-prints astro-ph/0701485 (2014)

P.J. Armitage, Physical processes in protoplanetary disks. ArXiv e-prints 1509.06382 [astro-ph] (2015)

P.J. Armitage, *Astrophysics of Planet Formation* (Cambridge University Press, Cambridge, 2010)

E. Asphaug, W. Benz, Size, density, and structure of comet Shoemaker-Levy 9 inferred from the physics of tidal breakup. Icarus **121**, 225–248 (1996)

C.A.L. Bailer-Jones, Close encounters of the stellar kind. Astron. Astrophys. **575**, 35 (2015)

M.E. Bailey, S.V.M. Clube, W.M. Napier, *The Origin of Comets* (Pergamon Press, Elmsford, 1990)

R.B. Baldwin, Was there a 'terminal lunar cataclysm' 3.9 to 4.0 billion years ago? Icarus **23**, 157–166 (1974)

P. Barge, J. Sommeria, Did planet formation begin inside persistent gaseous vortices? Astron. Astrophys. **295**, 1–4 (1995)

A.C. Barr, R.M. Canup, Origin of the Ganymede-Callisto dichotomy by impacts during the Late Heavy Bombardment. Nat. Geosci. **3**, 164–167 (2010)

J.A. Barranco, P.S. Marcus, Three-dimensional vortices in stratified protoplanetary disks. Astrophys. J. **623**, 1157–1170 (2005)

M.A. Barucci, I.N. Belskaya, M. Fulchignoni, M. Birlan, Taxonomy of Centaurs and Trans-Neptunian objects. Astron. J. **130**, 1291–1298 (2005)

C. Baruteau, A. Crida, S.-J. Paardekooper, F. Masset, J. Guilet, B. Bitsch, R. Nelson, W. Kley, J. Papaloizou, Planet-disk interactions and early evolution of planetary systems, in *Protostars and Planets VI*, ed. by H. Beuther, R.S. Klessen, C.P. Dullemond, T. Henning (Univ. Arizona Press, Tucson, 2014), pp. 667–689

F. Bash, The present, past and future velocity of nearby stars—the path of the Sun in 10^8 years, in *The Galaxy and the Solar System*, ed. by R. Smoluchowski, J.M. Bahcall, M.S. Matthews (Univ. Arizona Press, Tucson, 1986), pp. 35–46

K. Batygin, M.E. Brown, H. Betts, Instability-driven dynamical evolution model of a primordially five-planet outer Solar System. Astrophys. J. Lett. **744**, 3 (2012)

K. Batygin, M.E. Brown, W.C. Fraser, Retention of a primordial cold classical Kuiper Belt in an instability-driven model of Solar System formation. Astrophys. J. **738**, 13 (2011)

K. Batygin, A. Morbidelli, M.J. Holman, Chaotic disintegration of the inner Solar System. Astrophys. J. **799**, 120 (2015)

M.J.S. Belton, The mass disruption of Jupiter family comets. Icarus **245**, 87–93 (2015)

P.G. Benavidez, A. Campo Bagatin, Collisional evolution of trans-Neptunian populations: effects of fragmentation physics and estimates of the abundances of gravitational aggregates. Planet. Space Sci. **57**, 201–215 (2009)

G.M. Bernstein, D.E. Trilling, R.L. Allen, M.E. Brown, M. Holman, R. Malhotra, The size distribution of trans-Neptunian bodies. Astron. J. **128**, 1364–1390 (2004)

G.M. Bernstein, D.E. Trilling, R.L. Allen, M.E. Brown, M. Holman, R. Malhotra, Erratum: "the size distribution of trans-Neptunian bodies". Astron. J. **131**, 2364 (2006)

I. Bertini, Main belt comets: a new class of small bodies in the solar system. Planet. Space Sci. **59**, 365–377 (2011)

H. Beuther, R.S. Klessen, C.P. Dullemond, T. Henning, *Protostars and Planets VI* (Univ. Arizona Press, Tucson, 2014)

J.-P. Bibring, M.G.G.T. Taylor, C. Alexander, U. Auster, J. Biele, A.E. Finzi, F. Goesmann, G. Klingelhoefer, W. Kofman, S. Mottola, et al., Philae's first days on the comet. Science **349**, 493 (2015)

E.B. Bierhaus, L. Dones, Craters and ejecta on Pluto and Charon: anticipated results from the New Horizons flyby. Icarus **246**, 165–182 (2015)

E.B. Bierhaus, L. Dones, J.L. Alvarellos, K. Zahnle, The role of ejecta in the small crater populations on the mid-sized Saturnian satellites. Icarus **218**, 602–621 (2012)

B. Bitsch, M. Lambrechts, A. Johansen, The growth of planets by pebble accretion in evolving protoplanetary discs. Astron. Astrophys. **582**, 112 (2015)
B. Bitsch, A. Morbidelli, E. Lega, A. Crida, Stellar irradiated discs and implications on migration of embedded planets. II. Accreting-discs. Astron. Astrophys. **564**, 135 (2014)
J. Blum, B. Gundlach, S. Mühle, J.M. Trigo-Rodriguez, Comets formed in solar-nebula instabilities!—An experimental and modeling attempt to relate the activity of comets to their formation process. Icarus **235**, 156–169 (2014)
J. Blum, B. Gundlach, S. Mühle, J.M. Trigo-Rodriguez, Corrigendum to 'Comets formed in solar-nebula instabilities!—An experimental and modeling attempt to relate the activity of comets to their formation process' [Icarus 235 (2014) 156–169]. Icarus **248**, 135–136 (2015)
D. Bockelée-Morvan, J. Crovisier, M.J. Mumma, H.A. Weaver, The composition of cometary volatiles, in *Comets II*, ed. by M.C. Festou, H.U. Keller, H.A. Weaver, (Univ. Arizona Press, Tucson, 2004), pp. 391–423
D. Bockelée-Morvan, N. Biver, B. Swinyard, M. de Val-Borro, J. Crovisier, P. Hartogh, D.C. Lis, R. Moreno, S. Szutowicz, E. Lellouch, et al., Herschel measurements of the D/H and $^{16}O/^{18}O$ ratios in water in the Oort-cloud comet C/2009 P1 (Garradd). Astron. Astrophys. **544**, 15 (2012)
D. Bockelée-Morvan, U. Calmonte, S. Charnley, J. Duprat, C. Engrand, A. Gicquel, M. Hässig, E. Jehin, H. Kawakita, B. Marty, et al., Cometary isotopic measurements. Space Sci. Rev. (2015). doi:10.1007/s11214-015-0156-9
A.C. Boley, The two modes of gas giant planet formation. Astrophys. J. Lett. **695**, 53–57 (2009)
A. Bonsor, A.J. Mustill, M.C. Wyatt, Dynamical effects of stellar mass-loss on a Kuiper-like belt. Mon. Not. R. Astron. Soc. **414**, 930–939 (2011)
A. Bonsor, S.N. Raymond, J.-C. Augereau, C.W. Ormel, Planetesimal-driven migration as an explanation for observations of high levels of warm, exozodiacal dust. Mon. Not. R. Astron. Soc. **441**, 2380–2391 (2014)
W.F. Bottke, H.F. Levison, D. Nesvorný, L. Dones, Can planetesimals left over from terrestrial planet formation produce the lunar Late Heavy Bombardment? Icarus **190**, 203–223 (2007)
W.F. Bottke, D. Vokrouhlický, D. Minton, D. Nesvorný, A. Morbidelli, R. Brasser, B. Simonson, H.F. Levison, An Archaean heavy bombardment from a destabilized extension of the asteroid belt. Nature **485**, 78–81 (2012)
W.F. Bottke, D. Vokrouhlický, S. Marchi, T. Swindle, E.R.D. Scott, J.R. Weirich, H. Levison, Dating the Moon-forming impact event with asteroidal meteorites. Science **348**, 321–323 (2015)
T.S. Boyajian, D.M. LaCourse, S.A. Rappaport, D. Fabrycky, D.A. Fischer, D. Gandolfi, G.M. Kennedy, M.C. Liu, A. Moor, K. Olah, et al., Planet Hunters X. KIC 8462852—where's the flux? ArXiv e-prints 1509.03622 [astro-ph] (2015)
R. Brasser, M.J. Duncan, An analytical method to compute comet cloud formation efficiency and its application. Celest. Mech. Dyn. Astron. **100**, 1–26 (2008)
R. Brasser, M.-H. Lee, Tilting Saturn without tilting Jupiter: constraints on giant planet migration. Astron. J. (2015). doi:10.1088/0004-6256/150/5/157
R. Brasser, A. Morbidelli, Oort Cloud and Scattered Disc formation during a late dynamical instability in the Solar System. Icarus **225**, 40–49 (2013)
R. Brasser, M.E. Schwamb, Re-assessing the formation of the inner Oort Cloud in an embedded star cluster—II. Probing the inner edge. Mon. Not. R. Astron. Soc. **446**, 3788–3796 (2015)
R. Brasser, J.-H. Wang, An updated estimate of the number of Jupiter-family comets using a simple fading law. Astron. Astrophys. **573**, 102 (2015)
R. Brasser, M.J. Duncan, H.F. Levison, Embedded star clusters and the formation of the Oort Cloud. Icarus **184**, 59–82 (2006)
R. Brasser, M.J. Duncan, H.F. Levison, Embedded star clusters and the formation of the Oort Cloud. II. The effect of the primordial solar nebula. Icarus **191**, 413–433 (2007)
R. Brasser, M.J. Duncan, H.F. Levison, Embedded star clusters and the formation of the Oort Cloud. III. Evolution of the inner cloud during the Galactic phase. Icarus **196**, 274–284 (2008)
R. Brasser, A. Higuchi, N. Kaib, Oort Cloud formation at various galactic distances. Astron. Astrophys. **516**, 72 (2010)
R. Brasser, K.J. Walsh, D. Nesvorný, Constraining the primordial orbits of the terrestrial planets. Mon. Not. R. Astron. Soc. **433**, 3417–3427 (2013)
R. Brasser, A. Morbidelli, R. Gomes, K. Tsiganis, H.F. Levison, Constructing the secular architecture of the solar system II: the terrestrial planets. Astron. Astrophys. **507**, 1053–1065 (2009)
R. Brasser, M.E. Schwamb, P.S. Lykawka, R.S. Gomes, An Oort Cloud origin for the high-inclination, high-perihelion Centaurs. Mon. Not. R. Astron. Soc. **420**, 3396–3402 (2012a)
R. Brasser, M.J. Duncan, H.F. Levison, M.E. Schwamb, M.E. Brown, Reassessing the formation of the inner Oort Cloud in an embedded star cluster. Icarus **217**, 1–19 (2012b)

S. Breiter, R. Ratajczak, Vectorial elements for the Galactic disc tide effects in cometary motion. Mon. Not. R. Astron. Soc. **364**, 1222–1228 (2005)
G.D. Brin, D.A. Mendis, Dust release and mantle development in comets. Astrophys. J. **229**, 402–408 (1979)
M.E. Brown, The inclination distribution of the Kuiper Belt. Astron. J. **121**, 2804–2814 (2001)
M.E. Brown, The largest Kuiper Belt Objects, in *The Solar System Beyond Neptune*, ed. by M.A. Barucci, H. Boehnhardt, D.P. Cruikshank, A. Morbidelli, R. Dotson (Univ. Arizona Press, Tucson, 2008), pp. 335–344
M.E. Brown, C. Trujillo, D. Rabinowitz, Discovery of a candidate inner Oort Cloud planetoid. Astrophys. J. **617**, 645–649 (2004)
M.E. Brown, C.A. Trujillo, D.L. Rabinowitz, Discovery of a planetary-sized object in the scattered Kuiper Belt. Astrophys. J. Lett. **635**, 97–100 (2005)
D. Brownlee, The Stardust mission: analyzing samples from the edge of the Solar System. Annu. Rev. Earth Planet. Sci. **42**, 179–205 (2014)
D. Brownlee, P. Tsou, J. Aléon, C.M.O. Alexander, T. Araki, S. Bajt, G.A. Baratta, R. Bastien, P. Bland, P. Bleuet, et al., Comet 81P/Wild 2 under a microscope. Science **314**, 1711–1716 (2006)
S.J. Bus, M.F. A'Hearn, E. Bowell, S.A. Stern, (2060) Chiron: evidence for activity near aphelion. Icarus **150**, 94–103 (2001)
H. Campins, K. Hargrove, N. Pinilla-Alonso, E.S. Howell, M.S. Kelley, J. Licandro, T. Mothé-Diniz, Y. Fernández, J. Ziffer, Water ice and organics on the surface of the asteroid 24 Themis. Nature **464**, 1320–1321 (2010)
R.M. Canup, A giant impact origin of Pluto-Charon. Science **307**, 546–550 (2005)
R.W. Carlson, L.E. Borg, A.M. Gaffney, M. Boyet, Rb-Sr, Sm-Nd and Lu-Hf isotope systematics of the lunar Mg-suite: the age of the lunar crust and its relation to the time of Moon formation. Philos. Trans. R. Soc. Lond. A **372**, 30246 (2014)
D. Carrera, A. Johansen, M.B. Davies, How to form planetesimals from mm-sized chondrules and chondrule aggregates. Astron. Astrophys. **579**, 43 (2015)
A. Carusi, L'. Kresák, G.B. Valsecchi, Conservation of the Tisserand parameter at close encounters of interplanetary objects with Jupiter. Earth Moon Planets **68**, 71–94 (1995)
A. Carusi, L. Kresak, E. Perozzi, G.B. Valsecchi, Peculiar orbital histories of some short-period comets. Bull. Am. Astron. Soc. **16**, 704 (1984)
A. Carusi, L. Kresak, E. Perozzi, G.B. Valsecchi, *Long-Term Evolution of Short-Period Comets* (Adam Hilger, Bristol, 1985)
A. Carusi, L. Kresak, E. Perozzi, G.B. Valsecchi, High-order librations of Halley-type comets. Astron. Astrophys. **187**, 899–905 (1987)
J.E. Chambers, Making more terrestrial planets. Icarus **152**, 205–224 (2001)
J.E. Chambers, Planetesimal formation by turbulent concentration. Icarus **208**, 505–517 (2010)
J.E. Chambers, Giant planet formation with pebble accretion. Icarus **233**, 83–100 (2014)
S.L. Chapin, Early ideas about comets 1650–1700. Leafl., Astron. Soc. Pac. **6**(278), 221–228 (1952)
C.R. Chapman, B.A. Cohen, D.H. Grinspoon, What are the real constraints on the existence and magnitude of the Late Heavy Bombardment? Icarus **189**, 233–245 (2007)
J. Chen, D. Jewitt, On the rate at which comets split. Icarus **108**, 265–271 (1994)
Y.-T. Chen, J.J. Kavelaars, S. Gwyn, L. Ferrarese, P. Côté, A. Jordán, V. Suc, J.-C. Cuillandre, W.-H. Ip, Discovery of a new member of the inner Oort Cloud from the Next Generation Virgo Cluster Survey. Astrophys. J. Lett. **775**, 8 (2013)
E. Chiang, A.N. Youdin, Forming planetesimals in solar and extrasolar nebulae. Annu. Rev. Earth Planet. Sci. **38**, 493–522 (2010)
R. Cloutier, D. Tamayo, D. Valencia, Could Jupiter or Saturn have ejected a fifth giant planet? Astrophys. J. **813**, 8 (2015)
R. Cloutier, T. Currie, G.H. Rieke, S.J. Kenyon, Z. Balog, R. Jayawardhana, A deep Spitzer survey of circumstellar disks in the young double cluster, h and χ Persei. Astrophys. J. **796**, 127 (2014)
A.L. Cochran, A.-C. Levasseur-Regourd, M. Cordiner, E. Hadamcik, J. Lasue, A. Gicquel, D.G. Schleicher, S.B. Charnley, M.J. Mumma, L. Paganini, et al., The composition of comets. Space Sci. Rev. (2015). doi:10.1007/s11214-015-0183-6
B.A. Cohen, N.E. Petro, S.J. Lawrence, What do Nectaris Basin impact melt rocks look like and where can we find them? in *Early Solar System Impact Bombardment III*, (2015), p. 3019. http://ntrs.nasa.gov/archive/nasa/casi.ntrs.nasa.gov/20150015512.pdf
B.A. Cohen, T.D. Swindle, D.A. Kring, Support for the lunar cataclysm hypothesis from lunar meteorite impact melt ages. Science **290**, 1754–1756 (2000)
C.J. Cohen, E.C. Hubbard, Libration of the close approaches of Pluto to Neptune. Astron. J. **70**, 10–13 (1965)
M. Ćuk, Chronology and sources of lunar impact bombardment. Icarus **218**, 69–79 (2012)

M. Ćuk, B.J. Gladman, S.T. Stewart, Constraints on the source of lunar cataclysm impactors. Icarus **207**, 590–594 (2010)
M. Ćuk, B.J. Gladman, S.T. Stewart, Rebuttal to the comment by Malhotra and Strom on "Constraints on the source of lunar cataclysm impactors". Icarus **216**, 363–365 (2011)
J.N. Cuzzi, R.C. Hogan, W.F. Bottke, Towards initial mass functions for asteroids and Kuiper Belt Objects. Icarus **208**, 518–538 (2010)
J.N. Cuzzi, R.C. Hogan, K. Shariff, Toward planetesimals: dense chondrule clumps in the protoplanetary nebula. Astrophys. J. **687**, 1432–1447 (2008)
J.N. Cuzzi, R.C. Hogan, J.M. Paque, A.R. Dobrovolskis, Size-selective concentration of chondrules and other small particles in protoplanetary nebula turbulence. Astrophys. J. **546**, 496–508 (2001)
B.J.R. Davidsson, P.J. Gutiérrez, Estimating the nucleus density of comet 19P/Borrelly. Icarus **168**, 392–408 (2004)
B.J.R. Davidsson, P.J. Gutiérrez, Nucleus properties of comet 67P/Churyumov Gerasimenko estimated from non-gravitational force modeling. Icarus **176**, 453–477 (2005)
B.J.R. Davidsson, P.J. Gutiérrez, Non-gravitational force modeling of comet 81P/Wild 2. I. A nucleus bulk density estimate. Icarus **180**, 224–242 (2006)
B.J.R. Davidsson, P.J. Gutiérrez, H. Rickman, Nucleus properties of comet 9P/Tempel 1 estimated from non-gravitational force modeling. Icarus **187**, 306–320 (2007)
B. Davidsson, H. Sierks, C. Güttler, F. Marzari, M. Pajola, H. Rickman, A.-T. Auger, M.R. El-Maarry, The primordial nucleus of comet 67P/Churyumov-Gerasimenko. Astron. Astrophys. (2015), submitted
M.B. Davies, F.C. Adams, P. Armitage, J. Chambers, E. Ford, A. Morbidelli, S.N. Raymond, D. Veras, The long-term dynamical evolution of planetary systems, ed. by H. Beuther, R.S. Klessen, C.P. Dullemond, T. Henning (Univ. Arizona Press, 2014), pp. 787–808
A.M. Davis, K.D. McKeegan, Short-lived radionuclides and early Solar System chronology, in *Treatise on Geochemistry*, 2nd edn., ed. by K.K. Turekian, H.D. Holland (Elsevier, Amsterdam, 2014), pp. 361–395
A.M. Davis, C.M.O. Alexander, F.J. Ciesla, M. Gounelle, A.N. Krot, M.I. Petaev, T. Stephan, Samples of the Solar System: recent developments, in *Protostars and Planets VI*, ed. by H. Beuther, R.S. Klessen, C.P. Dullemond, T. Henning (Univ. Arizona Press, Tucson, 2014), pp. 809–831
R.I. Dawson, R. Murray-Clay, Neptune's wild days: constraints from the eccentricity distribution of the classical Kuiper Belt. Astrophys. J. **750**, 43 (2012)
R. Deienno, T. Yokoyama, E.C. Nogueira, N. Callegari, M.T. Santos, Effects of the planetary migration on some primordial satellites of the outer planets. I. Uranus' case. Astron. Astrophys. **536**, 57 (2011)
R. Deienno, D. Nesvorný, D. Vokrouhlický, T. Yokoyama, Orbital perturbations of the Galilean satellites during planetary encounters. Astron. J. **148**, 25 (2014)
A. Delsanti, D. Jewitt, The Solar System beyond the planets, in *Solar System Update*, ed. by P. Blondel, J.W. Mason (Springer, Berlin, 2006), pp. 267–294
A.H. Delsemme, D.C. Miller, The continuum of comet Burnham (1960 II): the differentiation of a short period comet. Planet. Space Sci. **19**, 1229–1257 (1971)
F.E. DeMeo, B. Carry, Solar System evolution from compositional mapping of the asteroid belt. Nature **505**, 629–634 (2014)
F.E. DeMeo, R.P. Binzel, B. Carry, D. Polishook, N.A. Moskovitz, Unexpected D-type interlopers in the inner main belt. Icarus **229**, 392–399 (2014)
R.P. Di Sisto, A. Brunini, The origin and distribution of the Centaur population. Icarus **190**, 224–235 (2007)
R.P. Di Sisto, J.A. Fernández, A. Brunini, On the population, physical decay and orbital distribution of Jupiter family comets: numerical simulations. Icarus **203**, 140–154 (2009)
L. Dones, P.R. Weissman, H.F. Levison, M.J. Duncan, Oort Cloud formation and dynamics, in *Comets II*, ed. by M.C. Festou, H.U. Keller, H.A. Weaver (Univ. Arizona Press, Tucson, 2004), pp. 153–174
L. Dones, C.R. Chapman, W.B. McKinnon, H.J. Melosh, M.R. Kirchoff, G. Neukum, K.J. Zahnle, Icy satellites of Saturn: impact cratering and age determination, in *Saturn from Cassini-Huygens*, ed. by M.K. Dougherty, L.W. Esposito, S.M. Krimigis (2009), pp. 613–635
B. Donn, J. Rahe, Structure and origin of cometary nuclei, in *IAU Colloq. 61: Comet Discoveries, Statistics, and Observational Selection*, ed. by L.L. Wilkening (1982), pp. 203–226
P.M. Doyle, K. Jogo, K. Nagashima, A.N. Krot, S. Wakita, F.J. Ciesla, I.D. Hutcheon, Early aqueous activity on the ordinary and carbonaceous chondrite parent bodies recorded by fayalite. Nat. Commun. **6**, 7444 (2015)
R. Duffard, D. Lazzaro, S. Pinto, J. Carvano, C. Angeli, A. Alvarez-Candal, S. Fernández, New activity of Chiron: results from 5 years of photometric monitoring. Icarus **160**, 44–51 (2002)
D. Dukes, M.R. Krumholz, Was the Sun born in a massive cluster? Astrophys. J. **754**, 56 (2012)
M. Duncan, T. Quinn, S. Tremaine, The formation and extent of the Solar System comet cloud. Astron. J. **94**, 1330–1338 (1987)

M.J. Duncan, H.F. Levison, A scattered comet disk and the origin of Jupiter family comets. Science **276**, 1670–1672 (1997)
M.J. Duncan, H.F. Levison, S.M. Budd, The dynamical structure of the Kuiper Belt. Astron. J. **110**, 3073–3081 (1995)
P.A. Dybczyński, Simulating observable comets. I. The effects of a single stellar passage through or near the Oort cometary cloud. Astron. Astrophys. **396**, 283–292 (2002)
P.A. Dybczyński, M. Królikowska, Where do long-period comets come from? Moving through the Jupiter-Saturn barrier. Mon. Not. R. Astron. Soc. **416**, 51–69 (2011)
P.A. Dybczyński, M. Królikowska, Near-parabolic comets observed in 2006–2010—II. Their past and future motion under the influence of the Galaxy field and known nearby stars. Mon. Not. R. Astron. Soc. **448**, 588–600 (2015)
P. Eberhardt, M. Reber, D. Krankowsky, R.R. Hodges, The D/H and $^{18}O/^{16}O$ ratios in water from comet P/Halley. Astron. Astrophys. **302**, 301 (1995)
S. Eggers, H.U. Keller, P. Kroupa, W.J. Markiewicz, Origin and dynamics of comets and star formation. Planet. Space Sci. **45**, 1099–1104 (1997)
D.J. Eicher, *Comets!: Visitors from Deep Space* (Cambridge University Press, Cambridge, 2013)
J.L. Elliot, S.D. Kern, K.B. Clancy, A.A.S. Gulbis, R.L. Millis, M.W. Buie, L.H. Wasserman, E.I. Chiang, A.B. Jordan, D.E. Trilling, K.J. Meech, The Deep Ecliptic Survey: a search for Kuiper Belt Objects and Centaurs. II. Dynamical classification, the Kuiper Belt plane, and the core population. Astron. J. **129**, 1117–1162 (2005)
V.V. Emel'yanenko, D.J. Asher, M.E. Bailey, A new class of trans-Neptunian objects in high-eccentricity orbits. Mon. Not. R. Astron. Soc. **338**, 443–451 (2003)
V.V. Emel'yanenko, D.J. Asher, M.E. Bailey, Centaurs from the Oort Cloud and the origin of Jupiter-family comets. Mon. Not. R. Astron. Soc. **361**, 1345–1351 (2005)
E. Everhart, Comet discoveries and observational selection. Astron. J. **72**, 716–726 (1967a)
E. Everhart, Intrinsic distributions of cometary perihelia and magnitudes. Astron. J. **72**, 1002–1011 (1967b)
E. Everhart, Change in total energy of comets passing through the Solar System. Astron. J. **73**, 1039–1052 (1968)
E. Everhart, The origin of short-period comets. Astrophys. Lett. **10**, 131–135 (1972)
P. Farinella, D.R. Davis, Short-period comets: primordial bodies or collisional fragments? Science **273**, 938–941 (1996)
F. Feng, C.A.L. Bailer-Jones, Finding the imprints of stellar encounters in long period comets. Mon. Not. R. Astron. Soc. **454**(3), 3267–3276 (2015)
J.A. Fernández, On the existence of a comet belt beyond Neptune. Mon. Not. R. Astron. Soc. **192**, 481–491 (1980)
J.A. Fernández, The formation of the Oort Cloud and the primitive galactic environment. Icarus **129**, 106–119 (1997)
J.A. Fernández, A. Brunini, The buildup of a tightly bound comet cloud around an early Sun immersed in a dense galactic environment: numerical experiments. Icarus **145**, 580–590 (2000)
J.A. Fernández, W.-H. Ip, Some dynamical aspects of the accretion of Uranus and Neptune—the exchange of orbital angular momentum with planetesimals. Icarus **58**, 109–120 (1984)
J.A. Fernández, A. Sosa, Magnitude and size distribution of long-period comets in Earth-crossing or approaching orbits. Mon. Not. R. Astron. Soc. **423**, 1674–1690 (2012)
J.A. Fernández, G. Tancredi, H. Rickman, J. Licandro, The population, magnitudes, and sizes of Jupiter family comets. Astron. Astrophys. **352**, 327–340 (1999)
J.A. Fernández, *Comets: Nature, Dynamics, Origin, and Their Cosmogonical Relevance*. Astrophysics and Space Science Library (Springer, Berlin, 2005)
Y.R. Fernández, D.C. Jewitt, S.S. Sheppard, Albedos of asteroids in comet-like orbits. Astron. J. **130**, 308–318 (2005)
Y.R. Fernández, M.S. Kelley, P.L. Lamy, I. Toth, O. Groussin, C.M. Lisse, M.F. A'Hearn, J.M. Bauer, H. Campins, A. Fitzsimmons, et al., Thermal properties, sizes, and size distribution of Jupiter-family cometary nuclei. Icarus **226**, 1138–1170 (2013)
M.C. Festou, H.U. Keller, H.A. Weaver (eds.), *Comets II* (Univ. Arizona Press, Tucson, 2004)
M. Fischer-Gödde, H. Becker, What is the age of the Nectaris basin? New Re-Os constraints for a pre-4.0 Ga bombardment history of the Moon, in *Lunar and Planetary Science Conference*, vol. 42 (2011), p. 1414
M. Fouchard, C. Froeschlé, J.J. Matese, G. Valsecchi, Comparison between different models of Galactic tidal effects on cometary orbits. Celest. Mech. Dyn. Astron. **93**, 229–262 (2005)
M. Fouchard, C. Froeschlé, J.J. Matese, G.B. Valsecchi, Erratum: Comparison between different models of Galactic tidal effects on cometary orbits. Celest. Mech. Dyn. Astron. **96**, 341–344 (2006a)
M. Fouchard, C. Froeschlé, G. Valsecchi, H. Rickman, Long-term effects of the Galactic tide on cometary dynamics. Celest. Mech. Dyn. Astron. **95**, 299–326 (2006b)

M. Fouchard, C. Froeschlé, H. Rickman, G.B. Valsecchi, The key role of massive stars in Oort Cloud comet dynamics. Icarus **214**, 334–347 (2011a)
M. Fouchard, H. Rickman, C. Froeschlé, G.B. Valsecchi, The last revolution of new comets: the role of stars and their detectability. Astron. Astrophys. **535**, 86 (2011b)
M. Fouchard, H. Rickman, C. Froeschlé, G.B. Valsecchi, Planetary perturbations for Oort Cloud comets. I. Distributions and dynamics. Icarus **222**, 20–31 (2013)
M. Fouchard, H. Rickman, C. Froeschlé, G.B. Valsecchi, Planetary perturbations for Oort Cloud comets: II. Implications for the origin of observable comets. Icarus **231**, 110–121 (2014)
P.J. Francis, The demographics of long-period comets. Astrophys. J. **635**, 1348–1361 (2005)
W.C. Fraser, The collisional divot in the Kuiper Belt size distribution. Astrophys. J. **706**, 119–129 (2009)
W.C. Fraser, J.J. Kavelaars, The size distribution of Kuiper Belt Objects for $D \gtrsim 10$ km. Astron. J. **137**, 72–82 (2009)
W.C. Fraser, M.E. Brown, M.E. Schwamb, The luminosity function of the hot and cold Kuiper Belt populations. Icarus **210**, 944–955 (2010)
W.C. Fraser, M.E. Brown, A. Morbidelli, A. Parker, K. Batygin, The absolute magnitude distribution of Kuiper Belt Objects. Astrophys. J. **782**, 100 (2014)
C.I. Fuentes, M.J. Holman, A Subaru archival search for faint trans-Neptunian objects. Astron. J. **136**, 83–97 (2008)
E.J. Gaidos, Paleodynamics: Solar System formation and the early environment of the Sun. Icarus **114**, 258–268 (1995)
J. García-Sánchez, P.R. Weissman, R.A. Preston, D.L. Jones, J.-F. Lestrade, D.W. Latham, R.P. Stefanik, J.M. Paredes, Stellar encounters with the Solar System. Astron. Astrophys. **379**, 634–659 (2001)
R. Genzel, F. Eisenhauer, S. Gillessen, The Galactic center massive black hole and nuclear star cluster. Rev. Mod. Phys. **82**, 3121–3195 (2010)
A.M. Gilbert, P.A. Wiegert, Searching for main-belt comets using the Canada-France-Hawaii Telescope Legacy Survey. Icarus **201**, 714–718 (2009)
A.M. Gilbert, P.A. Wiegert, Updated results of a search for main-belt comets using the Canada-France-Hawaii Telescope Legacy Survey. Icarus **210**, 998–999 (2010)
B. Gladman, B.G. Marsden, C. Vanlaerhoven, Nomenclature in the outer Solar System, in *The Solar System Beyond Neptune*, ed. by M.A. Barucci, H. Boehnhardt, D.P. Cruikshank, A. Morbidelli, R. Dotson (Univ. Arizona Press, Tucson, 2008), pp. 43–57
B. Gladman, J.J. Kavelaars, P.D. Nicholson, T.J. Loredo, J.A. Burns, Pencil-beam surveys for faint trans-Neptunian objects. Astron. J. **116**, 2042–2054 (1998)
B. Gladman, J.J. Kavelaars, J.-M. Petit, A. Morbidelli, M.J. Holman, T. Loredo, The structure of the Kuiper Belt: size distribution and radial extent. Astron. J. **122**, 1051–1066 (2001)
B. Gladman, M. Holman, T. Grav, J. Kavelaars, P. Nicholson, K. Aksnes, J.-M. Petit, Evidence for an extended scattered disk. Icarus **157**, 269–279 (2002)
B. Gladman, J. Kavelaars, J.-M. Petit, M.L.N. Ashby, J. Parker, J. Coffey, R.L. Jones, P. Rousselot, O. Mousis, Discovery of the first retrograde transneptunian object. Astrophys. J. Lett. **697**, 91–94 (2009)
B. Gladman, S.M. Lawler, J.-M. Petit, J. Kavelaars, R.L. Jones, J.W. Parker, C. Van Laerhoven, P. Nicholson, P. Rousselot, A. Bieryla, M.L.N. Ashby, The resonant trans-Neptunian populations. Astron. J. **144**, 23 (2012)
P. Goldreich, S. Tremaine, Disk-satellite interactions. Astrophys. J. **241**, 425–441 (1980)
P. Goldreich, W.R. Ward, The formation of planetesimals. Astrophys. J. **183**, 1051–1062 (1973)
R.S. Gomes, The origin of the Kuiper Belt high-inclination population. Icarus **161**, 404–418 (2003)
R.S. Gomes, A. Morbidelli, H.F. Levison, Planetary migration in a planetesimal disk: why did Neptune stop at 30 AU? Icarus **170**, 492–507 (2004)
R.S. Gomes, E.C. Nogueira, R. Brasser, The fate of regular satellites during the Nice model's planetary close encounters, in *AAS/Division for Planetary Sciences Meeting Abstracts*, vol. 44 (2012). 415.02
R.S. Gomes, T. Gallardo, J.A. Fernández, A. Brunini, On the origin of the high-perihelion Scattered Disk: the role of the Kozai mechanism and mean motion resonances. Celest. Mech. Dyn. Astron. **91**, 109–129 (2005a)
R. Gomes, H.F. Levison, K. Tsiganis, A. Morbidelli, Origin of the cataclysmic Late Heavy Bombardment period of the terrestrial planets. Nature **435**, 466–469 (2005b)
M. Gounelle, G. Meynet, Solar system genealogy revealed by extinct short-lived radionuclides in meteorites. Astron. Astrophys. **545**, 4 (2012)
J.S. Greaves, M.C. Wyatt, W.S. Holland, W.R.F. Dent, The debris disc around τ Ceti: a massive analogue to the Kuiper Belt. Mon. Not. R. Astron. Soc. **351**, 54–58 (2004)
J.S. Greaves, W.S. Holland, M.C. Wyatt, W.R.F. Dent, E.I. Robson, I.M. Coulson, T. Jenness, G.H. Moriarty-Schieven, G.R. Davis, H.M. Butner, et al., Structure in the ϵ Eridani debris disk. Astrophys. J. Lett. **619**, 187–190 (2005)

J.M. Greenberg, Comet Halley—a carrier of interstellar dust chemical evolution. Adv. Space Res. **7**, 33–44 (1987)

J.M. Greenberg, J.I. Hage, From interstellar dust to comets—a unification of observational constraints. Astrophys. J. **361**, 260–274 (1990)

P. Gronkowski, M. Wesołowski, A model of cometary outbursts: a new simple approach to the classical question. Mon. Not. R. Astron. Soc. **451**, 3068–3077 (2015)

O. Groussin, G. Hahn, P.L. Lamy, R. Gonczi, G.B. Valsecchi, The long-term evolution and initial size of comets 46P/Wirtanen and 67P/Churyumov-Gerasimenko. Mon. Not. R. Astron. Soc. **376**, 1399–1406 (2007)

O. Groussin, L. Jorda, A.-T. Auger, E. Kührt, R. Gaskell, C. Capanna, F. Scholten, F. Preusker, P. Lamy, S. Hviid, et al., Gravitational slopes, geomorphology, and material strengths of the nucleus of comet 67P/Churyumov-Gerasimenko from OSIRIS observations. ArXiv e-prints 1509.02707 [astro-ph] (2015)

M.S. Gudipati, N. Abou Mrad, J. Blum, S.B. Charnley, T. Chiavassa, M.A. Cordiner, O. Mousis, G. Danger, F. Duvernay, B. Gundlach, et al., Laboratory studies towards understanding comets. Space Sci. Rev. (2015). doi:10.1007/s11214-015-0192-5

A. Guilbert-Lepoutre, S. Besse, O. Mousis, M. Ali-Dib, S. Höfner, D. Koschny, P. Hager, On the evolution of comets. Space Sci. Rev. (2015). doi:10.1007/s11214-015-0148-9

C. Güttler, J. Blum, A. Zsom, C.W. Ormel, C.P. Dullemond, The outcome of protoplanetary dust growth: pebbles, boulders, or planetesimals? I. Mapping the zoo of laboratory collision experiments. Astron. Astrophys. **513**, 56 (2010)

M. Guzzo, E. Lega, A study of the past dynamics of comet 67P/Churyumov-Gerasimenko with fast Lyapunov indicators. Astron. Astrophys. **579**, 79 (2015)

F. Habashi, Hevel, Johannes, in *Biographical Encyclopedia of Astronomers*, ed. by T. Hockey, V. Trimble, T.R. Williams, K. Bracher, R.A. Jarrell, J.D. Marché II, J. Palmeri, D.W.E. Green (2014) pp. 967–970

G. Hahn, H. Rickman, Asteroids in cometary orbits. Icarus **61**, 417–442 (1985)

J.M. Hahn, R. Malhotra, Orbital evolution of planets embedded in a planetesimal disk. Astron. J. **117**, 3041–3053 (1999)

J.M. Hahn, R. Malhotra, Neptune's migration into a stirred-up Kuiper Belt: a detailed comparison of simulations to observations. Astron. J. **130**, 2392–2414 (2005)

A. Halle, P. Di Matteo, M. Haywood, F. Combes, Quantifying stellar radial migration in an N-body simulation: blurring, churning, and the outer regions of galaxy discs. Astron. Astrophys. **578**, 58 (2015)

D.P. Hamilton, W.R. Ward, Tilting Saturn. II. Numerical model. Astron. J. **128**, 2510–2517 (2004)

M.S. Hanner, A comparison of the dust properties in recent periodic comets. Adv. Space Res. **4**, 189–196 (1984)

B.M.S. Hansen, Formation of the terrestrial planets from a narrow annulus. Astrophys. J. **703**, 1131–1140 (2009)

J.K. Harmon, M.C. Nolan, S.J. Ostro, D.B. Campbell, Radar studies of comet nuclei and grain comae, in *Comets II*, ed. by M.C. Festou, H.U. Keller, H.A. Weaver (Univ. Arizona Press, Tucson, 2004), pp. 265–279

W.K. Hartmann, Lunar 'cataclysm'—a misconception. Icarus **24**, 181–187 (1975)

W.K. Hartmann, Megaregolith evolution and cratering cataclysm models—lunar cataclysm as a misconception (28 years later). Meteorit. Planet. Sci. **38**, 579–593 (2003)

W.K. Hartmann, History of the terminal cataclysm concept: a cataclysm that never happened? in *AAS/Division for Planetary Sciences Meeting Abstracts*, vol. 46 (2014). 106.03

W.K. Hartmann, D.J. Tholen, D.P. Cruikshank, The relationship of active comets, "extinct" comets, and dark asteroids. Icarus **69**, 33–50 (1987)

W.K. Hartmann, G. Ryder, L. Dones, D. Grinspoon, The time-dependent intense bombardment of the primordial Earth/Moon system, in *Origin of the Earth and Moon*, ed. by R.M. Canup, K. Righter (Univ. Arizona Press, Tucson, 2000), pp. 493–512

P. Hartogh, D.C. Lis, D. Bockelée-Morvan, M. de Val-Borro, N. Biver, M. Küppers, M. Emprechtinger, E.A. Bergin, J. Crovisier, M. Rengel, et al., Ocean-like water in the Jupiter-family comet 103P/Hartley 2. Nature **478**, 218–220 (2011)

I. Hasegawa, S. Nakano, Orbit of periodic comet 153P/Ikeya-Zhang. Mon. Not. R. Astron. Soc. **345**, 883–888 (2003)

L.A. Haskin, The Imbrium impact event and the thorium distribution at the lunar highlands surface. J. Geophys. Res. **103**, 1679–1689 (1998)

L.A. Haskin, R.L. Korotev, K.M. Rockow, B.L. Jolliff, The case for an Imbrium origin of the Apollo Th-rich impact-melt breccias. Meteorit. Planet. Sci. **33**, 959–975 (1998)

C. Hayashi, Structure of the solar nebula, growth and decay of magnetic fields and effects of magnetic and turbulent viscosities on the nebula. Prog. Theor. Phys. Suppl. **70**, 35–53 (1981)

M.R. Hayden, J. Bovy, J.A. Holtzman, D.L. Nidever, J.C. Bird, D.H. Weinberg, B.H. Andrews, S.R. Majewski, C. Allende Prieto, F. Anders, et al., Chemical cartography with APOGEE: metallicity distribution functions and the chemical structure of the Milky Way disk. Astrophys. J. **808**, 132 (2015)
T. Heidarzadeh, *A History of Physical Theories of Comets, from Aristotle to Whipple* (Springer, Berlin, 2008)
J. Heisler, Monte Carlo simulations of the Oort comet cloud. Icarus **88**, 104–121 (1990)
J. Heisler, S. Tremaine, The influence of the galactic tidal field on the Oort comet cloud. Icarus **65**, 13–26 (1986)
J. Heisler, S. Tremaine, C. Alcock, The frequency and intensity of comet showers from the Oort Cloud. Icarus **70**, 269–288 (1987)
A. Higuchi, E. Kokubo, Effect of stellar encounters on comet cloud formation. Astron. J. **150**, 26 (2015)
A. Higuchi, E. Kokubo, H. Kinoshita, T. Mukai, Orbital evolution of planetesimals due to the galactic tide: formation of the comet cloud. Astron. J. **134**, 1693–1706 (2007)
L.A. Hillenbrand, J.M. Carpenter, J.S. Kim, M.R. Meyer, D.E. Backman, A. Moro-Martín, D.J. Hollenbach, D.C. Hines, I. Pascucci, J. Bouwman, The complete census of 70 μm-bright debris disks within "The Formation and Evolution of Planetary Systems" Spitzer Legacy Survey of Sun-like stars. Astrophys. J. **677**, 630–656 (2008)
J.G. Hills, Comet showers and the steady-state infall of comets from the Oort cloud. Astron. J. **86**, 1730–1740 (1981)
J.R. Hind, *The Comets: A Descriptive Treatise upon Those Bodies with a Condensed Account of the Numerous Modern Discoveries Respecting Them, and a Table of All the Calculated Comets, from the Earliest Ages to the Present Time* (John W. Parker and Son, London, 1852)
M. Hirabayashi, D.J. Scheeres, D.P. Sánchez, T. Gabriel, Constraints on the physical properties of main belt comet P/2013 R3 from its breakup event. Astrophys. J. Lett. **789**, 12 (2014)
P.-Y. Ho, Ancient and medieval observations of comets and novae in Chinese sources. Vistas Astron. **5**, 127–225 (1962)
R.C. Hogan, J.N. Cuzzi, Cascade model for particle concentration and enstrophy in fully developed turbulence with mass-loading feedback. Phys. Rev. E **75**(5), 056305 (2007)
M.D. Hopkins, S.J. Mojzsis, A protracted timeline for lunar bombardment from mineral chemistry, Ti thermometry and U-Pb geochronology of Apollo 14 melt breccia zircons. Contrib. Mineral. Petrol. **169**, 30 (2015)
P.F. Hopkins, Jumping the gap: the formation conditions and mass function of pebble-pile planetesimals. ArXiv e-prints 1401.2458 [astro-ph] (2014)
J. Horner, N.W. Evans, Biases in cometary catalogues and Planet X. Mon. Not. R. Astron. Soc. **335**, 641–654 (2002)
A.W. Howard, Observed properties of extrasolar planets. Science **340**, 572–576 (2013)
H.H. Hsieh, The Hawaii trails project: comet-hunting in the main asteroid belt. Astron. Astrophys. **505**, 1297–1310 (2009)
H.H. Hsieh, D.C. Jewitt, Y.R. Fernández, The strange case of 133P/Elst-Pizarro: a comet among the asteroids. Astron. J. **127**, 2997–3017 (2004)
H.H. Hsieh, K.J. Meech, J. Pittichová, Main-belt comet 238P/Read revisited. Astrophys. J. Lett. **736**, 18 (2011)
H.H. Hsieh, D. Jewitt, P. Lacerda, S.C. Lowry, C. Snodgrass, The return of activity in main-belt comet 133P/Elst-Pizarro. Mon. Not. R. Astron. Soc. **403**, 363–377 (2010)
H.H. Hsieh, L. Denneau, R.J. Wainscoat, N. Schörghofer, B. Bolin, A. Fitzsimmons, R. Jedicke, J. Kleyna, M. Micheli, P. Vereš, et al., The main-belt comets: the Pan-STARRS1 perspective. Icarus **248**, 289–312 (2015)
W.F. Hübner, Über die Gasproduktion der Kometen. Z. Astrophys. **63**, 22 (1965)
D.W. Hughes, Cometary magnitude distribution and the ratio between the numbers of long- and short-period comets. Icarus **73**, 149–162 (1988)
D.W. Hughes, The magnitude distribution, perihelion distribution and flux of long-period comets. Mon. Not. R. Astron. Soc. **326**, 515–523 (2001)
J.R. Hurley, O.R. Pols, S.J. Aarseth, C.A. Tout, A complete N-body model of the old open cluster M67. Mon. Not. R. Astron. Soc. **363**, 293–314 (2005)
D.M. Hurwitz, D.A. Kring, Differentiation of the South Pole-Aitken basin impact melt sheet: implications for lunar exploration. J. Geophys. Res., Planets **119**, 1110–1133 (2014)
B.A. Ivanov, G. Neukum, W.F. Bottke, W.K. Hartmann, The comparison of size-frequency distributions of impact craters and asteroids and the planetary cratering rate, in *Asteroids III*, ed. by W.F.J. Bottke, A. Cellino, P. Paolicchi, R.P. Binzel (Univ. Arizona Press, Tucson, 2002), pp. 89–101
A. Izidoro, N. Haghighipour, O.C. Winter, M. Tsuchida, Terrestrial planet formation in a protoplanetary disk with a local mass depletion: a successful scenario for the formation of Mars. Astrophys. J. **782**, 31 (2014)

A. Izidoro, S.N. Raymond, A. Morbidelli, O.C. Winter, Terrestrial planet formation constrained by Mars and the structure of the asteroid belt. Mon. Not. R. Astron. Soc. **453**, 3619–3634 (2015)

S.A. Jacobson, K.J. Walsh, Earth and terrestrial planet formation. ArXiv e-prints 1502.03852 [astro-ph] (2015)

J.H. Jeans, The origin of binary systems. Mon. Not. R. Astron. Soc. **79**, 408–416 (1919)

D. Jewitt, From comets to asteroids: when hairy stars go bald. Earth Moon Planets **72**, 185–201 (1996)

D. Jewitt, The active asteroids. Astron. J. **143**, 66 (2012)

D. Jewitt, Properties of near-Sun asteroids. Astron. J. **145**, 133 (2013)

D. Jewitt, Color systematics of comets and related bodies. ArXiv e-prints 1510.07069 [astro-ph] (2015)

D. Jewitt, J. Li, Activity in Geminid parent (3200) Phaethon. Astron. J. **140**, 1519–1527 (2010)

D. Jewitt, J. Luu, Discovery of the candidate Kuiper Belt Object 1992 QB_1. Nature **362**, 730–732 (1993)

D. Jewitt, H. Hsieh, J. Agarwal, The active asteroids, in *Asteroids IV*, ed. by P. Michel, F. DeMeo, W. Bottke, (2015). ArXiv e-prints 1502.02361 [astro-ph]

D. Jewitt, J. Luu, C. Trujillo, Large Kuiper Belt Objects: the Mauna Kea 8K CCD survey. Astron. J. **115**, 2125–2135 (1998)

D.C. Jewitt, From Kuiper Belt Object to cometary nucleus: the missing ultrared matter. Astron. J. **123**, 1039–1049 (2002)

D. Jewitt, H. Weaver, J. Agarwal, M. Mutchler, M. Drahus, A recent disruption of the main-belt asteroid P/2010 A2. Nature **467**, 817–819 (2010)

D. Jewitt, H. Weaver, M. Mutchler, S. Larson, J. Agarwal, Hubble Space Telescope observations of main-belt comet (596) Scheila. Astrophys. J. Lett. **733**, 4 (2011)

D. Jewitt, J. Agarwal, J. Li, H. Weaver, M. Mutchler, S. Larson, Disintegrating asteroid P/2013 R3. Astrophys. J. Lett. **784**, 8 (2014)

D. Jewitt, J. Li, J. Agarwal, H. Weaver, M. Mutchler, S. Larson, Nucleus and mass loss from active asteroid 313P/Gibbs. Astron. J. **150**, 76 (2015)

A. Johansen, H. Klahr, Planetesimal formation through streaming and gravitational instabilities. Earth Moon Planets **108**, 39–43 (2011)

A. Johansen, A.N. Youdin, Y. Lithwick, Adding particle collisions to the formation of asteroids and Kuiper belt objects via streaming instabilities. Astron. Astrophys. **537**, 125 (2012)

A. Johansen, J.S. Oishi, M.-M. Mac Low, H. Klahr, T. Henning, A. Youdin, Rapid planetesimal formation in turbulent circumstellar disks. Nature **448**, 1022–1025 (2007)

A. Johansen, J. Blum, H. Tanaka, C. Ormel, M. Bizzarro, H. Rickman, The multifaceted planetesimal formation process, in *Protostars and Planets VI*, ed. by H. Beuther, R.S. Klessen, C.P. Dullemond, T. Henning (Univ. Arizona Press, Tucson, 2014), pp. 547–570

A. Johansen, M.-M. Mac Low, P. Lacerda, M. Bizzarro, Growth of asteroids, planetary embryos, and Kuiper Belt Objects by chondrule accretion. Sci. Adv. **1**, 15109 (2015a)

A. Johansen, E. Jacquet, J.N. Cuzzi, A. Morbidelli, M. Gounelle, New paradigms for asteroid formation, in *Asteroids IV*, ed. by P. Michel, F. DeMeo, W. Bottke (2015b). ArXiv e-prints 1505.02941 [astro-ph]

L. Jones, M. Brown, Solar System science with the Large Synoptic Survey Telescope, in *AAS/Division for Planetary Sciences Meeting Abstracts*, vol. 47 (2015). 312.22

L. Jorda, R. Gaskell, C. Capanna, S. Hviid, P. Lamy, G. Faury, O. Groussin, P. Gutiérrez, C. Jackman, H.U. Keller, et al., The global shape and density of comet 67P/Churyumov-Gerasimenko from pre-perihelion Rosetta/OSIRIS observations. Icarus (2015), in preparation

Y.C. Joshi, Displacement of the Sun from the Galactic plane. Mon. Not. R. Astron. Soc. **378**, 768–776 (2007)

P.C. Joss, On the origin of short-period comets. Astron. Astrophys. **25**, 271–273 (1973)

M. Jura, An upper bound to the space density of interstellar comets. Astron. J. **141**, 155 (2011)

M. Jutzi, E. Asphaug, The shape and structure of cometary nuclei as a result of low-velocity accretion. Science **348**, 1355–1358 (2015)

N.A. Kaib, J.E. Chambers, The fragility of the terrestrial planets during a giant planet instability. ArXiv e-prints 1510.08448 [astro-ph] (2015)

N.A. Kaib, N.B. Cowan, The feeding zones of terrestrial planets and insights into Moon formation. Icarus **252**, 161–174 (2015)

N.A. Kaib, T. Quinn, The formation of the Oort Cloud in open cluster environments. Icarus **197**, 221–238 (2008)

N.A. Kaib, T. Quinn, Reassessing the source of long-period comets. Science **325**, 1234–1236 (2009)

N.A. Kaib, R. Roškar, T. Quinn, Sedna and the Oort Cloud around a migrating Sun. Icarus **215**, 491–507 (2011)

P. Kalas, J.R. Graham, M.P. Fitzgerald, M. Clampin, STIS coronagraphic imaging of Fomalhaut: main belt structure and the orbit of Fomalhaut b. Astrophys. J. **775**, 56 (2013)

J. Kavelaars, L. Jones, B. Gladman, J.W. Parker, J.-M. Petit, The orbital and spatial distribution of the Kuiper Belt, in *The Solar System Beyond Neptune*, ed. by M.A. Barucci, H. Boehnhardt, D.P. Cruikshank, A. Morbidelli, R. Dotson (Univ. Arizona Press, Tucson, 2008), pp. 59–69

H.U. Keller, Comets: dirty snowballs or icy dirtballs, in *Physics and Mechanics of Cometary Materials*, ed. by J.J. Hunt, T.D. Guyenne. ESA Special Publication, vol. 302 (1989), pp. 39–45

H.U. Keller, C. Arpigny, C. Barbieri, R.M. Bonnet, S. Cazes, M. Coradini, C.B. Cosmovici, W.A. Delamere, W.F. Huebner, D.W. Hughes, et al., First Halley multicolour camera imaging results from Giotto. Nature **321**, 320–326 (1986)

M.S.P. Kelley, C.E. Woodward, D. Bodewits, T.L. Farnham, M.S. Gudipati, D.E. Harker, D.C. Hines, M.M. Knight, L. Kolokolova, A. Li, I. de Pater, S. Protopapa, R.W. Russell, M.L. Sitko, D.H. Wooden, Cometary science with the James Webb Space Telescope. ArXiv e-prints 1510.05878 [astro-ph] (2015a)

M.S.P. Kelley, D.J. Lindler, D. Bodewits, M.F. A'Hearn, C.M. Lisse, L. Kolokolova, J. Kissel, B. Hermalyn, Erratum to "a distribution of large particles in the coma of comet 103P/Hartley 2" [Icarus 222 (2013) 634–652]. Icarus **262**, 187–189 (2015b)

M.S. Kelley, D.J. Lindler, D. Bodewits, M.F. A'Hearn, C.M. Lisse, L. Kolokolova, J. Kissel, B. Hermalyn, A distribution of large particles in the coma of comet 103P/Hartley 2. Icarus **222**, 634–652 (2013)

G.M. Kennedy, M.C. Wyatt, Do two-temperature debris discs have multiple belts? Mon. Not. R. Astron. Soc. **444**, 3164–3182 (2014)

S.J. Kenyon, B.C. Bromley, Prospects for detection of catastrophic collisions in debris disks. Astron. J. **130**, 269–279 (2005)

S.J. Kenyon, J.X. Luu, Accretion in the early Kuiper Belt. I. Coagulation and velocity evolution. Astron. J. **115**, 2136–2160 (1998)

S.J. Kenyon, J.X. Luu, Accretion in the early outer Solar System. Astrophys. J. **526**, 465–470 (1999)

F. Kiefer, A. Lecavelier des Etangs, J. Boissier, A. Vidal-Madjar, H. Beust, A.-M. Lagrange, G. Hébrard, R. Ferlet, Two families of exocomets in the β Pictoris system. Nature **514**, 462–464 (2014)

Y. Kim, M. Ishiguro, F. Usui, Physical properties of asteroids in comet-like orbits in infrared asteroid survey catalogs. Astrophys. J. **789**, 151 (2014)

D.R. Kirsh, M. Duncan, R. Brasser, H.F. Levison, Simulations of planet migration driven by planetesimal scattering. Icarus **199**, 197–209 (2009)

J. Kissel, D.E. Brownlee, K. Buchler, B.C. Clark, H. Fechtig, E. Grün, K. Hornung, E.B. Igenbergs, E.K. Jessberger, F.R. Krueger, et al., Composition of comet Halley dust particles from Giotto observations. Nature **321**, 336–337 (1986)

M.M. Knight, M.F. A'Hearn, D.A. Biesecker, G. Faury, D.P. Hamilton, P. Lamy, A. Llebaria, Photometric study of the Kreutz comets observed by SOHO from 1996 to 2005. Astron. J. **139**, 926–949 (2010)

C. Koeberl, The Late Heavy Bombardment in the inner Solar System: is there any connection to Kuiper Belt Objects? Earth Moon Planets **92**, 79–87 (2003)

W. Kofman, A. Herique, Y. Barbin, J.-P. Barriot, V. Ciarletti, S. Clifford, P. Edenhofer, C. Elachi, C. Eyraud, J.-P. Goutail, et al., Properties of the 67P/Churyumov-Gerasimenko interior revealed by CONSERT radar. Science **349**(2), 020639 (2015)

E. Kokubo, S. Ida, Oligarchic growth of protoplanets. Icarus **131**, 171–178 (1998)

M.B.N. Kouwenhoven, S.P. Goodwin, R.J. Parker, M.B. Davies, D. Malmberg, P. Kroupa, The formation of very wide binaries during the star cluster dissolution phase. Mon. Not. R. Astron. Soc. **404**, 1835–1848 (2010)

L. Kresák, Dynamics, interrelations and evolution of the systems of asteroids and comets. Moon Planets **22**, 83–98 (1980)

L. Kresák, Comet discoveries, statistics, and observational selection, in *Comets*, ed. by L.L. Wilkening (Univ. Arizona Press, Tucson, 1982), pp. 56–82

L. Kresák, Dormant phases in the aging of periodic comets. Astron. Astrophys. **187**, 906–908 (1987)

L. Kresák, M. Kresáková, The absolute magnitudes of periodic comets. I—catalogue. Bull. Astron. Inst. Czechoslov. **40**, 269–284 (1989)

K.A. Kretke, H.F. Levison, Challenges in forming the Solar System's giant planet cores via pebble accretion. Astron. J. **148**, 109 (2014)

K.A. Kretke, H.F. Levison, Evidence for pebbles in comets. Icarus **262**, 9–13 (2015)

D.A. Kring, B.A. Cohen, Cataclysmic bombardment throughout the inner solar system 3.9–4.0 Ga. J. Geophys. Res., Planets **107**, 5009 (2002)

M. Królikowska, A study of the original orbits of "hyperbolic" comets. Astron. Astrophys. **376**, 316–324 (2001)

M. Królikowska, 67P/Churyumov-Gerasimenko—potential target for the Rosetta mission. Acta Astron. **53**, 195–209 (2003)

M. Królikowska, Non-gravitational effects in long-period comets and the size of the Oort Cloud. Acta Astron. **56**, 385–412 (2006)

M. Królikowska, Warsaw catalogue of cometary orbits: 119 near-parabolic comets. Astron. Astrophys. **567**, 126 (2014)

M. Królikowska, P.A. Dybczyński, Where do long-period comets come from? 26 comets from the non-gravitational Oort spike. Mon. Not. R. Astron. Soc. **404**, 1886–1902 (2010)

M. Królikowska, S. Szutowicz, Non-gravitational motion of the Jupiter-family comet 81P/Wild 2. I. The dynamical evolution. Astron. Astrophys. **448**, 401–409 (2006)

M. Królikowska, G. Sitarski, E.M. Pittich, S. Szutowicz, K. Ziołkowski, H. Rickman, R. Gabryszewski, B. Rickman, New catalogue of one-apparition comets discovered in the years 1901–1950. I. Comets from the Oort spike. Astron. Astrophys. **571**, 63 (2014)

G.W. Kronk, *Cometography: A Catalog of Comets, vol. 1: Ancient–1799* (Cambridge University Press, Cambridge, 1999)

G.W. Kronk, *Cometography: A Catalog of Comets, vol. 2: 1800–1899* (Cambridge University Press, Cambridge, 2003)

G.W. Kronk, *Cometography: A Catalog of Comets, vol. 3, 1900–1932*, vol. 3 (Cambridge University Press, Cambridge, 2007)

G.W. Kronk, *Cometography: A Catalog of Comets, vol. 4, 1933–1959*, vol. 4 (Cambridge University Press, Cambridge, 2009)

G.W. Kronk, M. Meyer, *Cometography: A Catalog of Comets, vol. 5, 1960–1982*, vol. 5 (Cambridge University Press, Cambridge, 2010)

M.R. Krumholz, M.R. Bate, H.G. Arce, J.E. Dale, R. Gutermuth, R.I. Klein, Z.-Y. Li, F. Nakamura, Q. Zhang, Star cluster formation and feedback, in *Protostars and Planets VI*, ed. by H. Beuther, R.S. Klessen, C.P. Dullemond, T. Henning (Univ. Arizona Press, Tucson, 2014), pp. 243–266

E. Kuehrt, H.U. Keller, The formation of cometary surface crusts. Icarus **109**, 121–132 (1994)

G. Kuiper, On the origin of the Solar System, in *Astrophysics: A Topical Symposium*, ed. by J.A. Hynek (McGraw–Hill, New York, 1951), pp. 357–424

M. Küppers, L. O'Rourke, D. Bockelée-Morvan, V. Zakharov, S. Lee, P. von Allmen, B. Carry, D. Teyssier, A. Marston, T. Müller, et al., Localized sources of water vapour on the dwarf planet (1) Ceres. Nature **505**, 525–527 (2014)

C.J. Lada, The physics and modes of star cluster formation: observations. Philos. Trans. R. Soc. Lond. A **368**, 713–731 (2010)

C.J. Lada, E.A. Lada, Embedded clusters in molecular clouds. Annu. Rev. Astron. Astrophys. **41**, 57–115 (2003)

G. Laibe, J.-F. Gonzalez, S.T. Maddison, Revisiting the "radial-drift barrier" of planet formation and its relevance in observed protoplanetary discs. Astron. Astrophys. **537**, 61 (2012)

M. Lambrechts, A. Johansen, Rapid growth of gas-giant cores by pebble accretion. Astron. Astrophys. **544**, 32 (2012)

M. Lambrechts, A. Johansen, Forming the cores of giant planets from the radial pebble flux in protoplanetary discs. Astron. Astrophys. **572**, 107 (2014)

H.J.G.L.M. Lamers, M. Gieles, N. Bastian, H. Baumgardt, N.V. Kharchenko, S. Portegies Zwart, An analytical description of the disruption of star clusters in tidal fields with an application to Galactic open clusters. Astron. Astrophys. **441**, 117–129 (2005)

P.L. Lamy, A. Herique, I. Toth, The subsurface structure and density of cometary nuclei. Space Sci. Rev. (2015). doi:10.1007/s11214-015-0160-0

P.L. Lamy, I. Toth, Y.R. Fernandez, H.A. Weaver, The sizes, shapes, albedos, and colors of cometary nuclei, in *Comets II*, ed. by M.C. Festou, H.U. Keller, H.A. Weaver (Univ. Arizona Press, Tucson, 2004), pp. 223–264

D. Lardner, On the classification of comets and the distribution of their orbits in space. Mon. Not. R. Astron. Soc. **13**, 188 (1853)

J. Laskar, Large scale chaos and the spacing of the inner planets. Astron. Astrophys. **317**, 75–78 (1997)

J. Laskar, M. Gastineau, Existence of collisional trajectories of Mercury, Mars and Venus with the Earth. Nature **459**, 817–819 (2009)

G. Laughlin, J.J. Lissauer, Exoplanetary geophysics—an emerging discipline. ArXiv e-prints 1501.05685 [astro-ph] (2015)

S.M. Lawler, C.A. Beichman, G. Bryden, D.R. Ciardi, A.M. Tanner, K.Y.L. Su, K.R. Stapelfeldt, C.M. Lisse, D.E. Harker, Explorations beyond the snow line: Spitzer/IRS spectra of debris disks around solar-type stars. Astrophys. J. **705**, 89–111 (2009)

Z.M. Leinhardt, S.T. Stewart, Full numerical simulations of catastrophic small body collisions. Icarus **199**, 542–559 (2009)

H.F. Levison, Comet taxonomy, in *Completing the Inventory of the Solar System*, ed. by T. Rettig, J.M. Hahn. Astronomical Society of the Pacific Conference Series, vol. 107 (1996), pp. 173–191

H.F. Levison, M.J. Duncan, From the Kuiper Belt to Jupiter-family comets: the spatial distribution of ecliptic comets. Icarus **127**, 13–32 (1997)
H.F. Levison, A. Morbidelli, The formation of the Kuiper Belt by the outward transport of bodies during Neptune's migration. Nature **426**, 419–421 (2003)
H.F. Levison, S.A. Stern, On the size dependence of the inclination distribution of the main Kuiper Belt. Astron. J. **121**, 1730–1735 (2001)
H.F. Levison, G.R. Stewart, Remarks on modeling the formation of Uranus and Neptune. Icarus **153**, 224–228 (2001)
H.F. Levison, L. Dones, M.J. Duncan, The origin of Halley-type comets: probing the inner Oort Cloud. Astron. J. **121**, 2253–2267 (2001)
H.F. Levison, K.A. Kretke, M.J. Duncan, Growing the gas-giant planets by the gradual accumulation of pebbles. Nature **524**, 322–324 (2015)
H.F. Levison, J.J. Lissauer, M.J. Duncan, Modeling the diversity of outer planetary systems. Astron. J. **116**, 1998–2014 (1998)
H.F. Levison, E. Thommes, M.J. Duncan, Modeling the formation of giant planet cores. I. Evaluating key processes. Astron. J. **139**, 1297–1314 (2010)
H.F. Levison, L. Dones, C.R. Chapman, S.A. Stern, M.J. Duncan, K. Zahnle, Could the lunar "Late Heavy Bombardment" have been triggered by the formation of Uranus and Neptune? Icarus **151**, 286–306 (2001)
H.F. Levison, D. Terrell, P.A. Wiegert, L. Dones, M.J. Duncan, On the origin of the unusual orbit of comet 2P/Encke. Icarus **182**, 161–168 (2006a)
H.F. Levison, M.J. Duncan, L. Dones, B.J. Gladman, The Scattered Disk as a source of Halley-type comets. Icarus **184**, 619–633 (2006b)
H.F. Levison, A. Morbidelli, R. Gomes, D. Backman, Planet migration in planetesimal disks, in *Protostars and Planets V*, ed. by B. Reipurth, D. Jewitt, K. Keil (Univ. Arizona Press, Tucson, 2007), pp. 669–684
H.F. Levison, A. Morbidelli, C. Van Laerhoven, R. Gomes, K. Tsiganis, Origin of the structure of the Kuiper Belt during a dynamical instability in the orbits of Uranus and Neptune. Icarus **196**, 258–273 (2008)
H.F. Levison, W.F. Bottke, M. Gounelle, A. Morbidelli, D. Nesvorný, K. Tsiganis, Contamination of the asteroid belt by primordial trans-Neptunian objects. Nature **460**, 364–366 (2009)
H.F. Levison, M.J. Duncan, R. Brasser, D.E. Kaufmann, Capture of the Sun's Oort Cloud from stars in its birth cluster. Science **329**, 187–190 (2010)
H.F. Levison, A. Morbidelli, K. Tsiganis, D. Nesvorný, R. Gomes, Late orbital instabilities in the outer planets induced by interaction with a self-gravitating planetesimal disk. Astron. J. **142**, 152 (2011)
H.F. Levison, K.A. Kretke, K.J. Walsh, W.F. Bottke, Growing the terrestrial planets from the gradual accumulation of submeter-sized objects. Proc. Natl. Acad. Sci. USA (2015b). doi:10.1073/pnas.1513364112
A.R. Lewis, T. Quinn, N.A. Kaib, The influence of outer Solar System architecture on the structure and evolution of the Oort Cloud. Astron. J. **146**, 16 (2013)
G. Li, F.C. Adams, Cross-sections for planetary systems interacting with passing stars and binaries. Mon. Not. R. Astron. Soc. **448**, 344–363 (2015)
Z.-Y. Li, R. Banerjee, R.E. Pudritz, J.K. Jørgensen, H. Shang, R. Krasnopolsky, A. Maury, The earliest stages of star and planet formation: core collapse, and the formation of disks and outflows, in *Protostars and Planets VI*, ed. by H. Beuther, R.S. Klessen, C.P. Dullemond, T. Henning (Univ. Arizona Press, Tucson, 2014), pp. 173–194
D.N.C. Lin, J. Papaloizou, On the tidal interaction between protoplanets and the protoplanetary disk. III—orbital migration of protoplanets. Astrophys. J. **309**, 846–857 (1986)
D.N.C. Lin, P. Bodenheimer, D.C. Richardson, Orbital migration of the planetary companion of 51 Pegasi to its present location. Nature **380**, 606–607 (1996)
D.C. Lis, N. Biver, D. Bockelée-Morvan, P. Hartogh, E.A. Bergin, G.A. Blake, J. Crovisier, M. de Val-Borro, E. Jehin, M. Küppers, et al., A Herschel study of D/H in water in the Jupiter-family comet 45P/Honda-Mrkos-Pajdušáková and prospects for D/H measurements with CCAT. Astrophys. J. Lett. **774**, 3 (2013)
J.J. Lissauer, R.I. Dawson, S. Tremaine, Advances in exoplanet science from Kepler. Nature **513**, 336–344 (2014)
J.J. Lissauer, O. Hubickyj, G. D'Angelo, P. Bodenheimer, Models of Jupiter's growth incorporating thermal and hydrodynamic constraints. Icarus **199**, 338–350 (2009)
C.M. Lisse, M.C. Wyatt, C.H. Chen, A. Morlok, D.M. Watson, P. Manoj, P. Sheehan, T.M. Currie, P. Thebault, M.L. Sitko, Spitzer evidence for a Late-Heavy Bombardment and the formation of ureilites in η Corvi at $\sim$1 Gyr. Astrophys. J. **747**, 93 (2012)
S.H. Lubow, S. Ida, Planet migration, in *Exoplanets*, ed. by S. Seager (Univ. Arizona Press, Tucson, 2011), pp. 347–371
J. Luu, B.G. Marsden, D. Jewitt, C.A. Trujillo, C.W. Hergenrother, J. Chen, W.B. Offutt, A new dynamical class of object in the outer Solar System. Nature **387**, 573–575 (1997)

P.S. Lykawka, Trans-Neptunian Objects as natural probes to the unknown Solar System. Monogr. Environ. Earth Planets **1**, 121–186 (2012)

R. Malhotra, The origin of Pluto's peculiar orbit. Nature **365**, 819–821 (1993)

R. Malhotra, The origin of Pluto's orbit: implications for the Solar System beyond Neptune. Astron. J. **110**, 420–429 (1995)

R. Malhotra, R.G. Strom, Comment on "constraints on the source of lunar cataclysm impactors" (Cuk et al., 2010, Icarus 207, 590–594). Icarus **216**, 359–362 (2011)

R. Malhotra, M.J. Duncan, H.F. Levison, Dynamics of the Kuiper Belt, in *Protostars and Planets IV*, ed. by V. Mannings, S.S. Russell, A. Boss (Univ. Arizona Press, Tucson, 2000), pp. 1231–1245

D. Malmberg, F. de Angeli, M.B. Davies, R.P. Church, D. Mackey, M.I. Wilkinson, Close encounters in young stellar clusters: implications for planetary systems in the solar neighbourhood. Mon. Not. R. Astron. Soc. **378**, 1207–1216 (2007)

K.E. Mandt, O. Mousis, B. Marty, T. Cavalié, W. Harris, P. Hartogh, K. Willacy, Constraints from comets on the formation and volatile acquisition of the planets and satellites. Space Sci. Rev. (2015). doi:10.1007/s11214-015-0161-z

L. Maquet, The recent dynamical history of comet 67P/Churyumov-Gerasimenko. Astron. Astrophys. **579**, 78 (2015)

S. Marchi, W.F. Bottke, D.A. Kring, A. Morbidelli, The onset of the lunar cataclysm as recorded in its ancient crater populations. Earth Planet. Sci. Lett. **325**, 27–38 (2012)

S. Marchi, W.F. Bottke, B.A. Cohen, K. Wünnemann, D.A. Kring, H.Y. McSween, M.C. de Sanctis, D.P. O'Brien, P. Schenk, C.A. Raymond, C.T. Russell, High-velocity collisions from the lunar cataclysm recorded in asteroidal meteorites. Nat. Geosci. **6**, 303–307 (2013)

S. Marchi, W.F. Bottke, L.T. Elkins-Tanton, M. Bierhaus, K. Wuennemann, A. Morbidelli, D.A. Kring, Widespread mixing and burial of Earth's Hadean crust by asteroid impacts. Nature **511**, 578–582 (2014)

B. Marsden, G. Williams, *Catalogue of Cometary Orbits*, 15th edn. (Smithsonian Astrophysical Observatory, Cambridge, 2003)

B.G. Marsden, G.V. Williams, *Catalogue of Cometary Orbits*, 8th edn. (Smithsonian Astrophysical Observatory, Cambridge, 1993)

B.G. Marsden, G.V. Williams, *Catalogue of Cometary Orbits*, 17th edn. (Smithsonian Astrophysical Observatory, Cambridge, 2008)

B.G. Marsden, Z. Sekanina, E. Everhart, New osculating orbits for 110 comets and analysis of original orbits for 200 comets. Astron. J. **83**, 64–71 (1978)

B.G. Marsden, Z. Sekanina, D.K. Yeomans, Comets and nongravitational forces. V. Astron. J. **78**, 211–225 (1973)

C.A. Martínez-Barbosa, A.G.A. Brown, S. Portegies Zwart, Radial migration of the Sun in the Milky Way: a statistical study. Mon. Not. R. Astron. Soc. **446**, 823–841 (2015)

F. Masset, M. Snellgrove, Reversing type II migration: resonance trapping of a lighter giant protoplanet. Mon. Not. R. Astron. Soc. **320**, 55–59 (2001)

J. Matese, D. Whitmire, Tidal imprint of distant Galactic matter on the Oort comet cloud. Astrophys. J. Lett. **472**, 41–43 (1996)

B.C. Matthews, A.V. Krivov, M.C. Wyatt, G. Bryden, C. Eiroa, Observations, modeling, and theory of debris disks, in *Protostars and Planets VI* (Univ. Arizona Press, Tucson, 2014), pp. 521–544

M. Mayor, D. Queloz, A Jupiter-mass companion to a solar-type star. Nature **378**, 355–359 (1995)

J.A.M. McDonnell, J. Kissel, E. Gruen, R.J.L. Grard, Y. Langevin, R.E. Olearczyk, C.H. Perry, J.C. Zarnecki, Giotto's Dust Impact Detection System (DIDSY) and Particulate Impact Analyzer (PIA): interim assessment of the dust distribution and properties within the coma, in *ESLAB Symposium on the Exploration of Halley's Comet*, ed. by B. Battrick, E.J. Rolfe, R. Reinhard. ESA Special Publication, vol. 250 (1986), pp. 25–38

T.A. McGlynn, R.D. Chapman, On the nondetection of extrasolar comets. Astrophys. J. Lett. **346**, 105–108 (1989)

K.D. McKeegan, J. Aléon, J. Bradley, D. Brownlee, H. Busemann, A. Butterworth, M. Chaussidon, S. Fallon, C. Floss, J. Gilmour, et al., Isotopic compositions of cometary matter returned by Stardust. Science **314**, 1724 (2006)

T.R. Medupe, Indigenous astronomy in Southern Africa, in *Handbook of Archaeoastronomy and Ethnoastronomy*, ed. by C.L.N. Ruggles (Springer, Berlin, 2015), pp. 1031–1036

K.J. Meech, O.R. Hainaut, B.G. Marsden, Comet nucleus size distributions from HST and Keck telescopes. Icarus **170**, 463–491 (2004)

K.J. Meech, M.F. A'Hearn, J.A. Adams, P. Bacci, J. Bai, L. Barrera, M. Battelino, J.M. Bauer, E. Becklin, B. Bhatt, N. Biver, et al., EPOXI: comet 103P/Hartley 2 observations from a worldwide campaign. Astrophys. J. Lett. **734**, 1 (2011)

S.N. Milam, J.A. Stansberry, G. Sonneborn, C. Thomas, The James Webb Space Telescope's plan for operations and instrument capabilities for observations in the Solar System. ArXiv e-prints 1510.04567 [astro-ph] (2015)
D.A. Minton, H.F. Levison, Planetesimal-driven migration of terrestrial planet embryos. Icarus **232**, 118–132 (2014)
D.A. Minton, R. Malhotra, A record of planet migration in the main asteroid belt. Nature **457**, 1109–1111 (2009)
S. Molinari, J. Bally, S. Glover, T. Moore, A. Noriega-Crespo, R. Plume, L. Testi, E. Vázquez-Semadeni, A. Zavagno, J.-P. Bernard, P. Martin, The Milky Way as a star formation engine, ed. by H. Beuther, R.S. Klessen, C.P. Dullemond, T. Henning (Univ. Arizona Press, 2014), pp. 125–148
S.L. Montgomery, B.Y. Welsh, Detection of variable gaseous absorption features in the debris disks around young A-type stars. Publ. Astron. Soc. Pac. **124**, 1042–1056 (2012)
A. Morbidelli, Origin and dynamical evolution of comets and their reservoirs. ArXiv e-prints astro-ph/0512256 (2005)
A. Morbidelli, A. Crida, The dynamics of Jupiter and Saturn in the gaseous protoplanetary disk. Icarus **191**, 158–171 (2007)
A. Morbidelli, H.F. Levison, Scenarios for the origin of the orbits of the trans-Neptunian objects 2000 CR_{105} and 2003 VB_{12} (Sedna). Astron. J. **128**, 2564–2576 (2004)
A. Morbidelli, H. Rickman, Comets as collisional fragments of a primordial planetesimal disk. ArXiv e-prints 1504.04512 [astro-ph] (2015)
A. Morbidelli, H.F. Levison, K. Tsiganis, R. Gomes, Chaotic capture of Jupiter's Trojan asteroids in the early Solar System. Nature **435**, 462–465 (2005)
A. Morbidelli, K. Tsiganis, A. Crida, H.F. Levison, R. Gomes, Dynamics of the giant planets of the Solar System in the gaseous protoplanetary disk and their relationship to the current orbital architecture. Astron. J. **134**, 1790–1798 (2007)
A. Morbidelli, W.F. Bottke, D. Nesvorný, H.F. Levison, Asteroids were born big. Icarus **204**, 558–573 (2009a)
A. Morbidelli, H.F. Levison, W.F. Bottke, L. Dones, D. Nesvorný, Considerations on the magnitude distributions of the Kuiper Belt and of the Jupiter Trojans. Icarus **202**, 310–315 (2009b)
A. Morbidelli, R. Brasser, K. Tsiganis, R. Gomes, H.F. Levison, Constructing the secular architecture of the solar system. I. The giant planets. Astron. Astrophys. **507**, 1041–1052 (2009c)
A. Morbidelli, R. Brasser, R. Gomes, H.F. Levison, K. Tsiganis, Evidence from the asteroid belt for a violent past evolution of Jupiter's orbit. Astron. J. **140**, 1391–1401 (2010)
A. Morbidelli, S. Marchi, W.F. Bottke, D.A. Kring, A sawtooth-like timeline for the first billion years of lunar bombardment. Earth Planet. Sci. Lett. **355**, 144–151 (2012)
A. Morbidelli, M. Lambrechts, S. Jacobson, B. Bitsch, The great dichotomy of the Solar System: small terrestrial embryos and massive giant planet cores. Icarus **258**, 418–429 (2015)
A. Moro-Martín, E.L. Turner, A. Loeb, Will the Large Synoptic Survey Telescope detect extra-solar planetesimals entering the Solar System? Astrophys. J. **704**, 733–742 (2009)
A. Moro-Martín, J.M. Carpenter, M.R. Meyer, L.A. Hillenbrand, R. Malhotra, D. Hollenbach, J. Najita, T. Henning, J.S. Kim, J. Bouwman, et al., Are debris disks and massive planets correlated? Astrophys. J. **658**, 1312–1321 (2007)
A. Moro-Martín, J.P. Marshall, G. Kennedy, B. Sibthorpe, B.C. Matthews, C. Eiroa, M.C. Wyatt, J.-F. Lestrade, J. Maldonado, D. Rodriguez, et al., Does the presence of planets affect the frequency and properties of extrasolar Kuiper belts? Results from the Herschel Debris and Dunes surveys. Astrophys. J. **801**, 143 (2015)
N. Movshovitz, E. Asphaug, D. Korycansky, Numerical modeling of the disruption of comet D/1993 F2 Shoemaker-Levy 9 representing the progenitor by a gravitationally bound assemblage of randomly shaped polyhedra. Astrophys. J. **759**, 93 (2012)
N. Movshovitz, F. Nimmo, D. Korycansky, E. Asphaug, J. Owen, Disruption and reaccretion of midsized moons during an outer solar system late heavy bombardment. Geophys. Res. Lett. **42**, 256–263 (2015)
M.J. Mumma, S.B. Charnley, The chemical composition of comets—emerging taxonomies and natal heritage. Annu. Rev. Astron. Astrophys. **49**, 471–524 (2011)
C.D. Murray, S.F. Dermott, *Solar System Dynamics* (Cambridge University Press, Cambridge, 1999)
N. Murray, B. Hansen, M. Holman, S. Tremaine, Migrating planets. Science **279**, 69–72 (1998)
R.A. Murray-Clay, H.E. Schlichting, Using Kuiper Belt binaries to constrain Neptune's migration history. Astrophys. J. **730**, 132 (2011)
P.C. Myers, Filamentary structure of star-forming complexes. Astrophys. J. **700**, 1609–1625 (2009)
Y. Nakagawa, M. Sekiya, C. Hayashi, Settling and growth of dust particles in a laminar phase of a low-mass solar nebula. Icarus **67**, 375–390 (1986)

T. Nakamura, T. Noguchi, A. Tsuchiyama, T. Ushikubo, N.T. Kita, J.W. Valley, M.E. Zolensky, Y. Kakazu, K. Sakamoto, E. Mashio, et al., Chondrulelike objects in short-period comet 81P/Wild 2. Science **321**, 1664 (2008)
D. Nesvorný, Young Solar System's fifth giant planet? Astrophys. J. Lett. **742**, 22 (2011)
D. Nesvorný, Evidence for slow migration of Neptune from the inclination distribution of Kuiper Belt Objects. Astron. J. **150**, 73 (2015a)
D. Nesvorný, Jumping Neptune can explain the Kuiper Belt kernel. Astron. J. **150**, 68 (2015b)
D. Nesvorný, A. Morbidelli, Statistical study of the early Solar System's instability with four, five, and six giant planets. Astron. J. **144**, 117 (2012)
D. Nesvorný, D. Vokrouhlický, Chaotic capture of Neptune Trojans. Astron. J. **137**, 5003–5011 (2009)
D. Nesvorný, M. Broz, V. Carruba, Identification and dynamical properties of asteroid families, in *Asteroids IV*, ed. by P. Michel, F. DeMeo, W. Bottke (2015). ArXiv e-prints 1502.01628 [astro-ph]
D. Nesvorný, D. Vokrouhlický, R. Deienno, Capture of irregular satellites at Jupiter. Astrophys. J. Lett. **784**, 22 (2014)
D. Nesvorný, D. Vokrouhlický, A. Morbidelli, Capture of irregular satellites during planetary encounters. Astron. J. **133**, 1962–1976 (2007)
D. Nesvorný, D. Vokrouhlický, A. Morbidelli, Capture of Trojans by jumping Jupiter. Astrophys. J. **768**, 45 (2013)
D. Nesvorný, D. Vokrouhlický, W.F. Bottke, K. Noll, H.F. Levison, Observed binary fraction sets limits on the extent of collisional grinding in the Kuiper Belt. Astron. J. **141**, 159 (2011)
D. Nesvorný, D. Vokrouhlický, R. Deienno, K.J. Walsh, Excitation of the orbital inclination of Iapetus during planetary encounters. Astron. J. **148**, 52 (2014)
G. Neukum, B.A. Ivanov, W.K. Hartmann, Cratering records in the inner Solar System in relation to the lunar reference system. Space Sci. Rev. **96**, 55–86 (2001)
F. Nimmo, D.G. Korycansky, Impact-driven ice loss in outer Solar System satellites: consequences for the Late Heavy Bombardment. Icarus **219**, 508–510 (2012)
K.S. Noll, A.H. Parker, W.M. Grundy, All bright cold classical KBOs are binary, in *AAS/Division for Planetary Sciences Meeting Abstracts*, vol. 46 (2014). 507.05
K.S. Noll, W.M. Grundy, E.I. Chiang, J.-L. Margot, S.D. Kern, Binaries in the Kuiper Belt, in *The Solar System Beyond Neptune*, ed. by M.A. Barucci, H. Boehnhardt, D.P. Cruikshank, A. Morbidelli, R. Dotson (Univ. Arizona Press, Tucson, 2008a), pp. 345–363
K.S. Noll, W.M. Grundy, D.C. Stephens, H.F. Levison, S.D. Kern, Evidence for two populations of classical transneptunian objects: the strong inclination dependence of classical binaries. Icarus **194**, 758–768 (2008b)
M.D. Norman, A.A. Nemchin, A 4.2 billion year old impact basin on the Moon: U-Pb dating of zirconolite and apatite in lunar melt rock 67955. Earth Planet. Sci. Lett. **388**, 387–398 (2014)
J. Norwood, H. Hammel, S. Milam, J. Stansberry, J. Lunine, N. Chanover, D. Hines, G. Sonneborn, M. Tiscareno, M. Brown, P. Ferruit, Solar System observations with JWST. ArXiv e-prints 1403.6845 [astro-ph] (2014)
M. Ogihara, H. Kobayashi, S.-i. Inutsuka, T.K. Suzuki, Formation of terrestrial planets in disks evolving via disk winds and implications for the origin of the Solar System's terrestrial planets. Astron. Astrophys. **579**, 65 (2015)
A. Önehag, A. Korn, B. Gustafsson, E. Stempels, D.A. Vandenberg, M67-1194, an unusually Sun-like solar twin in M67. Astron. Astrophys. **528**, 85 (2011)
J.H. Oort, The structure of the cloud of comets surrounding the Solar System and a hypothesis concerning its origin. Bull. Astron. Inst. Neth. **11**, 91–110 (1950)
J.H. Oort, M. Schmidt, Differences between new and old comets. Bull. Astron. Inst. Neth. **11**, 259–269 (1951)
E. Öpik, Note on stellar perturbations of nearly parabolic orbits. Proc. Am. Acad. Arts Sci. **67**, 169–183 (1932)
C.W. Ormel, S. Ida, H. Tanaka, Migration rates of planets due to scattering of planetesimals. Astrophys. J. **758**, 80 (2012)
S. Ouzman, Flashes of brilliance: San rock paintings of heaven's things, in Seeing and Knowing: Understanding Rock Art with and Without Ethnography, ed. by G. Blundell, C. Chippindale, B. Smith (Wits Univ. Press, 2010), pp. 10–31
M. Pan, R. Sari, Shaping the Kuiper Belt size distribution by shattering large but strengthless bodies. Icarus **173**, 342–348 (2005)
D. Pankenier, Z. Xu, Y. Jiang, *Archaeoastronomy in East Asia: Historical Observational Records of Comets and Meteor Showers from China, Japan, and Korea* (Cambria Press, Amherst, 2008)
J. Papaloizou, D.N.C. Lin, On the tidal interaction between protoplanets and the primordial solar nebula. I—linear calculation of the role of angular momentum exchange. Astrophys. J. **285**, 818–834 (1984)

J.C.B. Papaloizou, C. Terquem, Planet formation and migration. Rep. Prog. Phys. **69**, 119–180 (2006)
A.H. Parker, The intrinsic Neptune Trojan orbit distribution: implications for the primordial disk and planet migration. Icarus **247**, 112–125 (2015)
A.H. Parker, J.J. Kavelaars, Destruction of binary minor planets during Neptune scattering. Astrophys. J. Lett. **722**, 204–208 (2010)
A.H. Parker, J.J. Kavelaars, J.-M. Petit, L. Jones, B. Gladman, J. Parker, Characterization of seven ultra-wide trans-Neptunian binaries. Astrophys. J. **743**, 1 (2011)
S.J. Peale, On the density of Halley's comet. Icarus **82**, 36–49 (1989)
N. Peixinho, A. Delsanti, A. Doressoundiram, Reanalyzing the visible colors of Centaurs and KBOs: what is there and what we might be missing. Astron. Astrophys. **577**, 35 (2015)
J.-M. Petit, O. Mousis, KBO binaries: how numerous were they? Icarus **168**, 409–419 (2004)
J.-M. Petit, J.J. Kavelaars, B.J. Gladman, J.L. Margot, P.D. Nicholson, R.L. Jones, J.W. Parker, M.L.N. Ashby, A. Campo Bagatin, P. Benavidez, et al., The extreme Kuiper Belt binary 2001 QW_{322}. Science **322**, 432–434 (2008)
J.-M. Petit, J.J. Kavelaars, B.J. Gladman, R.L. Jones, J.W. Parker, C. Van Laerhoven, P. Nicholson, G. Mars, P. Rousselot, O. Mousis, et al., The Canada-France Ecliptic Plane Survey—full data release: the orbital structure of the Kuiper Belt. Astron. J. **142**, 131 (2011)
S. Pfalzner, M.B. Davies, M. Gounelle, A. Johansen, C. Muenker, P. Lacerda, S. Portegies Zwart, L. Testi, M. Trieloff, D. Veras, The formation of the Solar System. Phys. Scr. **90**, 068001 (2015)
B. Pichardo, E. Moreno, C. Allen, L.R. Bedin, A. Bellini, L. Pasquini, The Sun was not born in M67. Astron. J. **143**, 73 (2012)
A. Pierens, R.P. Nelson, Constraints on resonant-trapping for two planets embedded in a protoplanetary disc. Astron. Astrophys. **482**, 333–340 (2008)
R.E. Pike, J.J. Kavelaars, J.M. Petit, B.J. Gladman, M. Alexandersen, K. Volk, C.J. Shankman, The 5:1 Neptune Resonance as Probed by CFEPS: dynamics and population. Astron. J. **149**, 202 (2015)
N. Piskunov, H. Rickman, B. Gustafsson, Introduction: Nobel Symposium 135: Physics of Planetary Systems (18–22 June 2007, Lidingö, Stockholm, Sweden). Phys. Scr. T **130**, 011001 (2008)
D. Prialnik, E.D. Rosenberg, Can ice survive in main-belt comets? Long-term evolution models of comet 133P/Elst-Pizarro. Mon. Not. R. Astron. Soc. **399**, 79–83 (2009)
D. Prialnik, J. Benkhoff, M. Podolak, Modeling the structure and activity of comet nuclei, in *Comets II*, ed. by M.C. Festou, H.U. Keller, H.A. Weaver (Univ. Arizona Press, Tucson, 2004), pp. 359–387
C. Qi, K.I. Öberg, D.J. Wilner, P. D'Alessio, E. Bergin, S.M. Andrews, G.A. Blake, M.R. Hogerheijde, E.F. van Dishoeck, Imaging of the CO snow line in a solar nebula analog. Science **341**, 630–632 (2013)
T. Quinn, S. Tremaine, M. Duncan, Planetary perturbations and the origins of short-period comets. Astrophys. J. **355**, 667–679 (1990)
N. Raettig, H. Klahr, W. Lyra, Particle trapping and streaming instability in vortices in protoplanetary disks. Astrophys. J. **804**, 35 (2015)
S.N. Raymond, A. Morbidelli, The Grand Tack model: a critical review, in *Complex Planetary Systems*, ed. by Z. Knežević, A. Lemaître. IAU Symposium, vol. 310 (Cambridge University Press, Cambridge, 2014), pp. 194–203
S.N. Raymond, D.P. O'Brien, A. Morbidelli, N.A. Kaib, Building the terrestrial planets: constrained accretion in the inner Solar System. Icarus **203**, 644–662 (2009)
M. Reyes-Ruiz, H. Aceves, C.E. Chavez, Stability of the outer planets in multiresonant configurations with a self-gravitating planetesimal disk. Astrophys. J. **804**, 91 (2015)
J.E. Richardson, H.J. Melosh, C.M. Lisse, B. Carcich, A ballistics analysis of the Deep Impact ejecta plume: determining comet Tempel 1's gravity, mass, and density. Icarus **191**, 176–209 (2007)
H. Rickman, Masses and densities of comets Halley and Kopff, in *Comet Nucleus Sample Return Mission*, ed. by O. Melita. ESA Special Publication, vol. 249 (1986), pp. 195–205
H. Rickman, The nucleus of comet Halley—surface structure, mean density, gas and dust production. Adv. Space Res. **9**, 59–71 (1989)
H. Rickman, The Oort Cloud and long-period comets. Meteorit. Planet. Sci. **49**, 8–20 (2014)
H. Rickman, C. Froeschlé, Thermal models for the nucleus of comet P/Halley, in *Cometary Exploration: Proceedings of the International Conference*, ed. by T.I. Gombosi (Akademiai Kiado, 1983), pp. 75–84
H. Rickman, L. Jorda, Comet 46P/Wirtanen, the target of the Rosetta mission. Adv. Space Res. **21**, 1491–1504 (1998)
H. Rickman, J.A. Fernandez, B.A.S. Gustafson, Formation of stable dust mantles on short-period comet nuclei. Astron. Astrophys. **237**, 524–535 (1990)
H. Rickman, L. Kamel, M.C. Festou, C. Froeschle, Estimates of masses, volumes and densities of short-period comet nuclei, in *Diversity and Similarity of Comets*, ed. by E.J. Rolfe, B. Battrick. ESA Special Publication, vol. 278 (1987), pp. 471–481

H. Rickman, L. Kamel, C. Froeschle, M.C. Festou, Nongravitational effects and the aging of periodic comets. Astron. J. **102**, 1446–1463 (1991)
H. Rickman, M. Fouchard, C. Froeschlé, G.B. Valsecchi, Injection of Oort Cloud comets: the fundamental role of stellar perturbations. Celest. Mech. Dyn. Astron. **102**, 111–132 (2008)
H. Rickman, S. Marchi, M.F. A'Hearn, C. Barbieri, M.R. El-Maarry, C. Güttler, W.-H. Ip, H.U. Keller, P. Lamy, F. Marzari, et al., Comet 67P/Churyumov-Gerasimenko: constraints on its origin from OSIRIS observations. ArXiv e-prints 1505.07021 [astro-ph] (2015)
A.S. Rivkin, J.P. Emery, Detection of ice and organics on an asteroidal surface. Nature **464**, 1322–1323 (2010)
A.S. Rivkin, F. Marchis, J.A. Stansberry, D. Takir, C. Thomas, the JWST Asteroids Focus Group, Asteroids and the James Webb Space Telescope. ArXiv e-prints 1510.08414 [astro-ph] (2015)
A. Roberge, Astronomy: Hurling comets around a planetary nursery. Nature **514**, 440–441 (2014)
K. Ros, A. Johansen, Ice condensation as a planet formation mechanism. Astron. Astrophys. **552**, 137 (2013)
R. Roškar, V.P. Debattista, T.R. Quinn, G.S. Stinson, J. Wadsley, Riding the spiral waves: implications of stellar migration for the properties of galactic disks. Astrophys. J. Lett. **684**, 79–82 (2008)
D.C. Rubie, S.A. Jacobson, A. Morbidelli, D.P. O'Brien, E.D. Young, J. de Vries, F. Nimmo, H. Palme, D.J. Frost, Accretion and differentiation of the terrestrial planets with implications for the compositions of early-formed Solar System bodies and accretion of water. Icarus **248**, 89–108 (2015)
J.A. Ruffner, Isaac Newton's Historia Cometarum and the quest for elliptical orbits. J. Hist. Astron. **41**, 425–451 (2010)
C.L.N. Ruggles, *Handbook of Archaeoastronomy and Ethnoastronomy* (Springer, Berlin, 2015)
G. Ryder, Lunar samples, lunar accretion and the early bombardment of the Moon. Eos **71**, 313 (1990)
V.S. Safronov, Evolution of the protoplanetary cloud and formation of the Earth and the planets (Evoliutsiia doplanetnogo oblaka i obrazovanie zemli i planet, Moscow, Izdatel'stvo Nauka. Jerusalem, Israel Program for Scientific Translations, Ltd., 1972), 1969
R.Z. Sagdeev, P.E. Elyasberg, V.I. Moroz, Estimate of the mass and density of the nucleus of comet Halley. Sov. Astron. Lett. **13**, 259–263 (1987)
H.E. Schlichting, E.O. Ofek, M. Wenz, R. Sari, A. Gal-Yam, M. Livio, E. Nelan, S. Zucker, A single sub-kilometre Kuiper Belt Object from a stellar occultation in archival data. Nature **462**, 895–897 (2009)
H.E. Schlichting, E.O. Ofek, R. Sari, E.P. Nelan, A. Gal-Yam, M. Wenz, P. Muirhead, N. Javanfar, M. Livio, Measuring the Abundance of sub-kilometer-sized Kuiper Belt Objects using stellar occultations. Astrophys. J. **761**, 150 (2012)
M.E. Schwamb, Solar System: stranded in no-man's-land. Nature **507**, 435–436 (2014)
M.E. Schwamb, M.E. Brown, W.C. Fraser, The small numbers of large Kuiper Belt Objects. Astron. J. **147**, 2 (2014)
M.E. Schwamb, M.E. Brown, D. Rabinowitz, B.G. Marsden, 2007 OR10. Minor Planet Electronic Circulars, 42 (2009)
M.E. Schwamb, M.E. Brown, D.L. Rabinowitz, D. Ragozzine, Properties of the distant Kuiper Belt: results from the Palomar Distant Solar System Survey. Astrophys. J. **720**, 1691–1707 (2010)
S. Seager (ed.), *Exoplanets*. Space Science Series (Univ. Arizona Press, Tucson, 2010)
Z. Sekanina, P.W. Chodas, Origin of the Marsden and Kracht Groups of sunskirting comets. I. Association with comet 96P/Machholz and its interplanetary complex. Astron. Astrophys. Suppl. Ser. **161**, 551–586 (2005)
J.A. Sellwood, J.J. Binney, Radial mixing in galactic discs. Mon. Not. R. Astron. Soc. **336**, 785–796 (2002)
C. Shankman, B.J. Gladman, N. Kaib, J.J. Kavelaars, J.M. Petit, A possible divot in the size distribution of the Kuiper Belt's scattering objects. Astrophys. J. Lett. **764**, 2 (2013)
A. Shannon, Y. Wu, Y. Lithwick, Forming the cold classical Kuiper Belt in a light disk. ArXiv e-prints 1510.01323 [astro-ph] (2015a)
A. Shannon, A.P. Jackson, D. Veras, M. Wyatt, Eight billion asteroids in the Oort Cloud. Mon. Not. R. Astron. Soc. **446**, 2059–2064 (2015b)
S.S. Sheppard, C.A. Trujillo, A thick cloud of Neptune Trojans and their colors. Science **313**, 511–514 (2006)
S.S. Sheppard, C.A. Trujillo, The size distribution of the Neptune Trojans and the missing intermediate-sized planetesimals. Astrophys. J. Lett. **723**, 233–237 (2010)
H. Sierks, C. Barbieri, P.L. Lamy, R. Rodrigo, D. Koschny, H. Rickman, H.U. Keller, J. Agarwal, M.F. A'Hearn, F. Angrilli, et al., On the nucleus structure and activity of comet 67P/Churyumov-Gerasimenko. Science **347**, 1044 (2015)
J.A. Simpson, D. Rabinowitz, A.J. Tuzzolino, L.V. Ksanfomaliti, R.Z. Sagdeev, The dust coma of comet P/Halley—measurements on the Vega-1 and Vega-2 spacecraft. Astron. Astrophys. **187**, 742–752 (1987)
B.A. Smith, R.J. Terrile, A circumstellar disk around Beta Pictoris. Science **226**, 1421–1424 (1984)

C. Snodgrass, A. Fitzsimmons, S.C. Lowry, P. Weissman, The size distribution of Jupiter Family comet nuclei. Mon. Not. R. Astron. Soc. **414**, 458–469 (2011)

M. Solontoi, Ž. Ivezić, M. Jurić, A.C. Becker, L. Jones, A.A. West, S. Kent, R.H. Lupton, M. Claire, G.R. Knapp, et al., Ensemble properties of comets in the Sloan Digital Sky Survey. Icarus **218**, 571–584 (2012)

A. Sosa, J.A. Fernández, Masses of long-period comets derived from non-gravitational effects—analysis of the computed results and the consistency and reliability of the non-gravitational parameters. Mon. Not. R. Astron. Soc. **416**, 767–782 (2011)

Space Studies Board, *Vision and Voyages for Planetary Science in the Decade 2013–2022* (National Academies Press, Washington, 2012)

P.D. Spudis, *The Once and Future Moon* (Smithsonian Institution Press, Washington, 1996)

P.D. Spudis, D.E. Wilhelms, M.S. Robinson, The sculptured hills of the Taurus Highlands: implications for the relative age of Serenitatis, basin chronologies and the cratering history of the Moon. J. Geophys. Res., Planets **116**, E00H03 (2011)

S.W. Stahler, F. Palla, *The Formation of Stars* (Wiley, New York, 2004)

D.I. Steel, D.J. Asher, On the origin of comet Encke. Mon. Not. R. Astron. Soc. **281**, 937–944 (1996)

D.C. Stephens, K.S. Noll, Detection of six trans-Neptunian binaries with NICMOS: a high fraction of binaries in the cold classical disk. Astron. J. **131**, 1142–1148 (2006)

F.R. Stephenson, K.K.C. Yau, Far Eastern observations of Halley's comet—240 BC to AD 1368. J. Br. Interplanet. Soc. **38**, 195–216 (1985)

S.A. Stern, On the number of planets in the outer solar system—evidence of a substantial population of 1000-km bodies. Icarus **90**, 271–281 (1991)

S.A. Stern, Collisional time scales in the Kuiper Disk and their implications. Astron. J. **110**, 856 (1995)

S.A. Stern, On the collisional environment, accretion time scales, and architecture of the massive, primordial Kuiper Belt. Astron. J. **112**, 1203 (1996)

S.T. Stewart, Z.M. Leinhardt, Velocity-dependent catastrophic disruption criteria for planetesimals. Astrophys. J. Lett. **691**, 133–137 (2009)

D. Stöffler, G. Ryder, Stratigraphy and isotope ages of lunar geologic units: chronological standard for the inner Solar System. Space Sci. Rev. **96**, 9–54 (2001)

R.G. Strom, R. Malhotra, T. Ito, F. Yoshida, D.A. Kring, The origin of planetary impactors in the inner Solar System. Science **309**, 1847–1850 (2005)

R.G. Strom, R. Malhotra, Z.-Y. Xiao, T. Ito, F. Yoshida, L.R. Ostrach, The inner Solar System cratering record and the evolution of impactor populations. Res. Astron. Astrophys. **15**, 407 (2015)

K.Y.L. Su, Y.-H. Chu, G.H. Rieke, P.J. Huggins, R. Gruendl, R. Napiwotzki, T. Rauch, W.B. Latter, K. Volk, A debris disk around the central star of the Helix Nebula? Astrophys. J. Lett. **657**, 41–45 (2007)

J.M. Sunshine, M.F. A'Hearn, O. Groussin, J.-Y. Li, M.J.S. Belton, W.A. Delamere, J. Kissel, K.P. Klaasen, L.A. McFadden, K.J. Meech, H.J. Melosh, et al., Exposed water ice deposits on the surface of comet 9P/Tempel 1. Science **311**, 1453–1455 (2006)

G. Tancredi, The dynamical memory of Jupiter family comets. Astron. Astrophys. **299**, 288 (1995)

G. Tancredi, Chaotic dynamics of planet-encountering bodies. Celest. Mech. Dyn. Astron. **70**, 181–200 (1998)

G. Tancredi, A criterion to classify asteroids and comets based on the orbital parameters. Icarus **234**, 66–80 (2014)

G. Tancredi, J.A. Fernández, H. Rickman, J. Licandro, Nuclear magnitudes and the size distribution of Jupiter family comets. Icarus **182**, 527–549 (2006)

H. Tang, N. Dauphas, Abundance, distribution, and origin of ^{60}Fe in the solar protoplanetary disk. Earth Planet. Sci. Lett. **359**, 248–263 (2012)

M.G.G.T. Taylor, C. Alexander, N. Altobelli, M. Fulle, M. Fulchignoni, E. Grün, P. Weissman, Rosetta begins its comet tale. Science **347**, 387 (2015)

S.C. Tegler, W. Romanishin, Extremely red Kuiper-Belt objects in near-circular orbits beyond 40 AU. Nature **407**, 979–981 (2000)

F. Tera, D.A. Papanastassiou, G.J. Wasserburg, Isotopic evidence for a terminal lunar cataclysm. Earth Planet. Sci. Lett. **22**, 1–21 (1974)

L. Testi, T. Birnstiel, L. Ricci, S. Andrews, J. Blum, J. Carpenter, C. Dominik, A. Isella, A. Natta, J.P. Williams, D.J. Wilner, Dust evolution in protoplanetary disks, in *Protostars and Planets VI*, ed. by H. Beuther, R.S. Klessen, C.P. Dullemond, T. Henning (Univ. Arizona Press, Tucson, 2014), pp. 339–361

E.W. Thommes, M.J. Duncan, H.F. Levison, The formation of Uranus and Neptune in the Jupiter-Saturn region of the Solar System. Nature **402**, 635–638 (1999)

M.S. Tiscareno, R. Malhotra, The dynamics of known Centaurs. Astron. J. **126**, 3122–3131 (2003)

B. Todorovic-Juchniewicz, When we may and need to use barycentric orbit of a comet. Acta Astron. **31**, 191–196 (1981)

K. Tomida, S. Okuzumi, M.N. Machida, Radiation magnetohydrodynamic simulations of protostellar collapse: nonideal magnetohydrodynamic effects and early formation of circumstellar disks. Astrophys. J. **801**, 117 (2015)

M.V. Torbett, Injection of Oort Cloud comets to the inner solar system by galactic tidal fields. Mon. Not. R. Astron. Soc. **223**, 885–895 (1986)

I. Toth, Impact-generated activity period of the asteroid 7968 Elst-Pizarro in 1996: identification of the asteroid 427 Galene as the most probable parent body of the impactors. Astron. Astrophys. **360**, 375–380 (2000)

S. Tremaine, The distribution of comets around stars, in *Planets Around Pulsars*, ed. by J.A. Phillips, S.E. Thorsett, S.R. Kulkarni. Astronomical Society of the Pacific Conference Series, vol. 36 (1993), pp. 335–344

C.A. Trujillo, S.S. Sheppard, A Sedna-like body with a perihelion of 80 astronomical units. Nature **507**, 471–474 (2014)

K. Tsiganis, R. Gomes, A. Morbidelli, H.F. Levison, Origin of the orbital architecture of the giant planets of the Solar System. Nature **435**, 459–461 (2005)

E.F. van Dishoeck, E.A. Bergin, D.C. Lis, J.I. Lunine, Water: from clouds to planets, in *Protostars and Planets VI*, ed. by H. Beuther, R.S. Klessen, C.P. Dullemond, T. Henning (Univ. Arizona Press, Tucson, 2014), pp. 835–858

J.-B. Vincent, D. Bodewits, S. Besse, H. Sierks, C. Barbieri, P. Lamy, R. Rodrigo, D. Koschny, H. Rickman, H.U. Keller, et al., Large heterogeneities in comet 67P as revealed by active pits from sinkhole collapse. Nature **523**, 63–66 (2015)

R. Visser, S.D. Doty, E.F. van Dishoeck, The chemical history of molecules in circumstellar disks. II. Gas-phase species. Astron. Astrophys. **534**, 132 (2011)

R. Visser, E.F. van Dishoeck, S.D. Doty, C.P. Dullemond, The chemical history of molecules in circumstellar disks. I. Ices. Astron. Astrophys. **495**, 881–897 (2009)

D. Vokrouhlický, D. Nesvorný, Tilting Jupiter (a bit) and Saturn (a lot) during planetary migration. Astrophys. J. **806**, 143 (2015)

D. Vokrouhlický, W.F. Bottke, S.R. Chesley, D.J. Scheeres, T.S. Statler, The Yarkovsky and YORP effects, in *Asteroids IV* (2015). ArXiv e-prints 1502.01249 [astro-ph]

K. Volk, R. Malhotra, The scattered disk as the source of the Jupiter Family Comets. Astrophys. J. **687**, 714–725 (2008)

S.K. Vsekhsvyatskij, Fizicheskie Kharakteristiki Komet. English translation (Add Izdatel'stvo Fiziko-Matematicheskoj Literatury, Moscow, 1958) (NASA TT F-80, 1964)

K. Wahlberg Jansson, A. Johansen, Formation of pebble-pile planetesimals. Astron. Astrophys. **570**, 47 (2014)

K.J. Walsh, A. Morbidelli, The effect of an early planetesimal-driven migration of the giant planets on terrestrial planet formation. Astron. Astrophys. **526**, 126 (2011)

K.J. Walsh, A. Morbidelli, S.N. Raymond, D.P. O'Brien, A.M. Mandell, A low mass for Mars from Jupiter's early gas-driven migration. Nature **475**, 206–209 (2011)

K.J. Walsh, A. Morbidelli, S.N. Raymond, D.P. O'Brien, A.M. Mandell, Populating the asteroid belt from two parent source regions due to the migration of giant planets—"the Grand Tack". Meteorit. Planet. Sci. **47**, 1941–1947 (2012)

W.R. Ward, Density waves in the solar nebula—differential Lindblad torque. Icarus **67**, 164–180 (1986)

W.R. Ward, Protoplanet migration by nebula tides. Icarus **126**, 261–281 (1997)

W.R. Ward, D.P. Hamilton, Tilting Saturn. I. Analytic model. Astron. J. **128**, 2501–2509 (2004)

A. Waszczak, E.O. Ofek, O. Aharonson, S.R. Kulkarni, D. Polishook, J.M. Bauer, D. Levitan, B. Sesar, R. Laher, J. Surace, PTF Team, Main-belt comets in the Palomar Transient Factory survey—I. The search for extendedness. Mon. Not. R. Astron. Soc. **433**, 3115–3132 (2013)

H.A. Weaver, P.D. Feldman, M.F. A'Hearn, N. Dello Russo, S.A. Stern, The carbon monoxide abundance in comet 103P/Hartley 2 during the EPOXI flyby. Astrophys. J. Lett. **734**, 5 (2011)

S.J. Weidenschilling, Aerodynamics of solid bodies in the solar nebula. Mon. Not. R. Astron. Soc. **180**, 57–70 (1977)

S.J. Weidenschilling, The origin of comets in the solar nebula: a unified model. Icarus **127**, 290–306 (1997)

S.J. Weidenschilling, From icy grains to comets, in *Comets II*, ed. by M.C. Festou, H.U. Keller, H.A. Weaver (2004), pp. 97–104

S.J. Weidenschilling, Accretion of planetary embryos in the inner and outer Solar System. Phys. Scr. T **130**(1), 014021 (2008)

S.J. Weidenschilling, Initial sizes of planetesimals and accretion of the asteroids. Icarus **214**, 671–684 (2011)

S.J. Weidenschilling, J.N. Cuzzi, Formation of planetesimals in the solar nebula, in *Protostars and Planets III*, ed. by E.H. Levy, J.I. Lunine (1993), pp. 1031–1060

M. Weiler, H. Rauer, C. Sterken, Cometary nuclear magnitudes from sky survey observations. Icarus **212**, 351–366 (2011)
P. Weissman, P. Hut, Dynamics of cometary showers, in *Lunar and Planetary Science Conference*. Lunar and Planetary Inst. Technical Report, vol. 17 (1986), pp. 935–936
P.R. Weissman, Stellar perturbations of the cometary cloud. Nature **288**, 242–243 (1980)
P.R. Weissman, The mass of the Oort Cloud. Astron. Astrophys. **118**, 90–94 (1983)
P.R. Weissman, Are cometary nuclei primordial rubble piles? Nature **320**, 242–244 (1986)
P.R. Weissman, The Oort Cloud, in *Completing the Inventory of the Solar System*, ed. by T. Rettig, J.M. Hahn. Astronomical Society of the Pacific Conference Series, vol. 107 (1996), pp. 265–288
P.R. Weissman, W.F. Bottke Jr., H.F. Levison, Evolution of comets into asteroids, in *Asteroids III*, ed. by W.F. Bottke Jr., A. Cellino, P. Paolicchi, R.P. Binzel (Univ. Arizona Press, Tucson, 2002), pp. 669–686
P.R. Weissman, M. A'Hearn, Accretion of cometary nuclei in the solar nebula: boulders, not pebbles, in *American Astronomical Society, DPS meeting*, vol. 47 (2015). 309.05
R.S. Westfall, *Never at Rest: A Biography of Isaac Newton* (Cambridge University Press, Cambridge, 1983)
G.W. Wetherill, Late Heavy Bombardment of the Moon and terrestrial planets, in *Lunar and Planetary Science Conference Proceedings*, vol. 6 (1975), pp. 1539–1561
F.L. Whipple, A comet model. I. The acceleration of comet Encke. Astrophys. J. **111**, 375–394 (1950)
F.L. Whipple, Cometary brightness variation and nucleus structure. Moon Planets **18**, 343–359 (1978)
P. Wiegert, S. Tremaine, The evolution of long-period comets. Icarus **137**, 84–121 (1999)
D.E. Wilhelms, J.F. McCauley, N.J. Trask, *The Geologic History of the Moon* (US Geological Survey, Washington, 1987)
K. Willacy, C. Alexander, M. Ali-Dib, C. Ceccarelli, S. Charnley, M. Doronin, Y. Ellinger, P. Gast, E. Gibb, S. Milam, et al., The composition of the protosolar disk and the formation conditions for comets. Space Sci. Rev. (2015). doi:10.1007/s11214-015-0167-6
J.G. Williams, G.S. Benson, Resonances in the Neptune-Pluto system. Astron. J. **76**, 167–176 (1971)
J.N. Winn, D.C. Fabrycky, The occurrence and architecture of exoplanetary systems. Annu. Rev. Astron. Astrophys. **53**, 409–447 (2015)
S. Wolff, R.I. Dawson, R.A. Murray-Clay, Neptune on tiptoes: dynamical histories that preserve the cold classical Kuiper Belt. Astrophys. J. **746**, 171 (2012)
J.T. Wright, K.M.S. Cartier, M. Zhao, D. Jontof-Hutter, E.B. Ford, The Ĝ search for extraterrestrial civilizations with large energy supplies. IV. The signatures and information content of transiting megastructures. ArXiv e-prints 1510.04606 [astro-ph] (2015)
M.C. Wyatt, Evolution of debris disks. Annu. Rev. Astron. Astrophys. **46**, 339–383 (2008)
S. Wyckoff, M. Kleine, B.A. Peterson, P.A. Wehinger, L.M. Ziurys, Carbon isotope abundances in comets. Astrophys. J. **535**, 991–999 (2000)
Z. Xu, D.W. Pankenier, Y. Jiang, *East-Asian Archaeoastronomy: Historical Records of Astronomical Observations of China, Japan and Korea* (CRC Press, Boca Raton, 2000)
D.K. Yeomans, J.D. Giorgini, S.R. Chesley, The history and dynamics of comet 9P/Tempel 1. Space Sci. Rev. **117**, 123–135 (2005)
D.K. Yeomans, J. Rahe, R.S. Freitag, The history of comet Halley. J. R. Astron. Soc. Can. **80**, 62–86 (1986)
D.K. Yeomans, *Comets. A Chronological History of Observation, Science, Myth, and Folklore* (Wiley, New York, 1991)
A.N. Youdin, J. Goodman, Streaming instabilities in protoplanetary disks. Astrophys. J. **620**, 459–469 (2005)
E.D. Young, Inheritance of solar short- and long-lived radionuclides from molecular clouds and the unexceptional nature of the solar system. Earth Planet. Sci. Lett. **392**, 16–27 (2014)
K. Zahnle, P. Schenk, H. Levison, L. Dones, Cratering rates in the outer Solar System. Icarus **163**, 263–289 (2003)
J.-Q. Zheng, M.J. Valtonen, L. Valtaoja, Capture of comets during the evolution of a star cluster and the origin of the Oort Cloud. Celest. Mech. Dyn. Astron. **49**, 265–272 (1990)
M. Ziółkowski, Observations of comets and eclipses in the Andes, in *Handbook of Archaeoastronomy and Ethnoastronomy*, ed. by C.L.N. Ruggles (Springer, Berlin, 2015), pp. 913–920
A. Zsom, C.W. Ormel, C. Güttler, J. Blum, C.P. Dullemond, The outcome of protoplanetary dust growth: pebbles, boulders, or planetesimals? II. Introducing the bouncing barrier. Astron. Astrophys. **513**, 57 (2010)
B. Zuckerman, I. Song, A 40 Myr old gaseous circumstellar disk at 49 Ceti: massive CO-rich comet clouds at young A-type stars. Astrophys. J. **758**, 77 (2012)

Space Sci Rev (2015) 197:271–296
DOI 10.1007/s11214-015-0148-9

On the Evolution of Comets

A. Guilbert-Lepoutre[1,2] · S. Besse[1] · O. Mousis[3] · M. Ali-Dib[2] · S. Höfner[4,5] · D. Koschny[1] · P. Hager[6,7]

Received: 7 November 2014 / Accepted: 25 March 2015 / Published online: 15 April 2015

Abstract Studying comets is believed to bring invaluable clues on the formation and evolution of our planetary system. In comparison to planets, they have undergone much less alteration, and should have therefore retained a relatively pristine record of the conditions prevailing during the early phases of the solar system. However, comets might not be entirely pristine. As of today, we have not been able to determine which of the observed physical, chemical and orbital characteristics of comets, after they have evolved for more than 4 Gyr in a time-varying radiative and collisional environment, will provide the best clues to their origin. Comet physical characteristics as inherited from their formation stage may be very diverse, both in terms of composition and internal structure. The subsequent evolution of comet nuclei involves some possible processing from radiogenic heating, space weathering and large- and small-scale collisions, which might have modified their primordial structures and compositions with various degrees. When comets enter the inner solar system and become active, they start to lose mass at a very high rate. The effects of activity on comet nuclei involve a layering of the composition, a substantial non-even erosion and modification of their size and shape, and may eventually result in the death of comets. In this review, we present the dominating processes that might affect comet physical and chemical properties at different stages of their evolution. Although the evolutionary track may be specific to each comet, we can focus on long-lasting modifications which might be common to all

✉ A. Guilbert-Lepoutre
aurelie.guilbert-lepoutre@utinam.cnrs.fr

1 European Space Agency, ESTEC, 2200 AG Noordwijk, Netherlands

2 Institut UTINAM, UMR 6213–CNRS/Université de Franche Comté, 25000 Besancon, France

3 LAM, UMR 7326—CNRS/Université Aix Marseille, 13388, Marseille, France

4 Max-Planck-Institut für Sonnensystemforschung, 37077 Göttingen, Germany

5 Institut für Geophysik und extraterrestrische Physik, Technische Universität Braunschweig, 38106 Braunschweig, Germany

6 Institute of Astronautics, Technische Universität München, 85748 Garching b. München, Germany

7 SpaceTech GmbH, 88090 Immenstaad, Germany

nuclei after their formation stage, during their storage in reservoirs in the outer solar system, and once comets enter the inner solar system and become active objects.

Keywords Comets · Evolution · Thermal processing · Collisions · Space weathering · Origins

1 Introduction

The solar system as we know it today displays a very complex architecture, in which three major reservoirs of small bodies can be identified. In the inner solar system, the Main Belt of Asteroids between the orbits of Mars and Jupiter was recently identified as a possible source of comets, as mass loss was observed for some of these objects. Although the source for such activity is not fully constrained yet (Jewitt 2012), the "Grand Tack" model, which examines the dynamical evolution of the solar system a few million years after the formation of the first solids (Walsh et al. 2011), suggests that some primitive icy objects formed beyond the giant planets could have populated this region. In the outer solar system, two main reservoirs can be identified: the transneptunian region holding two possible sources of comets, namely the Kuiper Belt and the Scattered Disk, and the Oort Cloud, an inferred structure at the edge of the solar system. Multiple models try to explain how the different dynamical features encountered in the outer solar system were formed, including the orbits of the giant planets or the complex dynamical structures in the transneptunian region. For instance, the Nice model addresses the dynamical evolution of the solar system after the formation of giant planets, and manages to reproduce many of these observed features (Gomes et al. 2005; Morbidelli et al. 2005; Tsiganis et al. 2005; Levison et al. 2008). However, it is still very difficult to explain the formation of all three sources of comets in the outer solar system in a consistent way (Charnoz and Morbidelli 2007). As of today, it is generally accepted that the Oort Cloud should be populated by objects formed in the 5–40 AU region, mostly in the 20–40 AU zone (Dones et al. 2004), that have been scattered out by the giant planets. The Scattered Disk should be populated by objects formed between 25 and 35 AU, which have been scattered mostly by Neptune (Duncan and Levison 1997; Morbidelli et al. 2004; Dones et al. 2004). Finally, the Kuiper Belt should contain objects mostly formed in situ ($\sim$30–50 AU) or moderately pushed during the chaotic dynamical evolution of the giant planets (Gomes 2003; Levison and Morbidelli 2003). Once stored in these reservoirs, comets can be perturbed by gravitational interactions and can be handed down through the giant planet region, toward the inner solar system. Jupiter Family Comets (JFCs) for instance are believed to come from the Scattered Disk (Levison and Duncan 1997; Tiscareno and Malhotra 2003; Gomes et al. 2008), while Long Period Comets (LPCs) and Halley-Type Comets (HTCs) could originate from the Oort Cloud (Dones et al. 2004) and the inner Oort Cloud (Levison et al. 2001) respectively.

Studying the composition of comets is believed to provide crucial clues on the physical and chemical processes which shaped the solar system. The composition of the protosolar nebula (PSN) and protoplanetary disk can be inferred from measurements of the Sun itself. Variations in composition, measured in comets for instance, can therefore be ascribed to physical or chemical processes which took place in the early stages of the solar system formation. Comets exhibit different physical and chemical characteristics (for example size, shape, or composition) as reviewed by Bockelée-Morvan et al., Cochran et al. and Lamy et al. in companion papers. While the range of sizes observed may be attributed to systematic differences among comet groups (Lamy et al. 2004), there may not be any compositional difference between JFCs and LPCs-HTCs (A'Hearn et al. 2012). However, within one dynamical group, comets might display a range of composition (A'Hearn et al. 2012), and nuclei with both uniform and non-uniform compositions have been observed (Dello Russo et al. 2007; Feaga et al. 2007; A'Hearn 2008). In order to link comets characteristics as observed today to their origin, we must try to distinguish whether the variety of physical and chemical characteristics is primordial or the product of evolutionary processes. The early physical evolution of comets, during the planet formation stage and before they reach their storage location either in the Oort Cloud or the Kuiper Belt, is a matter of strong debate. This evolution would be dominated by thermal processing and collisions, with a wide range of possible outcomes. During their 4.5 Gyr storage phase, comet surface layers would be altered mainly by space weathering, while interiors might have undergone modifications from residual radiogenic heating (if any), and thermal processing before they even enter the inner solar system.

Once a comet enters the inner solar system it evolves rapidly. The active-comet phase is dominated by the thermal processing of nuclei due to solar radiation, which has been studied by many authors (e.g. Huebner et al. (2006) and references therein). We gained a better understanding of the active-comet phase owing to great developments in laboratory experiments and numerical modeling, and few spacecraft missions. The first cometary mission, the ESA Giotto mission to 1P/Halley in 1986, has revealed a dark and porous nucleus, with active, less active, and completely inactive areas in its surface, the later being covered of non-volatile material quenching the ice sublimation (Keller et al. 1986). A few other comets were more recently studied by spacecraft missions, such as 19P/Borrelly by Deep Space 1 (Soderblom et al. 2002), 81P/Wild 2 by Stardust (Brownlee et al. 2004), 9P/Tempel 1 by Deep Impact (A'Hearn et al. 2005) and Stardust-Next (Veverka et al. 2013), and 103P/Hartley 2 by EPOXI (A'Hearn et al. 2011). Currently, the ESA Rosetta mission is orbiting comet 67P/Churyumov-Gerasimenko and is performing a detailed study never achieved before (see Sierks et al. 2015 or Thomas et al. 2015 for example). On a more general perspective, some comets can show a sporadic activity and sometimes outbursts, of which causes are not certain. Other comets show low levels of activity attributed to the progressive loss of near-surface volatiles. Eventually, some comet nuclei might become dead or dormant (see Jewitt 2004 for example), either from running out of volatiles or from catastrophic disruptions. In this paper, we present some long-lasting modifications of comets physical characteristics that would result from various processes, typically expected to affect the different stages of a comet's life.

2 Properties Inherited from the Formation Stage

2.1 Material Incorporated

2.1.1 Refractories

Cometary dust appears to be a mixture of silicate minerals, such as olivine and pyroxenes, both amorphous and crystalline (Hanner and Zolensky 2010), while a significant fraction of the carbon is expected to be in the form of organic refractory material. Such organic material was recently observed at the surface of 67P/Churyumov-Gerasimenko (Capaccioni et al. 2015). Minerals formed at high temperatures, including crystalline silicates that presumably condensed in the 1200–1400 K temperature range in the solar nebula (Hanner 1999), have been detected with ground-based telescopes in a large number of comets

(Campins and Ryan 1989; Crovisier et al. 2000; Sitko et al. 2004; Wooden et al. 2000, 2004, 2010). Cometary dust has been brought back to Earth from comet 81P/Wild 2 by the Stardust mission (Brownlee et al. 2006), which provided a direct way of studying refractory cometary material. The dust found in the Stardust samples consists of amorphous and crystalline silicates, metals and organics. All of these components been seen in primitive meteorites before, although never in the particular combination revealed by the Stardust samples. Possible aqueous alteration phases have been reported in the Stardust samples (Stodolna et al. 2012), although the most common products of aqueous alteration, phyllosilicates, have not been seen yet (Zolensky et al. 2006; Zolensky et al. 2008; Joswiak et al. 2012). The most intriguing result of the Stardust mission is the evidence of a large-scale transport of dust grains across the protoplanetary disk, given by the presence of minerals formed both in high and low temperature environments. For instance, the discovery of CAIs is puzzling because it is believed that CAIs were the first solids to form, close to the Sun during the earliest stages of the disk evolution (Grossman 1972; Jones et al. 2000; Simon et al. 2008). Therefore, the history of comet 81P/Wild 2, and by extrapolation of comets in general, must be more complex than previously envisioned.

A number of mechanisms has been proposed to explain the origin of these high-temperature minerals in comets. For example, the presence of shock waves triggered by gravitational instabilities in the outer PSN may anneal the amorphous silicates to crystallinity in situ prior to their incorporation in comets (Harker and Desch 2002). In these conditions, thermal annealing of submicron- and micron-sized silicate dust grains may occur in the formation region of comets, obviating the need for large-scale radial transport. However, the isotopic composition, minor element composition, and even the range of Fe/Si ratios measured in the dust that was returned by the Stardust spacecraft from Comet 81P/Wild 2 seem to rule out this scenario. There is indeed no model or experiment suggesting the possibility to form such compositions from plausible amorphous interstellar materials (Brownlee et al. 2006).

An alternative possibility is that small dust particles formed in the inner part of the PSN would have been redistributed outward during the dynamical evolution of the disk. In a classical model, the viscous stresses serve to drive mass inward with time and allow the disk to spread in the radial direction in order to preserve angular momentum (Shakura and Sunyaev 1973; Lynden-Bell and Pringle 1974). Hence, most of the disk gas moved inwards and was accreted by the protosun but some gas and gas-coupled particles would have been transported outwards. Because of its turbulent nature, gas follows random motions and leads to diffusion of material within it, allowing dust particles to be redistributed in a way that smooths out the concentration gradients (Ciesla 2009). Turbulence favors the rapid diffusion of the different gaseous compounds and gas-coupled solids throughout the nebula. One-dimensional (vertically averaged) diffusive transport of particles in the disk (Bockelée-Morvan et al. 2002), or two-dimensional transport through its surrounding layers (Ciesla 2007, 2009) and at various epochs of its evolution (Yang and Ciesla 2012) have therefore been proposed to account for the presence of hot temperature minerals in the formation zone of comets. It is uncertain however whether turbulent transport suffices to explain the observations, or whether alternative physical processes are also needed. Indeed, Hugues and Armitage (2010) investigated the outward transport of particles in the nebula via a combination of advection (inward drift of particles though interaction with gas) and turbulent diffusion in an evolving disk. These authors found that the advection of solids within the gas flow significantly reduces the outward transport efficiency for larger particles (typically a few millimeters), thereby limiting the extent of mixing uniformity that is achievable within the disk via turbulent diffusion.

Another proposed transport mechanism of dust within the solar nebula is photophoresis (Krauss and Wurm 2005; Wurm and Krauss 2006; Krauss et al. 2007; Mousis et al. 2007; Wurm et al. 2010; Moudens et al. 2011; Wurm et al. 2013). This effect is based on a radiation-induced temperature gradient on the surface of a particle and the subsequent non uniform interaction with surrounding gas. When the existence of an inner gap is assumed in the disk, this latter becomes optically thin enough for particles to see the protosun, but still has a reasonable gas content, which enables the photophoretic force to push dust grains outward (Mousis et al. 2007; Moudens et al. 2011). This process provides a mechanism to transport high-temperature material from the inner solar system to the regions in which the comets were forming. Eventually, the dust driven outward in this manner will reach a region where the gas pressure and irradiation are so low that the combined outward forces of radiation pressure and photophoresis can only balance the inward drift of particles, inducing the formation of a ring of dust at this location. So far, simulations of photophoretic transport require the presence of an inner gap in the protosolar nebula to enable the drift of particles towards the formation regions of comets (Moudens et al. 2011). Further studies will take into account models depicting the time evolution of the Sun's illumination from its earliest phases to present. Indeed, the assumption of an inner gap could possibly be avoided if one takes into account the increase of UV light from the early Sun.

2.1.2 Volatiles

Compared to refractories, volatiles condensed in form of ices are more sensitive to changes in temperature, whether in the protosolar nebula or later in comet nuclei, so that they offer a better diagnostic as to where comets might have formed. Therefore, understanding the composition of the gaseous and solid phase in the protoplanetary disk is a key step toward assessing the bulk composition of comets. Molecular hydrogen and helium gas contribute to most of the mass in the PSN. Refractories (material with high (>1000 K) sublimation temperatures) are thought to represent up to 0.6 % of the PSN (Asplund et al. 2009). Volatiles (molecular and atomic species with relatively low sublimation temperatures, in this context species like H_2O, CO, CO_2, NH_3 etc) may hold 1.4 % of the PSN mass (Lodders 2003; Asplund et al. 2009). Chemical-thermodynamic equilibrium calculations show that CO and N_2 may be the major C- and N-bearing species at temperatures higher than a few hundred K, and that CH_4 and NH_3 should dominate at lower temperatures (Cyr et al. 1999). However, non-equilibrium calculations indicate that the reaction producing CH_4 and NH_3 from CO and N_2 respectively is kinetically inhibited, so that CO and N_2 could have remained the dominant species at low temperatures (Lewis and Prinn 1980). This is supported by the observations of CO and/or CO_2 as being the main C-bearing volatiles in comets, as reported for example by the observations from the Deep Impact spacecraft of comets 9P/Tempel 1 and 103P/Hartley 2 and the AKARI satellite (Feaga et al. 2007; A'Hearn et al. 2011; Ootsubo et al. 2012).

Volatiles are supposed to be present in gaseous as well as solid form, depending on the temperature distribution across the protoplanetary disk. The distance at which a volatile specie can condense is referred to as the snowline. The locations of snowlines of molecules such as H_2O, CO or CO_2 should have dictated the bulk composition of comets. Determining the positions of such snowlines is key to probing the composition of planetesimals across the protoplanetary disk. The study of snowlines in the PSN is very uncertain (Pontoppidan et al. 2014), so that solar analogs are being used to perform this investigation (Qi et al. 2013). The H_2O snowline should be key to the formation of planetesimals in the Jupiter-Saturn region, whereas the CO snowline, estimated to be located at tens of AU from the

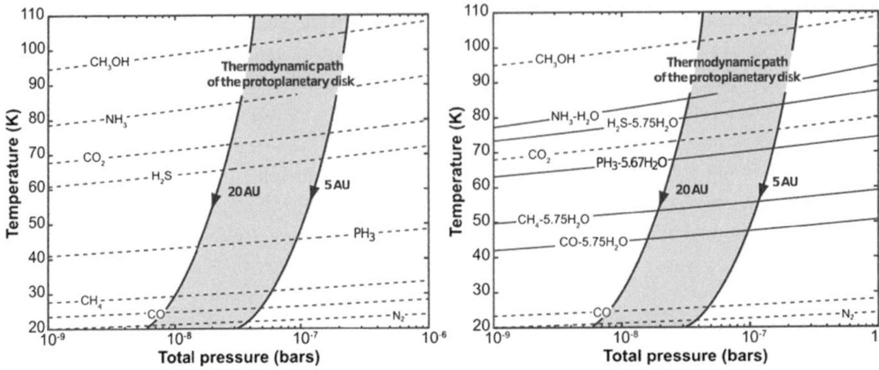

Fig. 1 Condensation sequence of the volatiles in the protosolar nebula. *Left*: Equilibrium curves of pure volatile condensates, along with the thermodynamic path of the nebula at 5 and 20 AU. *Right*: Equilibrium curves of hydrate (NH_3-H_2O), clathrates (X-5.75H_2O or X-5.67H_2O; *solid lines*), and pure condensates (*dotted lines*), and cooling curve of the solar nebula at 5 and 20 AU, assuming a full efficiency of clathration. Adapted from Madhusudhan et al. (2011)

protosun, should be key to the formation of planetesimals in the Uranus-Neptune region (Qi et al. 2013). However, the locations of snowlines may evolve with time. Indeed, in most classical one-dimensional and two-dimensional PSN models (Cassen 1994; D'Alessio et al. 1998; Hueso and Guillot 2005, for example), the disks midplanes cool down with time as their material essentially falls onto the parent star. This causes an inward migration of the snowline positions with time, as shown in Fig. 1. Recent layered-PSN models, including a midplane deadzone, predict more complex snowline evolutions (Martin and Livio 2012). Dodson-Robinson et al. (2009) suggests that between 3×10^4 yr and 10^6 yr, the CO snowline could have moved from 12 AU to 8 AU, while the H_2O line could have moved from 5 AU to 2 AU. Because comets exhibit a range of CO abundances, the CO snowline may have been located in the outer part of their formation region (Mumma and Charnley 2011). A'Hearn et al. (2012) suggests that the CO and CO_2 abundances could be explained if comets were formed between the CO_2 and CO snowlines. However, given the small number of comets investigated, there remains a possibility that the heterogeneity of volatile abundances among comets could be explained as a result of evolution in the inner solar system (Belton and Melosh 2009), or a combination of both formation and subsequent evolution.

In addition to their incorporation as pure condensates in the ~20–30 K range in the PSN, volatiles may be trapped into the lattice of water ice grains. These icy grains may be formed from amorphous ice either condensed at low temperature (<70 K) in the outer layers of the protoplanetary disk (Ciesla 2014), or originating from interstellar medium where extremely low temperature conditions (~10 K) prevail (Gibb et al. 2004). We know from laboratory experiments that amorphous water ice has the ability to trap large amounts of volatiles, which can be expelled from the lattice upon crystallization (Bar-Nun et al. 1985, Owen and Bar-Nun 1998, Yokochi et al. 2012). The formation temperature range of amorphous ice appears consistent with the spin temperatures measured in H_2O and NH_3 in several comets and that cluster near 30 K (Mumma and Charnley 2011). On the other hand, volatiles can also be trapped in crystalline water ice in the form of clathrates during the cooling of the protosolar nebula. These clathrates have been hypothesized to contribute to the observed compositions of solar system bodies like icy moons of the giant planets (Lunine and Stevenson 1985; Mousis et al. 2010). Figure 1 shows the thermodynamic conditions needed for such trapping to occur. In this case, volatiles are trapped at temperatures higher than for amorphous water ice, and should also be released from the water ice matrix at higher temperatures than their respective sublimation temperatures.

While evidence from the study of refractories indicate a large scale mixing of dust grains within the protoplanetary disk, it is not clear yet whether such mixing occurred for icy grains. In particular, one key question is whether any large scale mixing of cometesimals (the building blocks of comets) may have taken place early in the evolution of the solar system. A'Hearn et al. (1995) argue that observational data are consistent with a mixing of water ice and other volatile species at the level of grains, rather than at the level of molecules, and that they appear to require the mixing of cometesimals of different compositions into individual comet nuclei. If so, this would indeed require that cometesimals were scattered among different regions of the PSN at the time comets were accreted. The extent of such scattering remains to be understood, since some nuclei might contain the full range of compositions observed within the whole population of comets (A'Hearn et al. 1995) and present a heterogeneous composition like 9P/Tempel 1 (Mumma et al. 2005; Feaga et al. 2007), while some other comets like 73P/Schwassmann-Wachmann 3 appear homogeneous on a macroscopic scale (Dello Russo et al. 2007; Kobayashi et al. 2007). However, for the volatile species studied by A'Hearn et al. (1995), the sample of observed comets seem quite uniform and only two main taxonomic classes may exist for the overall population: one typical class and a second depleted in C-bearing compounds.

2.2 Accretion and Collisions

There are currently two opposite views of the formation of comet nuclei: they could either be relatively unprocessed aggregates of cometesimals formed directly in the PSN, or collisional debris of larger bodies. A superposition of the two nuclei types might alternatively exist. In the later case (collisional debris), we need to consider comets along with Transneptunian Objects (TNOs) and Centaurs (all referred to as icy bodies), believed to be dynamically connected to JFCs (Levison and Duncan 1997; Tiscareno and Malhotra 2003). The largest TNOs, like Pluto, Haumea, Makemake or Eris, are defined as dwarf planets. JFCs on the other hand are smaller km-sized objects. There is therefore a wide range of icy bodies in the solar system, going from km to 10^3-km diameter objects. However, the size distribution of these objects is difficult to establish, since small objects are difficult to observe in the outer regions of the solar system, and large comets are lacking in the inner regions. Nonetheless, the size distribution of TNOs shows similarities with asteroids. Asteroids in the Main Belt have undergone some substantial dynamical depletion and collisional erosion. Using dynamical evolution models, we can link their current size distribution to their primordial birth size distribution. Weidenschilling (2011) suggests that a primordial population of 100 m bodies can reproduce the current asteroid population, including the knee at 100 km in their size distribution. However, Bottke et al. (2005) proposed that asteroids larger than 120 km should be primordial, and that the smaller objects should result from the collisional grinding of the larger ones. This is consistent with Morbidelli et al. (2009) who tested different formation models to determine that asteroids should have formed big. Similarly, the knee at 100 km in the TNO size distribution suggests that the typical cometesimal birth size should be about 100 km.

In addition, Stern and Weissman (2001) suggested that collisions between cometesimals in the giant-planet region would be catastrophic for most of the population, so that comet nuclei ejected into the Oort Cloud could mostly be collisional debris of larger objects. Stern (1995) and Davis and Farinella (1997) also argued that most JFCs are likely collisional fragments of disrupted parent bodies with initial sizes up to $\sim$100 km. However, recent calculations by Belton (2014) show that many comets would be primordial planetesimals relatively unaffected by impacts, rather than collisional debris of larger objects. The outcomes of such collisions on the physical characteristics of comets remain to be constrained. Comet nuclei visited by spacecrafts are lacking impact crater features that could help us identify physical and chemical characteristics linked to collisions (Soderblom et al. 2002; A'Hearn et al. 2011; Belton et al. 2013). However, the surface of these nuclei may have been refreshed due to subsequent cometary activity, and all traces of past collisions may have been removed (see for example Weissman et al. 2004). The observation of some bi-lobate shapes might indicate that smooth collisions could have produced the nuclei of comets 19P/Borrelly, 103P/Hartley 2 and 67P/Churyumov-Gerasimenko. Among the comets targeted by space missions, 9P/Tempel 1 and 67P/Churyumov-Gerasimenko display a layering in their composition (Veverka et al. 2013; Sierks et al. 2015), which could be the result some thermal processing (see following sections), or rather an argument in favor of early collisions during which a nucleus may accrete material of various composition (Belton et al. 2007a, 2007b). Although the predicted decrease in layer thickness is not observed on 67P, different sets of lineaments between its two lobes may indicate that these two parts may have been accreted by slow velocity collision at some point (Sierks et al. 2015).

Comet internal structures may help us understand the accretion process behind their formation. However, inferring this structure from the current available observations is challenging, although the results from the CONSERT experiment on Rosetta will provide the first determination of a comet's internal structure. Several models exist, that can explain many of the observed features of cometary nuclei and activity (see Weissman et al. 2004 and A'Hearn 2001 for a review). The current consensus may favor the rubble pile model (either primordial by Weissman (1986) or collisional by Weissman et al. (2004), and the Talps model (Belton et al. 2007a, 2007b). If so, the size of cometesimal incorporated into individual nuclei may be inferred, as reviewed by A'Hearn (2011). The layering observed on comet 9P/Tempel 1 by the Deep Impact mission does not appear like the pattern expected from most evolutionary models (A'Hearn et al. 2011) but rather like primordial cometesimals: Belton et al. (2007a, 2007b) inferred that these cometesimals should be ~400-m objects. Some hints of the primordial size of cometesimals may also be inferred from splitting comets. Indeed, comets are known to fragment spontaneously with no obvious reason (see following section). The size of their fragments may tell us about the size of their constituting cometesimals (see Boehnhardt 2004 and Fernandez 2008 for reviews on splitting comets, and Knight et al. 2010 for a review on the size distribution of sun-grazing comets). These data are generally consistent with a typical size of cometesimals on the order of 50–100 m.

Once ejected into their storage regions, comet nuclei might not be significantly modified by collisions, in particular in the Oort Cloud which constitutes a dilute and cryogenic environment in which modifications of comets' physical and chemical properties are expected to be only limited. Indeed, collisions within the Oort Cloud, with an orbital speed of $\sim 0.2\ \mathrm{km\,s^{-1}}$, have effectively no influence on the evolution of comets. Stern (1988) found however that impacts would be more significant in the Inner Oort Cloud (Duncan et al. 1987) so that comets stored in this region should have undergone some surface modification. Studies of the physical and dynamical evolution of comets suggest that collisions may play an important role in the Kuiper Belt (Stern 1995, 1996; Farinella and Davis 1996; Stern and Weissman 2001, Duncan et al. 2004).

2.3 Early Thermal Processing

It is not clear whether impact-induced or accretional heating would have actually played a significant role in the global evolution of comet compositions in this early stage of their life. The thermal effect of collisions on comets is poorly known, although it might potentially be inferred from what we know of the collisional processing of asteroids. We therefore consider it useful to review the thermal consequences of impacts among asteroids, for which many studies have been performed. Impacts have been invoked to explain the metamorphism and melting seen in meteorites, especially when such events occurred lately (>5 Myr after CAI formation) in the history of meteorite parent bodies, when short-lived radioactive nuclides had already decayed (Rubin 2004; Schulz et al. 2012 and references therein). However, impacts would have been unable to drive a global-scale thermal processing of asteroids, since impact heating remains localized and can only affect a small volume of the target body. Keil et al. (1997) suggested that the global temperature increase after impact would be approximately 10 K. Davison et al. (2010) studied the influence of porosity on impact heating, and showed that collisions on porous objects would result in higher temperatures than on non-porous objects. However, they estimated that the globally-averaged temperature increase would also be of the order of 10K. Considering the speed of collisions in comet forming regions, the temperature increase would be expected to be quite small too (see Huebner et al. 2006).

Comets are however made of volatile species, more sensitive to variations of the temperature, so that small increases of the temperature can be relevant for the evolution of the composition of subsurface layers. Laboratory experiments show for example that collisions under present day dynamical conditions would result in the melting of water ice at the point of impact (Stewart and Ahrens 2004). The production of craters filled with melts that subsequently rapidly solidified as glass have been observed in the lab (Koschny et al. 2001; Love et al. 1993). In addition, Ciesla et al. (2013) showed that the post-impact structure of asteroids would be strongly influenced by their pre-impact internal structure, in particular with respect to the layered structure resulting from radiogenic heating. They found that heating would be mostly localized around the point of impact, and that radiogenically-heated internal material, which would have remained buried under the surface for a non-impacted object, would flow to the surface. This process would lead to an overall increased cooling rate of the impacted asteroid, implying that collisions on targets affected by radiogenic heating would help cooling them instead of further heating them. Therefore, if comets were formed big like asteroids, they could have been affected by an impact-induced thermal processing in the early stages of their physical evolution, yet to be constrained, in addition to the thermal processing described below.

The decay of ^{26}Al has long been recognized as a potentially powerful heat source, capable of melting rocky body interiors. Its actual effect on the thermal evolution of comets remains to be fully constrained though. In particular, whether this nuclide was actually accreted in comet nuclei however remains debatable. The study of meteorites and Stardust samples should be helpful in understanding the early evolution of comets. The dynamics of meteorite delivery to the Earth are well understood (Morbidelli and Gladman 1998; Vokrouhlicky and Farinella 2000), and all meteorites for which a precise orbit has been calculated originate from the asteroid belt (Gounelle et al. 2006 and references therein). There is thus a large consensus that asteroids are the source of all meteorites, except for a few which are lunar or martian meteorites. To understand if comets can produce meteorites worth studying in the laboratory, we should understand whether comets can impact the Earth. Gounelle et al. (2008) estimated that impacts of JFCs on Earth should be very

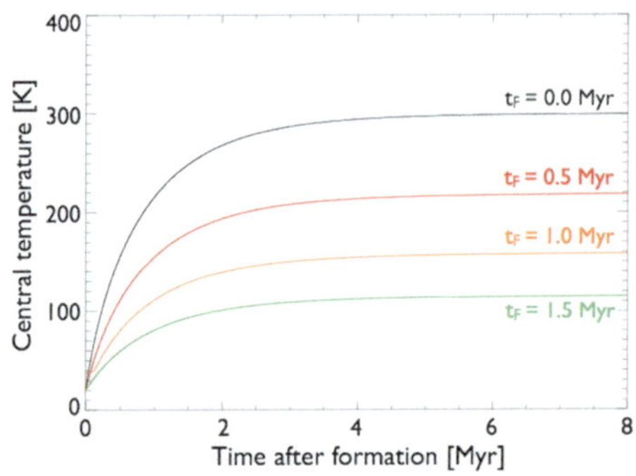

Fig. 2 Evolution of the central temperature for a 30 km comet parent body, with realistic physical characteristics (density of 700 kg m^{-3}, thermal conductivity of 10^{-2} Wm^{-1} K^{-1}), under the influence of radiogenic heating due to the decay of short-lived radioactive nuclides. The different lines correspond to different formation times after CAI formation: the longer the formation time, the smaller the amount of decaying radioactive nuclides available for heating comet interiors. Based on Mousis et al. (2012)

rare, and—given the very short lifetime of small comets in the inner solar system—could be considered as non-existent in practice. Despite the intense modeling effort, the impact rate of comets versus asteroids with Earth is still not constrained (see Gounelle et al. 2008, for a review). Although pathways from the outer solar system should be extremely rare, they can exist. In fact, Gounelle et al. (2006) suggested that the orbit of the Orgueil meteorite could be more compatible with that of a JFC, than with that of an asteroid. Impact probabilities with the Earth and entry velocities suggest that the proportion of cometary meteorites compared to asteroidal meteorites should be small but not zero. The study of Stardust samples shows that they contain minerals that were all previously seen in primitive meteorites, though the combination found in comet 81P/Wild 2 could not be matched to any known meteorite type. Although Ca–Al rich Inclusions (CAIs, believed to be the first solids to condensate in the solar system) were detected in the Stardust samples, the measurements of the decay product of short-lived radionuclide ^{26}Al, ^{26}Mg, do not show the former presence of ^{26}Al.

The ^{26}Al/^{27}Al ratio, with a canonical value of 5×10^{-5} (MacPherson et al. 1995), appears consistent in CAIs and meteorites of different classes which argues for a uniform distribution of ^{26}Al within the solar system (Huss et al. 2001). CAIs with low ratio ($\sim 5 \times 10^{-6}$) have been reported (Kunihiro et al. 2004; Makide et al. 2009), so that the existence of such inclusions, or the inclusions with no radiogenic excess of ^{26}Mg found in the Stardust samples, is perceived as a reason to invoke a non-uniform distribution of ^{26}Al within the solar system (Makide et al. 2011; Krot et al. 2012; Ogliore et al. 2012). This constitutes the first impediment to our understanding of the early thermal processing of comets. The second impediment comes from our lack of knowledge of the comet formation process itself, both in term of size (see previous section) and timescale (see Fig. 2). Indeed, given its short lifetime, the effectiveness of ^{26}Al in heating a comet interior strongly depends on the nucleus formation time with respect to the CAIs formation. Prialnik et al. (1987) found that with an initial ^{26}Al/^{27}Al ratio of 5×10^{-5}, the heat released upon decay would have resulted in the complete crystallization of amorphous ice, and even melting of comet cores for objects larger than 6 km. Haruyama et al. (1993) showed that comet thermal histories are very sensitive to thermal conductivity, leading to very high temperatures and full crystallization of the comet interiors, or alternatively very limited temperatures not higher than 100 K. According to Prialnik and Podolak (1995), the early thermal evolution of comets could have lead to distinct configurations, depending on their size, thermal conductivity and formation time.

Table 1 Estimated irradiation dose in eV/16-amu molecule for 4.6 Gyr, with an ice density of 1.0 g/cm^3, from Hudson et al. (2008)

Distance from the Sun (AU)	At 1-μm depth	At 100-μm depth	At 1-m depth
5–35 (Giant planet region)	100–10,000	100–200	30
40–∼1,000 (Transneptunian region)	100–500,000	100–30,000	30–50
∼40,000–100,000 (Oort Cloud)	500,000	30,000	50

A pristine structure could be thoroughly preserved, or the interior could be partly to completely crystallized except for a negligible outer layer which might remain primitive. The occurrence of high internal temperatures and production of liquid water would be strengthened when accounting for the effect of accretional heating, affecting the early evolution of comets concurrently with radiogenic heating during the short lifetime of ^{26}Al. Merk and Prialnik (2006) showed that the occurrence of liquid water in 2 to 32-km radius bodies may be a very common phenomenon. For example, they find that for a given set of initial parameters, all accreting objects with a final radius above 4 km could produce liquid water cores, extending from 10 to 90 % of the overall interior.

3 Processing During the Storage Phase

3.1 Effect of Space Weathering

Stern (2003) argued that even during the quiescent storage phase of comet nuclei, some evolutionary processes like thermal processes or radiation might play a role. Space weathering describes processes affecting the surface of airless bodies, such as bombardment by micrometeorites, irradiation by UV photons, solar wind particles and cosmic ray ions. A particle passing through an ice layer looses its energy by producing secondary particles, which in turn lead to many excitations in the surface material. These events result in the breakage of molecule chemical bonds, and the rearrangement of molecule fragments to produce new complex molecules. From laboratory experiments (Brunetto et al. 2006; Brunetto and Roush 2008; Hudson et al. 2008; Modica et al. 2012), we know that the simple molecules found in comets and their parent bodies, such as CH_4 and CH_3OH, are easily destroyed by space weathering, so to produce more complex organic molecules leading to the formation of a crust of refractory material. Indeed, such an organic-rich surface has recently been observed for comet 67P/Churuymov-Gerasimenko (Capaccioni et al. 2015). After prolonged irradiation, this crust might evolve into a layer of amorphous carbon (Strazzulla and Baratta 1992). Irradiation of cometary surfaces by high energy particles and photons is therefore a dominant evolutionary process during the storage phase of comet's life, whether in the Oort Cloud or the Kuiper Belt, as shown by the study of the radiation distribution in the outer solar system, performed both by theoretical modeling and in situ spacecraft measurements (see Hudson et al. 2008 for a review and Table 1).

Space weathering has a number of effects relevant in understanding the evolution of comets since it can cause significant variations in the surface properties such as color, composition, or other physical and optical characteristics. The progressive destruction of

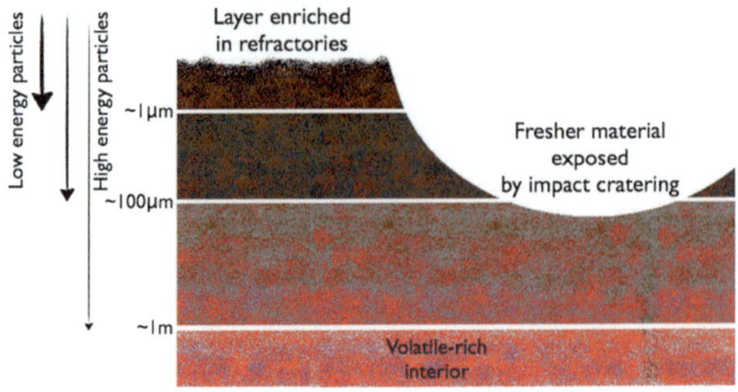

Fig. 3 Schematic illustrating the effects of irradiation by high energy particles on subsurface layer of icy objects in the outer solar system, and the effect of localized impact cratering. Based on Guilbert et al. (2009)

molecules leads to their signatures being progressively erased from comet spectra. Additionally, laboratory studies have shown that initially bright and flat spectra evolve into redder and darker spectra with increasing irradiation dose (Brunetto et al. 2006 for example). These characteristics are typically observed among comets, since they usually display a very low albedos (see Lamy et al. 2004), and red visual spectral slopes. As an example, we may mention the recent results of Capaccioni et al. (2015) who measured an albedo of 6 % and a visual slope of 5 to 25 %/100 nm for comet 67P/Churyumov-Gerasimenko. However, there is a superposition of reddening and darkening that needs to be accounted for, since initially red spectra can evolve into flat dark spectra (Kanuchova et al. 2012). Ultimately, in the presence of organics, space weathering trends are not well established yet. The end-result chemistry depends on whether the ice is initially pure, or if several ices are mixed together, i.e. on the initial composition of the mixture present at the surface of comet nuclei. In addition, irradiation by energetic particles can induce sputtering of surface molecules. This process is especially important for icy satellites, exposed to energetic ions trapped into the magnetosphere of giant planets. However, sputtering can also occur on cometary surfaces, due to irradiation by energetic solar wind ions. Among the potential effects of sputtering, we can expect some changes in composition, depletion in volatiles, erosion of the surface, variations of the surface texture, albedo and porosity (Palumbo 2006; Dartois et al. 2013). In particular, sputtering is believed to participate to the formation of comet mantles in the Oort Cloud (see Johnson 1995 for a review).

In summary, space weathering leads to the formation of an irradiation layer on the upper few meters of a comet nucleus. This mantle is expected to be carbon-rich and depleted in volatiles i.e. chemically transformed compared to the initial composition of cometary surfaces. From laboratory experiments, Strazzulla et al. (1991) suggested that in the Oort cloud the external 0.1 to 0.5 m thick layer of comet nuclei would be exposed to an irradiation dose sufficient to produce a substantial nonvolatile layer. Local removal of this layer due to impact cratering (illustrated by Fig. 3) could induce variations in the overall observed color, albedo and composition of the surface, and lead to very complex inhomogeneous surface properties, occurring before the comets enter the inner solar system. This external layer could survive the subsequent sublimation of underlying ices (Strazzulla et al. 1991; Strazzulla and Palumbo 2001). However, it is generally assumed that irradiation layers would be blown off or be buried by subsequent outgassing, and would not survive a comet first entry in the inner solar system.

3.2 Thermal Processing

The thermal processing of comets during their storage phase is usually assumed to be limited, since the equilibrium temperature in the outer solar system reservoirs is very low. However, when it comes to the survival of supervolatile species like CO or N_2, all potential heating needs to be accounted for. The processing of comet nuclei by radiogenic heating, as described previously, is relevant here since some residual ^{26}Al might still be decaying at the time comets enter their storage phase, either in the Oort Cloud or the transneptunian region. Long-lived isotopes like ^{40}K, ^{235}U, ^{238}U or ^{232}Th would inevitably decay over the age of the solar system, generating a possible more moderate but more extended heating of TNO interiors, and therefore play a significant role in differentiating the largest parent bodies. Between 30 and 50 AU, radiogenic heat can in fact be comparable to solar radiation. However, the actual outcome of radiogenic heating, both in the Oort Cloud and in the Kuiper Belt, depends on the amount of radiogenic elements accreted in comets and on the physical parameters, such as thermal conductivity, porosity, or the size of the nucleus. Thermal evolution modeling performed for porous comet nuclei with $R > 10$ km shows that these objects may have completely preserved their initial stratigraphy, or have completely crystallized (assuming they were initially composed of amorphous water ice), or developed a 'differentiated' structure, with a crystallized core, a layer of condensed volatiles, and pristine outer layers made of unaltered material (Prialnik and Podolak 1995). In addition to this radiogenic heating, Stern and Shull (1988) proposed that up to 20 % Oort cloud comets could have been heated to at least 30 K to a depth of several dozen meters, due to the passage of luminous stars. They suggested that most of them might have been heated as high as 45 K to a depth of 1 m from stochastic supernovae events. This would lead to the formation of a surface layer where super volatiles would be absent in their pure condensate form (they could have survived as molecules trapped in water ice). Stern (2003) summarized that passing stars and supernovae heating events may mask the primordial composition of comet nuclei down to 5 to 50 m for stars, and 0.1 to 2 m for supernovae.

After leaving their storage location, comet nuclei may enter the inner solar system following chaotic orbits, owing to the gravitational perturbations of the four giant planets. They can be injected into the inner solar system either on a direct path or a multistage path, sometimes extremely hard to track back from their current observed orbit. We consider here the effect of thermal processing before comets perform their first passage close to the Sun. We stress that the multistage process is not efficient in injecting comets into the inner solar system, since a minority of comet nuclei are found to reach short-period orbits without being ejected of the solar system (Everhart 1969). If we consider the coupling between thermal and dynamical evolution, a complex multistage dynamical evolution may have resulted in a complex and drastic modification of comets internal composition and structure, different from the ones resulting from a more direct injection. While a direct-injection orbit would most likely lead to volatile-depleted layers protecting internal ones, thereby preserving the pristine composition, multistage-injection orbits can be such that surface layers are strongly ablated and the underlying pristine material exposed. Huebner et al. (2006) performed a comparison between the "final" stratigraphy of a comet nucleus being injected via a direct and a multistage process. In the multistage-injection nucleus, the CO interface is located at a depth of more than a hundred meters, while it remains close to the surface for the direct-injection nucleus. Similarly, the interface between amorphous and crystalline water ice (assuming water ice is initially amorphous) is found more than a hundred meters below the surface in the multistage-injection nucleus, but remains close to the surface in the direct-injection nucleus. They also found that the subsequent CO emission from the multistage-injection nucleus would be almost ten times less than from the direct-injection one, and that its pattern would not follow the water emission. From the comparison, they conclude that the behavior of volatile gases could be used as a diagnostic for earlier evolution. A sudden outburst of volatile gases can indeed be a diagnostic of a recent orbital change, as suggested in the case of 46P/Wirtanen (Jorda and Rickman 1995).

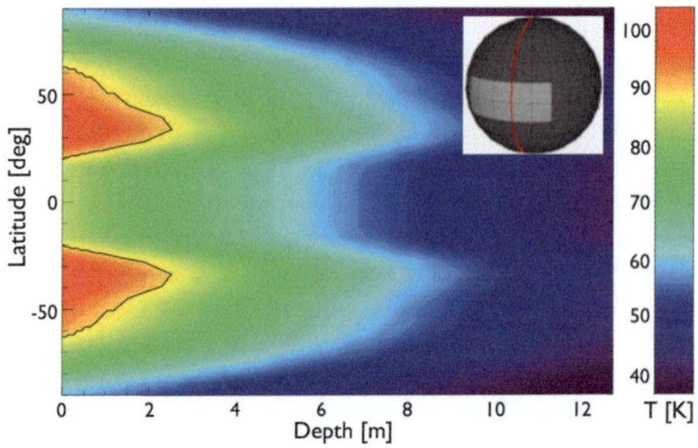

Fig. 4 Temperature distribution across a meridian cut of a comet subsurface layer. The surface holds a patch of higher albedo, casting a long-lived thermal shadow in the subsurface. *Black solid line* gives the location of the crystallization front. Adapted from Guilbert-Lepoutre and Jewitt (2011)

In addition to the radial layering described above, Guilbert-Lepoutre and Jewitt (2011) suggested that a lateral chemical differentiation would be expected from comet nuclei with non-uniform surface properties. They studied the evolution of the internal composition of comet nuclei orbiting on Centaur-like orbits, following the idea that a dynamical cascade between Kuiper Belt objects, Centaurs and Jupiter Family Comets might exist (Levison and Duncan 1997; Tiscareno and Malhotra 2003). They found that surface inhomogeneities, due to local variations of the albedo for example, would cast long-lived thermal shadows in subsurface layers. This would result in strong lateral modifications of the composition at a given depth, with regions in the thermal shadows being enriched in volatiles and surrounded by crystallized and volatile-depleted material (see Fig. 4). The superposition of the two processes, radial layering and lateral chemical differentiation, could therefore lead to some very complex internal structures, possibly mimicking the structures inherited from a complex formation mechanism such as described in the Talps model (Belton et al. 2007a, 2007b), before comets even entered the inner solar system. Recent collisions could also produce some localized variations of physical properties such as composition, color, albedo, porosity or thermal inertia, which would in turn increase the importance of thermal shadowing described before. Koschny et al. (2001) and Burchell et al. (2002) suggested that dense impactors may end up embedded into the surface of their targets, leading to possible strongly inhomogeneous surfaces. For instance, in an extreme case, a volatile-free projectile might get trapped into a volatile-rich cometary surface. In addition, Housen et al. (1999) and Housen and Holsapple (2003) suggested that collisions would result in the local compaction of the surface material, rather than in the traditional crater excavation flow, leading to strong variations of the thermal conductivity across comet surfaces. We stress again that observations of comets from spacecrafts have revealed that nuclei are lacking impact crater features that could help us identify physical and chemical characteristics linked to collisions, although craters would likely be removed by erosion due to subsequent cometary activity on a short timescale, and not expected to be observed (Weissman et al. 2004).

4 Active-Comet Phase

4.1 Loss of Volatiles

The active phase has been studied by many authors. We refer to extensive reviews by Prialnik et al. (2004) or Huebner et al. (2006) for detailed discussions on this intense period of a comet's life during which changes are expected to occur on its shape, size, or rotation properties (Jewitt 2004). Several mechanisms, like sublimation, clathrate devolatilization or crystallization of amorphous ice, can be identified to explain the mass loss observed during the active-comet phase. Sublimation of ices is the primary driver for cometary activity in the inner solar system. While volatile species like CO or CO_2 would start to be released at large heliocentric distances, depending on their sublimation temperature, water ice sublimation becomes important inside $\sim$3 AU (Meech and Svoren 2004). Clathrates devolatilize at specific temperatures, lower than the sublimation temperature of water ice. Once a clathrate has devolatilized, water ice remains crystalline until it reaches its own sublimation temperature. In addition, if amorphous water ice is present, its transition to crystalline water ice may be initiated. In this case, all volatiles released upon crystallization would escape the comet nucleus together, independently of their sublimation temperature. However, these mobilized volatile molecules can diffuse toward colder regions inside the nucleus where they can refreeze. Whether amorphous water ice can be preserved inside comets is a matter of debate, since it has never been observed directly. There could be some indirect evidence that amorphous water ice might have been able to survive close to the surface of comets though. For instance, the phase transition between amorphous and crystalline water ice has been invoked to explain the activity at large heliocentric distance of Centaurs (Jewitt 2009; Guilbert-Lepoutre 2012) and comets beyond 5 AU (Cochran et al. 1992). The phase transition is also one of the processes considered to explain cometary outbursts. For example, crystallization of amorphous ice, accompanied by the release of occluded volatiles and dust was invoked to interpret the outbursts observed for 1P/Halley (Espinasse et al. 1991; Prialnik and Bar-Nun 1992), C/1995 O1 (Hale-Bopp) (Prialnik 1999, 2002, Capria et al. 2002), and more recently 17P/Holmes (Hillman and Prialnik 2012). Therefore, coma observations alone are insufficient in constraining the composition of comets, since the abundances of volatiles in the nucleus are not mirrored directly by the observed abundances in the coma (Huebner and Benkhoff 1999).

The fraction of solar radiation that is not reflected or used for the different phase transitions at the surface of comets may be conducted into the nucleus (see Fig. 5). The depth reached by this process depends many unknown or poorly-constrained thermophysical properties such as porosity, pore size distribution, inertia or local geomorphological features. If the heat wave reaches volatile-rich layers, the same phase transitions may occur. We note that below the thermal skin depth (determined by the surface thermal inertia), the material is not sensitive to day-night variations of the surface temperature. Volatiles mobilized from subsurface layers could flow outward or diffuse inward through a complex network of pores where they would eventually refreeze (Benkhoff and Boice 1996). Since transition rates are strongly temperature dependent and vary from one volatile specie to an other, several distinct sublimation and condensation fronts and are expected to develop inside the nucleus. This process also leads to a chemical differentiation of the subsurface layers, with near-surface layers being depleted in volatiles over a few orbital periods, and the formation of a stratified structure (Prialnik et al. 2008; DeSanctis et al. 2007). We note however that the evidence from the Deep Impact mission to 9P/Tempel 1 suggest that the layering observed does not appear like the stratified structure expected from most thermal evolution models (A'Hearn 2001). They also suggest that the internal composition, as inferred from the few meters of subsurface layer studied, is not strongly differentiated, or that the differentiation expected from solar radiation processing proceeded below 20 m. The case of 9P/Tempel 1 may illustrate the fact that surface erosion and chemical differentiation are in competition, so that surfaces may erode as rapidly as differentiation fronts proceed into the nucleus. In addition, Jewitt (2004) argues that given

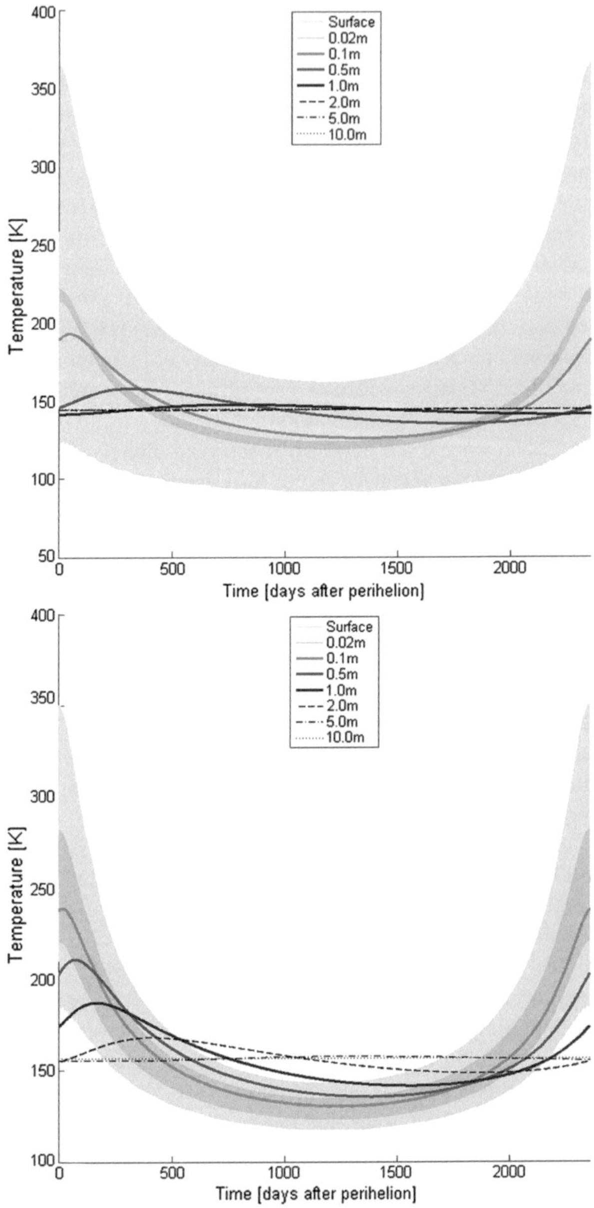

Fig. 5 Evolution of the temperature at selected depths and for different porosities over one orbital period of a Jupiter Family Comet (semi major axis of 3.5 AU and eccentricity of 0.64). Upper layers are affected by day-night variations of the temperature (*grey areas*), while below the skin depth the material is not sensitive to such variations. A higher surface porosity (*upper image*, thermal inertia of $20\ \mathrm{J\,m^{-2}\,K^{-1}\,s^{-1/2}}$) is linked to a lower conductivity, and results in a shallower penetration of the heat wave. Less porous material (lower image, thermal inertia of $100\ \mathrm{J\,m^{-2}\,K^{-1}\,s^{-1/2}}$), with higher conductivity, allows a deeper propagation of the heat wave. The effect of sublimation is not considered here, so to keep the problem simple. Sublimation would delay the propagation of the heat wave to deeper layers, as it consumes a significant fraction of the transported energy

the typical heat-conduction timescale for comets, deep interiors (below 1 km) should be thermally decoupled from surfaces. This process is complicated by the presence of dust, and the unknown internal structure, primordial and inherited from the evolutionary track followed before comet nuclei entered the inner solar system. Indeed, Rosenberg and Prialnik (2010) showed that internal inhomogeneities would affect the activity pattern of a comet and possibly produce outbursts at large heliocentric distance.

When solar radiation is dominant, orbital properties such as the spin rate or the inclination of the spin axis start to play an important role in the evolution of comets. They trigger a non-uniform heating of cometary surfaces which, combined with a low thermal inertia and an initially specific compositional structure, can result in strong modifications of local thermo-physical properties as well as uneven erosion, and in strong changes in the nucleus shape (Cohen et al. 2003). This shape would lead to distinct illumination functions for various areas at the surface of a comet, with some regions being almost permanently shadowed. In about a hundred orbits, the variable sublimation rate would lead to changes in the moment of inertia and spin properties, thus repeating the cycle of thermo-physical, compositional and structural changes. As summarized by Jewitt (2004), the various side effects of activity, such as changes in shape, size or rotation period are all closely intertwined. Space missions have revealed that comet nuclei have active, less active and completely inactive areas on their surface. Areas with higher material strength can become dead zones with respect to sublimation (Blum et al. 2014), so would areas covered by a layer of dust quenching ice sublimation (see following section). Ground-based and spacecraft observations have identified dust jets in the inner coma of many comets, which could be linked to confined areas on the nucleus (e.g. Keller et al. 1988; Belton 2010). Observational data suggest that active and non-active areas may be distinguished not only by illumination conditions, but also by topography and composition. It is not clear yet whether the geomorphological features observed at the surface of comets visited by spacecrafts are the result of past uneven outgassing, or the cause for such localized activity. For example, two types of terrains have been observed at the surface of 9P/Tempel 1 and 103P/Hartley 2: pitted terrains with a rough appearance, and smooth terrain where erosion and deposition took place (Thomas et al. 2013a, 2013b; Bruck-Syal et al. 2013). A scarp of higher albedo was observed, and interpreted as being due to some activity-driven surface reshaping (Farnham et al. 2013).

4.2 Dust Surface-Layer Formation and Death of Comets

The problem of dust layer formation has been studied since Brin and Mendis (1979). The existence of the mantle of dust at the surface of comets has been inferred from ground-based observations (Jewitt 1992, A'Hearn et al. 1995), and confirmed by recent space mission observations (see Gulkis et al. 2015; Thomas et al. 2015). When ice on a comet nucleus sublimates, it lifts dust grains previously embedded within the ice matrix, leaving at the surface those too heavy relative to their effective cross section to escape the nucleus weak gravitational field. Although initially large dust particles are isolated from each other, the surface gets covered with dust as particles accumulate. The gas flow through such a mantle can be modeled by considering gas diffusion through the porous medium. Jewitt (2002) estimated that a dust mantle would grow in $\sim 10^3$ years at 5 AU and 1 year at 1 AU. If the gas pressure is high enough, the dust layer may be blown off and the process will start again. Kuehrt and Keller (1994) suggested that eventually, the pore size of this dust mantle would be too small to allow particles to escape, leading to a very stable mantle, with a cohesive strength larger than the vapor pressure of sublimating ice underneath. When the mantle becomes thicker than the diurnal thermal skin depth, it seals the interior so that cometary activity would be quenched, as observed in the KOSI experiments (Grün et al. 1993). Any gas would be driven toward the interior, thus forming an ice layer of increased density (Spohn et al. 1989; Prialnik and Mekler 1991). Skorov and Blum (2012) reassessed the problem of dust mantle formation in the light of recent spacecraft observations showing ice sublimation coming from underneath a dry hot cohesive dust layer, and dust emission being correlated with gas activity (e.g. A'Hearn et al. 2011), and water ice being found ~ 1 m below a volatile-poor surface layer (A'Hearn et al. 2008). They studied the formation and removal of a dust layer at the surface of comets by evaluating its tensile strength. They showed that the sublimation of water ice would not be able to remove the dust mantle, while the sublimation of CO would, within a repetitive cycle of dust mantle formation and destruction, since gas pressure below the crust increases as the crust thickness increases. This seems to be in agreement with the results from the EPOXI mission toward 103P/Hartley 2, where water vapor was observed to come from the dust-covered region in the neck (A'Hearn et al. 2011).

Meteoroid observations are relevant to the study of dust mantle properties, since the bulk of the meteoroid population originates from comets, being blown away from the nucleus as ices sublimate close to the Sun. For instance, 1P/Halley is associated with the Aquarids and Orionids, and comets 21P/Giacobini-Zinner and 109P/Swift-Tuttle with the Draconids and Perseids respectively. Some meteoroids can also come from asteroids: 3200 Phaeton is indeed believed to be the parent body of the Geminids (thermal fracture has been invoked to explain the mass loss of Phaeton, Jewitt and Li 2010). How much information we can gain on comets and cometary activity from a meteoroid stream remains to be understood, since detailed orbital analysis of dust particles ejected from comets started only recently (e.g. Brown and Jones 1998; McNaught and Asher 1999; Lyytinen and Van Flandern 2000; Vaubaillon et al. 2005). The light curve and magnitude of a meteoroid can be used to assess dust properties such as density or strength. For instance, the Draconids contain extremely friable meteoroids with a very low density (Ceplecha 1968). More generally, the cometary material observed in meteoroids exhibits a low strength, of less than 1 kPa, the strength of fresh snow (Borovicka et al. 2007). In situ measurements performed from the Deep Impact spacecraft yielded an extremely low value (<65 Pa, A'Hearn et al. 2005) although any strength between 0 and 12 kPa has been shown to fit the observations (Holsapple and Housen 2007). More generally, the estimation of comet tensile strengths leads to low values, as expected for very porous bodies (Davidsson 2001; Toth and Lisse 2006). The study of the tidal breakup of comet D/1993 F2 (Shoemaker-Levy 9) led to an extremely low value of the tensile strength (Greenberg et al. 1995; Asphaug and Benz 1996). These low values are in agreement with laboratory measurements of the tensile strength of loosely packed micron-sized dust particles (Blum and Schrapler 2004; Blum et al. 2006).

Therefore, once a comet is injected into the inner solar system, it progressively decays. The median dynamical lifetime of a JFC is of the order of 10^5 years (Levison and Duncan 1994), although they follow a chaotic trajectory among planets, so that they retain a dynamical memory of only $\sim$1000 years (Tancredi 1995). A comet is considered active when it is producing a detectable coma, and inactive when the coma is not detectable, generally in the outer part of its orbit due to low insolation at large heliocentric distance. A comet is considered dead or dormant when no detectable coma can be observed at any point of its orbit: this was for example the case of comet 107P/Wilson-Harrington which was lost in 1942, and rediscovered as an asteroid-like object finally named (4015) Wilson-Harrington. Following the aforementioned description of the dust layer formation, one obvious reason for becoming a dead, or rather dormant, is the lack of volatiles available near the surface, due to previous cometary activity (comets would become dead when all volatiles would be consumed). The timescale for thermal devolatilization is of the order of $10^5 \times r$ years, with r the nucleus size in km (Jewitt 2004). This is comparable to the dynamical timescale for JFCs of km-sized objects, so that it is reasonable to expect that some comets might become dead or dormant, in the sense that they might have lost all or part of their initial volatile content respectively. Many comets display a very low level of activity, which is usually interpreted as a sign of their progressive loss of volatiles in the subsurface layers (Luu and Jewitt 1992). In addition, following the idea of a dynamical cascade between Scattered Disk Objects beyond

Neptune and JFCs, Belton (2014) suggested that in order to reproduce the size distribution of TNOs as constrained by recent surveys (Fuentes et al. 2010; Schlichting et al. 2012; Zhang et al. 2013), the current population of JFCs must contain a large fraction of defunct comets.

If comets can die from running out of volatiles, other mechanisms such as splitting can strongly limit their lifetime. A compilation of data on splitting comets by Weissman (1980) and Boehnhardt (2004) suggests that splitting would statistically occur of 10 % of Oort Cloud comets, 4 % of long-period comets and 1 % of short-period comets. Levison et al. (2002) also showed that 99 % of Oort Cloud comets would disaggregate on their first passage in the inner solar system. These numbers might reflect different internal properties, either primordial or evolutionary. Comets might be disrupted by tidal interaction with the Sun or a giant planet. The best example could be comet D/1993 F2 (Shoemaker-Levy 9), which was tidally disrupted by a close encounter with Jupiter in 1994, or the parent body that gave birth to the Kreutz group of sungrazing comets (Marsden 1989; Knight et al. 2010). Other comets may disrupt with no obvious reason, like D/1999 S4 which fragmented as it passed through perihelion in 2000 (Weaver et al. 2001). Boehnhardt (2004) suggested that over a lifetime as a comet, a nucleus might loose a mass equivalent to a 500 to 1000 m radius body, i.e. the typical size of a JFC. Therefore splitting and fragmenting should be an important process to be accounted for in the evolution of comets. In absence of tidal interaction, the mechanism behind the fracture of splitting comets is still unknown. Several other mechanisms can explain the splitting of comets, such as rotational spin-up (Weissman and Lowry 2003), gas pressure release (Samarasinha 1999). However, when the event is associated with a burst of activity, the process may be connected to internal stresses or pressure build-up. Indeed, if gas is being accumulated inside the nucleus more rapidly than it can be removed by flowing through porous ice or dust on the surface, large stresses may be experiences by the cometary subsurface material. If the pressure is not released rapidly enough to prevent an instability, this mechanism could lead to a major outburst, such as the one recently observed for comet 17P/Holmes. This comet experienced the largest outburst event in recorded history (Moreno et al. 2008; Li et al. 2011), during which the nucleus lost $\sim$5 % of its mass at most (Reach et al. 2010). Similarly, comet 73P/Schwassmann-Wachmann 3 shed fragments twice in past recorded history, but is still returning as an active comet, hence not disrupted. In the worst cases, comet nuclei could be completely disrupted by such an event. The temporal gap between the outburst and the splitting observed in many non-tidally split comets was interpreted as being due to the sustained activity that is required to fracture the material (Sekanina et al. 2002). If true, this constitutes a strong argument against the idea that comets should be strengthless rubble-pile objects. Other splitting mechanisms include internal frictions due to changes in the moment of inertia caused by uneven outgassing, resulting in a nucleus fragmenting at its weakest structural parts.

5 Discussion

We study comets because we believe that they hold invaluable clues on the formation and evolution of our planetary system. In comparison to planets, they have undergone much less alteration, having been exposed to fewer evolutionary processes that acted preferentially at their surface. Therefore, they should have retained a relatively pristine record of the conditions prevailing during the early phases of the solar system. Interpreting this record via observations, laboratory experiments and numerical modeling can thus tell us a lot about how our planetary system formed. However, we still have not been able to determine which of the observed physical, chemical and orbital characteristics of comets will provide the best clues to their origin, after they have evolved for more than 4 Gyr in a time-varying radiative and collisional environment. We have exposed the major processes that dominate the processing of comets at different stages of their evolution. Ultimately, we need to emphasize that the evolutionary track may be specific to each comet. Comet physical characteristics as inherited from their formation stage may be very diverse. Nuclei could be formed of a single component, either relatively uniform in composition, or heterogeneous if comets accreted material from different regions when migrating through the protoplanetary disk. Alternatively, comet nuclei may be made of multiple components loosely bound together. These sub-units may similarly be relatively uniform, or could originate from different regions of the protoplanetary disk. So far, direct observations by spacecraft missions have revealed that both types of nuclei, single- or multiple-components, exist. Furthermore, some of the observed nuclei might be collisional debris of larger, possibly chemically differentiated parent bodies. The subsequent evolution of comet nuclei involves some processing from radiogenic heating, space weathering and large- and small-scale collisions, which modified their primordial structures and compositions with various degrees. When comets enter the inner solar system and become active, they start to loose mass at a very high rate. The effects of activity on comet nuclei involve a layering of the composition, a substantial non-even erosion and modification of their size and shape, and may eventually result in the death of comets.

Because evolutionary processes may lead to so many different outcomes, we need to focus on their long-lasting effects. The initial population of icy bodies in the early solar system might have included a significant number of objects with sufficient thermal processing to alter their original composition. Depending mainly on the formation time of comets, but also on initial parameters such as internal structure, composition, porosity, or thermal conductivity, the outcomes of the early evolution of comets due to thermal processes could be very diverse. Pristine structures could have been preserved, but physical and chemical differentiation could have occurred, leading to unknown modifications of comets initial thermophysical properties. The current presence of highly volatile species like CO and N_2 seems to argue against a full scale modification of the original chemical composition, but rather tends toward limited alteration. The decay of radioactive nuclides could have nonetheless lead to the depletion of volatiles from the innermost layers of comet nuclei. Even if no liquid phase was produced, cometary material is such a poor heat conductor that residual heat from ^{26}Al or long-lived nuclides might have lead to the formation of volatile-depleted cores. If the population of icy objects was dominated by large cometesimals, so that current comet nuclei would mostly result from the collisional grinding of larger objects, then these parent bodies should have formed slowly and/or late with respect to CAI formation so to decrease the amount of ^{26}Al available to heat them, or they should have formed in a ^{26}Al poor environment. Comets could have also formed small, although the wide spectrum of alteration results described before is applicable even for km-sized objects. The study of meteorites is not clearly helpful in constraining the possible early thermal processing of comets yet. Although the Stardust samples do contain some CAIs, the lack of former ^{26}Al seems to argue against any strong early processing due to radiogenic heating. However, the samples do show hints of aqueous alteration, but lack the most commonly formed minerals. This suggests that (i) 81P/Wild 2 did not suffer from significant internal heating, (ii) the aqueous alteration minerals were debris from a different parent-body, then re-accreted at a later stage in this comet, or (iii) these minerals were formed locally for example in impact sites. In turn, this suggests that 81P/Wild 2 was presumably formed small and/or late with respect to CAI formation, but it is difficult to extrapolate this conclusion to the overall comet population, and argues in favor of sample-return spacecraft mission toward a comet.

As of today, little is known of the actual role of collisions in the evolution of comets, since impact craters have never been undoubtedly observed at the surface of comets visited by spacecrafts: pits seen on the surface of 81P/Wild 2 (Brownlee et al. 2004), 9P/Tempel 1 (Belton et al. 2013) and very recently 67P/Churyumov-Gerasimenko (Sierks et al. 2014) could be interpreted as impact craters (Basilevsky and Keller 2007), but could also be caused by outbursting activity (Belton 2010). Possible reasons include: (i) these comets were never impacted, (ii) impact cratering was not efficient, as suggested by experiments adding species more volatile than water (Burchell et al. 1998; Burchell and Johnson 2005), or (iii) the inevitable subsequent erosion due to cometary activity removed craters and other impact-related features from their surface. There are however supporting evidence for recent collisions among populations dynamically related to comets. In the Kuiper Belt, we know of one dynamical family which was produced by impact (the Haumea family, Brown et al. 2007). In the giant planet region, a ring system was recently detected around Centaur (10199) Chariklo (Braga-Ribas et al. 2014), which was interpreted as the result of a recent collision (Duffard et al. 2014). In summary, thermal processing during the early and late stages of comet evolution, and collisional processing lead to the superposition of two types of chemical differentiation, both radial (layering) and lateral. They are therefore expected to lead to some very complex internal structures, possibly mimicking the structures inherited from a complex formation mechanism such as described in the talps model (Belton et al. 2007a, 2007b), before comets even entered the inner solar system. Deep-seated global-scale layering of the composition has been observed on 9P/Tempel 1 and 67P/Churyumov-Gerasimenko (Thomas et al. 2007; Thomas et al. 2013a; Sierks et al. 2014). Whether their origin is primordial, evolutionary, or a mixture of both remains to be understood via additional observational constraints. We can thus expect that the coming ESA/Rosetta results, with the first determination of a comet internal structure, and the monitoring of surface features and activity, will shed some new light on the evolution of comet nuclei as well as a better understanding of their formation.

Acknowledgement This paper was triggered by a workshop initiated by A. Guilbert-Lepoutre on surface-interior interactions on small bodies, and funded by the ESTEC Faculty Council. The authors thus acknowledge J. Benkhoff, J. Blum, F. Cipriani, J. Deller, B. Gundlach, D. Heisselman, E. Lellouch and J.-B. Vincent for the fruitful discussions that lead to this manuscript. A. Guilbert-Lepoutre and S. Besse were supported by the European Space Agency Research Fellowship program. O. Mousis acknowledges support from CNES. This work has been supported by the A∗MIDEX project (n^0 ANR-11-IDEX-0001-02) funded by the "Investissements d'Avenir" French Government program, managed by the French National Research Agency (ANR). M. Ali-Dib was supported by a PhD grant from the city of Besancon, France.

References

M.F. A'Hearn, Annu. Rev. Astron. Astrophys. **49**, 281 (2001)

M.F. A'Hearn, Space Sci. Rev. **138**, 237 (2008)

M.F. A'Hearn, R.C. Millis, D.O. Schleicher, D.J. Osip, P.V. Birch, Icarus **118**, 223 (1995)

M.F. A'Hearn, M.J.S. Belton, W.A. Delamere, J. Kissel, K.P. Klaasen, L.A. McFadden, K.J. Meech, H.J. Melosh et al., Science **310**, 258–264 (2005)

M.F. A'Hearn, M.J.S. Belton, S.M. Collins, T.L. Farnham, L.M. Feaga, O. Groussin, C.M. Lisse, K.J. Meech et al., Earth Planets Space **60**, 61–66 (2008)

M.F. A'Hearn, M.J.S. Belton, W.A. Delamere, L.M. Feaga, D. Hampton, J. Kissel, K.P. Klaasen, L.A. McFadden et al., Science **332**, 1396 (2011)

M.F. A'Hearn, L.M. Feaga, H.U. Keller, H. Kawakita, D.L. Hampton, J. Kissel, K.P. Klaasen, L.A. McFadden et al., Astrophys. J. **758**, 29 (2012)

E. Asphaug, W. Benz, Icarus **121**, 225 (1996)
M. Asplund, N. Grevesse, A.J. Sauval, P. Scott, Annu. Rev. Astron. Astrophys. **47**, 481–522 (2009)
A. Bar-Nun, G. Herman, D. Laufer, M.L. Rappaport, Icarus **63**, 317 (1985)
A.T. Basilevsky, H.U. Keller, Sol. Syst. Res. **41**, 109–117 (2007)
M.J.S. Belton, Icarus **210**, 881 (2010)
M.J.S. Belton, Icarus **231**, 168 (2014)
M.J.S. Belton, J. Melosh, Icarus **200**, 280 (2009)
M.J.S. Belton, P. Thomas, J. Veverka, P. Schulz, M.F. A'Hearn, L. Feaga, T. Farnham, O. Groussin et al., Icarus **187**, 332 (2007a)
M.J.S. Belton, P. Thomas, J. Veverka, P. Schulz, M.F. A'Hearn, L. Feaga, T. Farnham, O. Groussin et al., Icarus **191**, 573 (2007b)
M.J.S. Belton, P. Thomas, B. Carcich, A. Quick, J. Veverka, J.H. Melosh, M.F. A'Hearn, J.-Y. Li et al., Icarus **222**, 477 (2013)
J. Benkhoff, D.C. Boice, Planet. Space Sci. **44**, 665 (1996)
J. Blum, B. Gundlach, S. Muhle, J.M. Trigo-Rodrigues, Icarus **235**, 156 (2014)
J. Blum, R. Schrapler, Phys. Rev. Lett. **93**, 115503 (2004)
J. Blum, R. Schrapler, B.J.R. Davidsson, J.M. Trigo-Rodrigues, Astrophys. J. **652**, 1768 (2006)
D. Bockelée-Morvan, D. Gautier, F. Hersant, J.-M. Huré, F. Robert, Astron. Astrophys. **384**, 1107–1118 (2002)
H. Boehnhardt, in *Comets II*, ed. by M.C. Festou, H.U. Keller, H.A. Weaver (University of Arizona Press, Tucson, 2004), pp. 301–316
J. Borovicka, P. Spurny, P. Koten, Astron. Astrophys. **473**, 661 (2007)
W.F. Bottke, D.D. Durda, D. Nesvorny, R. Jedicke, A. Morbidelli, D. Vokrouhlicky, H. Levison, Icarus **175**, 111 (2005)
F. Braga-Ribas, B. Sicardy, J.L. Ortiz, C. Snodgrass, F. Roques, R. Vieira-Martins, J.I.B. Camargo, M. Assafin et al., Nature **508**, 72 (2014)
G.D. Brin, D.A. Mendis, Astrophys. J. **229**, 402 (1979)
P. Brown, J. Jones, Icarus **133**, 36 (1998)
M.E. Brown, K.M. Barkume, D. Ragozzine, E.L. Schaller, Nature **446**, 294 (2007)
D.E. Brownlee, F. Horz, R.L. Newburn, M. Zolensky, T.C. Duxbury, S. Sandford, Z. Sekanina, P. Tsou et al., Science **304**, 1764 (2004)
D.E. Brownlee, P. Tsou, J. Aléon, C.M.O.D. Alexander, T. Araki, S. Bajt, G.A. Baratta, R. Bastien et al., Science **314**, 1711 (2006)
M. Bruck-Syal, P.H. Schulz, J.M. Sunshine, M.F. A'Hearn, T.L. Farnham, D.S.P. Dearborn, Icarus **222**, 610 (2013)
R. Brunetto, T.L. Roush, Astron. Astrophys. **481**, 879 (2008)
R. Brunetto, M.A. Barucci, E. Dotto, G. Strazzulla, Astrophys. J. **644**, 646 (2006)
M.J. Burchell, E. Johnson, Mon. Not. R. Astron. Soc. **360**, 769 (2005)
M.J. Burchell, M.J. Willis, S.P. Armes, M.A. Khan, M.J. Percy, C. Perruchot, Planet. Space Sci. **50**, 1025 (2002)
M.J. Burchell, W. Brooke-Thomas, J. Leliwa-Kopystynski, J.C. Zarnecki, Icarus **131**, 210 (1998)
H. Campins, E.V. Ryan, Astrophys. J. **341**, 1059–1066 (1989)
F. Capaccioni, A. Coradini, G. Filacchione, S. Erard, G. Arnold, P. Drossart, M.C. De Sanctis, D. Bockelée-Morvan et al., Science **347** (2015). doi:10.1126/science.aaa0628
M.T. Capria, A. Coradini, M.C. De Sanctis, Earth Moon Planets **90**, 217 (2002)
P. Cassen, Icarus **112**, 405–429 (1994)
Z. Ceplecha, SAO Special Report #279 (1968)
S. Charnoz, A. Morbidelli, Icarus **188**, 468 (2007)
F.J. Ciesla, Science **318**, 613 (2007)
F.J. Ciesla, Icarus **200**, 655–671 (2009)
F.J. Ciesla, Astrophys. J. **784**, L1 (2014)
F.J. Ciesla, T.M. Davison, G.S. Collins, D.P. O'Brien, Meteorit. Planet. Sci. **48**, 2559 (2013)
A.L. Cochran, E.S. Barker, T.F. Ramseyer, A.D. Storrs, Icarus **98**, 151 (1992)
M. Cohen, D. Prialnik, M. Podolak, New Astron. **8**, 179 (2003)
J. Crovisier, T.Y. Brooke, K. Leech, D. Bockelée-Morvan, E. Lellouch, M.S. Hanner, B. Altieri, H.U. Keller et al., in *Thermal Emission Spectroscopy and Analysis of Dust, Disks, and Regoliths*, vol. 196 (2000), pp. 109–117
K.E. Cyr, C.M. Sharp, J.I. Lunine, J. Geophys. Res. **104**, 19003–19014 (1999)
P. D'Alessio, J. Canto, N. Calvet, S. Lizano, Astrophys. J. **500**, 411–427 (1998)
E. Dartois, J.J. Ding, A.L.F. de Barros, P. Boduch, R. Brunetto, M. Chabot, A. Domaracka, M. Godard et al., Astron. Astrophys. **557**, 97 (2013)

D.R. Davis, P. Farinella, Icarus **125**, 50 (1997)
B.J.R. Davidsson, Icarus **149**, 375 (2001)
T.M. Davison, G.S. Collins, F.J. Ciesla, Icarus **208**, 468 (2010)
N. Dello Russo, R.J. Vervack, H.A. Weaver, N. Biver, D. Bockelée-Morvan, J. Crovisier, C.M. Lisse, Nature **448**, 172 (2007)
M.C. DeSanctis, M.T. Capria, A. Coradini, E. Ammannito, Astron. J. **133**, 1863 (2007)
S.E. Dodson-Robinson, K. Willacy, P. Bodenheimer, N.J. Turner, C.A. Beichman, Icarus **200**, 672 (2009)
D.R. Dones, P.R. Weissman, H.F. Levison, M.J. Duncan, in *Comets II*, ed. by M.C. Festou, H.U. Keller, H.A. Weaver (Univ. of Arizona Press, Tucson, 2004), p. 153
R. Duffard, N. Pinilla-Alonso, J.L. Ortiz, A. Alvarez-Candal, B. Sicardy, P. Santos-Sanz, N. Morales, C. Colazo et al., Astron. Astrophys. **568**, 79 (2014)
M.J. Duncan, H.F. Levison, Science **276**, 1670 (1997)
M. Duncan, T. Quinn, S. Tremaine, Astron. J. **94**, 1330 (1987)
M. Duncan, H. Levison, L. Dones, in *Comets II*, ed. by M.C. Festou, H.U. Keller, H.A. Weaver (University of Arizona Press, Tucson, 2004), p. 193
S. Espinasse, J. Klinger, C. Ritz, B. Schmitt, Icarus **92**, 350 (1991)
E. Everhart, Astron. J. **74**, 735 (1969)
P. Farinella, D.R. Davis, Science **273**, 938 (1996)
T.L. Farnham, D. Bodewits, J.-Y. Li, J. Veverka, P. Thomas, M.J.S. Belton, Icarus **222**, 540 (2013)
L.M. Feaga, M.F. A'Hearn, J.M. Sunshine, O. Groussin, T.L. Farnham, Icarus **190**, 345 (2007)
J.A. Fernandez, Space Sci. Rev. **138**, 27–42 (2008)
C.I. Fuentes, M.J. Holman, D.E. Trilling, P. Protopapas, Astrophys. J. **722**, 1290 (2010)
E.L. Gibb, D.C.B. Whittet, A.C.A. Boogert, A.G.G.M. Tielens, Astrophys. J. Suppl. Ser. **151**, 35–73 (2004)
R. Gomes, Nature **426**, 393 (2003)
R. Gomes, H.F. Levison, K. Tsiganis, A. Morbidelli, Nature **435**, 466 (2005)
R. Gomes, J.A. Fernandez, T. Gallardo, A. Brunini, in *The Solar System Beyond Neptune*, ed. by M.A. Barucci, H. Boehnhardt, D.P. Cruikshank, A. Morbidelli (University of Arizona Press, Tucson, 2008), p. 259
M. Gounelle, P. Spurny, P.A. Bland, Meteorit. Planet. Sci. **41**, 135 (2006)
M. Gounelle, A. Morbidelli, P.A. Bland, P. Spurny, E.D. Young, M. Sephton, in *The Solar System Beyond Neptune*, ed. by M.A. Barucci, H. Boehnhardt, D.P. Cruikshank, A. Morbidelli (University of Arizona Press, Tucson, 2008), pp. 525–541
J.M. Greenberg, H. Mitzutani, T. Yamamoto, Astron. Astrophys. **295**, 35 (1995)
L. Grossman, Geochim. Cosmochim. Acta **36**, 597–619 (1972)
E. Grün, J. Gebhard, A. Bar-Nun, J. Benkhoff, H. Dueren, G. Eich, R. Hische, W.F. Huebner et al., J. Geophys. Res. **98**, 15091–15104 (1993)
A. Guilbert, M.A. Barucci, R. Brunetto, A. Delsanti, F. Merlin, A. Alvarez-Candal, S. Fornasier, C. de Bergh, G. Sarid, Astron. Astrophys. **501**, 777 (2009)
A. Guilbert-Lepoutre, Astron. J. **144**, 97 (2012)
A. Guilbert-Lepoutre, D. Jewitt, Astrophys. J. **743**, 31 (2011)
S. Gulkis, M. Allen, P. von Allmen, G. Beaudin, N. Biver, D. Bockelée-Morvan, M. Choukroun, J. Crovisier et al., Science **347**, 0709 (2015)
M.S. Hanner, Space Sci. Rev. **90**, 99–108 (1999)
M.S. Hanner, M.E. Zolensky, Astromineralogy **815**, 203–232 (2010)
D.E. Harker, S.J. Desch, Astrophys. J. **565**, 109–112 (2002)
J. Haruyama, T. Yamamoto, H. Mizutani, J.M. Greenberg, J. Geophys. Res. **98**, 15079–15090 (1993)
Y. Hillman, D. Prialnik, Icarus **221**, 147 (2012)
K.A. Holsapple, K.R. Housen, Icarus **187**, 345 (2007)
K.R. Housen, K.A. Holsapple, Icarus **163**, 102 (2003)
K.R. Housen, K.A. Holsapple, M.E. Voss, Nature **402**, 155 (1999)
R.L. Hudson, M.E. Palumo, G. Strazzulla, M.H. Moore, J.F. Cooper, S.J. Sturner, in *The Solar System Beyond Neptune*, ed. by M.A. Barucci, H. Boehnhardt, D.P. Cruikshank, A. Morbidelli (University of Arizona Press, Tucson, 2008), pp. 507–523
R. Hueso, T. Guillot, Astron. Astrophys. **442**, 703–725 (2005)
W.F. Huebner, J. Benkhoff, Space Sci. Rev. **90**, 117 (1999)
W.F. Huebner, J. Benkhoff, M.T. Capria, A. Coradini, C. DeSanctis, R. Orosei, D. Prialnik, Heat an gas diffusion in comet nuclei. ISSI Scientific Report, ISBN 1608-280X (2006)
A.L.H. Hugues, P.J. Armitage, Astrophys. J. **719**, 1633–1653 (2010)
G.R. Huss, G.J. MacPherson, G.J. Wasserburg, S.S. Russell, G. Srinivasan, Meteorit. Planet. Sci. **36**, 975 (2001)

D. Jewitt, in *Observations and Physical Properties of Small Solar System Bodies*, ed. by A. Brahic, J.-C. Gerard, J. Surdej, I. d'Astrophys (Inst. d'Astrophys., Univ. Liege, Liege, 1992), p. 85
D. Jewitt, Astron. J. **123**, 1039 (2002)
D. Jewitt, in *Comets II*, ed. by M.C. Festou, H.U. Keller, H.A. Weaver (Univ. of Arizona Press, Tucson, 2004), p. 659
D. Jewitt, Astron. J. **137**, 4296 (2009)
D. Jewitt, Astron. J. **143**, 66 (2012)
D. Jewitt, J. Li, Astron. J. **140**, 1519 (2010)
R.E. Johnson, Rev. Mod. Phys. **68**, 305–312 (1995)
R.H. Jones, T. Lee, H.C. Connolly Jr., S.G. Love, H. Shang, *Protostars and Planets IV* (2000), p. 927
L. Jorda, H. Rickman, Planet. Space Sci. **43**, 575 (1995)
D.J. Joswiak, D.E. Brownlee, G. Matrajt, A.J. Westphal, C.J. Snead, Z. Gainsforth, Meteorit. Planet. Sci. **47**, 471 (2012)
Z. Kanuchova, R. Brunetto, M. Melita, G. Strazzulla, Icarus **221**, 12 (2012)
K. Keil, D. Stoeffler, S.G. Love, E.R.D. Scott, Meteorit. Planet. Sci. **32**, 349 (1997)
H.U. Keller, C. Arpigny, C. Barbieri, R.M. Bonnet, S. Cazes, M. Coradini, C.B. Cosmovici, W.A. Delamere et al., Nature **321**, 320 (1986)
H.U. Keller, R. Kramm, N. Thomas, Nature **331**, 227 (1988)
M.M. Knight, M.F. A'Hearn, D.A. Biesecker, G. Faury, D.P. Hamilton, P. Lamy, A. Llebaria, Astron. J. **139**, 926 (2010)
H. Kobayashi, H. Kawakita, M.J. Mumma, B.P. Bonev, J. Watanabe, T. Fuse, Astrophys. J. **668**, 75 (2007)
D. Koschny, G. Kargl, M. Rott, Adv. Space Res. **28**, 1533–1537 (2001)
O. Krauss, G. Wurm, Astrophys. J. **630**, 1088–1092 (2005)
O. Krauss, G. Wurm, O. Mousis, J.-M. Petit, J. Horner, Y. Alibert, Astron. Astrophys. **462**, 977–987 (2007)
A.N. Krot, K. Makide, K. Nagashima, G.R. Huss, R.C. Ogliore, F.J. Ciesla, L. Yang, E. Hellebrand, E. Gaidos, Meteorit. Planet. Sci. **47**, 1948 (2012)
T. Kunihiro, A.E. Rubin, K.D. McKeegan, J.T. Wasson, Geochim. Cosmochim. Acta **68**, 2947–2957 (2004)
E. Kuehrt, H.U. Keller, Icarus **109**, 121 (1994)
P.L. Lamy, I. Toth, Y.R. Fernandez, H.A. Weaver, in *Comets II*, ed. by M.C. Festou, H.U. Keller, H.A. Weaver (University of Arizona Press, Tucson, 2004), p. 223
H.F. Levison, M.J. Duncan, Icarus **108**, 18 (1994)
H.F. Levison, M.J. Duncan, Icarus **127**, 13 (1997)
H.F. Levison, A. Morbidelli, Nature **426**, 419 (2003)
H.F. Levison, L. Dones, M.J. Duncan, Astron. J. **121**, 2253 (2001)
H.F. Levison, A. Morbidelli, L. Dones, R. Jedicke, P.A. Wiegert, W.F. Bottke, Nature **296**, 2212 (2002)
H.F. Levison, A. Morbidelli, C. Van Laerhoven, R. Gomes, K. Tsiganis, Icarus **196**, 258 (2008)
J.S. Lewis, R.G. Prinn, Astrophys. J. **238**, 357–364 (1980)
J. Li, D. Jewitt, J.M. Clover, B.V. Jackson, Astrophys. J. **728**, 31 (2011)
K. Lodders, Astrophys. J. **591**, 1220–1247 (2003)
S.G. Love, F. Hortz, D.E. Brownlee, Icarus **105**, 216 (1993)
J.I. Lunine, D.J. Stevenson, Astrophys. J. Suppl. Ser. **58**, 493–531 (1985)
J.X. Luu, D. Jewitt, Astron. J. **104**, 2243 (1992)
D. Lynden-Bell, J.E. Pringle, Mon. Not. R. Astron. Soc. **168**, 603–637 (1974)
E.J. Lyytinen, T. Van Flandern, Earth Moon Planets **82**, 149 (2000)
G.J. MacPherson, A.M. Davis, E.K. Zinner, Meteoritics **30**, 365 (1995)
N. Madhusudhan, O. Mousis, T.V. Johnson, J.I. Lunine, Astrophys. J. **743**, 191 (2011)
K. Makide, K. Nagashima, A.N. Krot, G.R. Huss, I.D. Hutcheon, A. Bischoff, Geochim. Cosmochim. Acta **73**, 5018 (2009)
K. Makide, K. Nagashima, A.N. Krot, G.R. Huss, F.J. Ciesla, E. Heebrand, E. Gaidos, L. Yang, Astrophys. J. **733**, 31 (2011)
B.G. Marsden, Astron. J. **98**, 2306 (1989)
R.G. Martin, M. Livio, Mon. Not. R. Astron. Soc. **425**, 6–9 (2012)
R.H. McNaught, D.J. Asher, WGN **27**, 85–102 (1999)
K.J. Meech, J. Svoren, in *Comets II*, ed. by M.C. Festou, H.U. Keller, H.A. Weaver (University of Arizona Press, Tucson, 2004), p. 317
R. Merk, D. Prialnik, Icarus **183**, 283 (2006)
P. Modica, M.E. Palumbo, G. Strazzulla, Planet. Space Sci. **73**, 425 (2012)
A. Morbidelli, B. Gladman, Meteorit. Planet. Sci. **33**, 999 (1998)
A. Morbidelli, V.V. Emel'yanenko, H.F. Levison, Mon. Not. R. Astron. Soc. **355**, 935 (2004)
A. Morbidelli, H.F. Levison, K. Tsiganis, R. Gomes, Nature **435**, 462 (2005)
A. Morbidelli, W.F. Bottke, D. Nesvorny, H.F. Levison, Icarus **204**, 558 (2009)

F. Moreno, J.L. Ortiz, P. Santos-Sanz, N. Morales, M.J. Vidal-Nunez, L.M. Lara, P.J. Gutierrez, Astrophys. J. **677**, 63 (2008)
A. Moudens, O. Mousis, J.-M. Petit, G. Wurm, D. Cordier, S. Charnoz, Astron. Astrophys. **531**, 106 (2011)
O. Mousis, J.-M. Petit, G. Wurm, O. Krauss, Y. Alibert, J. Horner, Astron. Astrophys. **466**, 9–12 (2007)
O. Mousis, J.L. Lunine, S. Picaud, D. Cordier, Faraday Discuss. **147**, 509 (2010)
O. Mousis, A. Guilbert-Lepoutre, J.I. Lunine, A.l. Cochran, J.H. Waite, J.-M. Petit, P. Rousselot, Astrophys. J. **757**, 146 (2012)
M.J. Mumma, S.B. Charnley, Annu. Rev. Astron. Astrophys. **49**, 471 (2011)
M.J. Mumma, M.A. DiSanti, K. Magee-Sauer, B.P. Bonev, G.L. Villanueva, H. Kawakita, N. Dello Russo, E.L. Gibb et al., Science **310**, 270 (2005)
R.C. Ogliore, G.R. Huss, K. Nagashima, A.L. Butterworth, Z. Gainsforth, J. Stodolna, A.J. Westphal, D. Josniak, T. Tyliszczak, Astrophys. J. **745**, 19 (2012)
T. Ootsubo, H. Kawakita, S. Hamada, H. Kobayashi, M. Yamaguchi, F. Usui, T. Nakagawa, M. Ueno et al., Astrophys. J. **752**, 15 (2012)
T. Owen, A. Bar-Nun, Faraday Discuss. **109**, 453 (1998)
M.E. Palumbo, Astron. Astrophys. **453**, 903 (2006)
K.M. Pontoppidan, C. Salyk, E.A. Bergin, S. Brittain, B. Marty, O. Mousis, K.I. Oberg, in *Protostars and Planets VI*, ed. by H. Beuther, R.S. Klesser, C.P. Dullemond, T. Henning (University of Arizona Press, Tucson, 2014), pp. 363–385
D. Prialnik, Earth Moon Planets **77**, 223 (1999)
D. Prialnik, Earth Moon Planets **89**, 27–52 (2002)
D. Prialnik, A. Bar-Nun, Astron. Astrophys. **258**, 9 (1992)
D. Prialnik, Y. Mekler, Astrophys. J. **366**, 318 (1991)
D. Prialnik, M. Podolak, Icarus **117**, 420 (1995)
D. Prialnik, A. Bar-Nun, M. Podolak, Astrophys. J. **319**, 993 (1987)
D. Prialnik, J. Benkhoff, M. Podolak, in *Comets II*, ed. by M.C. Festou, H.U. Keller, H.A. Weaver (University of Arizona Press, Tucson, 2004), p. 359
D. Prialnik, G. Sarid, E.D. Rosenberg, R. Merk, Space Sci. Rev. **138**, 147 (2008)
C. Qi, K.I. Oberg, D.J. Wilner, P. D'Alession, E. Bergin, S.M. Andrews, G.A. Blake, M.R. Hogerheijde, E.F. van Dishoeck, Science **341**, 630 (2013)
W.T. Reach, J. Vaubaillon, C.M. Lisse, M. Holloway, J. Rho, Icarus **208**, 276 (2010)
E.D. Rosenberg, D. Prialnik, Icarus **209**, 753 (2010)
A.E. Rubin, Geochim. Cosmochim. Acta **68**, 673 (2004)
N.H. Samarasinha, Icarus **154**, 540–544 (1999)
H.E. Schlichting, E.O. Ofek, R. Sari, E.P. Nelan, A. Gal-Yam, M. Wenz, P. Muirhead, N. Javanfar, M. Livio, Astrophys. J. **761**, 150 (2012)
T. Schulz, D. Upadhyay, C. Munker, K. Mezger, Geochim. Cosmochim. Acta **85**, 200 (2012)
N.I. Shakura, R.A. Sunyaev, Astron. Astrophys. **24**, 337–355 (1973)
Z. Sekanina, E. Jehin, H. Boehnhardt, X. Bonfils, O. Schuetz, D. Thomas, Astrophys. J. **572**, 679 (2002)
H. Sierks et al., DPS Abstract (2014)
H. Sierks, C. Barbieri, P. Lamy, R. Rodrigo, D. Koschny, H. Rickman, H.U. Keller, J. Agarwal et al., Science **347** (2015). doi:10.1126/science.aaa1044
S.B. Simon, D.J. Joswiak, H.A. Ishii, J.P. Bradley, M. Chi, L. Grossman, J. Aleon, D.E. Brownlee et al., Meteorit. Planet. Sci. **43**, 1861 (2008)
M.L. Sitko, D.K. Lynch, R.W. Russell, M.S. Hanner, Astrophys. J. **612**, 576–580 (2004)
Y. Skorov, J. Blum, Icarus **221**, 1 (2012)
L.A. Soderblom, T.L. Becker, G. Bennett, D.C. Boice, D.T. Britt, R.H. Brown, B.J. Buratti, C. Isbell et al., Science **296**, 1087 (2002)
T. Spohn, J. Benkhoff, J. Klinger, E. Grun, H. Kochan, Adv. Space Res. **9**, 127 (1989)
S.A. Stern, Icarus **73**, 499 (1988)
S.A. Stern, Astron. J. **110**, 856 (1995)
S.A. Stern, Astron. Astrophys. **310**, 999 (1996)
S.A. Stern, Nature **424**, 639 (2003)
S.A. Stern, J.M. Shull, Nature **332**, 407 (1988)
S.A. Stern, P. Weissman, Nature **409**, 589 (2001)
S.A. Stewart, P.J. Ahrens, in *AIP Conference Proceedings on Shock Compression of Condensed Matter—2003*, vol. 706, ed. by M.D. Furnish, Y.M. Gupta, J.W. Forbes (American Institute of Physics, New York, 2004), pp. 1478–1483
J. Stodolna, D. Jacob, H. Leroux, Geochim. Cosmochim. Acta **87**, 35 (2012)
G. Strazzulla, G.A. Baratta, Astron. Astrophys. **266**, 434 (1992)
G. Strazzulla, G.A. Baratta, R.E. Johnson, B. Donn, Icarus **91**, 101 (1991)

G. Strazzulla, M.E. Palumbo, Adv. Space Res. **27**, 237 (2001)
G. Tancredi, Astron. Astrophys. **299**, 288 (1995)
M.S. Tiscareno, R. Malhotra, Astron. J. **126**, 3122 (2003)
P.C. Thomas, J. Veverka, M.J.S. Belton, A. Hidy, M.F. A'Hearn, T.L. Farnham, O. Groussin, J.-Y. Li et al., Icarus **187**, 4 (2007)
P. Thomas, M. A'Hearn, M.J.S. Belton, D. Brownlee, B. Carcich, B. Hermalyn, K. Klaasen, S. Sackett et al., Icarus **222**, 453 (2013a)
P. Thomas, M.F. A'Hearn, J. Veverka, M.J.S. Belton, J. Kissel, K.P. Klaasen, L.A. McFadden, H.J. Melosh et al., Icarus **222**, 550 (2013b)
N. Thomas, H. Sierks, C. Barbieri, P. Lamy, R. Rodrigo, H. Rickman, D. Koschny, H.U. Keller et al., Science **347**, 0440 (2015)
I. Toth, C.M. Lisse, Icarus **181**, 162 (2006)
K. Tsiganis, R. Gomes, A. Morbidelli, H.F. Levison, Nature **435**, 459 (2005)
J. Vaubaillon, F. Colas, L. Jorda, Astron. Astrophys. **439**, 751 (2005)
J. Veverka, K. Klaasen, M. A'Hearn, M. Belton, D. Brownlee, S. Chesley, B. Clark, T. Economou et al., Icarus **222**, 424 (2013)
D. Vokrouhlicky, P. Farinella, Nature **407**, 606 (2000)
K.J. Walsh, A. Morbidelli, S.N. Raymond, D.P. O'Brien, A.M. Mandell, Nature **475**, 206 (2011)
H.A. Weaver, Z. Sekanina, I. Toth, C.E. Delahodde, O.R. Hainaut, P.L. Lamy, J.M. Bauer, M.F. A'Hearn et al., Science **292**, 1329 (2001)
S.J. Weidenschilling, Icarus **214**, 671 (2011)
P. Weissman, Astron. Astrophys. **85**, 191 (1980)
P.R. Weissman, Nature **320**, 242 (1986)
P.R. Weissman, S.C. Lowry, in *Lunar and Planetary Science XXXIV*, #2003 (2003)
P.R. Weissman, E. Asphaug, S.C. Lowry, in *Comets II*, ed. by M.C. Festou, H.U. Keller, H.A. Weaver (University of Arizona Press, Tucson, 2004), p. 337
D.H. Wooden, D.E. Harker, C.E. Woodward, in *Thermal Emission Spectroscopy and Analysis of Dust, Disks, and Regoliths*, vol. 196 (2000), pp. 99–108
D.H. Wooden, C.E. Woodward, D.E. Harker, Astrophys. J. **612**, 77–80 (2004)
D.H. Wooden, S.S. Lindsay, D.E. Harker, M.S. Kelley, C.E. Woodward, D.T. Richard, L. Kolokolova, F. Moreno, Bull. Am. Astron. Soc. **42**, 960 (2010)
G. Wurm, O. Krauss, Icarus **180**, 487–495 (2006)
G. Wurm, J. Teiser, A. Bischoff, H. Haack, Rosziar, Icarus **208**, 482–495 (2010)
G. Wurm, M. Trieloff, H. Rauer, Astrophys. J. **769**, 78 (2013)
L. Yang, F.J. Ciesla, Meteorit. Planet. Sci. **47**, 99 (2012)
R. Yokochi, U. Marboeuf, E. Quirico, B. Schmitt, Icarus **218**, 760 (2012)
Z.-W. Zhang, M.J. Lehner, J.-H. Wang, S.-Y. Wang, S.-K. King, A.P. Granados, C. Alcock, T. Axelrod et al., Astron. J. **146**, 14 (2013)
M.E. Zolensky, T.J. Zega, H. Yano, S. Wirick, A.J. Westphal, M.K. Weisberg, I. Weber, J.L. Warren et al., Science **314**, 1735 (2006)
M.E. Zolensky, K. Nakamura-Messenger, F. Rietmeijer, H. Leroux, T. Mikouchi, K. Ohsumi, S. Simon, L. Grossman et al., Meteorit. Planet. Sci. **43**, 261 (2008)

Space Sci Rev (2015) 197:297–342
DOI 10.1007/s11214-015-0161-z

Constraints from Comets on the Formation and Volatile Acquisition of the Planets and Satellites

K.E. Mandt[1] · O. Mousis[2] · B. Marty[3] · T. Cavalié[4] · W. Harris[5] · P. Hartogh[4] · K. Willacy[6]

Received: 14 November 2014 / Accepted: 11 May 2015 / Published online: 21 May 2015

Abstract Comets play a dual role in understanding the formation and evolution of the solar system. First, the composition of comets provides information about the origin of the giant planets and their moons because comets formed early and their composition is not expected to have evolved significantly since formation. They, therefore serve as a record of conditions during the early stages of solar system formation. Once comets had formed, their orbits were perturbed allowing them to travel into the inner solar system and impact the planets. In this way they contributed to the volatile inventory of planetary atmospheres. We review here how knowledge of comet composition up to the time of the Rosetta mission has contributed to understanding the formation processes of the giant planets, their moons and small icy bodies in the solar system. We also discuss how comets contributed to the volatile inventories of the giant and terrestrial planets.

Keywords Solar system formation · Comets · Atmospheres · Giant planets · Terrestrial planets · Moon formation

Abbreviations

AU	Astronomical Unit
CG	Churyumov-Gerasimenko
CIRS	Cassini Infrared Spectromer
dex	logarithmic abundance
DM	Depleted Mantle

✉ K.E. Mandt
kmandt@swri.edu

1 Southwest Research Institute, San Antonio, TX, USA

2 Aix Marseille Université, CNRS, LAM (Laboratoire d'Astrophysique de Marseille) UMR 7326, 13388, Marseille, France

3 CRPG-CNRS, Nancy-Université, Vandoeuvre-lès-Nancy, France

4 Max Planck Institute for Solar System Research, Göttingen, Germany

5 University of Arizona, Tucson, AZ, USA

6 Jet Propulsion Laboratory, Pasadena, CA, USA

ESA	European Space Agency
Ga	Billion years
GPMS	Galileo Probe Mass Spectrometer
HIFI	Heterodyne Instrument for the Far Infrared
HST	Hubble Space Telescope
INMS	Ion Neutral Mass Spectrometer
IRIS	Infrared Spectrometer
ISO	Infrared Space Observatory
JCMT	James Clerk Maxwell Telescope
JFC	Jupiter Family Comet
JUICE	JUpiter ICy moons Explorer
NRAO	National Radio Astronomy Observatory
NUV	Near Ultraviolet
OCC	Oort Cloud Comet
OPR	Ortho-to-Para Ratio
ppb	parts per billion
ppm	parts per million
PSN	Protosolar Nebula
SIS	superconductor insulator superconductor
STIS	Space Telescope Imaging Spectrograph
SWAS	Submillimeter Wave Astronomy Satellite
TLHB	Terrestrial Late Heavy Bombardment
UV	Ultraviolet
VSMOW	Vienna Standard Mean Ocean Water

1 Introduction

Understanding the formation and evolution of the solar system requires evaluation of measurements that have been made in various solar system bodies that include the Sun, terrestrial planet mantles and atmospheres, giant planet atmospheres, giant planet satellites, meteorites and comets. These measurements are compared to models for solar system formation and evolution in an iterative manner where new measurements improve models and models provide guidelines for future measurements that are needed.

Measurements of cometary composition provide constraints in two areas. First, cometary composition provides a tracer of chemical conditions during solar system formation. Comets are small bodies that formed from ices and dust in the protosolar nebula (PSN) and are presumed to have evolved very little since their initial formation. Their composition is representative of conditions in the region in which they formed. After formation, the orbits of many comets were perturbed by migration of the giant planets causing them to impact the planets and contribute to planetary volatile inventories. Our goal is to review current state of knowledge—including some early Rosetta mission results—on the role of comets in understanding the formation and evolution of other solar system bodies.

In Sect. 2 we outline the current understanding of the formation of the solar system and the relevant composition of the PSN. We then discuss what is known about the formation of the giant planets in Sect. 3 and their satellites in Sect. 4. The role of comets in producing the volatile inventories of the terrestrial planets is discussed in Sect. 5, and Pluto and Kuiper Belt objects are covered in Sect. 6. We summarize the current state of knowledge, what we expect to learn from future measurements and what future measurements are needed in Sect. 7.

2 Formation of the Solar System

In order to evaluate how comets contribute to understanding the formation and evolution of other bodies in the solar system, we must first understand how the solar system formed and the composition of the PSN during the time of formation of comets and other solar system bodies. There are two types of models used to study the formation and evolution of the solar system: chemical models that describe the composition and chemistry in the PSN (for a more detailed description see the review by Willacy et al. 2015, this issue), and dynamical models that evaluate the physical processes determining where objects formed and how they ended up where they are today (for a more detailed description see the review by Dones et al. 2015, this issue).

Measurements made in various solar system bodies are used to determine the bulk composition of the PSN and the variability of its composition with distance from the young Sun. This "taxonomy" of solar system measurements can provide a basis for investigating processes within the PSN and processes that led to the evolution of solar system bodies after their formation. The most valuable measurements include noble gas abundances, noble gas isotope ratios, relative abundances of hydrogen, nitrogen, carbon and oxygen and their stable isotope ratios.

2.1 The Formation Processes of the Solar System and Conditions in the Solar Nebula

2.1.1 Initial Stages and Chemistry

The formation of the solar system began about 4.6 billion years ago with the collapse of a molecular cloud core as illustrated in Fig. 1. Prior to collapse, this core had been evolving over a period of a few million years. Despite very cold temperatures (10 K), a rich chemistry takes place within molecular clouds—almost 200 molecules have currently been identified in interstellar clouds. Additionally, because of the low temperatures, any molecules colliding with dust grains stick to them, forming icy mantles. These ices are also chemically active and more molecules can form by reactions within them. Indeed the most abundant ice species, water, is thought to form mainly by the hydrogenation of oxygen atoms on grain surfaces.

Core collapse occurs from the inside out (Shu 1977), with the protostar forming at the center where the densities are highest (Fig. 1b). Once the star switches on it generates powerful bipolar outflows that begin to clear away the parent cloud. During this stage, infall is still occurring (Fig. 1c) and a disk forms. The infalling material experiences an accretion shock at the surface of the disk, which could alter it chemically, for example by desorption of ices or dissociation of molecules. The importance of the shock in influencing the composition of material that enters the disk is still a matter of debate (e.g. Lunine et al. 1991; Visser et al. 2009).

In the *T-Tauri* phase (Fig. 1d) the infall is reduced, outflows continue and the disk can now be observed directly. Figure 2 shows the chemical structure of a typical *T-Tauri* disk around a Sun-like star. The wide range of physical conditions (temperature, density and ultraviolet or UV field) generates very different chemistries, from the cold midplane where the majority of molecules are frozen out as ices, to the highly irradiated, photon-dominated surface. Between these two layers is a region where desorption is efficient enough to maintain a population of molecules in the gas and where the UV field is low enough for these molecules to survive. This is the molecular layer that is detected in many observations of protostellar disks. Turbulent mixing is very important at this stage of evolution, with both

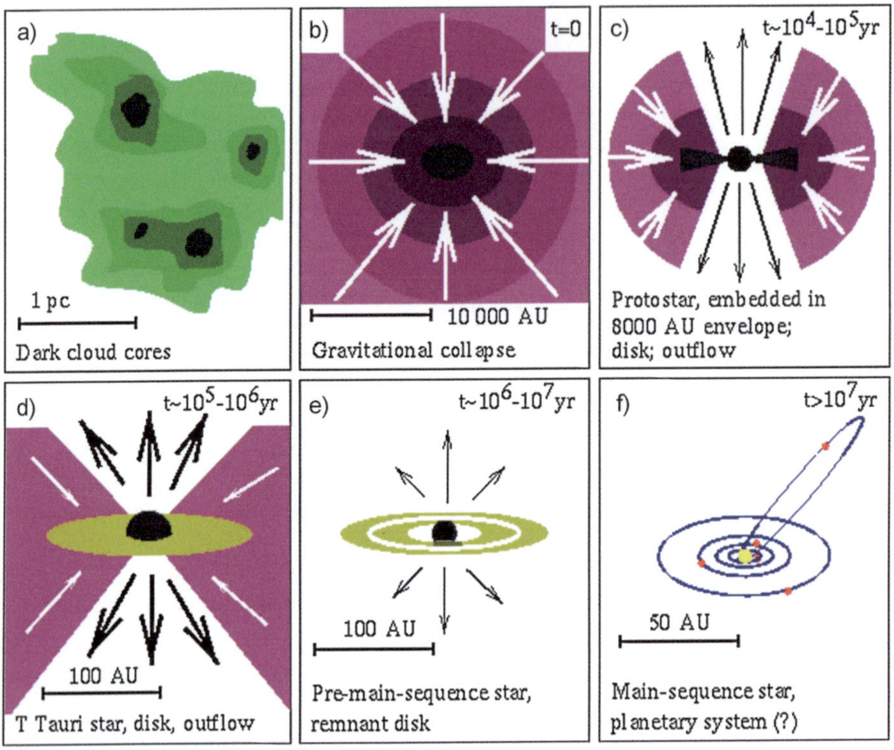

Fig. 1 The low mass star formation process (based on Shu et al. 1987). Reprinted from Hogerheijde (1998) with permission

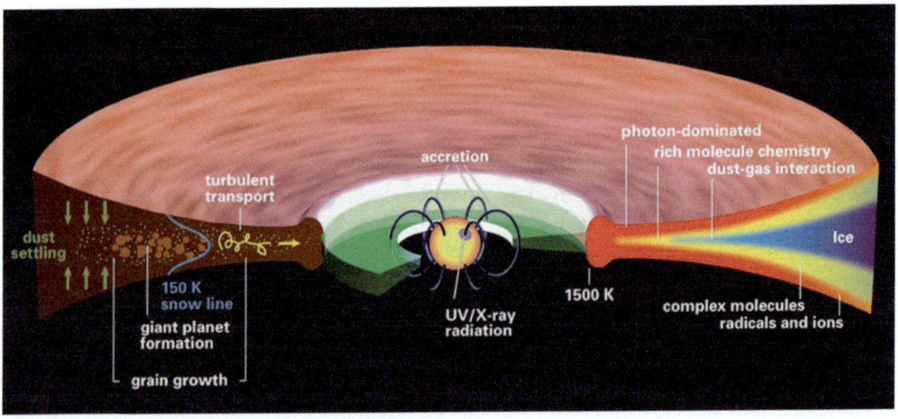

Fig. 2 Physical and chemical structure of a 1–5 Myrs protostellar disk around a Sun-like star. Reprinted from Henning and Semenov (2013) with permission. Copyright © 2013 American Chemical Society

gas and dust being transported between regions with different physical conditions driving changes in composition (see Willacy et al. 2015, this issue). Grain growth is another crucial process as grains coagulate to form larger particles that decouple from the gas motions and sink towards the midplane. The removal of small grains from the surface layers of the disk changes the opacity and hence the degree to which UV photons can penetrate, thus inducing more chemical changes. All of these chemical changes could be reflected in the composition of the planetesimals, and eventually the planetary bodies that form in the disk (Figs. 1e and 1f).

2.1.2 Formation of Planets, Moons and Small Bodies

The planets, moons and small bodies (including comets) are commonly believed to have formed through the process of accretion, where dust grains form clumps that collide and accumulate into objects known as planetesimals, which are tens of kilometers in size (Goldreich and Ward 1973). Planetesimals that formed closer to the Sun were depleted in volatiles because of high temperatures, while beyond the "snow line" (~ 4 astronomical units or AU, illustrated in Fig. 2) planetesimals formed with significant amounts of water ice and other volatiles. After formation, the inner planets are thought to have migrated inward from their formation location (e.g. Goldreich and Tremaine 1980). On the other hand, the gas giants, Jupiter, Saturn, Uranus and Neptune, are believed to have formed between 5.5 and 17 AU and then migrated to their present positions of ~ 5, ~ 10, ~ 20 and ~ 30 AU, respectively (Gomes et al. 2005; Tsiganis et al. 2005; Morbidelli et al. 2005).

There are a variety of methods through which the moons in the solar system formed. Many of the moons of the giant planets are believed to have formed in a subnebula surrounding the planet that they now orbit (e.g. Prinn and Fegley 1989), although some may have formed from building blocks formed earlier in the PSN that migrated into the subnebula (e.g. Mousis et al., 2009a, 2009b; Mandt et al. 2014). Others are thought to be planetesimals that were captured by the planet (e.g. Agnor and Hamilton 2006). Finally, some moons are thought to have formed due to a major collision, including the Earth's moon (Canup and Asphaug 2001) and Pluto's moon Charon (Canup 2005). Asteroids and comets are planetesimals left over from the process of forming the planets and the moons.

Based on what is known of their orbital parameters, comets are believed to originate primarily from two regions of the solar system—the Kuiper Belt and scattered disc, which are located 30 to 100 AU from the Sun, and the Oort Cloud which is located beyond 50000 AU. The formation regions of comets have been studied for some time by dynamical models, but many questions remain (see Dones et al. 2015, this issue, for a complete review). A key question for our study is to evaluate whether the comets from the Kuiper belt and Oort Cloud would have compositions that are distinct from each other, or if there is little connection between composition and their current location.

Oort Cloud comets (OCCs) originate from the Oort Cloud and may have formed at a distance of 5–30 AU from the Sun and then scattered to their current location (e.g. Duncan et al. 1987), or they could have been captured from nearby stars in the Sun's birth cluster (Levison et al. 2010). Jupiter Family Comets (JFCs) are believed to originate in the Kuiper Belt and are suggested to have formed at distances 30–35 AU from the Sun and been scattered outwards (Levison et al. 2008). It would be reasonable to presume that if Kuiper Belt and Oort Cloud objects formed at distinct distances from the Sun with unique temperature and composition conditions in the PSN they would represent two groups of comets with unique compositions (Mumma et al. 2003; Mumma and Charnley 2011). However, another model suggests that all comets formed in a region 15–35 AU from the Sun (Tsiganis et al. 2005; O'Brien et al. 2006) and were scattered into both the Oort Cloud and Kuiper Belt regions by the inward migration of Saturn (Crovisier 2007). In this case comets from both the Kuiper

Table 1 Elemental abundances in the Sun and protosun (adapted from Lodders et al. 2009)

Element	Solar dex	Protosolar dex	Δ dex	Protosolar X/H	Protosolar X/H_2	Δ (X/H_2)
He	10.93	10.99	0.02	9.68×10^{-2}	1.94×10^{-1}	9.13×10^{-3}
C	8.39	8.44	0.04	2.77×10^{-4}	5.55×10^{-4}	5.35×10^{-5}
N	7.86	7.91	0.12	8.18×10^{-5}	1.64×10^{-4}	5.21×10^{-5}
O	8.73	8.78	0.07	6.07×10^{-4}	1.21×10^{-3}	2.12×10^{-4}
Ne	8.05	8.10	0.10	1.27×10^{-4}	2.54×10^{-4}	6.56×10^{-5}
P	5.46	5.51	0.04	3.26×10^{-7}	6.52×10^{-7}	6.29×10^{-8}
S	7.14	7.19	0.01	1.56×10^{-5}	3.12×10^{-5}	7.27×10^{-7}
Ar	6.50	6.55	0.10	3.57×10^{-6}	7.15×10^{-6}	1.85×10^{-6}
Kr	3.28	3.33	0.08	2.15×10^{-9}	4.31×10^{-9}	8.71×10^{-10}
Xe	2.27	2.32	0.08	2.10×10^{-10}	4.21×10^{-10}	8.51×10^{-11}

Belt and Oort Cloud would display a variety of compositions with no distinction between the two groups. Therefore, it is unclear at this time whether different comet families can be expected to have different compositions given current uncertainties in their dynamical evolutions.

2.2 Defining "Solar Composition"

The first step in understanding formation of the solar system is to constrain the bulk composition of the PSN. The standard reference for this is the chemical composition of the Sun corrected for the evolution of the Sun over the past 4.6 billion years (Ga). Solar chemical composition has been investigated for several decades and knowledge improves regularly thanks to progress made in measurements of the primitive matter of the Solar System, in atomic and molecular data, and in solar atmospheric modeling, allowing better inference of the photospheric abundances. Table 1 represents a list of solar and protosolar abundances of elements derived from the recent compilation of Lodders et al. (2009), which provides the most representative list of abundances of volatiles that potentially existed in the protosolar nebula. In order to compare data to elemental abundances in the early Sun or in the PSN, it is preferable to use the protosolar abundances listed in the table because the Sun's evolution reduced the abundances of elements heavier than H_2 by more than 13 % in the photosphere over the last 4.56 Ga (Lodders et al. 2009; Asplund et al. 2009). The conversion of present-day to protosolar abundances is made via a correction of +0.061 logarithmic abundances (dex) for He and +0.053 dex for all other elements except H. The abundances and associated uncertainties are given in dex and converted in ratios relative to H and H_2.

The isotopes of main volatile elements and their relative proportions are listed in Table 2. These data derive from Lodders et al. (2009) who updated the compilation of isotopes published by Lodders (2003).

2.3 Volatiles and Noble Gases in the PSN

Having constrained the bulk elemental composition of the PSN, the next step in understanding the histories of the planets and moons is to constrain the molecular composition and variability of this composition within the PSN. In this way comets are highly valuable as

Table 2 Percent contributions of the isotopes of main volatile elements. These values are adapted from Lodders et al. (2009) with the exception of the nitrogen (from Marty et al. 2011) and carbon (from Hashizume et al. 2004)

Element	Atomic mass	Atom (%)
H	1	99.9981
H	2	0.00194
		100
He	3	0.0166
He	4	99.9834
		100
C	12	98.998
C	13	1.001
		100
N	14	99.774
N	15	0.226
		100
O	16	99.763
O	17	0.037
O	18	0.200
		100
Ne	20	92.9431
Ne	21	0.2228
Ne	22	6.8341
		100
P	31	100
S	32	95.018
S	33	0.75
S	34	4.215
S	36	0.017
		100
Ar	36	84.595
Ar	38	15.381
Ar	40	0.024
		100
Kr	78	0.362
Kr	80	2.326
Kr	82	11.655
Kr	83	11.546
Kr	84	56.903
Kr	86	17.208
		100
Xe	124	0.129
Xe	126	0.112
Xe	128	2.234
Xe	129	27.463
Xe	130	4.378
Xe	131	21.802
Xe	132	26.355
Xe	134	9.661
Xe	136	7.868
		100

their composition provides a record of the composition of the PSN in the region where they formed (for a more detailed review of comet composition see Cochran et al. 2015, this issue and Bockelée-Morvan et al. 2015, this issue). In addition to comets, meteorites can help to constrain composition within the PSN. Meteorites are presumed to have initially been part of larger planetesimals, of which there are two general categories—differentiated and undifferentiated. Chondritic meteorites are presumed to originate from undifferentiated planetesimals and thus represent the bulk composition of those source planetesimals (Lodders et al. 2009). Chondrites are broken into nine different classes and CI-chondrites demonstrate the closest abundance pattern to solar abundance with the exception of the most volatile elements H, C, N and O (Lodders et al. 2009). Key measurements are noble gas abundances, relative abundances of H, C, N and O and the stable isotope ratios of these elements.

2.3.1 C, H, O and N

The relative abundances of C, H, N and O give a record of volatile history in the PSN and in solar system bodies. The bulk abundance of C, H, N and O in the giant planets is of particular interest as a tracer of their formation processes, and will be discussed further in Sect. 3.

Stable isotope ratios of C, H, N and O are valuable for studying the volatile history of different bodies. They serve as a tracer for temperature conditions and dominant reaction pathways within the PSN as well as for the history of chemical processing of an atmosphere, surface or interior. Furthermore, they can be used to determine the primordial form of volatiles delivered to a solar system body (e.g. Mandt et al. 2014).

Measurements of isotopic ratios for the most abundant elements in cometary and planetary volatiles is still a developing discipline with only a limited number of targets and source species having been studied to date. The observations themselves are challenging, with both high-efficiency and high-spectral resolution needed to detect the weaker isotopic signatures in Earth-based remote sensing. Direct detection from mass spectroscopy can be more effective, but the opportunities in comets are limited to the few JFCs that have been visited by spacecraft (e.g. Balsiger et al. 1995; Eberhardt et al. 1995; Altwegg and Bockelée-Morvan 2003; Altwegg et al. 2015).

The D/H ratio in water depends significantly on temperature of formation while the isotopic ratios of elements heavier than water depend on temperature as well as the mechanism of chemical storage and the current state of the body at the time of release. For molecular radicals in comets, both the source molecule (e.g. HCN for CN) and the mechanism of storage (e.g. ice, clathrate, PAH, silicate grain, etc.) can affect the observed ratio of the same molecular species in different objects formed under nearly identical circumstances.

2.3.2 Water and D/H

The variability of the D/H ratio throughout the solar system provides important clues to solar system formation conditions. Of particular interest is how this ratio varies in water. In prestellar cores the very low temperature conditions cause most volatiles to freeze out and condense onto grains. An upper limit of the gaseous water abundance relative to hydrogen of only 1.3 ppb was recently reported in starless cores (Caselli et al. 2010), while the fraction of water in ice mantles covering dusts grains is about 100 ppm relative to hydrogen (e.g. Whittet and Duley 1991). The formation process of water on ice grains involving chemical reactions favors heavier water isotopologues. Ceccarelli et al. (2005) and Butner et al. (2007) derived D/H ratios in the water of molecular clouds and protostellar envelopes of about

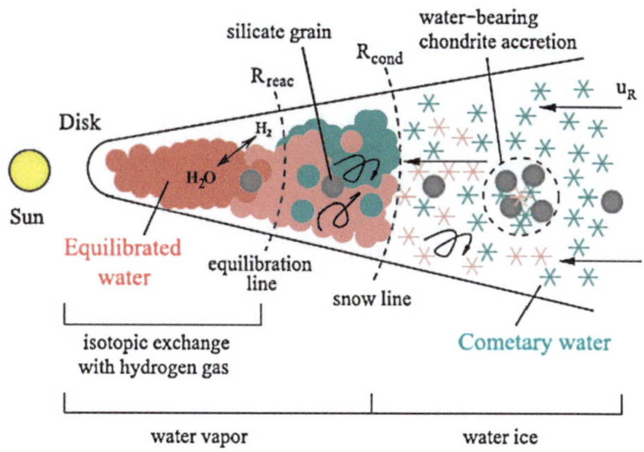

Fig. 3 Illustration of the formation of a D/H gradient with heliocentric distance. *Arrows* symbolize motions of gas. Reprinted from Jacquet and Robert (2013) with permission. Copyright © 2013 Elsevier

1.0×10^{-2} and 1.0×10^{-3} respectively. This is an enrichment of 2–3 orders of magnitude relative to the D/H ratio of protosolar hydrogen, 2.1×10^{-6}, or the local interstellar medium, 1.6×10^{-6} (Linsky et al. 2006). These values may be considered as upper and lower limits of the D/H ratio in water in our solar system, taking into account the possibility of isotopic exchange between water and hydrogen.

Figure 3 illustrates how the origin of water in our solar system may be traced back by determination of the water D/H ratio.

In the region of the disk near the sun under pressures in the 10 μbar to 1 mbar range and temperatures between 600 and 1300 K (compare e.g. Yang et al. 2013, Fig. 1), gaseous water equilibrates with hydrogen within a few hundred years. This means that the D/H ratio in water, which was initially highly enriched, will be defractionated to the value of hydrogen. With increasing heliocentric distance, temperature and density drops and the efficiency of the isotopic exchange process decreases and finally stops at an intermediate distance from the sun where water is still present in gaseous form. Here, equilibrated water transported from the inner zone by turbulent diffusion and sublimating water from highly D-enriched ices drifting inwards mix and establish a radial gradient of D/H, becoming fixed beyond the snow line.

Observational constraints on the D/H ratio in water in comets became available for the first time with the Giotto mission. Balsiger et al. (1995) and Eberhardt et al. (1995) derived D/H ratios of about 3.0×10^{-4} for 1P/Halley, which originated in the Oort Cloud. This value, about twice as high as the Vienna Standard Mean Ocean Water (VSMOW) of 1.56×10^{-4} was confirmed (within error bars) in OCCs by a number of ground-based detections in C/1996 B2 (Hyakutake) (Bockelée-Morvan et al. 1998), C/1995 O1 (Hale-Bopp) (Meier et al. 1998), C/2002 T7 (LINEAR) (Hutsemékers et al. 2008) and 8P/Tuttle (Villanueva et al. 2009). Biver et al. (2006) provided an upper limit for 153P/Ikeya-Zhang of 2.50×10^{-4}. Waite et al. (2009) added the D/H found in the Enceladus plume, which was in line with D/H of Oort cloud comets.

Simulations of the D/H gradient using upper limits for the ice D/H from the analysis of meteorite samples, managed to verify these observations (e.g. Drouart et al. 1999; Mousis et al. 2000; Horner et al. 2007 and Kavelaars et al. 2011). Therefore, models appear to be compatible with the assumption that OCCs formed in the vicinity of the gas giants. The models indeed show an increase of the water D/H ratio with heliocentric distance. Kavelaars et al. (2011) predicted a higher D/H ratio of at least 5.00×10^{-4} for JFCs, believed to have formed at larger heliocentric distances than the OCCs, i.e. beyond Neptune in the Kuiper Belt or scattered disk.

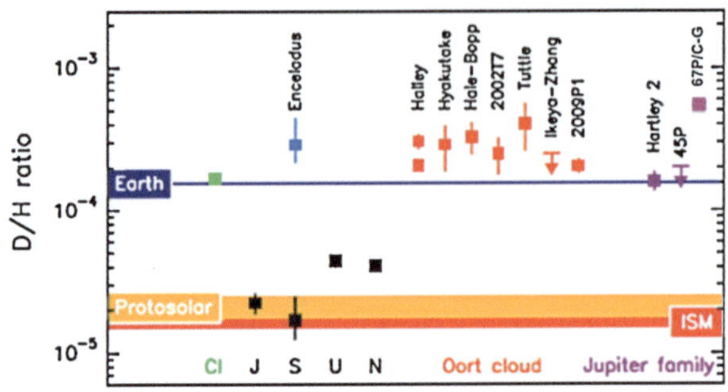

Fig. 4 Summary of measured D/H values in the solar system. *Orange symbols*: Oort cloud comets, *purple symbols*: Jupiter family comets. *Black symbols*: measurements in molecular hydrogen in Giant planets. *Blue* and *green symbols* are values in the Enceladus plume and in CI carbonaceous chondrites. ISM and protosolar values are in H_2 and Earth value is in water. Error bars are 1 sigma. Adapted with permission from Lis et al. (2013). Copyright © 2013 American Astronomical Society

The determination of the D/H ratio in JFCs is a greater challenge because they are generally much fainter than OCCs. The first constraint of D/H in water of a JFC was provided by observations of the Heterodyne Instrument for the Far Infrared (HIFI) (De Graauw et al. 2010) on Herschel (Pilbratt et al. 2010). HIFI is characterized by very high spectral resolution ($> 10^6$), near quantum limit sensitivity of its superconductor insulator superconductor (SIS) mixers and the largest collecting area ever flown, i.e. the most sensitive sensor in space for the detection of the D/H ratio thus far. On 17 November 2010, about three weeks after its perihelion, a D/H ratio of $1.61 \pm 0.24 \times 10^{-4}$ was derived for the JFC 103P/Hartley 2 from HIFI observations (Hartogh et al. 2011a). This value is compatible with VSMOW and more than 3 times lower than predicted by Kavelaars et al. (2011). Lis et al. (2013) then determined an upper limit for the D/H ratio of the JFC 45P/Honda-Mrkos-Pajdusakova (with HIFI) of 2.00×10^{-4}, compatible with the result from 103P. Herschel also determined the D/H ratio in the Oort cloud comet C/2009 P1 (Garradd) to be $2.06 \pm 0.22 \times 10^{-4}$ (Bockelée-Morvan et al. 2012). Finally, early Rosetta measurements made by the Rosetta Orbiter Spectrometer for Ion and Neutral Analysis (ROSINA) Double-Focusing Mass Spectrometer (DFMS; Balsiger et al. 2007) provide the highest D/H yet measured in a comet, $5.3. \pm 0.7 \times 10^{-4}$ (Altwegg et al. 2015). Figure 4 summarizes measurements of the D/H ratio in water and hydrogen throughout the solar system.

The initial finding that the D/H ratio measured in two JFCs did not agree with model predictions followed by the much higher ratio measured by ROSINA in a JFC challenge the existing D/H models, and raise questions as to whether the "classical picture" on the formation regions of OCCs and JFCs is correct. Levison et al. (2010) suggest based on numerical simulations that only 10 % of the OCCs originate in the solar system and perhaps more than 90 % are from protoplanetary disks of other stars. Based on this model, the D/H determined in all OCCs would be expected to show greater scatter from the average value of $\sim 3.00 \times 10^{-4}$. Brasser and Morbidelli (2013) suggest, however, that JFCs and OCCs originated from the same extended outer region of the PSN. This would mean that comets like Hartley 2, which has a lower D/H, formed closer to the Sun and comets with a higher D/H, like 67P/Churyumov-Gerasimenko (hereafter 67P/CG) formed farther out (Altwegg et al. 2015). This is consistent with the scenario suggesting that the migration of the young Jupiter and Saturn towards the inner solar system and back, caused mixing of material over large heliocentric distances (Walsh et al. 2011) causing the original D/H gradient with heliocentric distance to completely vanish and allow similar D/H ratios in both comet families.

Hogerheijde et al. (2011) point in a similar direction, concluding based on considerations on the ortho-to-para ratio (OPR) of water in planet forming disks that comets should contain heterogeneous ice mixtures collected across the entire solar nebula during the early stages of planetary birth. OPR measurements can, within certain limits, be used to determine the spin temperature of water within a cometary coma (Bonev et al. 2007). These spin temperatures are suggested to have been preserved from the time of comet formation (e.g. Crovisier 1984; Mumma et al. 1987), but little is known about the conditions that permit nuclear spin conversion and change the spin temperature of water in comets. Although this conclusion is based on the OPRs determined only in OCCs (e.g. Bonev et al. 2007), measurements for three JFCs (Crovisier et al. 1999; Bonev et al. 2008; Paganini et al. 2012) are consistent with what is observed in OCCs.

More recent D/H models provide alternative scenarios. Albertsson et al. (2014) present results of a laminar model and a model with two-dimensional turbulent mixing taking into account gas-grain chemistry including multiply deuterated species and nuclear spin-states. They find an overlap in the possible formation location for OCCs and JFCs. The model of Yang et al. (2013) does not result in a monotonic increase of the D/H ratio with distance from the Sun, but shows a decrease in the outermost region of the disk. They assume that water equilibrates in the inner disk already during the disk formation phase, when infall of material has not yet stopped. Due to conservation of angular momentum, water that was incorporated early into the disk near the young star would have been pushed outward, carrying low D/H ratios. Since crystalline silicates also formed under high temperature conditions in the inner disk, they predict a correlation between low D/H and a high crystalline/amorphous silicates ratio of comets.

2.3.3 Carbon Isotopes

The $^{12}C/^{13}C$ ratio is the best studied of all cometary isotopic ratios with measurements from more than 25 comets that include observations of C_2, CN, and HCN. Of these, the CN and HCN measurements are the most extensive (e.g. Manfroid et al. 2009), while C_2 measurements have the longest history (e.g. Danks et al. 1974; Vanysek 1973; Owen 1973; Lambert and Danks 1983), extending from comet Ikeya (1963a) (Stawikowski and Greenstein 1964) through comet C/2002 T7 (LINEAR) (Rousselot et al. 2012). The most recent compilations of data (Jehin et al. 2009; Manfroid et al. 2009 and references therein) include measurements of 23 comets including 8 JFCs and 15 OCCs at heliocentric distances from 1–3 AU and covering multiple compositional classes (e.g. A'Hearn et al. 1995; Fink 2009).

The third carbon species studied is HCN (e.g. Jewitt et al. 1997), which is observed in a different spectral range with different systematic issues than visible CN and C_2. HCN has been observed at radio wavelengths from the James Clerk Maxwell Telescope (JCMT, Jewitt et al. 1997) and the 12-meter National Radio Astronomy Observatory (NRAO) telescope on Kitt Peak (Ziurys et al. 1999). HCN is the likely parent of CN at large heliocentric distances (Rauer et al. 2003), which introduces a formation bias into the relative isotopic ratios. However, at smaller distances there is at least one additional parent that contributes over a different range of cometocentric distances (e.g. Woodney et al. 2002). Several potential candidates have been suggested ranging from more complex molecular species to direct liberation from dust (e.g. Klavetter and A'Hearn 1994; Woodney et al. 2002; A'Hearn et al. 1986), but to date no definitive source has been identified.

Despite significant systematic and theoretical difficulties associated with the measurements, a remarkable consistency has been obtained in the measurements. Figure 5 summa-

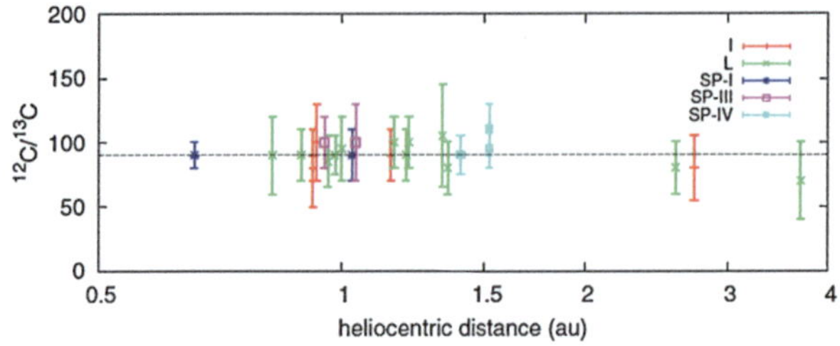

Fig. 5 The measured isotopic ratios for carbon are plotted as a function of heliocentric distance and orbital classification—Intermediate-period (I), Long-period (L) and Short-period Tisserand I (SP-I), III (SP-III) and IV (SP-IV). The relative uncertainties overlap for the entire set to a value between 90 and 110, which is consistent with the both the terrestrial value of 89 and the protosolar value of 99.8 (Hashizume et al. 2004). Reprinted from Manfroid et al. (2009) with permission from Astronomy and Astrophysics. Copyright © 2009 ESO

rizes the merged dataset of Manfroid et al. (2009), which includes all of the CN measurements. While there are hints of a potential trend toward a reduction in the ratio with increasing heliocentric distance, within the relative uncertainties of the various measurements obtained from a wide variety of objects with different facilities over several decades is a consistent value for $^{12}C/^{13}C$ between 90 and 110 with uncertainties between 10 and 20 %. The C_2 measurements are similar. All of the measurements are within their relative precision of each other. The majority of objects are slightly above the terrestrial value of 89 and in good agreement with the protosolar value of 99.8 determined from solar wind measurements (Hashizume et al. 2004). However the differences are not statistically significant.

2.3.4 Nitrogen Isotopes

Of the volatile elements other than hydrogen, only nitrogen has an isotopic ratio that is significantly lower in comets compared to the protosolar $^{14}N/^{15}N$ value of ~ 445 (Marty et al. 2011) and also lower than the terrestrial value of 272 (Anders and Grevesse 1989). It has been observed with approximately the same frequency as the carbon ratio since it is present in both CN and HCN, which are the primary species used for carbon studies. The early measurements of the $^{14}N/^{15}N$ ratio were obtained from comet C/1995 O1 (Hale-Bopp) using the JCMT and NRAO telescopes (Jewitt et al. 1997; Ziurys et al. 1999). These measurements, 323 ± 46 and 330 ± 98, were both reported as being close to the terrestrial value. However, successive study of CN B-X (0,0) band in the NUV consistently yielded a value for $^{14}N/^{15}N$ that was half (i.e. heavily *enriched* in ^{15}N) that obtained for HCN (e.g. Arpigny et al. 2003; Manfroid et al. 2009; Hutsemékers et al. 2005; Bockelée-Morvan et al. 2008). This differential between radio and visible measurements extended to Hale-Bopp (Arpigny et al. 2003), which had already been shown to have a consistent isotopic ratio for carbon, as discussed above. Arpigny et al. (2003) suggested that this difference was due to the known changing parentage of CN with heliocentric distance with the contributing parents at small distances having potentially been enriched in ^{15}N via an NH_3 ice-dust grain fractionation process in dense molecular clouds (Charnley and Rodgers 2002).

The discrepancy in $^{14}N/^{15}N$ may have been resolved by observations of the outburst of 17P/Holmes that were obtained using both the NUV (CN) and radio JCMT (HCN). These observations produced similar results for both species with $^{14}N/^{15}N$ in HCN of 139 ± 26, which was only slightly *enriched* relative to $^{14}N/^{15}N$ in CN of 165 ± 40 (Bockelée-Morvan et al. 2008). It is important to note that Hale-Bopp and 17/P Holmes are objects in completely different classes (OCC vs. JFC), and the outburst nature of the Holmes event may

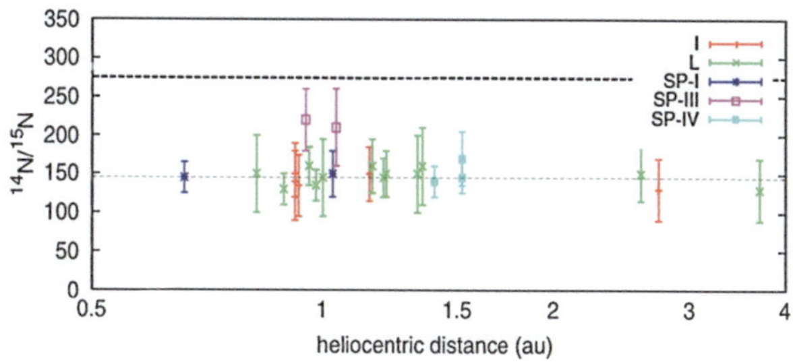

Fig. 6 The $^{14}N/^{15}N$ ratios of the comets listed in Manfroid et al. (2009) reveals a consistent value for the CN derived number as a function of comet class—Intermediate-period (I), Long-period (L) and Short-period Tisserand I (SP-I), III (SP-III) and IV (SP-IV)—and heliocentric distance. The two HCN values from Hale-Bopp (*pink*) are the least enriched of the entire set. The *thick dashed line* above $^{14}N/^{15}N = 250$ represents the terrestrial value of 272, which does overlap with HCN at the level of their uncertainties. Reprinted from Manfroid et al. (2009) with permission from Astronomy and Astrophysics. Copyright © 2009 ESO

not have followed the heliocentric distance-parentage relationship observed in other comets. Either factor may have affected the extent of any difference in the $^{14}N/^{15}N$ ratio. As a check, Bockelée-Morvan et al. (2008) performed an updated analysis of the original JCMT and NRAO Hale-Bopp data. They identified several contaminating factors that could have biased the original reduction and, after correction, obtained new ratios of 205 ± 70 and 207 ± 48, respectively. While still less enriched than the ratios obtained for CN, the new values were closer to agreement, and to within their relative uncertainties, approximately equal (Fig. 6).

As Fig. 6 shows, the new values are as close to the terrestrial value of 272 as they are to the average of the other measurements. Hence, the possibility of an admixture of CN sources with different ^{15}N enrichment factors (Arpigny et al. 2003) cannot be ruled out. This suggests that considerable additional work is required to resolve differences between the terrestrial $^{14}N/^{15}N$ ratio and that derived from comets.

Two recent measurements of $^{14}N/^{15}N$ in cometary NH_3 add to the growing dataset: 127 ± 32 (Rousselot et al. 2014) using a summed spectrum of 12 comets, and 139 ± 38 (Shinnaka et al. 2014) from the outburst of comet C/2012 S1 (ISON). These values are in good agreement with the HCN trend of heavy cometary values for $^{14}N/^{15}N$.

2.3.5 Oxygen Isotopes

Oxygen has one of the most diverse set of isotopic measurements of the elements described here, but is also the most complex in terms of its interpretation. It is a major component of terrestrial rocks and interstellar grains with three stable isotopes that occur in ratios that are a strong function of the conditions of their formation (e.g. Webster and Mahaffy 2012; Dominguez 2010). Taken in isolation, the $^{16}O/^{18}O$ ratio can vary even if the number for the full system does not. The well-studied variation $\delta^{18}O$ of VSMOW as a function of global terrestrial glaciation, rainfall, and evaporation is a good example of this (e.g. Clark and Fritz 1997). The interaction between different stable isotopes in the formation of different types of oxygen bearing compounds produces a general trend that can be broken down by $\delta^{18}O$ and $\delta^{17}O$ values (Fig. 7) reflecting different processes in the formation and evolution of both the early and later solar system (Dominguez 2010).

The result of these processes presents a situation where comparison with the terrestrial VSMOW value does not provide clear answers on formation and evolution of the solar system. The observed features will reflect the process that formed them (e.g. water ice and

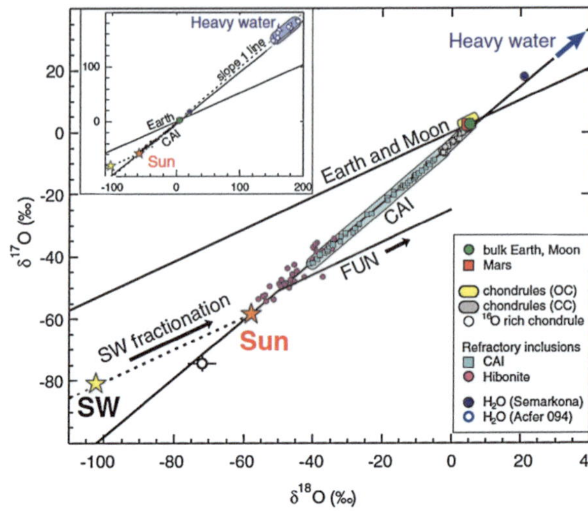

Fig. 7 A triple-isotope plot shows relative excesses and deficits relative to SMOW for a variety of astrophysical compounds. The differences come from multiple sources, including, for example, temperature, grain surface chemistry, shock effects, and molecular cloud density. Reprinted with permission from McKeegan et al. (2011). Copyright © 2011 American Association for the Advancement of Science

CHON grains may have different $^{16}O/^{18}O$ values even when obtained from the same body), with Earth itself representing only an emergent property derived from multiple mineralogical and chemical processes acting simultaneously. Because of this, the cometary values should be viewed in the context of whether they fit within the relational framework described in Fig. 7 as opposed to being an exact match to the terrestrial VSMOW.

Cometary $^{16}O/^{18}O$ ratios from water of 518 ± 45 and 470 ± 70 at Halley (Balsiger et al. 1995; Eberhardt et al. 1995) and 556 ± 70 at 67P/CG (Altwegg et al. 2015) have been obtained by mass spectroscopy and 530 ± 50 by radio band spectroscopic remote sensing (Lecacheux et al. 2003; Biver et al. 2007). Many other observations of ^{16}O and ^{18}O in water have also been made with Herschel observations, but are only in few cases converted to a $^{16}O/^{18}O$ ratio (e.g. Bockelée-Morvan et al. 2012) due to modeling challenges. Additional water-group spectroscopic measurements have been made of the OH daughter, 425 ± 55 (Hutsemékers et al. 2008). Non-water-group measurements have been made from direct sampling of the mineral grains returned from 81P/Wild2 by *Stardust*, 490–520 (McKeegan et al. 2006). Put within the context of Fig. 7, it is clear that the uncertainties of the measurements defy precision classification with respect to the VSMOW value of 499 in addition to lacking the $\delta^{17}O$ slope indicator. Although the general trend for cometary values is consistent with the range described between chondritic material and asteroidal water, more effective classification will require not only more measurements, but also more precise ones that include ^{17}O. Therefore, the oxygen isotope ratios that are currently available are not helpful in evaluating formation and evolution of solar system bodies.

2.3.6 Sulfur Isotopes

Like oxygen, sulfur has multiple stable isotopes and its isotopic state is very process dependent. Its mineralogical signature is very different however, with the bulk of its activity being chemical. On Earth, the modern sulfur cycle represents a mixture of biological processes, weathering, and anthropogenic sources (e.g. Farquhar et al. 2000). The best direct comparison with cometary sampling will come from older samples that follow $\delta^{33}S$, $\delta^{34}S$, and $\delta^{36}S$ from the modern mean as shown in Fig. 8. Even when focusing on the oldest features, the comparison with cometary values is not straightforward, and given the paucity of measurements, unclear in its significance.

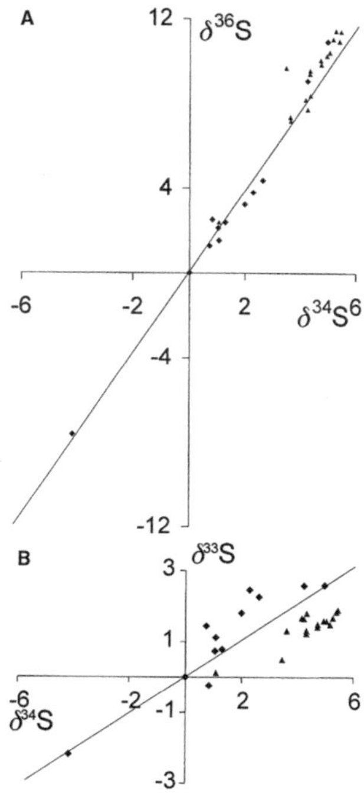

Fig. 8 Three isotope plots of δ^{33}S, δ^{34}S, and δ^{36}S are shown for terrestrial samples older than 3 Gyr. The reported cometary values here are generally consistent with the zero point on the δ^{34}S axis and the lone sample at a significant negative value. Reprinted from Farquhar et al. (2000) with permission. Copyright © 2000 American Association for the Advancement of Science

Sulfur isotopes have been sampled in multiple species including plasma (S^+ mass spectroscopy, Altwegg 1996) and radio-band remote sensing of CS (Jewitt et al. 1997) and H_2S (Crovisier et al. 2004). All are focused on the $^{32}S/^{34}S$ and its comparison with the modern value of ~ 23. The resulting values are nominally consistent with the terrestrial value, but do have certain inconsistencies. In particular, the $^{32}S/^{34}S$ ratios in CS and H_2S from Hale-Bopp of 27 ± 3 and 16 ± 3, respectively, are highly disparate. However, since they come from different species they may be process specific.

The $^{32}S/^{34}S$ ratio in CS has also been observed to vary, with the value from Hale-Bopp of 27 ± 3 being very different from that obtained during the outburst of 17P/Holmes of 16 ± 3. Aside from the previously noted difference in both the evolutionary state and population (dynamically new OCC vs. JFC) and activity cycle (heliocentric driven vs. outburst) it is difficult to make any quantitative conclusions about this difference, particularly given the small sample size of sulfur measurements. The similarity between the H_2S value from Hale-Bopp and the CS value from Holmes may be indicative of an underlying process, but until a larger sample size is obtained that targets multiple isotopes from several comets of different classes and heliocentric distances, any conclusions outside of a general agreement with terrestrial averages is premature. Therefore, as with the oxygen isotope ratios the sulfur isotope ratios that are currently available are not helpful in evaluating formation and evolution of solar system bodies.

2.3.7 Noble Gases

Noble gases and their isotopologues are inert and not subject to chemical processes. Their abundance in comets, meteorites and atmospheres can serve as a valuable tracer of conditions within the PSN, processes that deliver volatiles to solar system bodies and processes

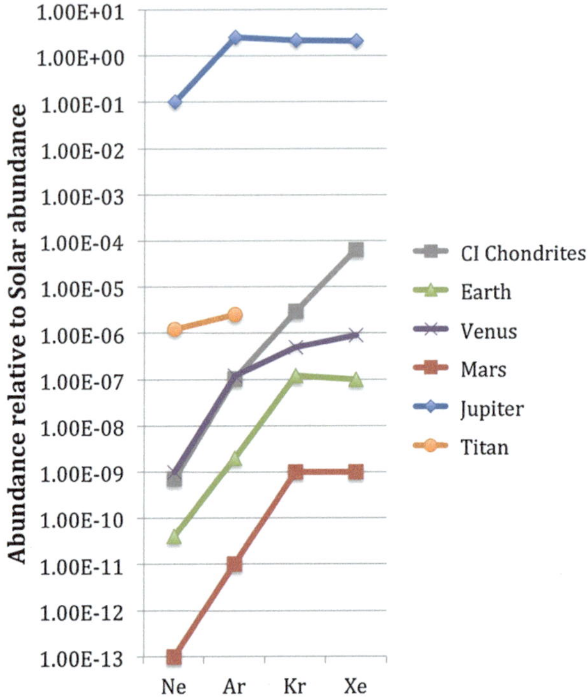

Fig. 9 Noble gas abundances relative to solar noble gas abundances for Jupiter, CI Chondrites, Titan and the terrestrial planets

that change atmospheres with time. Unfortunately at the present time the only available measurements of noble gases in comets are upper limits and a tentative detection of Argon (Stern et al. 2000). Noble gas measurements in chondrites and terrestrial planet atmospheres compared to solar abundances have been used to evaluate the origin of volatiles on the terrestrial planets (e.g. Owen and Barnun 1995) and the evolution of the terrestrial planet atmospheres (e.g. Pepin 1991).

Figure 9 illustrates the noble gas measurements available for planets, for CI chondrites and for one of the moons in the solar system relative to the solar abundance of noble gases. Titan's abundance is determined based on noble gas abundance relative to nitrogen compared to solar abundance relative to nitrogen assuming that the ratio of ^{22}Ne to ^{20}Ne is solar. The remaining ratios are determined based on the abundance of noble gases to 10^6Si atoms.

Jupiter's noble gas abundances are greater than the solar abundances while the terrestrial planets, CI chondrites and Titan are significantly depleted in noble gases. All of these bodies are most depleted in Ne compared to solar values and a trend of increased depletion with decreasing mass is observed in Titan, the terrestrial planets and CI chondrites. These depletion patterns could provide information about the potential sources of volatiles for these bodies, although the present lack of measurements in comets is very limiting.

3 The Giant Planets

Here we review the atmospheric elemental and isotopic compositions of Jupiter, Saturn, Uranus and Neptune. We then discuss the constraints that can be derived on the formation conditions of these giant planets in the protosolar nebula from their inferred compositions.

Table 3 Elemental and isotopic abundances in the four giant planets

Element	Jupiter/Sun	Saturn/Sun	Uranus/Sun	Neptune/Sun
He	0.8[a]	0.7 ± 0.1[g]	0.9 ± 0.2[n]	1.2 ± 0.2[r]
Ne	0.1[b]	–	–	–
O	0.4 ± 0.1[c]	$(1.6 \pm 0.29) \times 10^{-4}$[h]	–	–
C	4.3 ± 1.1[c]	9.6 ± 1.0[i]	41.5 ± 16.7[o]	72.1 ± 19.3[o]
N	4.1 ± 2.0[c]	2.8 ± 1.1[j]	–	–
S	2.9 ± 0.7[c]	12.05[k]	22.5 ± 11.3[p]	22.5 ± 11.3[p]
P	3.3 ± 0.4[d]	11.2 ± 1.3[d]	–	–
Ar	2.5 ± 0.8[b]	–	–	–
Kr	2.2 ± 0.6[b]	–	–	–
Xe	2.1 ± 0.6[b]	–	–	–
Isotope	**Jupiter**	**Saturn**	**Uranus**	**Neptune**
D/H (in H_2)	$(2.60 \pm 0.7) \times 10^{-5}$[e]	$1.70^{+0.75}_{-0.45} \times 10^{-5}$[l]	$(4.4 \pm 0.4) \times 10^{-5}$[q]	$(4.1 \pm 0.4) \times 10^{-5}$[q]
$^3He/^4He$	$(1.66 \pm 0.05) \times 10^{-4}$[e]	–	–	–
$^{12}C/^{13}C$ (in CH_4)	$92.6^{+4.5}_{-4.1}$[f]	$91.8^{+8.4}_{-7.8}$[i]	–	–
$^{14}N/^{15}N$ (in NH_3)	434.8^{+65}_{-50}[c]	> 500[m]	–	–

Error is defined as $(\Delta E/E)^2 = (\Delta X/X_{planet})^2 + (\Delta X/X_{Protosun})^2$. [a]von Zahn et al. (1998); [b]Mahaffy et al. (2000); [c]Wong et al. (2004); [d]Fletcher et al. (2009a); [e]Niemann et al. (1998); [f]Niemann et al. (1996); [g]Conrath and Gautier (2000); [h]de Graauw et al. (1997); [i]Fletcher et al. (2009b); [j]Fletcher et al. (2011a); [k]Briggs and Sackett (1989); [l]Lellouch et al. (2001); [m]Fletcher et al. (2014); [n]Conrath et al. (1987); [o]Karkoschka and Tomasko (2009, 2011); [p]de Pater et al. (1991); [q]Feuchtgruber et al. (2013); [r]Conrath et al. (1991).

3.1 Atmospheric Composition

The abundances and isotopic ratios of most significant volatiles measured at Jupiter, Saturn, Uranus and Neptune are summarized in Table 3. Elemental abundances have been expressed as a function of the protosolar abundances derived from Lodders et al. (2009). The abundances of CH_4, NH_3, H_2O, H_2S, Ne, Ar, Kr and Xe were measured by the Galileo Probe Mass Spectrometer (GPMS) in Jupiter's atmosphere (Mahaffy et al. 2000; Wong et al. 2004). The value of H_2O abundance reported for Jupiter corresponds to the deepest measurement made by the probe (at ~ 20 bar). It is probably smaller than the planet's bulk water abundance, which remains unknown (Atreya et al. 2003; Wong et al. 2004; Mousis et al. 2012). The He abundance in Jupiter has also been measured in situ by a Jamin-Mascart interferometer (Helium Abundance Detector) aboard the Galileo probe (von Zahn and Hunten 1992; von Zahn et al. 1998). PH_3 is the only species of our list of measurements in Jupiter whose abundance has been determined remotely by the Cassini Composite Infrared Spectrometer (CIRS) during the spacecraft encounter in 2000–2001 (Fletcher et al. 2009a). The GPMS instrument aboard the Galileo probe has also performed the isotopic measurements presented for Jupiter in Table 3 (Niemann et al. 1996, 1998; Mahaffy et al. 2000; Wong et al. 2004).

In the case of Saturn, only the abundances of CH_4, PH_3, NH_3, H_2O, He, and indirectly that of H_2S, have been determined. The abundance of CH_4 has been measured from the analysis of high spectral resolution Cassini/CIRS data (Fletcher et al. 2009b). Similar to Jupiter,

PH_3 has been determined remotely in Saturn from Cassini/CIRS observations (Fletcher et al. 2009a). The NH_3 abundance corresponds to the deepest value derived by Fletcher et al. (2011a) who examined Saturn's tropospheric composition from Cassini/VIMS thermal emission spectroscopy. Tropospheric H_2O has been inferred in Saturn via the Short Wavelength Spectrometer Instrument onboard the Infrared Space Observatory (ISO) (de Graauw et al. 1997). However, H_2O is unsaturated at this altitude (~ 3 bar level), implying that its bulk abundance should be higher than the measured one. The H_2S abundance is quoted from the indirect determination of Briggs and Sackett (1989) who investigated the influence of models of NH_3-H_2S-H_2O cloud decks on Saturn's atmospheric opacity at microwave wavelengths. The He abundance in Saturn's atmosphere comes from a reanalysis of Voyager's infrared spectrometer (IRIS) measurements (Conrath and Gautier 2000). The only isotopic ratios measured in Saturn are D/H in H_2 (Lellouch et al. 2001) and $^{12}C/^{13}C$ in CH_4 (Fletcher et al. 2009b).

Due to their higher heliocentric distances, the measurements of molecular abundances and isotopic ratios are scarce in Uranus and Neptune. CH_4 has been measured in the atmospheres of the two planets with the Hubble Space Telescope (HST) Space Telescope Imaging Spectrograph (STIS) (Karkoschka and Tomasko 2009, 2011). The H_2S abundance has been indirectly inferred in Uranus and Neptune from the fits of thermal spectra with models including the microwave absorption of this molecule (de Pater et al. 1991). N has not been measured in the atmospheres of the two ice giants, but its equilibrium form should be NH_3 in the troposphere, as is the case for Jupiter and Saturn (Fegley and Prinn 1986). The He abundance in the two planets has been inferred from the combination of Voyager radio-occultation and infrared spectroscopy measurements (Conrath et al. 1987, 1991).

3.2 Interpretations of the Volatile Enrichments in the Atmospheres of Jupiter, Saturn, Uranus and Neptune

Table 3 shows that C, N, P, S, Ar, Kr and Xe are found to be uniform and enriched by a factor from ~ 2 to 4 in Jupiter. In contrast, C, N and P (the only heavy elements reliably measured) are found to be enriched by factors of ~ 10, 3 and 12 respectively in Saturn. Helium is depleted compared to the protosolar value in Jupiter and Saturn because of its condensation into droplets that precipitate in the deep interiors of the two giants (Fortney and Hubbard 2003). Interestingly, the Galileo probe entry site in Jupiter's atmosphere was an unusually dry meteorological system. This implies that the probe did not measure the deep, well-mixed water abundance (Wong et al. 2004).

An interpretation of the volatile enrichments measured in the four giant planets is that their building blocks agglomerated from a mixture of rocks and crystalline ices that condensed during the cooling period of the protosolar nebula (Gautier et al. 2001; Hersant et al. 2004, 2008; Alibert et al. 2005a, 2005b; Gautier and Hersant 2005; Mousis et al. 2006, 2009c, 2010, 2012; Marboeuf et al. 2008, 2014a, 2014b). The ices consist of a mixture of clathrates or hydrates (case of NH_3) and pure condensates whose relative proportions were fixed by the availability of crystalline water at the time of volatiles trapping in the clathrate phase and the efficiency of the clathration kinetics in the protosolar nebula (Mousis et al. 2009c). In order to explain the uniform enrichment of volatiles in Jupiter's atmosphere, it has been proposed that the abundance of water ice was approximately twice as large as the protosolar O abundance (Gautier et al. 2001). In these conditions, essentially all volatiles, including noble gases, were trapped by water in the ~ 45–80 K range in the form of clathrates or hydrates instead of condensing as pure ices at lower temperatures in the protosolar nebula

(Gautier et al. 2001). The volatile enrichments in Jupiter can then be explained by the accretion and the vaporization in its envelope of the icy planetesimals formed in its feeding zone. Alternative models suggest that the building blocks of Jupiter may have formed in a cooler disk (down to ~ 20 K), implying the relaxing of the condition of a supersolar O abundance that is needed to form clathrates at higher temperatures (Mousis et al. 2009c, 2012). In this case, the assumption of a protosolar O abundance in the nebula would imply that the fraction of pure condensates would be higher than that of clathrates in planetesimals.

However, more recent work suggests that both Jupiter and Saturn were formed in environments where the O abundance was supersolar. Indeed, the scenario of full clathration of volatiles explains their observed enrichments in the two giants and is also found consistent with the high $^{14}N/^{15}N$ ratios measured in their atmospheres (Mousis et al. 2014a). Consequently, these values require significant amounts of N_2 to have been trapped in the building blocks of the two planets at higher temperatures in the disk than those required for the condensation of this volatile (~ 22 K). The scenario of clathration of volatiles also matches the observational data of Uranus and Neptune (Gautier and Hersant 2005) but the known constraints on these two planets are too scarce and can also be explained by alternative scenarios. For example, it has been proposed that Uranus and Neptune could have formed between the CO and N_2 ice lines, assuming that the disk was stationary (Ali-Dib et al. 2014). This model explains the D/H value found in both planets, matches well the heavy C enrichment in the two planets and predicts that N is protosolar in the envelopes. However, it requires a homogenous envelope while many internal structure models suggest a non-homogenous envelope. Clathration models predict that nitrogen is moderately enriched in Uranus and Neptune (Gautier and Hersant 2005).

3.3 Contribution of Comets to the Giant Planet Atmospheres

While comets provide us with invaluable information on the Solar System formation conditions from their composition, they are also thought more and more to be active players in the temporal evolution of giant planet atmospheric composition. This section reviews the current knowledge regarding the contribution of comets to the composition of giant planet atmospheres.

3.3.1 An Emblematic Case: The Shoemaker-Levy 9 Comet Impacts of 1994 with Jupiter

The Shoemaker-Levy 9 comet (SL9; Shoemaker et al. 1993), also known as the "String of Pearls" comet because of its fragmented form, collided with Jupiter in July 1994 and left visible scars on the Jovian disk for several weeks. This spectacular event was the first direct observation of an extraterrestrial collision in the Solar System and was followed worldwide by professional and amateur astronomers.

Observations of the impacts over a wide range of wavelengths led to the detection of several species, including H_2O, CO, OCS, CS, CS_2, S_2, HCN, and NH_3, in Jupiter's stratosphere at the impact sites (e.g., Bjoraker et al. 1996; Lellouch et al. 1995; Atreya et al. 1995; Noll et al. 1995; Marten et al. 1995; Orton et al. 1995). While the presence of NH_3 was likely due to the updraft of air parcels from the NH_3-rich upper troposphere following the impacts (Orton et al. 1995), the formation of most of these species was explained by shock chemistry occurring at extreme temperatures during the impacts (Zahnle 1996). Post-impact high temperatures were still observable a few days after the impacts (Orton et al. 1995; Bézard et al. 1997a, 1997b; Lellouch et al. 1997; Moreno et al. 2001).

Comets are best known for being water-rich bodies (e.g., Bockelée-Morvan 2011). Therefore, the detection of water vapor in the stratospheres of the giant planets (i.e., above the water ice cloud that prevents any transport of water vapor from their water-rich deep atmospheres up to observable levels) by the ISO (Feuchtgruber et al. 1997) was less of a surprise in Jupiter than in the other giant planets: SL9 was indeed a perfect candidate source as its fragments had disintegrated in Jupiter's stratosphere a few years earlier. However, other possible sources for external water exist in the outer solar system: interplanetary dust particles (Prather et al. 1978), which result from comet activity and asteroid collisions, and icy rings and satellites (Strobel and Yung 1979). Another complication comes from the results of shock chemistry simulations, according to which most of the cometary H_2O had converted into CO during the impact (Zahnle 1996). So, while H_2O was directly detected at some impact sites by Bjoraker et al. (1996), subsequent ISO Submillimeter Wave Astronomy Satellite (SWAS), and Odin observations did not lead to a direct proof that Jupiter's stratospheric H_2O had been delivered by SL9, because of limitations in spatial resolution, spectral resolution, and sensitivity (Lellouch et al. 2002; Bergin et al. 2000; Cavalié et al. 2008, 2012). Only recently has Herschel been able to demonstrate that the spatial distribution of H_2O was a remnant of the SL9 impacts of 1994 (Cavalié et al. 2013).

The mid- and long-term consequences of such enrichment in the chemical inventory of Jupiter's stratosphere were investigated by means of photochemical models (Moses et al. 1995a, 1995b; Moses 1996). According to these simulations, the most stable species over the long-term (i.e., several years) after such impacts are CO, CS, HCN, and to a lesser extent H_2O. All of these species have been monitored since that time and used as atmospheric tracers to constrain vertical and horizontal transport, and Jovian oxygen photochemistry (Moreno et al. 2003; Griffith et al. 2004; Lellouch et al. 2006; Cavalié et al. 2008).

3.3.2 Beyond Jupiter

The comet impact rates in the Solar System have been studied since long before the SL9 events (e.g., Everhart 1969; Zimbelman 1984; Olsson-Steel 1987). Levison et al. (2000) and Zahnle et al. (2003) provided the latest estimates for the Giant Planets based on cratering rates. Work more focused on Jupiter has been published since that time (Sánchez-Lavega et al. 2010; Hueso et al. 2013).

CO is one of the most stable post-SL9 species in Jupiter's stratosphere with respect to its abundance (Moreno et al. 2003), and it is still monitored as of 2014. It is therefore no surprise that tentative evidence for comet impact signatures has been found in giant planets beyond Jupiter using observations of this species. The observability of cometary-derived species like CO is made plausible given the existing trade-space found between km-size comet impact rates at Jupiter, Saturn, Uranus, and Neptune (~ 0.01–0.001 year^{-1}; Zahnle et al. 2003), and stratospheric meridional/vertical mixing timescales (~ 100–1000 years; Moreno et al. 2003; Moses et al. 2005). This trade-space seems to ensure that comets of decent size can leave observable signatures on the long term in giant planet stratospheres.

In the past decade, observations of CO in the millimeter and submillimeter ranges at Saturn and Neptune have tentatively shown that the external source for this species may be ancient comet impacts (Lellouch et al. 2005, 2010a; Hesman et al. 2007; and Luszcz-Cook and de Pater 2013, for Neptune, and Cavalié et al. 2010 for Saturn). At Uranus, a cometary origin for CO is possible (Cavalié et al. 2014). These results are based on spectrally resolved ground- and space-based observations and vertical transport modeling.

To account for Neptune's observed external CO, the required CO flux is 10–500 times that of H_2O, i.e., the same order of magnitude as in Jupiter, thus favoring comet impact shock chemistry as the formation process for Neptune's external CO (Lellouch et al. 2005). In addition, HCN is also present in Neptune's stratosphere (Marten et al. 1993) and the CO/HCN ratio is similar to post-SL9 values. Lellouch et al. (2005) concluded that HCN and CO might therefore derive from the same comet.

At Saturn, there is as yet no additional clue (beyond CO) regarding the presence of cometary-derived material. HCN remains undetected and other oxygen species probably originate from the Enceladus geysers and subsequent transport in the Saturnian system (Waite et al. 2006; Cassidy and Johnson 2010; Hartogh et al. 2011b). At Uranus, the origin of other oxygen species is still under investigation (Orton et al. 2014).

In the cases of Saturn and Neptune, $\sim$ km-size comets impacting the planets a few centuries ago are required to fit the CO data. One should note that Bézard et al. (2002) had already demonstrated that Jupiter's CO vertical profile showed evidence for pre-SL9 comet impacts.

3.3.3 New Impacts?

In the last few years, amateur observations have led to the unexpected detection of new impacts in Jupiter's atmosphere in 2009, 2010, and 2012 (Sánchez-Lavega et al. 2010; Fletcher et al. 2010, 2011b; Hueso et al. 2010a, 2013). While the size of the 2009 impactor ($\sim$ 0.5 km) made the impact look like one of the main SL9 impacts, the high abundance of silica and the lack of spectroscopic evidence of cometary-derived species like CO, CS, and HCN, favor an asteroidal origin rather than a cometary origin for this impactor (Orton et al. 2011).

In the future, nearly continuous monitoring of Jupiter offered by the extensive amateur network (Hueso et al. 2010b; Mousis et al. 2014b), will undoubtedly lead to more impact detections and to a better understanding of the contribution of comets to the composition of giant planet atmospheres.

The study of the contributions of comets to the composition of giant planet atmospheres will not only benefit the Solar System atmosphere field but also the growing field of exoplanet atmosphere characterization (Turrini et al. 2014).

4 Titan, Enceladus and Giant Planet Satellites

The giant planet satellites are divided into two categories depending on their formation process: (1) regular satellites formed as a by-product of giant planet formation and (2) irregular satellites formed in the PSN that were captured by their planet. Captured satellites of giant planets are likely to have formed in conditions similar to objects in the Kuiper Belt and Oort Cloud. Their composition would provide important information about their source region and dynamical processes in the early solar system.

Understanding of the formation conditions of the regular satellites has progressed along with studies of giant planet formation. Initially, giant planets were thought to have formed within a subnebula that was significantly warmer than the PSN and dense enough to convert CO and N_2 from the PSN to CH_4 and NH_3 (Prinn and Fegley 1989). However, Canup and Ward (2002) found that the subnebula of Jupiter must have had a gas density that was orders of magnitudes lower than previously thought. Therefore, the subnebula was probably not dense enough to efficiently convert CO to CH_4 and N_2 to NH_3 (Mousis et al. 2002; Alibert and Mousis 2007). Canup and Ward (2006) then showed that the regular satellites formed within a circumplanetary disk that was actively supplied with gas and solids from the PSN, and that the earliest satellites to form were lost by collision with the giant planets. The regular satellites that remain today are proposed to have formed during the latest stages of planetary formation in a "gas-starved" disk (Canup and Ward 2006).

The composition of regular satellites is an important tool for constraining the effective temperature of the subnebula and thus the inflow properties during the formation of the giant planet (Canup and Ward 2006). Satellites that are formed early in the process of giant planet formation are predicted have a low ice-to-rock ratio compared to those that formed later, so the high ice-to-rock ratio of most of the giant planet satellites is interpreted to mean that they formed later in the process of planetary formation (Canup and Ward 2006). However, it is also possible that these satellites formed at a greater distance from their parent planet and migrated to their current position (Mousis et al. 2009a; Crida and Charnoz 2012). Comparing the primordial composition and isotopic ratios of noble gases and carbon, nitrogen, hydrogen and oxygen to solar composition and the composition of comets provides important clues as to the formation conditions of these satellites.

4.1 Enceladus

Enceladus is a small moon (radius of ~ 500 km) located within the E ring of Saturn. A thick layer of high albedo water ice covers the surface. The Cassini mission discovered a water-rich plume emanating from the south pole (Porco et al. 2006; Hansen et al. 2006; Spencer et al. 2006), the composition of which has been evaluated by the Cassini Ion Neutral Mass Spectrometer (INMS) (Waite et al. 2006; 2009). This composition includes volatiles that could either be primordial or produced by aqueous processes (Matson et al. 2007; Glein et al. 2008). Uncertainty in the source of some molecules found in the plume of Enceladus limits the use of its composition for direct comparison with formation processes. For example, the origin of CH_4 detected in the plume could be primordial or the methane may have been produced through serpentinization reactions in the interior.

One measurement from the plume that can be presumed to be primordial is the D/H ratio in H_2O. This was measured to be $2.9^{+1.5}_{-0.7} \times 10^{-4}$ (Waite et al. 2009), which is very close to the value found for OCCs (see Fig. 4), and serves as the most useful tool currently available for evaluating the formation of Enceladus.

When evaluating the formation of Saturn's moon Titan, Alibert and Mousis (2007) proposed that the D/H in water for Saturn's moons would show a gradient based on their formation region. Moons that formed closer to Saturn would have a lower D/H than moons that formed farther away or those that were captured. However, a D/H ratio for Enceladus that is in the range of OCCs strongly indicates that the building blocks for Enceladus must have formed in the PSN (Mousis et al. 2009b). These building blocks are proposed to have migrated inward through the Saturnian subnebula where they were partially devolatilized, losing primordial Ar, CO and N_2 in the process (Mousis et al. 2009b) in agreement with the model of Canup and Ward (2002, 2006).

4.2 Titan

Table 4 lists measurements made in Titan's atmosphere that are relevant to studying Titan's formation and evolution compared to measurements made in comets and estimates for PSN values. While each of these measurements provides insight into the evolution of Titan's atmosphere, the most important values for evaluating Titan's formation are the three that have been determined to represent primordial composition of the building blocks of Titan (italicized in Table 4). The D/H in water at Titan can be presumed to be in the same range as

Table 4 Measurements of Titan's current and *primordial* composition compared to what has been measured in comets and determined for the PSN

		Titan	Comets	PSN
$CO:CH_4$		$\sim 10^{-3a}$		70^{b}
D/H	H_2	$1.35 \pm 0.3 \times 10^{-4c}$	–	1.94×10^{-5d}
	H_2O	$\mathit{2.9^{+1.5}_{-0.7} \times 10^{-4}}$[e]	$1.5–5.3 \times 10^{-4f}$	variable
	CH_4	$1.6 \pm 0.3 \times 10^{-4g}$ $\mathit{0.95–1.6 \times 10^{-4}}$[h]	–	variable
$^{14}N/^{15}N$	N_2	167.7 ± 0.6^{c}	–	430 ± 50^{i}
	NH_3	$\mathit{\leq 190}$[j]	127 ± 32^{k} 139 ± 38^{l}	130 ± 40^{m}
	HCN	56 ± 8^{n}	157 ± 21°	157 ± 21°
$^{22}Ne/N$	Ne	$1.4 \pm 1.1 \times 10^{-7c}$	–	0.106^{d}
$^{36}Ar/N$	Ar	$1.1 \pm 0.4 \times 10^{-7c}$	–	3.67×10^{-2d}
Kr/N	Kr	$< 5 \times 10^{-9c}$	–	2.63×10^{-5d}
Xe/N	Xe	$< 5 \times 10^{-9c}$	–	2.58×10^{-6d}

[a]Gautier and Raulin (1997); [b]Lodders (2003); [c]Niemann et al. (2010); [d]Lodders et al. (2009); [e]Waite et al. (2009); [f]See Fig. 4; [g]Nixon et al. (2012); [h]Mandt et al. (2012); [i]Based on solar wind from Marty et al. (2011) and Jupiter's atmosphere from Owen et al. (2001); [j]Mandt et al. (2014); [k]Rousselot et al. (2014); [l]Shinnaka et al. (2014); [m]Based on comet measurements; [n]Vinatier et al. (2007); [o]Bockelée-Morvan et al. (2008).

was determined for Enceladus (Waite et al. 2009; Mousis et al. 2009a). The limited range of values for primordial D/H in methane is based on limits placed by the evolution of $^{12}C/^{13}C$ in Titan's methane (Mandt et al. 2009, 2012), while the upper limit for $^{14}N/^{15}N$ in Titan's nitrogen inventory is based on the maximum possible fractionation allowed due to escape processes (Mandt et al. 2014).

The primordial forms of carbon and nitrogen in the building blocks of Titan provide important clues to the conditions in which Titan formed. The ratio of CO to CH_4 in Titan's atmosphere (Gautier and Raulin 1997) is orders of magnitude lower than the ratio determined for the PSN (Lodders 2003), suggesting either that Titan's building blocks were initially enriched in CH_4 (Lunine 1989) or that the building blocks were rich in CO that was later converted to CH_4 through aqueous processes in the interior (Atreya et al. 2006). The nitrogen could have originated as N_2 or as NH_3 that was later converted to N_2 by photochemical processing in the early atmosphere (Atreya et al. 1978), impact shock heating (McKay et al. 1988), or endogenic processes (Glein et al. 2008).

The original thought that Titan formed in a warm subnebula that was dense enough to convert CO and N_2 from the PSN to CH_4 and NH_3 (Prinn and Fegley 1989) would have produced building blocks that preferentially trapped CH_4 and NH_3 for two reasons. The first reason is that CH_4 and NH_3 would have been more abundant than CO and N_2 in the Saturnian subnebula (Prinn and Fegley 1989). However, in a gas-starved disc (Canup and Ward 2002, 2006), CO and N_2 would be more abundant than CH_4 and NH_3 because the gas density is too low for conversion to be effective (Mousis et al. 2002; Alibert and Mousis 2007).

The second reason that the building blocks could preferentially trap CH_4 and NH_3 in the Saturnian subnebula is that the temperature could have been too high to trap CO and N_2, which are trapped at much lower temperatures than CH_4 and NH_3 (Bar-Nun et al., 1985, 1988). This possibility would still be valid in a gas-starved disc and is further supported by the measured ^{36}Ar abundance relative to N_2 of 1.1×10^{-7} (Niemann et al. 2010). Like CO and N_2, ^{36}Ar is trapped at very low temperatures. A measured abundance that is orders of magnitude lower than the solar abundance of ~ 0.04 suggests that Titan's building blocks may have formed in temperatures as high as 100 K (Owen 2008).

However, the $^{36}Ar/N$ ratio could also be explained by the removal of ^{36}Ar from the atmosphere by processes such as trapping in surface clathrates (e.g. Fortes and Stofan 2005; Osegovic and Max 2005; Mousis et al. 2011). Furthermore, if Titan had formed at such high temperatures, the D/H in water ice of Titan would be lower than that of OCCs (Owen 2008), which disagrees with the D/H measurement at Enceladus (Mousis et al. 2009a). As shown in Table 4, we can assume that Titan and Enceladus formed under similar conditions and that Titan's D/H in H_2O is also similar to that of OCCs (Mousis et al. 2009a). A D/H similar to that of OCCs requires very low formation temperatures, like those found in the PSN, and excludes the possibility of formation temperatures above 100 K. This suggests that N_2 and CO could have initially been trapped in the building blocks of Titan. Further constraints are, therefore, needed to determine the primordial forms of carbon and nitrogen in Titan's building blocks.

The primordial D/H in methane and $^{14}N/^{15}N$ in nitrogen provide important clues as to the origin of carbon and nitrogen on Titan. Based on the evolution of $^{12}C/^{13}C$ in methane, the primordial D/H in methane is distinctly lower than the water D/H, and must have been between 9.5×10^{-5} and 1.6×10^{-4} (Mandt et al. 2009, 2012; Nixon et al. 2012). If the methane in Titan's atmosphere had been formed by serpentinization reactions in the interior (e.g. Atreya et al. 2006) then the primordial D/H ratio in methane would be much higher due to the high D/H in Titan's water. A lower primordial D/H suggests, then, that Titan's methane is primordial (Mousis et al. 2009a). Furthermore, limitations on the evolution of $^{14}N/^{15}N$ in Titan's atmosphere place the primordial $^{14}N/^{15}N$ for Titan well within the range for primordial NH_3 in the PSN (Mandt et al. 2014), suggesting that Titan's nitrogen originated as NH_3. These constraints point to an origin of Titan's building blocks in the PSN (Mousis et al. 2009a; Mandt et al. 2014), but still require depletion in N_2, CO and possibly ^{36}Ar. This depletion can be explained by migration of the building blocks from the PSN into the warmer Saturnian subnebula (Canup and Ward 2002, 2006) where some partial devolatilization occurred (Mousis et al. 2009a; Mandt et al. 2014).

4.3 The Galilean Satellites

The four largest satellites of Jupiter—known as the Galilean satellites because of their initial discovery by Galileo in 1610—are Io, Europa, Ganymede and Callisto. Io, located closest to Jupiter, is subjected to significant tidal heating due to its orbital resonance with Jupiter and the other satellites, leading to extensive volcanic activity. The other Galilean satellites show evidence in their surface structures of thermal processing due to tidal heating, which decreases with increasing distance from Jupiter. Both Europa and Ganymede are found to be tectonically active due to tidal heating and have experienced extensive resurfacing. Callisto is subjected to the least amount of tidal heating and has a surface that is covered in impact craters suggesting little to no geological activity over its history (Estrada et al. 2009). The primordial composition of Io is difficult to constrain due to the extensive heating to which it has been subjected, but Europa, Ganymede and Callisto have all been able to retain substantial amounts of water ice throughout their history. It is unclear how much water ice Io may have initially contained.

As discussed earlier, the regular Jovian satellites are presumed to have formed in a circumplanetary nebula that was accreting material from the PSN (Canup and Ward 2002).

Models of the temperature and pressure profile initially suggested that water ice was accreted starting at or just beyond the orbit of Europa, the second closest Galilean satellite (e.g. Lunine and Stevenson 1982), and that the ice-to-rock ratios of Jupiter's satellites indicate their formation distance within the subnebula (e.g. Canup and Ward 2002; Mousis and Gautier 2004). However, the discovery that Amalthea, a small regular moon orbiting closer to Jupiter than Io, had a higher ice-to-rock ratio than the Galilean satellites (Anderson et al. 2005) challenged this assumption. This discovery brought about the possibility that all of the Jovian moons could have accreted significant amounts of ice, and that subsequent heating—either tidal heating or bombardment—could have driven off much of the water ice from Io and Europa (Estrada et al. 2009). However, this does not rule out the possibility that Amalthea formed later than the giant satellites or in a colder region of the subnebula and migrated to its current location (Anderson et al. 2005).

Other than the ice-to-rock ratio, the only available composition measurements for the Galilean satellites is surface composition, from which no isotopic or noble gas information is currently available. Galileo magnetic field measurements have found indications of subsurface oceans on Europa (Khurana et al. 1998; Kivelson et al. 1999), Ganymede (Kivelson et al. 2002) and Callisto (Khurana et al. 1998; Kivelson et al. 1999) where endogenic chemistry is likely to have influenced the composition of these internal oceans. The extensive resurfacing of Europa and Ganymede suggest exchange between an interior ocean and the surface. Therefore, it is difficult with the little information available on surface composition to determine the primordial forms of carbon and nitrogen in order to constrain formation scenarios for these satellites. Furthermore, the surfaces of all four satellites are subjected to extensive radiolitic chemistry caused by energetic particles in Jupiter's magnetosphere (Johnson et al. 2004) that modifies the surface composition. One thing to note is that the possible presence of liquid water below the surface suggests the possible presence of ammonia (e.g. Kargel 1992). However, it is difficult to know if either of this is primordial because NH_3 could be present as a result of endogenic chemistry. Therefore, little can be said about the formation processes of the Galilean satellites based on their composition beyond the fact that they must have accumulated significant amounts of water ice during their formation, as was predicted by Canup and Ward (2002, 2006).

4.4 Other Satellites of Interest

Data on the composition of the other outer planet satellites is very limited. CO_2 is frequently seen on the surface of outer planet moons, but many possibilities exist for *in situ* production of this molecule. It has been observed on the surface of Saturn's satellites Phoebe (Clark et al. 2005), Iapetus (Buratti et al. 2005) and Hyperion (Cruikshank et al. 2007) as well the Uranian moons Ariel, Umbriel and Titania (Grundy et al. 2006). As with Jupiter's icy moons, it is currently unclear if this CO_2 is primordial or produced through surface chemistry or endogenic processes so no conclusion can be made about the formation processes of these satellites based on this measurement.

Neptune's moon Triton has an orbit that is retrograde and highly inclined, suggesting that it is a captured moon presumed to have originated in the Kuiper Belt (Goldreich et al. 1989; Agnor and Hamilton 2006). Triton's surface is made up primarily ($\sim 55\ \%$) of N_2 ice with water and CO_2 ices making up the remaining $\sim 45\ \%$. Trace amounts of CH_4 and CO have also been detected on its surface (McKinnon and Randolph 2014) as well as in the atmosphere (Lellouch et al. 2010b). The source of the N_2 ices, whether N_2 or NH_3 in the PSN, can provide important information about the formation conditions of Triton and, by extension, Kuiper Belt objects. Nitrogen isotope ratio measurements, in particular, would be of high value.

Understanding of the composition of comets has played an important role in determining the origin of the Saturnian satellites based on the D/H measurement in water at Enceladus (Mousis et al. 2009b) and the $^{14}N/^{15}N$ in N_2 and D/H in CH_4 at Titan (Mousis et al. 2009a; Mandt et al. 2014). Further isotopic measurements in comets and in the outer planet satellites, as well as noble gas measurements in each of these solar system objects will play an important role in constraining the origins of other outer planet satellites as well as the giant planets.

5 The Terrestrial Planets

A cometary origin for the terrestrial oceans is often advocated in the literature. It makes sense to assume that terrestrial water originated from impacts of bodies like comets, which are otherwise known to be water-rich and which periodically cross the inner solar system. Modern versions of these scenarios arise from dynamical models advocating large-scale injection of small bodies formed beyond the snow line into the inner solar system following disruption of giant planet trajectories (Walsh et al. 2011). However, only Earth has a significant amount of water at its surface, while the other terrestrial bodies (Mercury, Venus, Moon, Mars) are dry or highly water-depleted. Such depletions could be the result of atmospheric escape processes and not necessarily the result of heterogeneous distribution of water contributors, but even the Earth contains a significantly small amount of water: the oceans are equivalent to about 260 ppm H_2O normalized to the mass of our planet, and the bulk Earth that includes deep reservoirs may contain no more than about 1000–3000 ppm (Marty 2012), which is still low compared to comets ($\sim$50 % water) or the other possible contributors that are wet asteroids.

Carbonaceous chondrites contain up to 15–20 % equivalent water in the form of hydrated minerals, which is not far from cometary concentrations. Thus a small contribution of either wet asteroidal material (e.g., 0.8–2 % carbonaceous chondrite-type; Marty 2012), or cometary matter (less than 0.5 %) could account for the water inventory of Earth. Alternatively the Earth and inner planets could have grown up from wet material from the beginning, but this possibility is at odds with the reduced character of Earth's building blocks that permitted formation of an iron-rich core. Thus current models advocate a proto-Earth (and inner planets) being built first from reduced, and supposedly dry, material, followed by increasing contributions of more oxidized contributors (e.g. Wänke and Dreibus 1988). This type of scenario is consistent with dynamical models in which growing inner planets become wet with time (e.g. Morbidelli 2002).

5.1 Asteroidal Versus Cometary Origin for Inner Planet Volatiles: Abundances

Two types of very volatile elements can be used as diagnostics of the origin(s) of the inner planet atmospheres. Here atmosphere means the inventory of volatile elements at the surface of planets, that is, atmosphere stricto sensu plus the oceans and sediments for Earth. On the one hand, the noble gas elemental and isotopic compositions are key tracers of origin and physical processes due to their chemical inertness. On the other hand, some of the stable isotope ratios of light elements, specifically those of H and N, show dramatic variations among solar system objects and reservoirs that make them unique cosmochemical tracers.

The elemental abundances of volatile elements (C, N, O and noble gases) elements and key isotope ratios for inner planet atmospheres, carbonaceous chondrites (thought to represent wet asteroids) and comets are given in Table 5 and displayed in Figs. 10 and 11.

Table 5 Elemental abundances of volatile elements and key isotope ratios in the atmospheres of Earth, Venus and Mars, compared to those of chondrites, of model cometary matter and of the PSN. The inner planet abundances are the surface inventories divided by the mass of the respective planets. The Earth atmosphere represents the sum of the oceans, sediments and atmosphere stricto sensu. Cometary abundances are those in amorphous ice for two end-member formation temperatures of 30 K and 70 K, respectively

Isotope	Earth atm.[a]	Venus atm.[a,b]	Mars atm.[a,c]	Chondritic[d]	Cometary @ 30 K[e]	Cometary @ 70 K[e]	PSN[f]
^{12}C	1.29×10^{-6}	2.17×10^{-6}	9.25×10^{-10}	2.28×10^{-3}	–	–	3.25×10^{-4}
^{14}N	5.98×10^{-8}	1.57×10^{-7}	5.21×10^{-11}	1.44×10^{-4}	–	–	3.36×10^{-3}
^{1}H	1.48×10^{-5}	–	–	6.14×10^{-3}	5.40×10^{-2}	5.40×10^{-2}	–
^{22}Ne	2.27×10^{-14}	1.07×10^{-12}	2.58×10^{-16}	1.89×10^{-12}	1.62×10^{-5}	2.28×10^{-9}	5.74×10^{-6}
^{36}Ar	9.58×10^{-13}	6.97×10^{-11}	6.00×10^{-15}	2.89×10^{-11}	1.30×10^{-4}	4.64×10^{-8}	2.49×10^{-6}
^{84}Kr	1.98×10^{-14}	5.60×10^{-14}	2.10×10^{-16}	3.61×10^{-13}	3.92×10^{-8}	1.03×10^{-8}	6.95×10^{-10}
^{130}Xe	1.08×10^{-16}	6.85×10^{-16}	1.60×10^{-18}	1.61×10^{-13}	3.13×10^{-10}	8.97×10^{-11}	6.21×10^{-12}
Isotope ratio							
$D/H \times 10^{-4}$	1.56	160 ± 20	8.6 ± 4.0[g]	1.7 ± 0.4	1.5–5.3[h]		0.21
$^{14}N/^{15}N$	272	270 ± 51	173 ± 11[g]	262 ± 29	130–160[i]		440[j]
$^{20}Ne/^{22}Ne$	9.80	–	~ 8(?)[g]	~ 8.5(Ne-A)	~ 10[k]		13.4–13.8[l]

[a]Pepin (1991); [b]Hoffman et al. (1980); [c]Bogard et al. (2001); [d]from Orgueil and Murchison carbonaceous chondrite data in Marty (2012) and references therein; [e]as computed by Dauphas (2003) with data from Bar-Nun and Owen (1998); [f]Lodders et al. (2009); [g]Garrison et al. (1998); [h]Bockelée-Morvan et al. (2004); Hartogh et al. (2011a); and Altwegg et al. (2015); [i]Rousselot et al. (2014) and Shinnaka et al. (2014); [j]Marty et al. (2011); [k]Marty et al. (2008) from Stardust cometary material; [l]Grimberg et al. (2006)

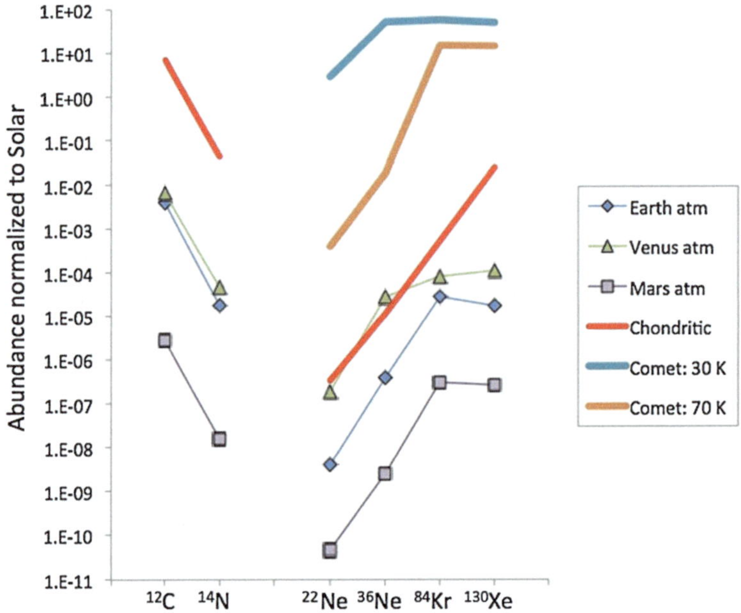

Fig. 10 Volatile element abundances (mol/g) normalized to Solar (Lodders 2003) for terrestrial planet atmospheres compared to chondritic and cometary abundances. See Table 5 references for sources of data. Volatile abundances of inner planet atmospheres fit better with chondritic than presumed cometary patterns, keeping in mind that for comets' real abundances are not yet measured. Heavy noble gases (e.g. Xe) appear nonetheless depleted relative to chondritic, indicating that a chondritic origin for planetary noble gases do not fully account for observed patterns (see the isotope subsection)

The abundances of volatile elements are reasonably well known in the Sun, our best representative for the PSN, as the Sun concentrates more than 99 % of the solar system mass at present. Carbonaceous chondrites are presumably representative of volatiles in wet asteroids from the outer part of the asteroid belt. The latter are potential candidates for the contributors of volatile elements to inner planets, which require addition of oxidized material to make the present-day chemistries of Earth and Mars (e.g., oxygen fugacity; Wänke and Dreibus 1988). It has been long recognized that the abundance patterns of noble gases in the terrestrial planets (except He which is not retained in inner planet atmospheres) differ from the solar abundance and resemble that of primitive meteorites (see Fig. 9). For this reason, these abundance patterns have been named "planetary" (e.g., Mazor et al. 1970) in contrast to "solar". This denomination is, however, misleading to some point since some of the key isotope ratios of planetary atmospheres (for instance those of Xe) differ significantly from those of primitive meteorites.

Nevertheless, the abundance ratios of noble gases, and by extension of C and N in the atmospheres of Earth, Venus and Mars are quite consistent each other and resemble indeed those of carbonaceous chondrites (Fig. 10). A notable exception is Xe, which is under-abundant in inner planet atmospheres by one order of magnitude relative to chondritic and other noble gases (e.g., Kr), an important difference to which we shall return later.

However, it is difficult to evaluate the possibility of a cometary origin based on noble gases simply because the composition of noble gases in comets has not yet been measured.

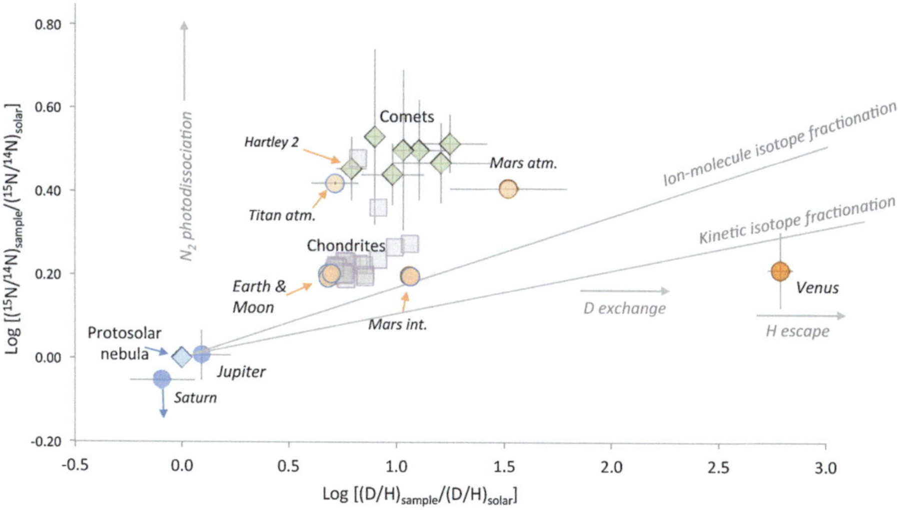

Fig. 11 Co-variations of the H and N isotope ratios among solar system reservoirs and objects. The D/H and $^{15}N/^{14}N$ ratios are normalized to the PSN values ($[2.1 \pm 0.5] \times 10^{-5}$ and $[2.27 \times 0.03] \times 10^{-3}$, respectively; after Aléon 2010). Data are consistent with different processes of isotope fractionation, e.g., ion-molecule isotope fractionation during low temperature isotope exchange between organics and the protosolar gas, kinetic isotope fractionation proportional to the square root of mass, or self-shielding enhancement of the isotope fractionation through photodissociation of protosolar N_2 by UV light (Clayton 2002; Chakraborty et al. 2013). The Earth shares H and N isotope signatures with bulk chondrites, whereas comets are richer in D and ^{15}N, thus suggesting an asteroidal, rather than cometary, origin for terrestrial volatiles (Alexander et al. 2012; Marty 2012). The Venusian atmosphere is extremely depleted in water and rich in D as a consequence of photodissociation of H_2O and subsequent loss (Grinspoon 1993). The enrichment in both D and ^{15}N of the Martian atmosphere is attributed to atmospheric escape processes. Sources—Moon: Füri et al. (2014), Kerridge et al. (1991); Venus: Grinspoon (1993), Hoffman et al. (1980); Mars: Leshin et al. (2013), Leshin (2000); chondrites: Deloule et al. (1998), Kerridge (1985), Robert (2003); Saturn: Fletcher et al. (2014), Macy and Smith (1978); Jupiter: Mahaffy et al. (1998), Owen et al. (2001); Titan atmosphere: Niemann et al. (2010); comets: Bockelée-Morvan et al. (2008); Ceccarelli et al. (2014), Hartogh et al. (2011a), Manfroid et al. (2009), Rousselot et al. (2014)

Attempts to measure noble gases by spectroscopy have failed, and it is hoped that mass spectrometers onboard the Rosetta spacecraft will have the sensitivity to detect and, hopefully, quantify noble gases (Balsiger et al. 2007). In the absence of real data, it has been proposed that noble gases in comets could have been trapped from the PSN gas during the growth of amorphous ice, so that results of experiments aimed at measuring the trapping efficiency of noble gases in amorphous ice (Bar-Nun and Owen 1998; Bar-Nun et al. 2007) have been used to estimate the noble gas content of comets (Dauphas 2003; Owen and Barnun 1995).

The trapping efficiency of noble gases in ice depends on ice formation temperature and on the noble gas mass, and in Fig. 10 we represent the cometary abundance patterns for two end-member temperatures of comet formation, using data from Bar-Nun and Owen (1998), Bar-Nun et al. (2007) and Dauphas (2003). In such a format, the absolute abundances are not as important as the relative ones, and one could conceivably find trapping conditions able to satisfy the planetary abundance pattern. Notably, a dual origin of the terrestrial atmosphere has been proposed by Dauphas (2003) in which the Earth first had a nebular gas atmosphere which was fractionated during escape, and then received its load of Xe-depleted cometary gases, explaining in this way the Xe deficiency of planetary atmospheres.

5.2 Asteroidal Versus Cometary Origin for Inner Planet Volatiles: H and N Isotopes

We know a little bit more about the isotopic composition of comets, thanks to spectroscopic measurements of the D/H and N isotopes (Bockelée-Morvan et al. 2008; Ceccarelli et al. 2014; Hartogh et al. 2011a; Manfroid et al. 2009; Rousselot et al. 2014; Shinnaka et al. 2014; Altwegg et al. 2015). As described in Sect. 2, the D/H and $^{14}N/^{15}N$ ratios display dramatic variations among solar system objects and reservoirs, thus allowing one to derive information on origins and relationships between reservoirs (Aléon 2010; Marty 2012; Marty et al. 2011). Figure 11 represents the co-variations of H and N isotope ratios for planetary atmospheres, comets, and meteorites normalized to the PSN values (see Fig. 11 caption for data sources). All reservoirs but the giant planets are enriched in D and ^{15}N compared to the PSN.

The causes of such enrichments are likely to result from extensive isotope fractionation during either ion-molecule reactions at low temperature, isotope exchange with D-rich molecules for H, isotope-selective photolysis, self shielding, or a combination of several of these processes. Notably, the Earth, the Moon and the interior of Mars and Venus all have comparable $^{15}N/^{14}N$ ratios that are also within the range of most values seen in primitive meteorites. The atmospheres of Titan and of Mars are exceptions (Mandt et al. 2015): for the Titan atmosphere, the enrichment in ^{15}N in N_2 (Niemann et al. 2010) is attributed to contribution of ^{15}N-rich material akin of cometary amines and nitrides (Mandt et al. 2014), whereas the enrichment in ^{15}N of the atmosphere of Mars is regarded as resulting from isotope-selective non-thermal escape processes (e.g., Bogard et al. 2001). In contrast to inner solar system bodies, all comets analyzed so far present strong enrichments (factors of about 2) in ^{15}N in HCN (Bockelée-Morvan et al. 2008; Manfroid et al. 2009) and NH_3 (Rousselot et al. 2014; Shinnaka et al. 2014). Although N_2 has been detected by ROSINA DFMS in 67P/CG, its abundance was found to be depleted by a factor of $\sim 25.4 \pm 8.9$ relative to CO (Rubin et al. 2015) and no isotope ratio has been measured. It is unclear if the depletion of N_2 in comets is primordial or due to loss after accretion (Mousis et al. 2012), so it is difficult to conclude based on this early measurement if comets contributed significantly to the Earth's nitrogen.

A similar, but somewhat more blurred, situation arises from D/H systematics. The Earth and the Moon and many meteorites share a common range of values about a factor of 6 enriched in D relative to the PSN. Venus is extremely rich in D, suggesting water loss through photodissociation of H_2O and isotope-selective hydrogen escape (Grinspoon 1993). The Martian atmosphere is also rich in D when compared to Earth and chondrites, an enrichment equally attributed to selective (charge-related) atmospheric escape as in the case of N (ibid.). Thus it appears that, based on these isotope H and N variations, asteroids fit better than comets for the origin of inner planet volatiles.

Recently, terrestrial-like D/H ratios have been reported for two comets presumed to be JFCs (Ceccarelli et al. 2014; Hartogh et al. 2011a), thus revitalizing the theory for a cometary origin for terrestrial oceans. However, the measurement of the D/H ratio of comet 67P/CG counters this argument (Altwegg et al. 2015). It is important to note that comet Hartley 2, for which an ocean-like D/H ratio was measured (Hartogh et al. 2011a), is nevertheless rich in ^{15}N compared to Earth, thus contradicting a cometary origin for terrestrial volatiles. The story may, however, evolve because we may not know fully the N isotope composition of comets. Indeed only CN, HCN and NH_2 (presumably from ammonia) could be analyzed by spectroscopy, and it is possible that comets could also contain N_2 from the PSN. If so, mixing of ^{15}N-poor PSN nitrogen with ^{15}N-rich nitriles and amines could conceivably yield

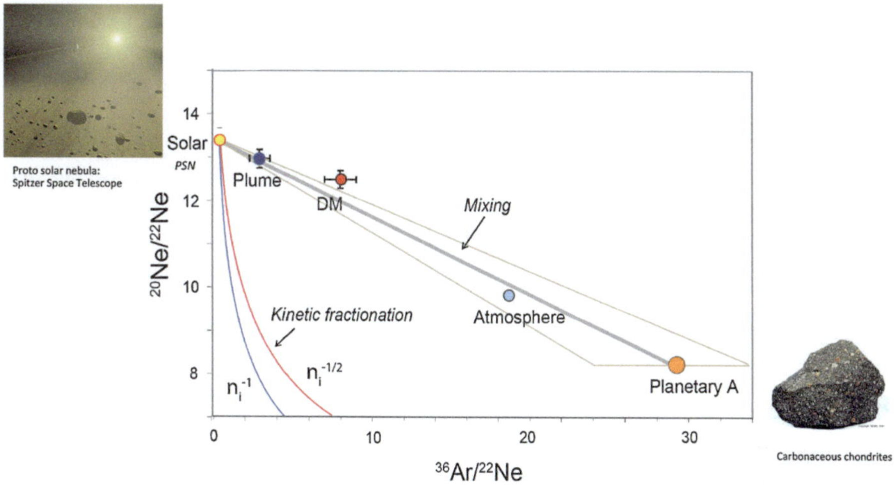

Fig. 12 Co-variations of the $^{20}Ne/^{22}Ne$ isotopic ratio vs. the $^{36}Ar/^{22}Ne$ ratio for terrestrial reservoirs compared to the PSN (Heber et al. 2012) and the carbonaceous chondrite (Planetary-A component as defined by Mazor et al. 1970) end-members. In this format, two end-member mixing results in a straight line connecting the end-member data points. The terrestrial shallow depleted mantle (DM—Moreira et al. 1998), the deep mantle, represented by mantle plume (plume) material (Mukhopadhyay 2012; Yokochi and Marty 2004), and the atmosphere fit remarkably well a mixing relationship between a PSN-like component, presumably retained deep in the Earth's interior, and a surficial reservoir—the atmosphere sensu largo—that was contributed by a wet asteroid end-member of carbonaceous chondrite composition. This diagram also illustrates the fact that the terrestrial atmosphere did not experience isotope-element fractionating escape processes, represented here by mass-dependent trajectories (*blue* and *red curves*). If atmospheric escape took place, it should have been non-fractionating, of the blown-off type (adapted from Marty 2012)

inner solar system-like N isotope compositions, provided a good mix in the cocktail of 1/3–2/3 approximately.

5.3 Asteroidal Versus Cometary Origin for Inner Planet Volatiles: Other Isotope Systems

Neon has three isotopes—^{20}Ne, ^{21}Ne, ^{22}Ne—which show significantly large variations in the solar system between a PSN $^{20}Ne/^{22}Ne$ end-member of 13.8 for the solar wind (possibly 13.4 for the PSN; Heber et al. 2012) and about 8.5 for the carbonaceous chondrite end-member (i.e., Ne-A; Mazor et al. 1970).

Solar-like neon has been identified in the terrestrial mantle (Mukhopadhyay 2012; Yokochi and Marty 2004), suggesting that the proto-Earth grew up in the presence of PSN gas, possibly under the form of a PSN, H_2-rich atmosphere that equilibrated with molten silicates of a magma ocean. Later on, this primary atmosphere would have been blown off and a secondary atmosphere would have developed from contribution of wet bodies. Ne isotope and Ar-Ne abundances fit very well into this scenario and further indicate carbonaceous chondrite-type material as the best progenitor for the secondary atmosphere (and oceans—Fig. 12). The important relationship depicted in Fig. 12 indicates that terrestrial noble gases are fully consistent with mixing between a deep component of PSN origin and a more surficial component, mostly in the atmosphere, that was contributed by material having a composition akin of carbonaceous chondrites, therefore presumably wet asteroidal matter.

A cometary contribution would not fit well into this scheme, but cannot be totally excluded since one does not know the noble gas composition of comets. The only measurement of noble gases in cometary material is from the analysis of grains returned by the Stardust mission (Brownlee et al. 2006), which showed the occurrence of extraterrestrial He and Ne (Marty et al. 2008). Only fragments of one grain that exploded in the collecting medium (aerogel) could be analyzed so far, which yielded a $^{3}He/^{4}He$ ratio intermediate between the post-deuterium burning composition of the solar wind, and that of the Jupiter, probably representing the PSN composition. It is therefore possible that comets have taken snapshots of the early evolution of the Sun in the composition of noble gases they trapped at different periods of time.

The Stardust Ne isotopic composition is surprisingly different from the solar composition, and instead close to that of the terrestrial atmosphere. It is also indistinguishable, within errors, from the composition of Phase Q Ne, an ubiquitous component found in meteorites and thought to be associated with primitive organic matter (Marty et al. 2008). These unique results suggest that comets may also contain noble gases trapped in their organics and not only frozen in ice. Furthermore, the concentration of Ne in the Stardust material was found to be extremely large, of the order of 5×10^{-7} mol/g. With such concentration, the contribution of comets to atmospheric Ne could have been significant, with roughly the required $^{20}Ne/^{22}Ne$ ratio.

For the purpose of illustration, the terrestrial late heavy bombardment (TLHB) around 3.8 Ga ago (about 1×10^{23} g, as scaled to the lunar cratering record) could have been made of approximately 50 % Kuiper Belt material, according to the Nice model (Gomes et al. 2005). With such a concentration, about 5×10^{16} moles of neon could have been contributed by the TLHB, one order of magnitude more than the atmospheric inventory of neon (3.2×10^{15} moles). Marty and Meibom (2007) have argued that the TLHB could not be made of 50 % cometary material, based on the noble gas content of comets (estimated from amorphous ice experiments, see above). They proposed an upper limit of about 0.5 % for cometary matter in the TLHB, the rest consisting of asteroidal matter. With such contribution, about 10 % atmospheric Ne could have been contributed by cometary matter.

Another intriguing isotope system relevant to the present discussion is that of the nine isotopes of xenon, the heaviest stable noble gas. Besides several isotopes produced by extinct and extant radioactivities which have their own interest for geochronological purpose, the stable isotopes of xenon in the terrestrial atmosphere as well as in the Martian atmosphere (unfortunately, Venus is not measured) are mass-fractionated relative to either solar, or meteoritic Xe. The extent of fractionation is so large (3–4 % per amu) for such a heavy element that no known kinetic mass fractionation can account for this difference. Together with the xenon elemental depletion relative to the other noble gases, this unique isotope fractionation could in fact provide the signature of cometary matter that would carry an exotic Xe component from the edge of the solar system not seen in its inner region.

This possibility is not without problems, since it requires the existence of a component not seen elsewhere in the solar system, and which must be different from Xe in the Sun. Instead several models call for early processing of solar-like Xe in the terrestrial atmosphere. Pepin (1991, 2006) proposed that terrestrial Xe was mass fractionated from an initial composition resembling the solar one during episodes of magma ocean degassing and atmospheric escape. One problem with such a possibility is that it would require the same suite of events occurring on Mars (whose atmosphere is equally fractionated for Xe isotopes), which is difficult to conceive for such drastically different planets.

Dauphas (2003) proposed a dual origin for the terrestrial atmosphere, in which a solar composition was fractionated during escape, leaving only isotopically fractionated xenon, and then contributed to by cometary gases depleted in xenon, leaving thus the fractionated Xe signature while providing the isotope compositions of the other noble gases. This attractive model may also suffer from the comparison with Mars—that is, requiring a similar suite of events for a planet very different from the Earth in terms of heliocentric distance and size.

More recently, Pujol et al. (2011) found that xenon trapped in Archean (3.5 Ga-old) sedimentary rocks is isotopically intermediate between chondritic/solar, and modern atmospheric Xe. They attributed this composition to the signature of the ancient atmosphere at that time. They proposed that the Xe isotope fractionation and Xe underabundance was due to an atmospheric escape process specific to Xe and not affecting other noble gases. This escape was progressive through time, and was not yet completed 3.5 Ga ago. The specific Xe atmospheric escape would be related to its electronic structure together with a higher far UV light flux in the distant past that would have enhanced selective loss of xenon from the atmosphere through time. If this were the case, the xenon "paradox" would no longer be a question of the nature of the extraterrestrial source (cometary versus asteroidal), but an aeronomic problem specific to Earth and Mars. All these models cannot be fully tested at present because we lack precise measurements of noble gases in comets.

5.4 Conclusion

With all of the data and information at hand, it appears that contribution from asteroidal material of the type of that sampled by carbonaceous chondrites can better explain the composition of the inner planet atmospheres than contribution of cometary matter. This appears to be the case for the abundances of noble gases, and for the isotopes of hydrogen and nitrogen. However, the situation is severely biased by the lack of cometary data: no heavy noble gases have been measured so far in these objects, and measurements of elemental and stable isotope compositions have considerable uncertainties that are inherent to the spectroscopic methods. Many of the considerations above are based on a very restricted number of data and on considerable assumptions, and they certainly highlight the need to have more data on the volatile content and isotopic composition of cometary volatiles.

6 Pluto and Kuiper Belt Objects

Pluto's mean density shows that it is poor in ice and suggests that it formed in a water-poor region of the solar nebula (Lunine 1989). It is also likely to have experienced further loss of volatiles by the impact formation of Charon (McKinnon and Mueller 1988). The surface of Pluto consists of a spatially heterogeneous mixture of N_2, CH_4, CO and C_2H_6 ices (Cruikshank et al. 2015). Molecular nitrogen is the most abundant ice on the surface and is presumed to be the primary constituent in Pluto's tenuous atmosphere (Owen et al. 1993) where CH_4 and CO have been detected (Lellouch et al. 2011). N_2 ices have also been detected on the surface of several Kuiper Belt objects, including Sedna (Barucci et al. 2010) and Quaoar (Dalle Ore et al. 2009).

The source of nitrogen for the surface ices and atmosphere of Pluto can provide important information about its formation conditions, particularly about the temperature and composition of the region of the PSN in which it formed. Like Triton, Pluto's nitrogen originated either as N_2 or NH_3 in the PSN. The abundance of N_2 is believed to have been greater in the PSN than NH_3 (Lewis and Prinn 1980), but would only have been accreted by Pluto if it formed at temperatures less than 38 K (Bar-Nun et al. 1985, 1988).

The formation conditions of comets provide important information relevant to the formation of Pluto. Based on measurements currently available, comets are found in general to be deficient in N_2 relative to NH_3. This means that temperature conditions in the region where comets (and Pluto) formed could have been too warm for N_2 ice to form, that they did not retain N_2 beyond their first pass through the solar system (Owen et al. 1993), or that detection methods for N_2 are too limited to determine its abundance in comets. The recent detection of N_2 in 67P/CG has been interpreted to mean that this comet formed at temperatures less than 30 K (Rubin et al. 2015), although the N_2/CO ratio is compatible with formation temperatures as high as 56 K if the nucleus agglomerated from ice grains made of clathrates (Mousis et al. 2012; Rubin et al. 2015). As with Titan (Mandt et al. 2014), if formation temperatures were low enough the source of Pluto's nitrogen was N_2 in the PSN, but if temperatures were higher than 38 K then the source of nitrogen for Pluto's surface and atmosphere would have been NH_3 in the PSN that was somehow converted to N_2. Nitrogen isotope ratios could provide important constraints on the origin on Pluto's nitrogen if the escape history of Pluto's atmosphere can be constrained as has been done with Titan (Mandt et al. 2014). Determining the origin of Pluto's nitrogen would have important implications for understanding the origin of Kuiper Belt objects and Neptune's moon Triton.

7 Discussion and Conclusions

We review here the state of knowledge on the formation of the planets, moons and small bodies in the solar system and the contributions that comets have made to this understanding. There are many limitations to this understanding, so we also highlight measurements that are needed to address these limitations. The European Space Agency (ESA) Rosetta mission (Glassmeier et al. 2007) is currently in orbit around comet 67P/CG and the NASA New Horizons spacecraft (Stern 2009) will fly by Pluto in July 2015. Both of these missions will take measurements that will help address these limitations as well as the upcoming ESA JUpiter ICy moons Explorer (JUICE) mission and a NASA Europa mission.

7.1 Current State of Knowledge and Limitations

The composition of giant planet atmospheres demonstrates that heavy elements are enriched in the giant planets. This is interpreted to mean that the building blocks of the giant planets were a mixture of rocks and crystalline ices (e.g. Gautier et al. 2001). The exact nature of these ices is still not determined, although most recent results seem to argue for full clathration of volatiles to explain the enrichment of volatiles in the giant planet atmospheres and the $^{14}N/^{15}N$ in Jupiter's (Mousis et al. 2009c, 2012) and Saturn's atmospheres (Mousis et al. 2014a). The formation scenarios of Uranus and Neptune are not as well constrained and could also be explained by formation in a region between the CO and N_2 ice lines (Ali-Dib et al. 2014). Questions remain as to the later contribution of comets to the volatile inventories of the giant planet atmospheres. The example of the SL9 impact with Jupiter demonstrated that CO and HCN produced by shock chemistry remain in a giant planet atmosphere long after a comet impact and can be used as tracers for recent comet contributions to the volatile inventory of giant planets.

The giant planet satellites are divided into two categories based on their origin. Captured satellites could be similar in composition to comets because they are likely to have formed in conditions similar to the formation conditions of Kuiper Belt and Oort Cloud objects while regular satellites formed as a byproduct of giant planet formation. However, the fact that comets appear to be deficient in nitrogen while Pluto and Triton, both presumed to be Kuiper Belt objects, contain significant nitrogen is a puzzle yet to be resolved. Current understanding of regular satellites is that they formed in a gas-starved disk around their planet and that they are expected to have large inventories of water ice and other volatiles (Canup and Ward 2002, 2006). Their composition is an important tool for providing further constraints on their origin. The D/H ratio in water in Enceladus demonstrates that the building blocks for Saturn's regular satellites must have formed in the PSN where conditions were cold enough to provide a D/H similar to that of OCCs and to trap N_2, CO and ^{36}Ar (Mousis et al. 2009b). The deficiency of CO and ^{36}Ar in Titan's atmosphere, as well as a $^{14}N/^{15}N$ in N_2 that indicates its origin as NH_3 in the PSN suggests that the building blocks for Saturn's regular satellites partially devolatilized prior to formation of the moons (Mousis et al. 2009a; Mandt et al. 2014). Knowledge about the composition of Jupiter's moons is limited primarily to their ice-to-rock ratio. This parameter is either an indication of their formation distance in Jupiter's subnebula (e.g. Canup and Ward 2002; Mousis and Gautier 2004) or they initially had high ice-to-rock ratios, and have lost part or all of their ice inventory as a result of tidal heating (Estrada et al. 2009). Due to extensive thermal processing and radiolitic chemistry on the surface, it is difficult to provide any further constraints on the origin of these moons based on what is currently known about their composition.

Even less is known about the moons of Uranus and Neptune beyond the fact that Triton is a captured moon with extensive surface ices composed primarily of N_2 with trace amounts of CH_4 and CO. This observation combined with similar observations of the atmosphere and surface composition of Pluto suggest that Kuiper Belt objects may have formed in conditions cold enough to trap and retain N_2 and CO, but more data are needed to confirm this.

The data available to evaluate volatile origins of the terrestrial planets point toward asteroidal material of the type of that sampled by carbonaceous chondrites to explain the composition of the inner planet atmospheres as opposed to contribution of cometary matter. This agrees with the abundances of noble gases, as well as the isotopes of hydrogen and nitrogen.

7.2 Measurements that Are Needed

Now that we have reviewed measurements that have implications for understanding the formation and evolution of the solar system, areas where more measurements are needed clearly stand out.

First, there are only ten measurements of D/H in water in comets currently available. Two of the three JFCs that were measured have D/H values in water in the range of D/H in water for Earth, while the D/H in water for 67P/CG and seven OCCs is a factor of 2–4 greater than the terrestrial ratio. This statistical sampling is far too low for any significant conclusions to be made. Further measurements of D/H in comets is desperately needed to enhance understanding of solar system formation processes and the role of comets in contributing volatiles to planetary atmospheres. The OPR measurements show trends that are also relevant, but only a few limits and three measurements are available for JFCs. More information about JFCs and more measurements in OCCs are needed.

Future work on carbon isotope ratios should include expanding number of C_2 measurements and inclusion of other carbon bearing species, particularly among the photochemical parents of the more commonly studied visible radicals. Combined with increasing the measurement and modeling precision of the existing methods and narrowing the selection of potential parents it will be possible to determine if there are any potential trends with heliocentric distance and to evaluate if a subtle shift toward higher values in comets relative to Earth is significant. Future work on nitrogen isotope ratios will require the acquisition of new HCN radio measurements in coordination with NUV CN, and expanding measurements of $^{14}N/^{15}N$ in cometary NH_3.

The oxygen and sulfur isotopic measurements in comets are not sufficient to provide any assistance in understanding formation and evolution of the solar system. Although the general trend for cometary $^{16}O/^{18}O$ ratios is consistent with the range described between chondritic material and asteroidal water, more effective classification will require not only more measurements, but also more precise ones that include ^{17}O. A larger sample size of sulfur isotope ratios is needed that targets multiple isotopes from several comets of different classes and heliocentric distances in order for these ratios to be compared to other solar system measurements.

Noble gas measurements have only been made in the atmospheres of Jupiter, Titan and the terrestrial planets. In order to better understand the formation processes in the solar system, the noble gas abundances need to be measured in several comets as well as in the atmospheres of Saturn, Uranus and Neptune and other satellites of giant planets.

To continue evaluation of the contribution of comets to giant planet atmospheric composition, further observations of CO and HCN in giant planet atmospheres are needed. This continuing study will not only benefit the Solar System atmosphere field but also the growing field of exoplanet atmosphere characterization (Turrini et al. 2014).

The origin of Kuiper Belt objects can be constrained through nitrogen isotope ratio measurements. Measuring the $^{14}N/^{15}N$ in the atmospheres of Triton, Pluto and several Kuiper Belt objects would significantly improve understanding of their formation conditions. A very high isotope ratio would suggest origin in very cold conditions, while a very low ratio would indicate origin as NH_3 (assuming that escape and photochemistry have not significantly changed the ratio over time).

Evaluation of the terrestrial planet atmospheres is severely biased by the lack of cometary data: no heavy noble gases have been measured so far in these objects, and measurements on elemental and stable isotope compositions have considerable uncertainties that are inherent to the spectroscopic methods. The conclusions made here are based on a very restricted number of data and on considerable assumptions, and they certainly highlight the need to have more data on the volatile content and isotopic composition of cometary volatiles.

7.3 Expected Contributions of Rosetta, New Horizons and Other Upcoming Missions

The Rosetta mission has now measured the D/H in another JFC and shown that the trend of a different D/H in different comet families is complicated. D/H ratios in CH_4 and HCN and $^{14}N/^{15}N$ in HCN and NH_3 are also anticipated along with the noble gas abundances. These measurements will not only add to the statistics on comets but also provide new measurements that may change perspectives on solar system formation. This mission is of high value and more missions like Rosetta are needed in the future.

The atmosphere of Saturn will continue to be studied by the Cassini mission. When this mission reaches the end of its operational life, the spacecraft will deorbit into the atmosphere of Saturn. This will provide a prime opportunity to measure *in situ* the noble gas abundances in Saturn's atmosphere as well as isotope ratios of some of the major atmospheric constituents.

The JUICE mission and an upcoming NASA-sponsored Europa mission may be able to provide improved composition measurements for the Galilean moons that could include isotopic composition, while the ALICE UV spectrometer (Stern et al. 2008) on the New

Horizons mission may be able to measure the $^{14}N/^{15}N$ in Pluto's atmosphere if the ratio is ≤ 330 (Jessup et al. 2013).

Although much has been learned by observations of comets and solar system bodies, a great deal of work remains.

Acknowledgements K.E.M. acknowledges support from JPL Subcontract 1345493 and NASA grant NNX13AQ99G. O.M. acknowledges support from CNES. The work contributed by O.M. was carried out thanks to the support of the A∗MIDEX project (no. ANR-11-IDEX-0001-02) funded by the "Investissements d'Avenir" French Government program, managed by the French National Research Agency (ANR). The work contributed by K.W. was carried out at the Jet Propulsion Laboratory/California Institute of Technology, under contract with the National Aeronautics and Space Administration with partial support from the NASA/Origins of Solar Systems program. B.M. acknowledges support from the European Research Council (grant agreement number 267255).

References

C.B. Agnor, D.P. Hamilton, Neptune's capture of its moon Triton in a binary-planet gravitational encounter. Nature **441**, 192–194 (2006)

M.F. A'Hearn, S. Hoban, P.V. Birch, C. Bowers, R. Martin, D.A. Klinglesmith III, Cyanogen jets in comet Halley. Nature **324**, 649–651 (1986)

M.F. A'Hearn, R.C. Millis, D.O. Schleicher, D.J. Osip, P.V. Birch, The ensemble properties of comets: Results from narrowband photometry of 85 comets, 1976–1992. Icarus **118**, 223–270 (1995)

T. Albertsson, D. Semenov, Th. Henning, Chemodynamical deuterium fractionation in the early solar nebula: The origin of water on earth and in asteroids and comets. Astrophys. J. **784**(1), 39 (2014)

J. Aléon, Multiple origins of nitrogen isotopic anomalies in meteorites and comets. Astrophys. J. **722**, 1342–1351 (2010)

C.M.O. Alexander, R. Bowden, M.L. Fogel, K.T. Howard, C.D.K. Herd, L.R. Nittler, The provenances of asteroids, and their contributions to the volatile inventories of the terrestrial planets. Science **337**, 721-3 (2012)

M. Ali-Dib, O. Mousis, J.-M. Petit, J.I. Lunine, The measured compositions of Uranus and Neptune from their formation on the CO ice line. Astrophys. J. **793**, 9 (2014)

Y. Alibert, O. Mousis, Formation of Titan in Saturn's subnebula: Constraints from Huygens probe measurements. Astron. Astrophys. **465**, 1051–1060 (2007)

Y. Alibert, O. Mousis, W. Benz, On the volatile enrichments and composition of Jupiter. Astrophys. J. **622**, L145–L148 (2005a)

Y. Alibert, O. Mousis, C. Mordasini, W. Benz, New Jupiter and Saturn formation models meet observations. Astrophys. J. **626**, L57–L60 (2005b)

K. Altwegg, Sulfur in the coma of comet Halley from in situ measurements. Ph.D. thesis, University of Bern (1996)

K. Altwegg, D. Bockelée-Morvan, Isotopic abundances in comets. Space Sci. Rev. **106**, 139–154 (2003)

K. Altwegg et al., 67P/Churyumov-Gerasimenko, a Jupiter family comet with a high D/H ratio. Science **347**, 1261952 (2015)

E. Anders, N. Grevesse, Abundances of the elements—Meteoritic and solar. Geochim. Cosmochim. Acta **53**, 197–214 (1989)

J.D. Anderson et al., Amalthea's density is less than that of water. Science **308**, 1291–1293 (2005)

C. Arpigny et al., Anomalous nitrogen isotope ratio in comets. Science **301**, 1522–1525 (2003)

M. Asplund, N. Grevesse, A.J. Sauval, P. Scott, The chemical composition of the Sun. Annu. Rev. Astron. Astrophys. **47**, 481–522 (2009)

S.K. Atreya, T.M. Donahue, W.R. Kuhn, Evolution of a nitrogen atmosphere on Titan. Science **201**, 611–613 (1978)

S.K. Atreya, S.G. Edgington, D. Gautier, T.C. Owen, Origin of the major planet atmospheres: Clues from trace species. Earth Moon Planets **67**, 71–75 (1995)

S.K. Atreya, P.R. Mahaffy, H.B. Niemann, M.H. Wong, T.C. Owen, Composition and origin of the atmosphere of Jupiter—An update, and implications for the extrasolar giant planets. Planet. Space Sci. **51**, 105–112 (2003)
S.K. Atreya et al., Titan's methane cycle. Planet. Space Sci. **54**, 1177–1187 (2006)
H. Balsiger, K. Altwegg, J. Geiss, D/H and O-18/O-16 ratio in the hydronium ion and in neutral water from in situ ion measurements in comet Halley. J. Geophys. Res. **100**, 5827–5934 (1995)
H. Balsiger et al., Rosina—Rosetta orbiter spectrometer for ion and neutral analysis. Space Sci. Rev. **128**, 745–801 (2007)
A. Bar-Nun, T. Owen, Trapping of gases in water ice and consequences to comets and the atmospheres of the inner planets, in *Solar System Ices*, ed. by B.B. Schmidt, C. De Bergh, M. Festou (Kluwer Academic, Norwell, 1998), pp. 353–366
A. Bar-Nun, G. Herman, D. Laufer, M.L. Rappaport, Trapping and release of gases by water ice and implications for icy bodies. Icarus **63**, 317–332 (1985)
A. Bar-Nun, I. Kleinfeld, E. Kochavi, Trapping of gas mixtures by amorphous water ice. Phys. Rev. B **38**, 7749–7754 (1988)
A. Bar-Nun, G. Notesco, T. Owen, Trapping of N_2, CO and Ar in amorphous ice—Application to comets. Icarus **190**, 655–659 (2007)
M.A. Barucci et al., (90377) Sedna: Investigation of surface compositional variation. Astron. J. **140**, 2095–2100 (2010)
E.A. Bergin et al., Submillimeter wave astronomy satellite observations of Jupiter and Saturn: Detection of 557-GHz water emission from the upper atmosphere. Astrophys. J. **539**, L147–150 (2000)
B. Bézard, C.A. Griffith, D.M. Kelly, Search for NH_3 in Jupiter's stratosphere ten months after SL9's collision. Icarus **125**, 331–339 (1997a)
B. Bézard, C.A. Griffith, D.M. Kelly, J.H. Lacy, T. Greathouse, G. Orton, Thermal infrared imaging spectroscopy of Shoemaker-Levy 9 impact sites: Temperature and HCN retrievals. Icarus **125**, 94–120 (1997b)
B. Bézard, E. Lellouch, D. Strobel, J.-P. Maillard, P. Drossart, Carbon monoxide on Jupiter: Evidence for both internal and external sources. Icarus **159**, 95–111 (2002)
N. Biver et al., Radio wavelength molecular observations of comets C/1999 T1 (McNaught-Hartley), C/2001 A2 (LINEAR), C/2000 WM_1 (LINEAR) and 153P/Ikeya-Zhang. Astron. Astrophys. **449**, 1255–1270 (2006)
N. Biver et al., Radio observations of Comet 9P/Tempel 1 before and after deep impact. Icarus **187**, 253–271 (2007)
G.L. Bjoraker, S.R. Stolovy, T.L. Herter, G.E. Gull, B.E. Pirger, Detection of water after the collision of fragments G and K of Comet Shoemaker-Levy 9 with Jupiter. Icarus **121**, 411–421 (1996)
D. Bockelée-Morvan, An overview of comet composition. The molecular universe, in *Proceedings of the International Astronomical Union*. IAU Symposium, vol. 280 (2011), pp. 261–274
D. Bockelée-Morvan et al., Deuterated water in Comet C/1996 B2 (Hyakutake) and its implications for the origin of comets. Icarus **133**, 147–162 (1998)
D. Bockelée-Morvan, J. Crovisier, M.J. Mumma, H.A. Weaver, The composition of cometary volatiles, in *Comets 2*, ed. by H.U. Keller, M.C. Festou, H.A. Weaver (University of Arizona Press, Tucson, 2004), pp. 391–423
D. Bockelée-Morvan et al., Large excess of heavy nitrogen in both hydrogen cyanide and cyanogen from Comet 17P/Holmes. Astrophys. J. Lett. **679**, L49–L52 (2008)
D. Bockelée-Morvan et al., Walker, H. Herschel measurements of the D/H and $^{16}O/^{18}O$ ratios in water in the Oort-cloud Comet C/2009 P1 (Garradd). Astron. Astrophys. **544**, L15 (2012)
D. Bockelée-Morvan et al., Cometary isotopic measurements. Space Sci. Rev. (2015, this issue). doi:10.1007/s11214-015-0156-9
D.D. Bogard, R.N. Clayton, K. Marti, T. Owen, G. Turner, Martian volatiles: Isotopic composition, origin, and evolution. Space Sci. Rev. **96**, 425–460 (2001)
B.P. Bonev et al., A search for variation in the H_2O ortho-para ratio and rotational temperature in the inner coma of Comet C/2004 Q2 (Machholz). Astrophys. J. **661**, L97–L100 (2007)
B.P. Bonev et al., IRCS/Subaru observations of water in the inner coma of Comet 73P-B/Schwassmann-Wachmann 3: Spatially resolved rotational temperatures and ortho-para ratios. Icarus **196**, 241–248 (2008)
R. Brasser, A. Morbidelli, Oort cloud and scattered disc formation during a late dynamical instability in the solar system. Icarus **225**, 40–49 (2013)
F.H. Briggs, P.D. Sackett, Radio observations of Saturn as a probe of its atmosphere and cloud structure. Icarus **80**, 77–103 (1989)
D. Brownlee et al., Comet 81P/Wild 2 under a microscope. Science **314**, 1711-6 (2006)

B.J. Buratti et al., Cassini visual and infrared mapping spectrometer observations of iapetus: Detection of CO_2. Astrophys. J. **622**, L149–L152 (2005)
H.M. Butner et al., Discovery of interstellar heavy water. Astrophys. J. **659**(2), L137–L140 (2007)
R.M. Canup, A giant impact origin of Pluto-Charon. Science **307**, 546–550 (2005)
R.M. Canup, E. Asphaug, Origin of the Moon in a giant impact near the end of the Earth's formation. Nature **412**, 708–712 (2001)
R.M. Canup, W.R. Ward, Formation of the Galilean satellites: Conditions of accretion. Astron. J. **124**, 3404–3423 (2002)
R.M. Canup, W.R. Ward, A common mass scaling for satellite systems of gaseous planets. Nature **441**, 834–839 (2006)
P. Caselli et al., Water vapor toward starless cores: The Herschel view. Astron. Astrophys. **521**, L29 (2010)
T. Cassidy, R. Johnson, Collisional spreading of Enceladus' neutral cloud. Icarus **209**, 696–703 (2010)
T. Cavalié et al., Observation of water vapor in the stratosphere of Jupiter with the Odin space telescope. Planet. Space Sci. **56**, 1573–1584 (2008)
T. Cavalié et al., A cometary origin for CO in the stratosphere of Saturn? Astron. Astrophys. **510**, A88 (2010)
T. Cavalié et al., Odin space telescope monitoring of water vapor in the stratosphere of Jupiter. Planet. Space Sci. **61**, 3–14 (2012)
T. Cavalié et al., Spatial distribution of water in the stratosphere of Jupiter from Herschel HIFI and PACS observations. Astron. Astrophys. **553**, A21 (2013)
T. Cavalié et al., The first submillimeter observation of CO in the stratosphere of Uranus. Astron. Astrophys. **562**, A33 (2014)
C. Ceccarelli et al., Discovery of deuterated water in a young protoplanetary disk. Astrophys. J. **631**(1), L81–L84 (2005)
C. Ceccarelli et al., Deuterium fractionation: The Ariadne's thread from the pre-collapse phase to meteorites and comets today, in *Protostars and Planets VI*, ed. by H. Beuther, R. Klessen, C. Dullemond, Th. Henning (University of Arizona Press, Tucson, 2014)
S. Chakraborty, T.L. Jackson, B.H. Muskatel, M. Ahmed, R.D. Levine, M.H. Thiemens, Huge isotopic effect in VUV photodissociation of N_2: Implications for meteorite data, in *44th Lunar Planet. Sci. Conf. Abstract*, vol. 1043 (2013)
S.B. Charnley, S.D. Rodgers, The end of interstellar chemistry as the origin of nitrogen in comets and meteorites. Astrophys. J. **569**, L133–L137 (2002)
I.D. Clark, P. Fritz, *Environmental Isotopes in Hydrogeology* (CRC Press, Boca Raton, 1997)
R.N. Clark et al., Compositional maps of Saturn's moon Phoebe from imaging spectroscopy. Nature **435**, 66–69 (2005)
R.N. Clayton, Solar system—Self-shielding in the solar nebula. Nature **415**, 860–861 (2002)
A. Cochran et al., The composition of comets: A pre-Rosetta look. Space Sci. Rev. (2015, this issue)
B.J. Conrath, D. Gautier, Saturn helium abundance: A reanalysis of voyager measurements. Icarus **144**, 124–134 (2000)
B. Conrath, R. Hanel, D. Gautier, A. Marten, G. Lindal, The helium abundance of Uranus from Voyager measurements. J. Geophys. Res. **92**, 15003–15010 (1987)
B.J. Conrath, D. Gautier, G.F. Lindal, R.E. Samuelson, W.A. Shaffer, The helium abundance of Neptune from Voyager measurements. J. Geophys. Res. **96**, 18907 (1991)
A. Crida, S. Charnoz, Formation of regular satellites from ancient massive rings in the solar system. Science **338**, 1196–1199 (2012)
J. Crovisier, The water molecule in comets—Fluorescence mechanisms and thermodynamics of the inner coma. Astron. Astrophys. **130**, 361–372 (1984)
J. Crovisier, Cometary diversity and cometary families, in *Proceedings of the XVIIIemes Rencontres de Blois: Planetary Science: Challenges and Discoveries*, 28th May–2nd June 2006, Blois, France (2007), available at astro-ph/0703785
J. Crovisier et al., ISO spectroscopic observations of short-period comets, in *The Universe as Seen by ISO*, vol. 427 (1999), p. 161
J. Crovisier et al., The composition of ices in Comet C/1995 O1 (Hale-Bopp) from radio spectroscopy. Further results and upper limits on undetected species. Astron. Astrophys. **418**, 1141–1157 (2004)
D.P. Cruikshank et al., The surface compositions of Pluto and Charon. Icarus **246**, 82–92 (2015)
D.P. Cruikshank et al., Surface composition of Hyperion. Nature **448**, 54–56 (2007)
C.M. Dalle Ore et al., Composition of KBO (50000) Quaoar. Astron. Astrophys. **501**, 349–357 (2009)
A.C. Danks, D.L. Lambert, C. Arpigny, The $^{12}C/^{13}C$ ratio in Comet Kohoutek /1973f/. Astrophys. J. **194**, 745–751 (1974)
N. Dauphas, The dual origin of the terrestrial atmosphere. Icarus **165**, 326–339 (2003)
T. de Graauw et al., First results of ISO-SWS observations of Saturn: Detection of CO_2, CH_3C_2H, C_4H_2 and tropospheric H_2O. Astron. Astrophys. **321**, L13–L16 (1997)

Th. De Graauw et al., The Herschel-Heterodyne instrument for the far-infrared (HIFI). Astron. Astrophys. **518**, L6 (2010)
I. de Pater, P.N. Romani, S.K. Atreya, Possible microwave absorption by H_2S gas in Uranus' and Neptune's atmospheres. Icarus **91**, 220–233 (1991)
E. Deloule, F. Robert, J.C. Doukhan, Interstellar hydroxyl in meteoritic chondrules: Implications for the origin of water in the inner solar system. Geochim. Cosmochim. Acta **62**, 3367–3378 (1998)
G. Dominguez, A heterogeneous chemical origin for the ^{16}O-enriched and ^{16}O-depleted reservoirs of the early solar system. Astrophys. J. **713**, L59–L63 (2010)
L. Dones et al., Origin and evolution of the cometary reservoirs. Space Sci. Rev. (2015, this issue)
A. Drouart, B. Dubrulle, D. Gautier, F. Robert, Structure and transport in the solar nebula from constraints on deuterium enrichment and giant planets formation. Icarus **140**(1), 129–155 (1999)
M. Duncan, T. Quinn, S. Tremaine, The formation and extent of the solar system comet cloud. Astron. J. **94**, 1330–1338 (1987)
P. Eberhardt, M. Reber, D. Krankowsky, R.R. Hodges, The D/H and $^{18}O/^{16}O$ ratios in water from Comet P/Halley. Astron. Astrophys. **302**, 301 (1995)
P.R. Estrada, I. Mosqueira, J.J. Lissauer, G. D'Angelo, D.P. Cruikshank, Formation of Jupiter and conditions for accretion of the Galilean satellites, in *Europa*, ed. by R.T. Pappalardo, W.B. McKinnon, K.K. Khurana (University of Arizona Press, Tucson, 2009), p. 27
E. Everhart, Close encounters of comets and planets. Astron. J. **74**, 735 (1969)
J. Farquhar, H. Bao, M. Thiemens, Atmospheric influence of Earth's earliest sulfur cycle. Science **289**, 756–759 (2000)
B. Fegley Jr., R.G. Prinn, Chemical models of the deep atmosphere of Uranus. Astrophys. J. **307**, 852–865 (1986)
H. Feuchtgruber, E. Lellouch, T. de Graauw, B. Bézard, T. Encrenaz, M. Griffin, External supply of oxygen in the giant planets. Nature **389**, 159–161 (1997)
H. Feuchtgruber et al., The D/H ratio in the atmospheres of Uranus and Neptune from Herschel-PACS observations. Astron. Astrophys. **551**, A126 (2013)
U. Fink, A taxonomic survey of comet composition 1985–2004 using CCD spectroscopy. Icarus **201**, 311–334 (2009)
L.N. Fletcher, G.S. Orton, N.A. Teanby, P.G.J. Irwin, Phosphine on Jupiter and Saturn from Cassini/CIRS. Icarus **202**, 543–564 (2009a)
L.N. Fletcher, G.S. Orton, N.A. Teanby, P.G.J. Irwin, G.L. Bjoraker, Methane and its isotopologues on Saturn from Cassini/CIRS observations. Icarus **199**, 351–367 (2009b)
L.N. Fletcher, G.S. Orton, I. de Pater, O. Mousis, Jupiter's stratospheric hydrocarbons and temperatures after the July 2009 impact from VLT infrared spectroscopy. Astron. Astrophys. **524**, 46 (2010)
L.N. Fletcher et al., The aftermath of the July 2009 impact on Jupiter: Ammonia, temperatures and particulates from Gemini thermal infrared spectroscopy. Icarus **211**, 568–586 (2011a)
L.N. Fletcher et al., The aftermath of the July 2009 impact on Jupiter: Ammonia, temperatures and particulates from Gemini thermal infrared spectroscopy. Icarus **211**, 568–586 (2011b)
L.N. Fletcher et al., The origin of nitrogen on Jupiter and Saturn from the $^{15}N/^{14}N$ ratio. Icarus **238**, 170–190 (2014)
A.D. Fortes, E.R. Stofan, Clathrate formation in the near-surface environment of Titan, in *36th Lunar Planet. Sci. Conf. Abstract*, vol. 1123 (2005)
J.J. Fortney, W.B. Hubbard, Phase separation in giant planets: Inhomogeneous evolution of Saturn. Icarus **164**, 228–243 (2003)
E. Füri, E. Deloule, A. Gurenko, B. Marty, New evidence for chondritic lunar water from combined D/H and noble gas analyses of single Apollo 17 volcanic glasses. Icarus **229**, 109–120 (2014)
D.H. Garrison et al., Isotopic composition of trapped and cosmogenic noble gases in several Martian meteorites. Meteorit. Planet. Sci. **33**, 721–736 (1998)
D. Gautier, F. Hersant, Formation and composition of planetesimals. Space Sci. Rev. **116**, 25–52 (2005)
D. Gautier, F. Raulin, Chemical composition of Titan's atmosphere, in *Huygens: Science, Payload and Mission*, vol. 1177 (1997), p. 359
D. Gautier, F. Hersant, O. Mousis, J.I. Lunine, Enrichments in volatiles in Jupiter: A new interpretation of the Galileo measurements. Astrophys. J. **550**, L227–L230 (2001)
K.-H. Glassmeier et al., The Rosetta mission: flying towards the origin of the solar system. Space Sci. Rev. **128**, 1–21 (2007)
C.R. Glein, M.Y. Zolotov, E.L. Shock, The oxidation state of hydrothermal systems on early Enceladus. Icarus **197**, 157–163 (2008)
P. Goldreich, S. Tremaine, Disk-satellite interactions. Astrophys. J. **241**, 425–441 (1980)
P. Goldreich, W.R. Ward, The formation of planetesimals. Astrophys. J. **183**, 1051–1062 (1973)
P. Goldreich, N. Murray, P.Y. Longaretti, D. Banfield, Neptune's story. Science **245**, 500–504 (1989)

R. Gomes, H.F. Levison, K. Tsiganis, A. Morbidelli, Origin of the cataclysmic late heavy bombardment period of the terrestrial planets. Nature **435**, 466–469 (2005)

C.A. Griffith, B. Bézard, T.G. Greathouse, E. Lellouch, J. Lacy, D. Kelly, M.J. Richter, Meridional transport of HCN from SL9 impacts on Jupiter. Icarus **170**, 58–69 (2004)

A. Grimberg et al., Solar wind neon from genesis: Implications for the lunar noble gas record. Science **314**, 1133–1135 (2006)

D. Grinspoon, Implications of the high D/H ratio for the sources of water in Venus' atmosphere. Nature **363**, 428–431 (1993)

W.M. Grundy et al., Distributions of H_2O and CO_2 ices on Ariel, Umbriel, Titania, and Oberon from IRTF/SpeX observations. Icarus **184**, 543–555 (2006)

C.J. Hansen et al., Enceladus' water vapor plume. Science **311**, 1422–1425 (2006)

P. Hartogh et al., Ocean-like water in the Jupiter-family Comet 103P/Hartley 2. Nature **478**(7368), 218–220 (2011a)

P. Hartogh et al., Direct detection of the Enceladus water torus with Herschel. Astron. Astrophys. **532**, L2 (2011b)

K. Hashizume, M. Chaussidon, B. Marty, K. Terada, Protosolar carbon isotopic composition: Implications for the origin of meteoritic organics. Astrophys. J. **600**, 480–484 (2004)

V.S. Heber et al., Isotopic mass fractionation of solar wind: Evidence from fast and slow solar wind collected by the genesis mission. Astrophys. J. **759**, 121 (2012)

Th. Henning, D. Semenov, Chemistry in protoplanetary disks. Chem. Rev. **113**, 9016 (2013)

F. Hersant, D. Gautier, J.I. Lunine, Enrichment in volatiles in the giant planets of the solar system. Planet. Space Sci. **52**, 623–641 (2004)

F. Hersant, D. Gautier, G. Tobie, J.I. Lunine, Interpretation of the carbon abundance in Saturn measured by Cassini. Planet. Space Sci. **56**, 1103–1111 (2008)

B.E. Hesman, G.R. Davis, H.E. Matthews, G.S. Orton, The abundance profile of CO in Neptune's atmosphere. Icarus **186**, 342–353 (2007)

J.H. Hoffman, V.I. Oyama, U. von Zahn, Measurements of the Venus lower atmosphere composition: A comparison of results. J. Geophys. Res. **85**, 7871 (1980)

M. Hogerheijde, The molecular environment of low mass protostars. Ph.D. thesis, University of Leiden (1998)

M.R. Hogerheijde et al., Detection of the water reservoir in a forming planetary system. Science **334**(6054), 338–340 (2011)

J. Horner, O. Mousis, F. Hersant, Constraints on the formation regions of comets from their D: H ratios. Earth Moon Planets **100**(1–2), 43–56 (2007)

R. Hueso et al., First Earth-based detection of a superbolide on Jupiter. Astrophys. J. **721**, L129–L133 (2010a)

R. Hueso, J. Legarreta, S. Pérez-Hoyos, J.F. Rojas, A. Sánchez-Lavega, A. Morgado, The international outer planets watch atmospheres node database of giant-planet images. Planet. Space Sci. **58**, 1152–1159 (2010b)

R. Hueso et al., Impact flux on Jupiter: From superbolides to large-scale collisions. Astron. Astrophys. **560**, A55 (2013)

D. Hutsemékers et al., Isotopic abundances of carbon and nitrogen in Jupiter-family and Oort cloud comets. Astron. Astrophys. **440**, L21–L24 (2005)

D. Hutsemékers, J. Manfroid, E. Jehin, J.-M. Zucconi, C. Arpigny, The $^{16}OH/^{18}OH$ and OD/OH isotope ratios in Comet C/2002 T7 (LINEAR). Astron. Astrophys. **490**, P31–L34 (2008)

E. Jacquet, F. Robert, Water transport in protoplanetary disks and the hydrogen isotopic composition of chondrites. Icarus **223**(2), 722–732 (2013)

E. Jehin et al., A multi-wavelength simultaneous study of the composition of the Halley family Comet 8P/Tuttle. Earth Moon Planets **105**, 343–349 (2009)

K.L. Jessup et al., $^{14}N/^{15}N$ detectability in Pluto's atmosphere. Icarus **226**, 1514–1526 (2013)

D. Jewitt, H.E. Matthews, T. Owen, R. Meier, The $^{12}C/^{13}C$, $^{14}N/^{15}N$ and $^{32}S/^{34}S$ isotope ratios in Comet Hale-Bopp (C/1995 O1). Science **278**, 90–93 (1997)

R.E. Johnson et al., Radiation effects on the surfaces of the Galilean satellites, in *Jupiter: The Planet, Satellites and Magnetosphere*, ed. by F. Bagenal, T.E. Dowling, W.B. McKinnon (Cambridge University Press, Cambridge, 2004)

J.S. Kargel, Ammonia-water volcanism on icy satellites—Phase relations at 1 atmosphere. Icarus **100**, 556–574 (1992)

E. Karkoschka, M. Tomasko, The haze and methane distributions on Uranus from HST-STIS spectroscopy. Icarus **202**, 287–309 (2009)

E. Karkoschka, M.G. Tomasko, The haze and methane distributions on Neptune from HST-STIS spectroscopy. Icarus **211**, 780–797 (2011)

J.J. Kavelaars, O. Mousis, J.-M. Petit, H.A. Weaver, On the formation location of Uranus and Neptune as constrained by dynamical and chemical models of comets. Astrophys. J. **734**(2), L30 (2011)

F. Kerridge, Carbon, hydrogen and nitrogen in carbonaceous chondrites: Abundances and isotopic compositions in bulk samples. Geochim. Cosmochim. Acta **49**, 1707–1714 (1985)

J.F. Kerridge, O. Eugster, J.S. Kim, K. Marti, Nitrogen isotopes in the 74001/74002 double-drive tube from Shorty Crater, in *Proc. 21st Lunar Planet. Sci. Conf.* (1991), pp. 291–299

K.K. Khurana et al., Induced magnetic fields as evidence for subsurface oceans in Europa and Callisto. Nature **395**, 777–780 (1998)

M.G. Kivelson et al., Europa and Callisto: Induced or intrinsic fields in a periodically varying plasma environment. J. Geophys. Res. **104**, 4609–4626 (1999)

M.G. Kivelson, K.K. Khurana, M. Volwerk, The permanent and inductive magnetic moments of Ganymede. Icarus **157**, 507–522 (2002)

J.J. Klavetter, M.F. A'Hearn, An extended source for CN jets in Comet P/Halley. Icarus **107**, 322–334 (1994)

D.L. Lambert, A.C. Danks, High-resolution spectra of C_2 Swan bands from comet West 1976 VI. Astrophys. J. **268**, 428–446 (1983)

A. Lecacheux et al., Observations of water in comets with Odin. Astron. Astrophys. **402**, L55–L58 (2003)

E. Lellouch et al., Chemical and thermal response of Jupiter's atmosphere following the impact of Comet Shoemaker-Levy 9. Nature **373**, 592–595 (1995)

E. Lellouch et al., Carbon monoxide in Jupiter after the impact of Comet Shoemaker-Levy 9. Planet. Space Sci. **45**, 1203–1212 (1997)

E. Lellouch et al., The deuterium abundance in Jupiter and Saturn from ISO-SWS observations. Astron. Astrophys. **370**, 610–622 (2001)

E. Lellouch et al., The origin of water vapor and carbon dioxide in Jupiter's stratosphere. Icarus **159**, 112–131 (2002)

E. Lellouch, R. Moreno, G. Paubert, A dual origin for Neptune's carbon monoxide? Astron. Astrophys. **430**, L37–L40 (2005)

E. Lellouch et al., On the HCN and CO2 abundance and distribution in Jupiter's stratosphere. Icarus **184**, 478–497 (2006)

E. Lellouch et al., First results of Herschel-PACS observations of Neptune. Astron. Astrophys. **518**, L152 (2010a)

E. Lellouch, C. de Bergh, B. Sicardy, S. Ferron, H.U. Käufl, Detection of CO in Triton's atmosphere and the nature of surface-atmosphere interactions. Astron. Astrophys. **512**, LL8 (2010b)

E. Lellouch, C. de Bergh, B. Sicardy, H.U. Käufl, A. Smette, High resolution spectroscopy of Pluto's atmosphere: Detection of the 2.3μm CH_4 bands and evidence for carbon monoxide. Astron. Astrophys. **530**, LL4 (2011)

L.A. Leshin, Insights into martian water reservoirs from analyses of martian meteorite QUE94201. Geophys. Res. Lett. **27**, 2017–2020 (2000)

L.A. Leshin et al., Volatile, isotope, and organic analysis of martian fines with the Mars Curiosity rover. Science **341**, 1238937 (2013)

H.F. Levison, M.J. Duncan, K. Zahnle, M. Holman, L. Dones, Planetary impact rates from ecliptic comets. Icarus **143**, 415–420 (2000)

H.F. Levison, A. Morbidelli, C. Van Laerhoven, R. Gomes, K. Tsiganis, Origin of the structure of the Kuiper belt during a dynamical instability in the orbits of Uranus and Neptune. Icarus **196**, 258–273 (2008)

H.F. Levison, M.J. Duncan, R. Brasser, D.E. Kaufmann, Capture of the Sun's Oort cloud from stars in its birth cluster. Science **329**(5988), 187–190 (2010)

J.S. Lewis, R.G. Prinn, Kinetic inhibition of CO and N2 reduction in the solar nebula. Astrophys. J. **238**, 357–364 (1980)

J.L. Linsky et al., What is the total deuterium abundance in the local galactic disk? Astrophys. J. **647**(2), 1106–1124 (2006)

D.C. Lis et al., Herschel study of D/H in water in the Jupiter-family Comet 45P/Honda-Mrkos-Pajdušáková and prospects for D/H measurements with CCAT. Astrophys. J. **774**(1), L3 (2013)

K. Lodders, Solar system abundances and condensation temperatures of the elements. Astrophys. J. **591**, 1220–1247 (2003)

K. Lodders, H. Palme, H.-P. Gail, Abundances of the elements in the solar system, in *Landolt Börnstein*, vol. 44 (2009)

J.I. Lunine, Origin and evolution of outer solar system atmospheres. Science **245**, 1417 (1989)

J.I. Lunine, D.J. Stevenson, Formation of the Galilean satellites in a gaseous nebula. Icarus **52**, 14–39 (1982)

J. Lunine, S. Engel, B. Rizk, M. Horanyi, Sublimation and reformation of icy grains in the primitive solar nebula. Icarus **94**, 333 (1991)

S. Luszcz-Cook, I. de Pater, Constraining the origins of Neptune's carbon monoxide abundance. with CARMA millimeter-wave observations. Icarus **222**, 379–400 (2013)

W. Macy, W.H. Smith, Detection of HD on Saturn and Uranus, and the D/H ratio. Astrophys. J. **222**, L73 (1978)
P.R. Mahaffy, T.M. Donahue, S.K. Atreya, T.C. Owen, H.B. Niemann, Galileo probe measurements of D/H and 3He/4He in Jupiter's atmosphere. Space Sci. Rev. **84**, 251–263 (1998)
P.R. Mahaffy et al., Noble gas abundance and isotope ratios in the atmosphere of Jupiter from the Galileo probe mass spectrometer. J. Geophys. Res. **105**, 15061–15072 (2000)
K.E. Mandt et al., Isotopic evolution of the major constituents of Titan's atmosphere based on Cassini data. Planet. Space Sci. **57**, 1917–1930 (2009)
K.E. Mandt et al., The $^{12}C/^{13}C$ ratio on Titan from Cassini INMS measurements and implications for the evolution of methane. Astrophys. J. **749**, 160 (2012)
K.E. Mandt, O. Mousis, J. Lunine, D. Gautier, Protosolar ammonia as the unique source of Titan's nitrogen. Astrophys. J. **788**, L24 (2014)
K.E. Mandt, O. Mousis, E. Chassefière, Comparative planetology of the history of nitrogen isotopes in the atmospheres of Titan and Mars. Icarus **254**, 259–261 (2015)
J. Manfroid et al., The CN isotopic ratios in comets. Astron. Astrophys. **503**, 613–624 (2009)
U. Marboeuf et al., Composition of ices in low-mass extrasolar planets. Astrophys. J. **681**, 1624–1630 (2008)
U. Marboeuf, A. Thiabaud, Y. Alibert, N. Cabral, W. Benz, From planetesimals to planets: Volatile molecules. Astron. Astrophys. **570**, A36 (2014a)
U. Marboeuf, A. Thiabaud, Y. Alibert, N. Cabral, W. Benz, From stellar nebula to planetesimals. Astron. Astrophys. **570**, A35 (2014b)
A. Marten et al., First observations of CO and HCN on Neptune and Uranus at millimeter wavelengths and the implications for atmospheric chemistry. Astrophys. J. **406**, 285–297 (1993)
A. Marten et al., The collision of Comet Shoemaker-Levy 9 with Jupiter: Detection and evolution of HCN in the stratosphere of the planet. Geophys. Res. Lett. **22**, 1589–1592 (1995)
B. Marty, The origins and concentrations of water, carbon, nitrogen and noble gases on Earth. Earth Planet. Sci. Lett. **313–314**, 56–66 (2012)
B. Marty, A. Meibom, Noble gas signature of the late heavy bombardment in the Earth's atmosphere. eEarth **2**, 43–49 (2007)
B. Marty et al., Helium and neon abundances and compositions in cometary matter. Science **319**, 75–78 (2008)
B. Marty, M. Chaussidon, R.C. Wiens, A.J.G. Jurewicz, D.S. Burnett, A^{15}N-poor isotopic composition for the solar system as shown by Genesis solar wind samples. Science **332**, 1533–1536 (2011)
D.L. Matson, J.C. Castillo, J. Lunine, T.V. Johnson, Enceladus' plume: Compositional evidence for a hot interior. Icarus **187**, 569–573 (2007)
E. Mazor, D. Heymann, E. Anders, Noble gases in carbonaceous chondrites. Geochim. Cosmochim. Acta **34**, 781–824 (1970)
C.P. McKay et al., High-temperature shock formation of N2 and organics on primordial Titan. Nature **332**, 520–522 (1988)
K.D. McKeegan et al., Isotopic compositions of cometary matter returned by Stardust. Science **314**, 1724–1728 (2006)
K.D. McKeegan et al., The oxygen isotopic composition of the Sun inferred from captured solar wind. Science **332**, 1528 (2011)
W.B. McKinnon, R.L. Kirk, Triton, in *Encyclopedia of the Solar System*, ed. by T. Spohn, D. Breuer, T. Johnson 3rd edn. (Elsevier, Amsterdam, 2014), pp. 861–882
W.B. McKinnon, S. Mueller, Pluto's structure and composition suggest origin in the solar, not a planetary, nebula. Nature **335**, 240–243 (1988)
R. Meier et al., Deuterium in Comet C/1995 O1 (Hale-Bopp): Detection of DCN. Science **279**, 6054, 942 (1998)
A. Morbidelli, Modern integrations of solar system dynamics. Annu. Rev. Earth Planet. Sci. **30**, 89–112 (2002)
A. Morbidelli, H.F. Levison, K. Tsiganis, R. Gomes, Chaotic capture of Jupiter's Trojan asteroids in the early solar system. Nature **435**, 462–465 (2005)
M. Moreira, J. Kunz, C.J. Allègre, Rare gas systematics in Popping Rock: Isotopic and elemental compositions in the upper mantle. Science **279**, 1178–1181 (1998)
R. Moreno, A. Marten, Y. Biraud, B. Bézard, E. Lellouch, G. Paubert, W. Wild, Jovian stratospheric temperature during the two months following the impacts of Comet Shoemaker-Levy 9. Planet. Space Sci. **49**, 473–486 (2001)
R. Moreno, A. Marten, H.E. Matthews, Y. Biraud, Long-term evolution of CO, CS and HCN in Jupiter after the impacts of Comet Shoemaker-Levy 9. Planet. Space Sci. **51**, 591–611 (2003)
J.I. Moses, SL9 impact chemistry: Long-term photochemical evolution, in *The Collision of Comet Shoemaker-Levy 9 and Jupiter*, ed. by K. Noll, H.A. Weaver, P.D. Feldman (Cambridge University Press, Cambridge, 1996), pp. 243–268

J.I. Moses, M. Allen, G.R. Gladstone, Nitrogen and oxygen photochemistry following SL9. Geophys. Res. Lett. **22**, 1601–1604 (1995a)

J.I. Moses, M. Allen, G.R. Gladstone, Post-SL9 sulfur photochemistry on Jupiter. Geophys. Res. Lett. **22**, 1597–1600 (1995b)

J.I. Moses, T. Fouchet, B. Bézard, G.R. Gladstone, E. Lellouch, H. Feuchtgruber, Photochemistry and diffusion in Jupiter's stratosphere: Constraints from ISO observations and comparisons with other giant planets. J. Geophys. Res. **110**, E08001 (2005). doi:10.1029/2005JE002411

O. Mousis et al., New insights on Saturn's formation from its nitrogen isotopic composition. Astrophys. J. **796**, L28 (2014a)

O. Mousis et al., Instrumental methods for professional and amateur collaborations in planetary astronomy. Exp. Astron. **38**, 91–191 (2014b)

O. Mousis, D. Gautier, Constraints on the presence of volatiles in Ganymede and Callisto from an evolutionary turbulent model of the Jovian subnebula. Planet. Space Sci. **52**, 361–370 (2004)

O. Mousis, D. Gautier, D. Bockelée-Morvan, F. Robert, B. Dubrulle, A. Drouart, Constraints on the formation of comets from D/H ratios measured in H_2O and HCN. Icarus **148**(2), 513–525 (2000)

O. Mousis, D. Gautier, D. Bockelée-Morvan, An evolutionary turbulent model of Saturn's subnebula: Implications for the origin of the atmosphere of Titan. Icarus **156**, 162–175 (2002)

O. Mousis, Y. Alibert, W. Benz, Saturn's internal structure and carbon enrichment. Astron. Astrophys. **449**, 411–415 (2006)

O. Mousis et al., A primordial origin for the atmospheric methane of Saturn's moon Titan. Icarus **204**, 749–751 (2009a)

O. Mousis et al., Formation conditions of enceladus and origin of its methane reservoir. Astrophys. J. **701**, L39–L42 (2009b)

O. Mousis et al., Determination of the minimum masses of heavy elements in the envelopes of Jupiter and Saturn. Astrophys. J. **696**, 1348–1354 (2009c)

O. Mousis et al., Impact regimes and post-formation sequestration processes: Implications for the origin of heavy noble gases in terrestrial planets. Astrophys. J. **714**, 1418–1423 (2010)

O. Mousis et al., Removal of Titan's atmospheric noble gases by their sequestration in surface clathrates. Astrophys. J. **740**, L9 (2011)

O. Mousis, J.I. Lunine, N. Madhusudhan, T.V. Johnson, Nebular water depletion as the cause of Jupiter's low oxygen abundance. Astrophys. J. **751**, LL7 (2012)

S. Mukhopadhyay, Early differentiation and volatile accretion recorded in deep-mantle neon and xenon. Nature **486**, 101–104 (2012)

M.J. Mumma et al., The ortho-para ratio and nuclear spin temperatures in Comets Halley and Wilson. Bull. Am. Astron. Soc. **19**, 866 (1987)

M.J. Mumma, S.B. Charnley, The chemical composition of comets—Emerging taxonomies and natal heritage. Annu. Rev. Astron. Astrophys. **49**, 471–524 (2011)

M.J. Mumma et al., Remote infrared observations of parent volatiles in comets: A window on the early solar system. Adv. Space Res. **31**, 2563–2575 (2003)

H.B. Niemann et al., The Galileo probe mass spectrometer: Composition of Jupiter's atmosphere. Science **272**, 846–849 (1996)

H.B. Niemann et al., The composition of the Jovian atmosphere as determined by the Galileo probe mass spectrometer. J. Geophys. Res. **103**, 22831–22846 (1998)

H.B. Niemann et al., Composition of Titan's lower atmosphere and simple surface volatiles as measured by the Cassini-Huygens probe gas chromatograph mass spectrometer experiment. J. Geophys. Res. **115**, E12006 (2010)

C.A. Nixon et al., Isotopic ratios in Titan's methane: Measurements and modeling. Astrophys. J. **749**, 159 (2012)

K.S. Noll et al., HST spectroscopic observations of Jupiter after the collision of Comet Shoemaker-Levy 9. Science **267**, 1307–1313 (1995)

D.P. O'Brien, A. Morbidelli, H.F. Levison, Terrestrial planet formation with strong dynamical friction. Icarus **184**, 39–58 (2006)

D. Olsson-Steel, Collisions in the solar system. IV—Cometary impacts upon the planets. Mon. Not. R. Astron. Soc. **227**, 501–524 (1987)

G.S. Orton et al., Mid-infrared spectroscopy of Uranus from the Spitzer infrared spectrometer: 2. Determination of the mean composition of the upper troposphere and stratosphere. Icarus **243**, 471–493 (2014)

G. Orton et al., Collision of Comet Shoemaker-Levy 9 with Jupiter observed by the NASA infrared telescope facility. Science **267**, 1277–1282 (1995)

G.S. Orton et al., The atmospheric influence, size and possible asteroidal nature of the July 2009 Jupiter impactor. Icarus **211**, 587–602 (2011)

J.P. Osegovic, M.D. Max, Compound clathrate hydrate on Titan's surface. J. Geophys. Res. **110**, E08004 (2005)
T. Owen, The isotope ratio $^{12}C/^{13}C$ in Comet Tago-Sato (1969g). Astrophys. J. **184**, 33–44 (1973)
T. Owen, The contributions of comets to planets, atmospheres, and life: Insights from Cassini-Huygens, Galileo, Giotto, and inner planet missions. Space Sci. Rev. **138**, 301–316 (2008)
T. Owen, A. Barnun, Comets, impacts, and atmospheres. Icarus **116**, 215–226 (1995)
T.C. Owen et al., Surface ices and the atmospheric composition of Pluto. Science **261**, 745–748 (1993)
T. Owen, P.R. Mahaffy, H.B. Niemann, S. Atreya, M. Wong, P. Nitrogen, Astrophys. J. **553**, L77–L79 (2001)
L. Paganini et al., The formation heritage of Jupiter Family Comet 10P/Tempel 2 as revealed by infrared spectroscopy. Icarus **218**, 644–653 (2012)
R.O. Pepin, On the origin and early evolution of terrestrial planetary atmospheres and meteoritic volatiles. Icarus **92**, 1–79 (1991)
R.O. Pepin, Atmospheres on the terrestrial planets: Clues to origin and evolution. Earth Planet. Sci. Lett. **252**, 1–14 (2006)
G.L. Pilbratt et al., Herschel space observatory. An ESA facility for far-infrared and submillimetre astronomy. Astron. Astrophys. **518**, L1 (2010)
C.C. Porco et al., Cassini observes the active South pole of enceladus. Science **311**, 1393–1401 (2006)
M.J. Prather, J.A. Logan, M.B. McElroy, Carbon monoxide in Jupiter's upper atmosphere: An extraplanetary source. Astrophys. J. **223**, 1072–1081 (1978)
R.G.P. Prinn, B. Fegley Jr., Solar nebula chemistry: Origins of planetary, satellite and cometary volatiles, in *Origin and Evolution of Planetary and Satellite Atmospheres* (University of Arizona Press, Tucson, 1989), pp. 78–136
M. Pujol, B. Marty, R. Burgess, Chondritic-like xenon trapped in Archean rocks: A possible signature of the ancient atmosphere. Earth Planet. Sci. Lett. **308**, 298–306 (2011)
H. Rauer et al., Long-term optical spectrophotometric monitoring of Comet C/1995 O1 (Hale-Bopp). Astron. Astrophys. **397**, 1109–1122 (2003)
F. Robert, The D/H ratio in chondrites. Space Sci. Rev. **106**, 87–101 (2003)
P. Rousselot, E. Jehin, J. Manfroid, D. Hutsemekers, The $^{12}C_2/^{12}C^{13}C$ isotopic ratio in comets C/2001 Q4 (NEAT) and C/2002 T7 (LINEAR). Astron. Astrophys. **545**, AA24 (2012)
P. Rousselot et al., Toward a unique nitrogen isotopic ratio in cometary ices. Astrophys. J. **780**, L17 (2014)
M. Rubin et al., Molecular nitrogen in Comet 67P/Churyumov-Gerasimenko indicates a low formation temperature. Science **348**, 232–235 (2015)
A. Sánchez-Lavega et al., The impact of a large object on Jupiter in 2009 July. Astrophys. J. **715**, L155–L159 (2010)
Y. Shinnaka, H. Kawakita, H. Kobayashi, M. Nagashima, D.C. Boice, $^{14}NH_2/^{15}NH_2$ ratio in Comet C/2012 S1 (ISON) observed during its outburst in 2013 November. Astrophys. J. **782**, L16 (2014)
C.S. Shoemaker, E.M. Shoemaker, D.H. Levy, J.V. Scotti, P. Bendjoya, J. Mueller, in *IAU Circ.*, vol. 5725 (1993), p. 1
F. Shu, Self similar collapse of isothermal spheres and star formation. Astrophys. J. **214**, 488 (1977)
F.H. Shu, F.C. Adams, S. Lizano, Star formation in molecular clouds—Observation and theory. Annu. Rev. Astron. Astrophys. **25**, 23–81 (1987)
J.R. Spencer et al., Cassini encounters Enceladus: Background and the discovery of a South polar hot spot. Science **311**, 1401–1405 (2006)
A. Stawikowski, J.L. Greenstein, The isotope ratio $^{12}C/^{13}C$ in a comet. Astrophys. J. **140**, 1280 (1964)
S.A. Stern, The New Horizons Pluto Kuiper belt mission: an overview with historical context, in *New Horizons* (Springer, New York, 2009), pp. 3–21
S.A. Stern et al., The discovery of argon in Comet C/1995 O1 (Hale-Bopp). Astrophys. J. **544**, L169 (2000)
S.A. Stern et al., ALICE: The ultraviolet imaging spectrograph aboard the New Horizons Pluto–Kuiper Belt mission. Space Sci. Rev. **140**, 155–187 (2008)
D.F. Strobel, Y.L. Yung, The Galilean satellites as a source of CO in the Jovian upper atmosphere. Icarus **37**, 256–263 (1979)
K. Tsiganis, R. Gomes, A. Morbidelli, H.F. Levison, Origin of the orbital architecture of the giant planets of the solar system. Nature **435**, 459–461 (2005)
D. Turrini, R.P. Nelson, M. Barbieri, The role of planetary formation and evolution in shaping the composition of exoplanetary atmospheres. Exp. Astron. **40**, 1–22 (2014)
V. Vanysek, Comet 1973f again. Rise Hvezd **55**, 81–84 (1973)
G.L. Villanueva et al., A sensitive search for deuterated water in Comet 8p/Tuttle. Astrophys. J. **690**(1), L3–L9 (2009)
S. Vinatier, B. Bezard, C.A. Nixon, The Titan $^{14}N/^{15}N$ and $^{12}C/^{13}C$ isotopic ratios in HCN from Cassini/CIRS. Icarus **191**, 712–721 (2007)

R. Visser, E.F. van Dishoeck, S.D. Doty, C.P. Dullemond, The chemical history of molecules in circumstellar disks I. Ices. Astron. Astrophys. **495**, 881 (2009)

U. von Zahn, D.M. Hunten, The Jupiter Helium interferometer experiment on the Galileo entry probe. Space Sci. Rev. **60**, 263–281 (1992)

U. von Zahn, D.M. Hunten, G. Lehmacher, Helium in Jupiter's atmosphere: Results from the Galileo probe helium interferometer experiment. J. Geophys. Res. **103**, 22815–22830 (1998)

J.H. Waite et al., Cassini ion and neutral mass spectrometer: Enceladus plume composition and structure. Science **311**, 1419–1422 (2006)

J.H. Waite et al., Liquid water on Enceladus from observations of ammonia and ^{40}Ar in the plume. Nature **460**(7254), 487–490 (2009)

K.J. Walsh, A. Morbidelli, S.N. Raymond, D.P. O'Brien, A.M. Mandell, A low mass for Mars from Jupiter's early gas-driven migration. Nature **475**(7355), 206–209 (2011)

H. Wänke, G. Dreibus, Chemical-composition and accretion history of terrestrial planets. Philos. Trans. R. Soc. Lond. Ser. A, Math. Phys. Sci. **325**, 545–557 (1988)

C.R. Webster, P.R. Mahaffy, Measuring isotope ratios across the solar system, in *International Workshop on Instrumentation for Planetary Missions, Abstract*, vol. 1683 (2012)

D.C.B. Whittet, W.W. Duley, Carbon monoxide frosts in the interstellar medium. Astron. Astrophys. Rev. **2**, 167–189 (1991)

Willacy et al., Space Sci. Rev. (2015, this issue)

M.H. Wong, P.R. Mahaffy, S.K. Atreya, H.B. Niemann, T.C. Owen, Updated Galileo probe mass spectrometer measurements of carbon, oxygen, nitrogen, and sulfur on Jupiter. Icarus **171**, 153–170 (2004)

L.M. Woodney et al., Morphology of HCN and CN in Comet Hale-Bopp (1995 O1). Icarus **157**, 193–204 (2002)

L. Yang, F.J. Ciesla, C.M.O.'D Alexander, The D/H ratio of water in the solar nebula during its formation and evolution. Icarus **226**(1), 256–267 (2013)

R. Yokochi, B. Marty, A determination of the neon isotopic composition of the deep mantle. Earth Planet. Sci. Lett. **225**, 77–88 (2004)

K. Zahnle, Dynamics and chemistry of SL9 plumes, in *The Collision of Comet Shoemaker-Levy 9 and Jupiter*, ed. by K. Noll, H.A. Weaver, P.D. Feldman (Cambridge University Press, Cambridge, 1996), pp. 183–212

K. Zahnle, P. Schenk, H. Levison, L. Dones, Cratering rates in the outer solar system. Icarus **163**, 263–289 (2003)

J. Zimbelman, Planetary impact probabilities for long-period comets. Icarus **57**, 48–54 (1984)

L.M. Ziurys et al., Cyanide chemistry in Comet Hale-Bopp (C/1995 O1). Astrophys. J. **527**, L67–L71 (1999)

MIX
Papier aus verantwortungsvollen Quellen
Paper from responsible sources
FSC® C105338

If you have any concerns about our products,
you can contact us on
ProductSafety@springernature.com

In case Publisher is established outside the EU,
the EU authorized representative is:
Springer Nature Customer Service Center GmbH
Europaplatz 3, 69115 Heidelberg, Germany

Printed by Libri Plureos GmbH
in Hamburg, Germany